Characterization of Crystal Growth Defects by X-Ray Methods

NATO ADVANCED STUDY INSTITUTES SERIES

A series of edited volumes comprising multifaceted studies of contemporary scientific issues by some of the best scientific minds in the world, assembled in cooperation with NATO Scientific Affairs Division.

Series B. Physics

Recent Volumes in this Series

Volume 53 – Atomic and Molecular Processes in Controlled Thermonuclear Fusion
edited by M. R. C. McDowell and A. M. Ferendeci

Volume 54 – Quantum Flavordynamics, Quantum Chromodynamics, and Unified Theories
edited by K. T. Mahanthappa and James Randa

Volume 55 – Field Theoretical Methods in Particle Physics
edited by Werner Rühl

Volume 56 – Vibrational Spectroscopy of Molecular Liquids and Solids
edited by S. Bratos and R. M. Pick

Volume 57 – Quantum Dynamics of Molecules: The New Experimental Challenge to Theorists
edited by R. G. Woolley

Volume 58 – Cosmology and Gravitation: Spin, Torsion, Rotation, and Supergravity
edited by Peter G. Bergmann and Venzo De Sabbata

Volume 59 – Recent Developments in Gauge Theories
edited by G. 't Hooft, C. Itzykson, A. Jaffe, H. Lehmann,
P. K. Mitter, I. M. Singer, and R. Stora

Volume 60 – Theoretical Aspects and New Developments in Magneto-Optics
Edited by Jozef T. Devreese

Volume 61 – Quarks and Leptons: *Cargèse 1979*
edited by Maurice Lévy, Jean-Louis Basdevant, David Speiser, Jacques Weyers, Raymond Gastmans, and Maurice Jacob

Volume 62 – Radiationless Processes
edited by Baldassare Di Bartolo

Volume 63 – Characterization of Crystal Growth Defects by X-Ray Methods
edited by Brian K. Tanner and D. Keith Bowen

This series is published by an international board of publishers in conjunction with NATO Scientific Affairs Division

A Life Sciences	Plenum Publishing Corporation
B Physics	London and New York
C Mathematical and Physical Sciences	D. Reidel Publishing Company Dordrecht, Boston and London
D Behavioral and Social Sciences	Sijthoff & Noordhoff International Publishers
E Applied Sciences	Alphen aan den Rijn, The Netherlands, and Germantown, U.S.A.

Characterization of Crystal Growth Defects by X-Ray Methods

Edited by
Brian K. Tanner
Durham University
Durham, United Kingdom

and
D. Keith Bowen
Warwick University
Coventry, United Kingdom

SPRINGER SCIENCE+BUSINESS MEDIA, LLC

Library of Congress Cataloging in Publication Data

Nato Advanced Study Institute on Characterization of Crystal Growth Defects by X-ray Methods, Durham, Eng., 1979.
Characterization of crystal growth defects by X-ray methods.
(NATO advanced study institutes series: Series B, Physics; v. 63)
"Published in cooperation with NATO Scientific Affairs Division."
Includes bibliographical references and index.
1. Crystals—Defects—Congresses. 2. X-ray crystallography—Congresses. I. Tanner, Brian Keith. II. Bowen, David Keith, 1940- III. Title. IV. Series.
QD921.N38 1979   548'.5   80-26509
ISBN 978-1-4757-1128-8   ISBN 978-1-4757-1126-4 (eBook)
DOI 10.1007/978-1-4757-1126-4

Proceedings of the NATO Advanced Study Institute on Characterization of Crystal Growth Defects by X-Ray Methods, held August 29-September 10, 1979, at Durham University, Durham, United Kingdom.

© 1980 Springer Science+Business Media New York
Originally published by Plenum Press, New York in 1980
Softcover reprint of the hardcover 1st edition 1980

All rights reserved

No part of this book may be reproduced, stored in a retrieval system, or transmitted, in any form or by any means, electronic, mechanical, photocopying, microfilming, recording, or otherwise, without written permission from the Publisher

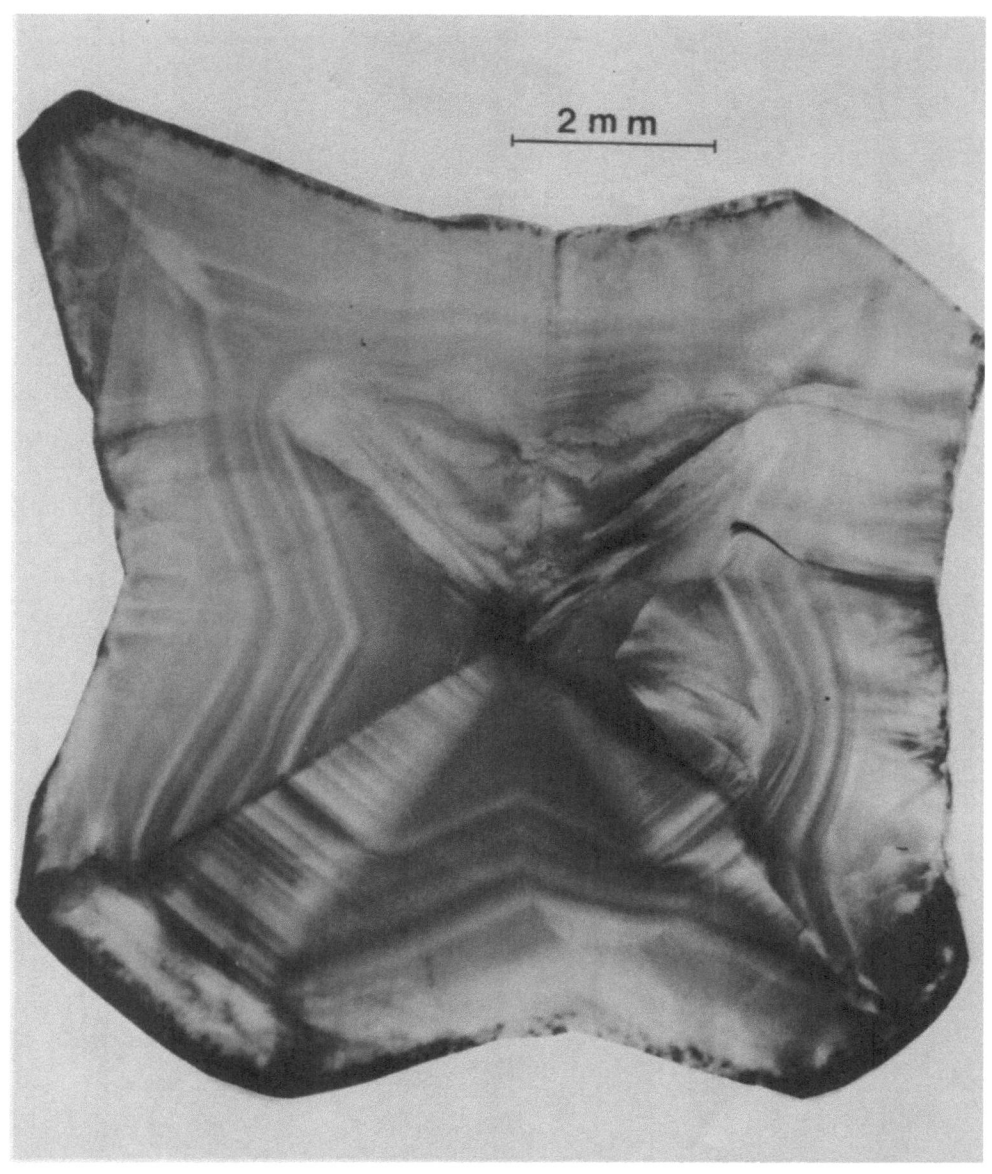

X-ray section topograph (MoKα$_1$ radiation, 440 reflexion) of a knobbly cuboid diamond of 12 carats, roughly 1 cm cube.

C.A. Fox and M. Moore   Royal Holloway College

FOREWORD

This book contains the proceedings of a NATO Advanced Study Institute entitled "Characterization of Crystal Growth Defects by X-ray Methods" held in the University of Durham, England from 29th August to 10th September 1979. The current interest in electronic materials, in particular silicon, gallium aluminium arsenide, and quartz, and the recent availability of synchrotron radiation for X-ray diffraction studies made this Advanced Study Institute particularly timely. Two main themes ran through the course:

1. A survey of the various types of defect occurring in different methods of crystal growth, the mechanism of their generation and their influence on the properties of relatively perfect crystals.

2. A detailed and advanced course on the observation and characterization of such defects by X-ray methods. The main emphasis was on X-ray topographic techniques but a substantial amount of time was spent on goniometric techniques such as double crystal diffractometry and gamma ray diffraction.

The presentation of material in this book reflects these twin themes. Section A is concerned with defects, Section C with techniques and in linking them. Section B provides a concise account of the basic theory necessary for the interpretation of X-ray topographs and diffractometric data. Although the sequence follows roughly the order of presentation at the Advanced Study Institute certain major changes have been made in order to improve the pedagogy. In particular, the first two chapters provide a vital, and seldom articulated, case for the need for characterization for crystals used in device technologies. It is against this pressing economic background, as well as the context of an intrinsic scientific interest in understanding the role of crystal lattice defects, that the book should be read. Quite deliberately, we turn full circle and end the exposition with a critical analysis of the cost effectiveness of the use of X-ray diffraction techniques for characterization.

During the Institute much emphasis was placed on the tutorial work which took place late each afternoon and the material presented in these sessions is included as seven substantial appendices. They provide a very useful set of problems from which anyone entering the field of X-ray topography will greatly benefit.

FOREWORD

The Advanced Study Institute itself, attended by 77 students and 18 lecturers from 17 different countries, was agreed by all to have been a great success. In part this was due to the atmosphere imparted by residence in a castle dating from the 11th Centruy overlooking a majestic cathedral of similar venerability and a miraculously fine spell of weather. In the main, however, its success was due to the way all the participants, lecturers and students alike, went out of their way to exchange ideas and information in informal contacts. The two informal poster displays in the bar, at which all participants were invited to present work reflecting the theme and tutorial function of the Institute, provided an excellent catalyst for such contacts.

The Director of the Institute, Professor A.R. Lang F.R.S., in his opening speech, drew attention to a subsidiary theme of the Institute: contrast. Much of the academic activity concerned the contrast of crystal defects, but the contrasts of ancient and modern buildings in Durham, of experienced and novice X-ray workers and of differing languages, countries and cultures were highlighted. The success of the Institute was in no small way a result of Professor Lang's enthusiasm and inspiration, in particular in relation to the tutorial programme.

The editors, responsible for detailed financial administration and programme organization respectively, wish to express their gratitude to the many people whose willing cooperation combined to make for very smooth running of the Institue. In particular, thanks are expressed to the organizing committee and the International Advisory panel for their constructive suggestions many of which they personally realized. We would also thank the Council of the University of Durham for acting as our hosts, Professor B.H. Bransden, Head of the Physics Department for permitting the use of the Department and laboratory facilities for practical classes, several colleagues in various Departments who loaned minibuses for local transport, Messrs G.F. Clark, D.G. Money, H. Davison and P.Cottle for driving them, Mr. C.F. Cleveland and his staff for their smooth local arrangements and Mrs. R.H. Tanner for running the accompanying persons' programme. Several people have collaborated in production of these proceedings and we gratefully thank Dr. S.T. Davies for proofreading; Mrs. M. Bradley, Miss J. Nelson and Mrs. V. Todd for typing services; Misses D. Dixon, J. Morgan, K. Gittings and Mr. W. Spalding for reprographic work; and the University of Warwick Computer Unit, in particular Mrs. V. Clayton and Mr. G. Sheridan for use of the word processing system. The quality of these proceedings is a tribute to their efforts.

B.K.Tanner
University of Durham

D.K. Bowen
University of Warwick

CONTRIBUTORS

Dr. E. S. Meieran, Intel Corporation, Santa Clara, .S.A

Dr. A.D. Milne, Wolfson Microelectrons Unit, Edinburgh University,U.K.

Dr. D.J. Hurle, R. S.R.E., Malvern, U.K.

Dr. B. Cockayne, R.S.R.E. Malvern, U.K.

Dr. J.R. Patel, Bell Laboratories, Murray Hill. N.J., U.S.A.

Professor G. Champier, E.N.S.I.M., Nancy, France

Dr. H. Klapper, Rhein-Westphalia Technische Hochschule, Aachen, W. Germany

Professor A.R. Lang, H.H. Wills Laboratory, Bristol University U.K.

Dr. J.R. Schneider, Hahn Meitner Institute, Berlin, W. Germany

Professor M. Hart, Department of Physics, Kings College, London University, U.K.

Professor N. Kato, Department of Crystallography, Nagoya University, Japan

Professor U. Bonse, Institute for Physics, Dortmund University, W. Germany

Dr. D.K.Bowen, Department of Engineering, Warwick University, Coventry, U.K.

Professor R. Armstrong, Department of Engineering, University of Maryland, College Park, U.S.A.

Dr. J.I. Chikawa, N.H.K. Research Laboratories, Tokyo, Japan

Dr. W. Hartmann, Max Planck Institute, Stuttgart, W. Germany

Dr. J. Miltat, Laboratory of Mineralogy and Crystallography, University of Paris VI, France

CONTRIBUTORS

Dr. M. Sauvage, Laboratory of Mineralogy and Crystallography, University of Paris VI, France

Dr. M. Moore, Department of Physics, Royal Holloway College, London University, U.K.

Dr. B.K. Tanner, Department of Physics, Durham University, U.K.

Dr. F. Balibar, Laboratory of Mineralogy and Crystallography, University of Paris VI, France

Dr. Y. Epelboin, Laboratory of Mineralogy and Crystallography, University of Paris VI, France.

CONTENTS

CHAPTER 1	Industrial Implications of Crystal Quality		1
	1.1	Introduction	1
	1.2	Production requirements	7
		1.2.1 Incoming inspection	8
		1.2.2 Yield	18
	1.3	Advances in technology	22
	1.4	Conclusion	26
		References	27
CHAPTER 2	The Technical Importance of Growth Defects		28
	2.1	Introduction	28
	2.2	Charge-coupled devices	30
		2.2.1 Applications	33
		2.2.2 Charge transfer inefficiency	35
		2.2.3 Dark current	36
	2.3	Magnetic bubble memories	37
		2.3.1 The manufacture of bubble memories	38
		2.3.2 Device architecture	39
		2.3.3 Bubble memory performance	40
	2.4	Surface acoustic wave devices	41
		2.4.1 Crystal properties	43
		Bibliography	44

CONTENTS

CHAPTER 3	<u>Defects and their Detectability in Melt-Grown Crystals</u>	46
	3.1 Introduction	46
	3.2 Solute striations	48
	3.3 Effects of facet formation	50
	3.4 Cellular and dendritic structures	55
	3.4.1 Introduction	55
	3.4.2 Morphology	56
	3.4.3 Chemical inhomogeneities	60
	3.4.4 Structural inhomogeneities	62
	3.5 Dislocations generated during cooling of the crystal	65
	3.6 Particulate matter and voids	69
	3.7 Summary and conclusions	71
	References	71
CHAPTER 4	<u>Defects and Their Detectability</u>	73
	4.1 Introduction	73
	4.2 Planar and line defects in crystals	73
	4.2.1 X-ray topography of stacking faults in crystals	73
	4.2.2 Dislocations in quartz	77
	4.3 Defects due to clustering or precipitation	79
	4.3.1 Anomalous X-ray transmission	79
	4.3.2 X-ray diffuse scattering	82
	4.4 Point defects	84
	4.4.1 Perfection of epitaxial layers	84
	4.4.2 Impurity lattice location studies	86
	References	95

CONTENTS

CHAPTER 5	Defect Generation in Metal Crystals		97
	5.1	Introduction	97
	5.2	Growth from liquid state	98
		5.2.1 Bridgman method	98
		5.2.2 Czochralski method	104
	5.3	Growth from the solid state	110
		5.3.1 Strain-anneal method	110
		5.3.2 Phase transformation method	117
	5.4	Growth from gas state and whiskers	117
		5.4.1 Growth from gas state	117
		5.4.2 Whiskers	118
	5.5	Stability of dislocation configurations at room temperature	119
	5.6	Stability of dislocation configurations at high temperature	123
		5.6.1 Annealing effect: observation at room temperature	123
		5.6.2 Observation at high temperature	126
	5.7	Conclusion	128
		References	130
CHAPTER 6	Defects in Non-metal Crystals		133
	6.1	Introduction	133
	6.2	Growth defects: crystals with planar growth faces	134
		6.2.1 General features of grown-in dislocations	134
		6.2.2 Origin of dislocations. Formation of inclusions	136
		6.2.3 Theory of preferred directions	138
		6.2.4 Growth bands and growth sector boundaries	145

CONTENTS

6.3	Growth defects: crystals grown on curved interfaces		146
	6.3.1	General remarks	146
	6.3.2	Main growth features	147
	6.3.3	Striations	147
	6.3.4	Facets	150
	6.3.5	Inclusions	150
	6.3.6	Grown-in dislocations	150
6.4	Growth defects: crystals grown under high thermal gradients		151
6.5	Post-growth defects		153
	6.5.1	Distinction between grown-in and post-growth dislocations	153
	6.5.2	Dislocation glide	154
	6.5.3	Dislocation climb	156
	6.5.4	Dislocation loops, stacking faults and precipitates	157
	References		157

CHAPTER 7 Defect Visualisation: Individual Defects 161

7.1	Introduction		161
	7.1.1	Experimental background	161
	7.1.2	Theoretical background	161
7.2	Production of X-ray topographs		163
	7.2.1	Section topographs	163
	7.2.2	Projection topographs	167
	7.2.3	Scanning reflection topographs	169
	7.2.4	Stereo methods	170
7.3	Interpretation of X-ray topographs		173
	7.3.1	Types of defect individually visualizable	173
	7.3.2	Effects of specimen thickness on defect contrast	176
	7.3.3	Dislocations and low-angle boundaries	178

CONTENTS

7.3.4	Surface damage	180
7.3.5	Precipitates and inclusions	181
7.3.6	Fault surfaces and growth bands	182

References 184

CHAPTER 8   Experimental Techniques for the Study of Statistically Distributed Defects 186

8.1   Introduction   186

8.2   Investigation of diffuse X-ray scattering due to point defects   187

    8.2.1   Huang diffuse scattering   189
    8.2.2   Zwischenreflex scattering   191
    8.2.3   Structure determination of self-interstitials in aluminium   193

8.3   Study of mosaic structure of large crystals by γ-rays   195

    8.3.1   The γ-ray diffractometer   196
    8.3.2   Interpretation of γ-ray rocking curves   198

8.4   Applications of γ-ray diffractometry   207

    8.4.1   Testing of as-grown Al single crystals   208
    8.4.2   Testing of neutron monochromator crystals   208
    8.4.3   Testing samples for neutron scattering experiments   209
    8.4.4   Study of structural phase transitions   213

References   214

CONTENTS

CHAPTER 9	Elementary Dynamical Theory		216
9.1	Introduction		216
9.2	The Laue condition		216
9.3	Formal definition of the crystal optical parameters		217
	9.3.1	The polarizability of a crystal	217
	9.3.2	Properties of the structure amplitude	219
9.4	Solution of Maxwell's equations		221
	9.4.1	Case 1: α_o and α_h both real	224
	9.4.2	Case 2: α_o and α_h are complex conjugates	224
	9.4.3	Summary of results for zero absorption	224
9.5	The Poynting vector and ray optics		226
	9.5.1	The ray approximation	227
	9.5.2	Phenomenology of absorption along the raypath	227
9.6	Boundary conditions linking external and crystal waves		229
	9.6.1	Amplitude matching	230
	9.6.2	Wavevector matching	230
	9.6.3	Spherical waves	231
9.7	Some demonstrations of the results of the dynamical theory		232
	9.7.1	The Borrmann effect	232
	9.7.2	Ray tracing in perfect crystals	235
	9.7.3	Crystal reflection profiles	236
	9.7.4	Pendellosung fringes	239
	9.7.5	Spatial distribution of intensity in section patterns	241
9.8	Wavefield propagation in distorted crystals		241

CONTENTS

	9.8.1	Ray propagation and the refractive index gradient ∇n	242
	9.8.2	The characteristic equation relating $d\xi$ to the deformation	243
	9.8.3	Ray paths in a homogeneously deformed crystal	245
	9.8.4	Intensity of rays in homogeneously deformed crystals	247
9.9	Experimental results from homogeneously deformed crystals		248
	9.9.1	The Borrmann effect in homogeneously deformed crystals	248
	9.9.2	Ray tracing experiments	250
	9.9.3	Pendellosung fringes in deformed crystals	250
	9.9.4	Spatial distribution of intensity in section patterns	252
9.10	Images of crystal defects in real crystals		254
	9.10.1	Qualitative interpretation of defect images	255
	9.10.2	Semi-quantitative interpretation of defect images	257
	References		258
	Appendix		261
9.11	A solution of Maxwell's equations		261
9.12	The Poynting vector and ray optics		262
9.13	Boundary conditions linking external and internal waves		262
9.14	Absorption of rays		263

CHAPTER 10	Perfect and Imperfect Crystals	264
10.1	Introduction	264
10.2	Ray optical considerations	265

xvii

CONTENTS

	10.2.1	Ray optics in vacuum; kinematical theory	266
	10.2.2	Plane wave theory and spherical wave theory	268
	10.2.3	Ray optics in the crystalline medium	269
	10.2.4	Perfect crystals	272
	10.2.5	Plate-like defects	275
	10.2.6	Long range distortion	278
10.3	Wave-optical considerations		283
	10.3.1	Wave equations of Takagi-Taupin type	283
	10.3.2	Perfect crystals	285
	10.3.3	The case of constant strain gradient	287
	10.3.4	Computer simulation	288
10.4	Statistical theory of dynamical diffraction		289
	10.4.1	Derivation of energy transfer equations	290
	References		295

CHAPTER 11 X-ray Sources   298

11.1	Definition of a 'powerful source'		298
11.2	X-rays generated by electron impact on solid targets (EI)		300
	11.2.1	Bremsspectrum	301
	11.2.2	Characteristic line spectrum	302
	11.2.3	Efficiency of X-ray production by EI	303
	11.2.4	X-ray tubes	304
11.3	Synchrotron X-ray sources		306
	11.3.1	General	306
	11.3.2	Calculation of the radiation properties of a SR source	307
	11.3.3	Time structure of SR from storage rings	310

CONTENTS

	11.3.4	SR from wigglers and undulators	310
11.4	Summary and conclusions		312
	References		318

CHAPTER 12 X-ray Detectors   320

12.1	Introduction		320
	12.1.1	X-ray photon counters	320
	12.1.2	Counting statistics and counting ratemeters	322
12.2	Photographic recording methods		323
	12.2.1	Photographic densitometry and sensitivity	323
	12.2.2	Statistical limitations on resolution	324
	12.2.3	Fast X-ray films	326
	12.2.4	Properties and processing of nuclear emulsions	327
	12.2.5	Handling and preservation of nuclear plates	328
	12.2.6	Ultra high resolution image recorders	328
	12.2.7	Photomicrography of X-ray topographs	329
12.3.	Electronic recording methods		330
	12.3.1	Basic problems	330
	References		332

CHAPTER 13 Sample Preparation   333

13.1	Introduction	333
13.2	Determining crystal orientation	334
13.3	Shaping and forming processes	338

xix

CONTENTS

13.4	Finishing processes: polishing and etching	342
13.5	Specimen mounting on the topographic camera	345
	References	348

CHAPTER 14 Laboratory Techniques for X-ray Reflection Topography 349

14.1	The nature of reflection topography	349
14.2	Diffraction contrast in reflection topographs	351
14.3	Dislocation subgrain boundaries	356
14.4	Applications to crystal growth processes	361
	References	366

CHAPTER 15 Laboratory Techniques for Transmission X-ray Topography 368

15.1	Introduction	368
15.2	Basic photographic techniques	368
	15.2.1 Topography by spherical waves	368
	15.2.2 Topography by the plane waves	372
	15.2.3 Analysis of defects	373
	15.2.4 Special techniques	379
15.3	Video display of topographic images	382
	15.3.1 Instrumentation	382
	15.3.2 Rapid survey of defects	388
	15.3.3 Live topography	390
	15.3.4 Future trends	396
15.4	Conclusion	398
	References	398

CONTENTS

CHAPTER 16	White Beam Synchrotron Radiation Topography		401
	16.1	Introduction	401
	16.2	Experimental set-up	401
	16.3	The perfect crystal: general properties of Laue topographs	403
		16.3.1 Geometrical resolution	403
		16.3.2 Integrated diffracted intensity	404
		16.3.3 Effects of polarization properties of Pendellosung fringes	406
	16.4	The imperfect crystal	408
		16.4.1 Sensitivisty to entrance surface misorientations	408
		16.4.2 Defect contrast	409
		16.4.2.1 Dislocation direct images	409
		16.4.2.2 Other type of contrast	412
		16.4.2.3 Influence of harmonics superposition	412
	16.5	Applications	413
		16.5.1 Recrystallization of Al polycrystals	414
		16.5.2 Misfit dislocations	415
		16.5.3 Plastic deformation of Fe-Si single crystals	416
		16.5.4 Wall movements in the antiferromagnet $KCoF_3$	416
	16.6	Conclusion	418
		References	419

CONTENTS

CHAPTER 17	Control of Wavelength, Polarization, Time-Structure and Divergence for Synchrotron Radiation Topography		421
	17.1	Introduction	421
	17.2	Summary of results from dynamical diffraction theory	421
	17.3	Control of wavelength	423
		17.3.1 Multiple Bragg reflections	424
		17.3.2 Harmonic-free systems for topography	424
	17.4	Polarizing optical systems	426
		17.4.1 Linear polarization	426
		17.4.2 Circular polarization	428
	17.5	Stroboscopy, modulation and timing	428
	17.6	Control of beam divergence	430
		References	431
CHAPTER 18	Monochromatic Synchrotron Radiation Topography		433
	18.1	Introduction	433
	18.2	Selection of the experimental conditions	433
		18.2.1 Integrated images	433
		18.2.2 Plane-wave imaging techniques	435
	18.3	Available and future facilities	436
	18.4	Analysis of an experiment	436
		18.4.1 Du Mond diagram	436
		18.4.2 Numerical application	439
	18.5	Examples of applications	441
		18.5.1 Stroboscopic observation of magnetic domain wall motion	441
		18.5.2 Plane-wave topography	444

CONTENTS

	18.6	Conclusion	453
		References	455
CHAPTER 19		Environmental Stages and Dynamic Experiments	456
	19.1	Introduction	456
	19.2	Simultaneous measurements	457
		19.2.1 Electric field	458
		19.2.2 Magnetic field	458
		19.2.3 Low temperature	459
		19.2.4 High temperature	461
		19.2.5 Stress	461
		19.2.6 Time lapse in a corrosive environment	464
		19.2.7 Crystal growth parameters	464
	19.3	Dynamic experiments	465
		19.3.1 conventional source experiments	465
		19.3.2 Synchrotron radiation experiments	466
	19.4	Instrumentation developments at SR sources	469
		19.4.1 White radiation camera	469
		19.4.2 Double crystal camera	470
		19.4.3 X-ray TV detector	470
		19.4.4 Environmental stage	470
	19.5	Future developments	470
		References	471
CHAPTER 20		Technology and Costs of X-Ray Diffraction Topography	474
	20.1	Introduction	474
		20.1.1 Cost estimates	475

CONTENTS

20.2	Experimental arrangements for transmission topography		476
	20.2.1	The X-ray source	476
	20.2.2.	Cameras for transmission topography	479
	20.2.3	Detectors for transmission topography	480
20.3	Double crystal experimental arrangements		482
	20.3.1	X-ray sources	482
	20.3.2	Double crystal cameras	483
	20.3.3	Detector systems for double crystal diffraction	485
20.4	Building your own apparatus		486
	20.4.1	X-ray sources	486
	20.4.2	Cameras for defect studies	486
	20.4.3	Electronics for X-ray diffraction topography	487
20.5	Synchrotron radiation sources		491
	20.5.1	Stanford Synchrotron Radiation Laboratory (SSRL)	491
	20.5.2	Laboratoire pour l'Utilization de Rayonnement Electromagnetique (LURE)	492
	20.5.3	Hamburg, Deutsches Electron-Synchrotron (DESY, DORIS)	492
	20.5.4	Cornell University (CHESS)	493
	20.5.5	Daresbury, storage ring source (SRS)	493
	20.5.6	National Synchrotron Light Source (Brookhaven)	494
	20.5.7	The Photon Factory (Japan)	495
	20.5.8	Costs and funding	495
	References		496

CONTENTS

CHAPTER 21	X-Ray TV Imaging and Real-Time Experiments		497
	21.1	Introduction	497
	21.2	Experimental arrangement	497
	21.3	Experiments	499
	21.4	Summary	501
		References	501
	21.5	Discussion	502
CHAPTER 22	Computer Modelling of Crystal Growth and Dissolution		503
	22.1	Crystal growth	503
	22.2	Growth horizons	503
	22.3	Crystal dissolution	504
		References	505
CHAPTER 23	Microradiography and Absorption Microscopy		506
	23.1	Introduction	506
	23.2	Experimental methods for microradiography	506
		23.2.1 White radiation method	506
		23.2.2 White radiation: self-monochromatisation	507
		23.2.3 Beam-conditioned radiography: difference maps	507
		23.2.4 Scanning methods	508
	23.3	Resolution	508
	23.4	Contrast in microradiography	509
	23.5	Detectors	510

CONTENTS

	23.6	Applications and Speculations	510
		References	511
CHAPTER 24		Reciprocal Lattice Spike Topography	512
		References	514
CHAPTER 25		Reflection Topography: Panel Discussion	515
	25.1	Introduction	515
	25.2	Discussion	515
APPENDIX 1		Designing a Topographic Experiment	520
APPENDIX 2		Defects and Artifacts	526
APPENDIX 3		Exercises in Diffraction Contrast	528
APPENDIX 4		Stereographic Projection Description for X-Ray Topography: Subgrain Boundaries and Stereo-Pairs	535
APPENDIX 5		Dispersion Surface Exercises	544
APPENDIX 6		Contrast of Stacking Faults	554
APPENDIX 7		Misfit Boundaries and Junctions of Purely Rotational Boundaries	565

Sponsors, Organising Committee, Advisory Panel	581
Subject Index	583
Chemical Formula Index	589

CHAPTER 1

INDUSTRIAL IMPLICATIONS OF CRYSTAL QUALITY

EUGENE S. MEIERAN

1.1 Introduction

The semiconductor and electronic industry consumption of single crystals is approximately doubling each year, and now accounts for a half billion dollar industry. While most of this volume is due to the silicon industry, other materials such as bubble memory garnets, sapphire for silicon-on-sapphire devices, III-V, and opto-electronic crystals are becoming increasingly important. In view of this economic driving force, it is not surprising that a high degree of competition exists between vendors of crystals eager to take advantage of the boom in sales, and that crystal users try to take advantage of such competition in order to obtain better quality material.

Cost, yield and reliability are the generally-accepted parameters compared by crystal users to decide between potential vendors. Since crystal costs are largely controlled by economic rather than technical factors, such as equipment cost, power, cost of raw materials, etc., the cost differential between vendors is easy to quantify and deal with. On the other hand, reliability factors are usually reasonably independent of the crystal itself, depending more on chip design, processing and packaging variables, and hence play a small role in deciding amongst crystal vendors. The third quantity, yields, is an entirely different matter, for unlike reliability, yield depends on crystal properties and can vary considerably from vendor to vendor, and the economics of yield variations may far overshadow cost differences.

Unfortunately, while yield trends are in principle easy to generate, there are several practical difficulties. First, yield trends require time, and problems with yield may arise long after material is started in a production line. Second, many different crystal properties affect yield, as do process variations, and the latter problems are the responsibility of the crystal user, not the crystal vendor. Finally, measurement of crystal quality is not usually quantitive in nature. Consequently, it is not easy to relate yield to vendor and to crystal quality.

Such problems lead to the topic for this chapter, namely, that there is a need to develop procedures that can be used in production to evaluate crystals. This is a subject related to that of relating specific device properties to specific crystal defects; the industrial need is to define simple quantifiable measurements which can be related to yield, so that incoming inspection decisions can be made. Specific cause and effect relationships between particular defects and device properties are important, but may be unnecessary in evaluating yield trends.

During the past 20 years, remarkable strides have been made in the art and science of growing single crystals. This has been achieved in large part as a response to the enormous economic driving force provided by the semiconductor and electronics industries. Not only has the quantity of material used increased during the past 20 years, but crystal quality and physical characteristics have likewise improved. For example, the silicon crystals used 20 years ago were about 1 cm in diameter and 15 cm long; the current crystals are 100 mm in diameter, and are over 1 metre long. Furthermore, the rate of diameter increase shows only moderate signs of levelling off, Fig. 1. Many factories are now looking at 120 or 125 mm diameter crystals for 1980/1981 usage. The trend towards larger diameters also appears to be followed by the garnet and sapphire based technologies, Fig. 1.

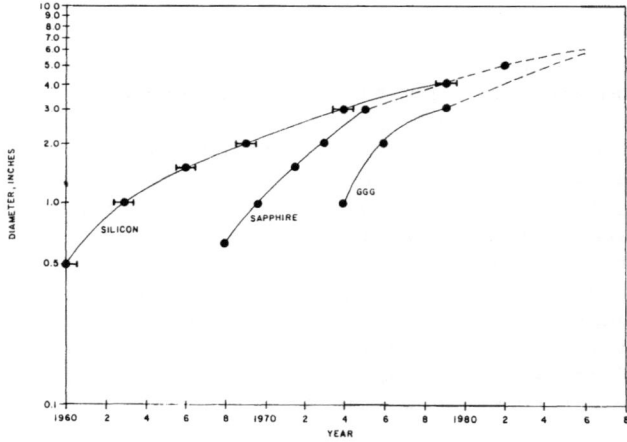

Figure 1. Wafer diameter vs. year.

As a second example, wafer flatness criteria have become tighter, due to the masking requirements needed for high resolution line geometries typical of modern devices. While 10 or 20 μm NTV (Non-Linear Thickness Variation) was acceptable for silicon ten years ago, current wafers must exhibit less than 5 μm NTV. Silicon easily meets this requirement; most other materials at present do not. Fortunately, because of the reasonably well identified requirements of new masking systems such as E-beam or wafer stepper lithography, the 5 μm NTV requirement appears adequate for next generation devices, and so the trend towards lower NTV has slowed.

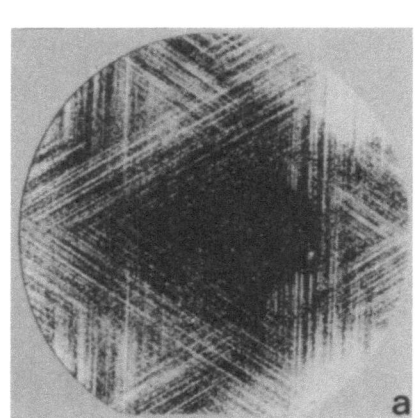 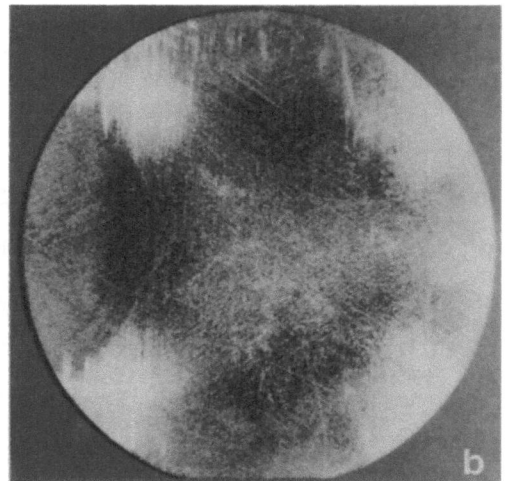

Figure 2. Dislocation etch pits in silicon x 1 (a) (111) silicon (b) (100) silicon

Less well defined is the trend in defect density. Originally, as-grown crystals were highly imperfect, containing many, many defects. Dislocation densities of 10^6 /cm^2 were common, Fig. 2, until about 1965, when the as-grown dislocation density dropped dramatically. In order to improve device yields, crystals were grown that contained zero dislocations, and few other macroscopic defects. However, as device technologies continued to mature, intentional defect densities were on occasion introduced to act as gettering centres for dissolved impurities, for example by controlled back side grinding procedures. At this time, it is not clear whether such defects are beneficial or detrimental to device performance – equally valid arguments have been made either way. While the trend in needs for larger crystals or for flatter crystals to increase device yield is well understood, the need for defect-free material to improve yields is more difficult to assess, due to its apparent process dependence, and is subject to much controversy. This topic is discussed in more detail in the chapter by Milne [1].

In order to evaluate the effects of crystal quality on device yields, many techniques have been developed for analyzing crystal perfection [2 - 4]. Among the more sophisticated of these techniques are those dependent on X-ray diffraction properties of crystals, such as X-ray topography, double crystal X-ray diffraction, section topography, anomalous transmission of X-rays, and others. While X-ray techniques have so far kept pace with research and development requirements, they are severely lagging behind production needs. To further aggravate the situation, in view of the rapid advances in device technology, it is not evident that the measurement technology advances will be satisfactory to meet future needs for better sensitivity to defects, even in research areas. The principles of the major X-ray techniques used to evaluate crystal perfection are listed in Table 1, and are illustrated in Fig. 3.

Table 1

X-RAY TECHNIQUES FOR QUALITY CONTROL

Method	$\delta d/d$	Mode	Defects
Transmission Topography	10^{-3}	Imaging Bulk	Dislocations Precipitates Damage Heavy Swirl
Reflection Topography	10^{-3}	Imaging Surface	Dislocations Damage Heavy Swirl
Section Topography	–	Imaging Cross Section	Precipitates General Perfection
Double Crystal Topography	10^{-8}	Imaging Surface	Strains Distortions Light Swirl
Anomalous Transmission	–	Non-imaging Bulk	General Perfection

It is this state of affairs to which this chapter is addressed. The explicit purposes of the chapter are to examine current production needs for crystal perfection measurement, and to look at trends in device and material technology, in order to anticipate future research and development needs. While most examples cited will be related to silicon technology, the principles are equally applicable to other uses of single crystals.

INDUSTRIAL IMPLICATIONS OF CRYSTAL QUALITY

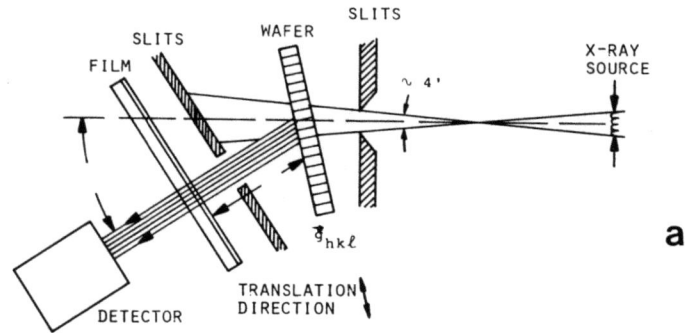

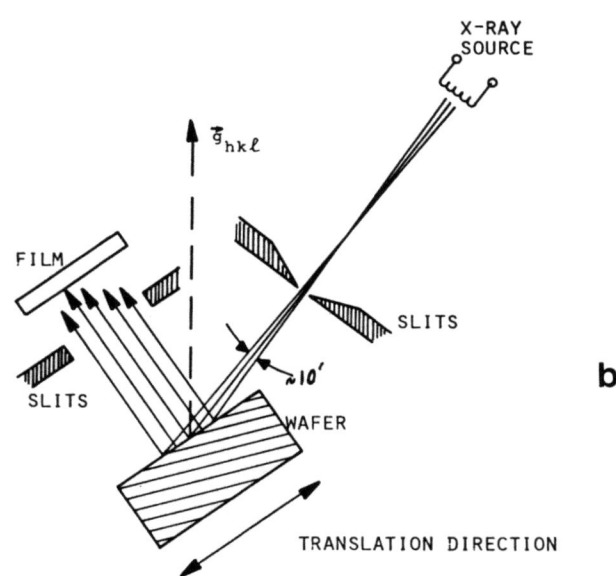

Figure 3. (a) Transmission topography, (b) reflection topography, (c) section topography, (d) double crystal topography, (e) anomalous transmission

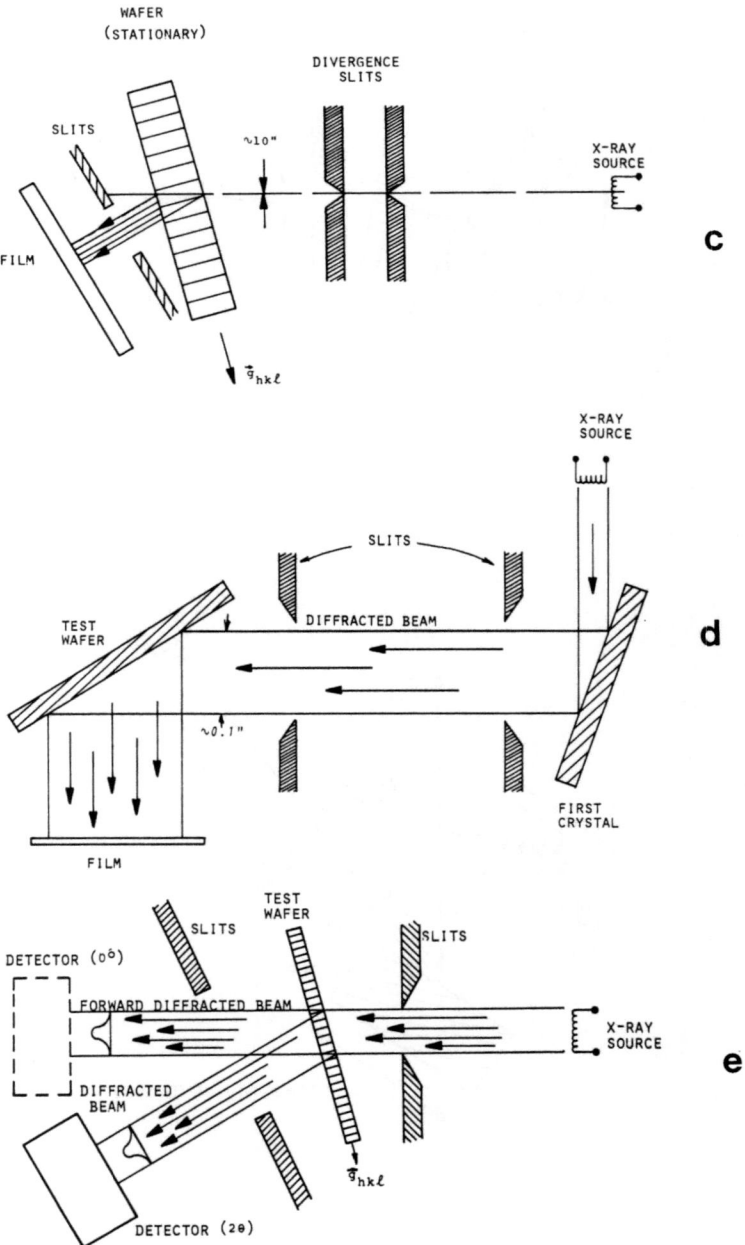

Figure 3 (continued)

INDUSTRIAL IMPLICATIONS OF CRYSTAL QUALITY

1.2 Production requirements

In order to aid in the discussion of production needs for defect analysis of single crystals, two quality indicators will be described. First is die yield, measured in percent of electrically sorted functional die per wafer completed (not per wafer started, which is a measure of wafer yield and is of less concern). Die yield is often expressed by means of Isodefect curves, in order to eliminate variations in yield due to variable factors such as device size, wafer size, or number of masking steps. An Isodefect curve gives the measure of a process in terms of equivalent 100 mil^2 die; i.e., the actual yield of a given device of a given size is adjusted to what the yield would be if the device were exactly 100 mil^2 (10,000 square mils). Isodefect analysis will be illustrated in detail later.

Secondly in order to maintain yields by preventing substandard quality material from entering the production line, incoming wafer lots are sample inspected. This inspection is of a sample size based on a statistical sampling plan, in most cases AQL (Acceptable Quality Level) sampling [5] (see Table 2.) In a situation where vendor shipments of material come in quantities of a few thousand or so wafers per lot, the AQL sampling plan calls for inspection of about 1% of the lot. For a wafer Fab area running at the 25,000 wafers per week level, this amounts to an inspection of about 1000 wafers per month, at a cost of several thousand dollars, since most inspection techniques are destructive. Incoming quality is measured in wafer percent defective – that is, the number of wafers with reject level defects, divided by the number of wafers inspected.

Table 2

REJECT/SAMPLE AQL SAMPLE PLAN

AQL:	0.65	1.0	1.5	2.5	4.0
LOT SIZE					
<500	1/20	2/50	3/50	4/50	6/50
501 - 1200	2/80	3/80	4/80	6/80	8/80
1201 - 3200	3/125	4/125	6/125	8/125	11/125

So, for the purpose of this chapter, the production environment relies on two inspection indicators – a destructive sampling of about 1% of the incoming wafers and a monitor of die yields in terms of Isodefect curves. The possible use of X-ray analysis in each of these areas is discussed below.

1.2.1 Incoming inspection

The role of incoming inspection is to provide confidence that incoming material quality meets or exceeds a specified level. In addition to the objective physical and dimensional parameters usually inspected, such as resistivity, type, wafer orientation, diameter, lifetime, flatness, thickness, flat orientation, etc., there are several somewhat subjective tests. In particular, although the precise effects of lattice defects on device yields may not be understood, enough general information is available to suggest defect screening. Some defects such as dislocations are universally accepted as deleterious to device performance, and at present, zero dislocation density is specified for all incoming silicon wafers. The effects of other defects, such as swirl and haze are less well documented and specifications are substantially less quantitative or objective. Finally, there are categories of defects such as point defect clusters, for which there are no defined procedures for measurement, nor any objective or subjective criteria for decision at this time.

According to the AQL table, Table 2, in order to achieve 95% confidence that a given incoming lot of a given defect level will pass an incoming sampleplan, a sample quantity is selected and inspected based on the lot size. Since most wafer lots consist of a few thousand wafers, for a typical 2.5% AQL plan (95% confidence that a given lot of 2.5% defective or less will be accepted) it is seen that 125 wafers must be inspected, with an accept level of 7 defective wafers out of 125 inspected. If 8 defective wafers are found, the lot is returned to the vendor.

A list of defects for which there are incoming inspection screens is given in Table 3, along with the present method used to detect the defects, and typical accept/reject criteria. As can be seen, wafers are usually etched, and are screened by visual inspection. Some of the defects are observed directly in as-received wafers; others must be decorated by undergoing a sequence of heat treatment steps prior to etching. Dislocation densities are measurable on an absolute scale - Fig. 4 shows examples of etch pits in silicon. Swirl is measured after a multistage heat treatment sequence - the inspection criteria are quite arbitrary and are subject to inspector sensitivity. Examples of several degrees of swirl are shown in Fig. 5. While most device manufacturers agree on the specification for dislocation density (i.e., zero dislocations) there is little agreement on swirl, and every company uses its own specification and procedure. Finally, haze, a term applied to a variety of defects, may be observed by etching either decorated or undecorated wafers. Unfortunately, while haze is comparatively easy to measure (in terms of etch pits per cm^2, much as dislocations are measured) it is by no means clear what is being measured. Haze has been attributed to a number of causes, so not only the effects of the defect on yields, but also the precise nature of the defect, is unknown. An example of haze is shown in Fig. 6.

Table 3

CURRENT INDUSTRIAL DEFECT DETECTION METHODS
AND ACCEPTANCE CRITERIA

DEFECT	PRE-TREATMENT	METHOD	CRITERION
Dislocations	None	Sirtl Etch	Zero dislocations
Scratches	None	Visual	Zero scratches
Haze	None	Seeco etch	"no haze"
Swirl	Multi-stage oxidation	Seeco Etch	No heavy swirl
Precipitates	None	Etch	Variable

Two major difficulties associated with incoming inspection are immediately apparent; the cost is large due to the destructive nature of the tests, and many tests are not objective. X-ray technology provides methods to overcome both difficulties. Firstly, in view of the fact that X-ray measurements may be non-destructive, there are potential cost savings of several thousands of dollars per month, if X-ray techniques replace destructive etching techniques. Secondly, the X-ray measurements are relatively quantitative and operator insensitive, thereby providing a reproducible method to inspect samples.

The non-destructive nature of X-ray techniques is unfortunately not all that assured. While the inspection itself does not damage silicon or other materials, the handling during inspection may introduce problems. Since inspected wafers should be returned to the lots from which they came, great care must be exercised to avoid contaminating the wafers; normal mounting such as used in X-ray cameras is unacceptable. The problem of contamination or damage effects can overshadow defect effects. This problem must be considered by both manufacturers and users of X-ray cameras if X-ray techniques are to be used as non-destructive inspections to replace etching tests. With regard to the second advantage of the use of X-ray diffraction techniques for evaluating wafers, the fact that a number of X-ray procedures are available is a great advantage.

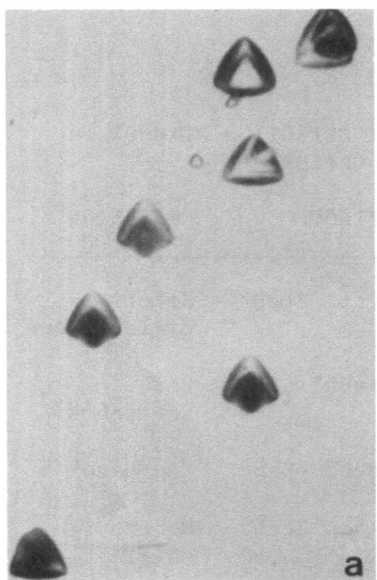 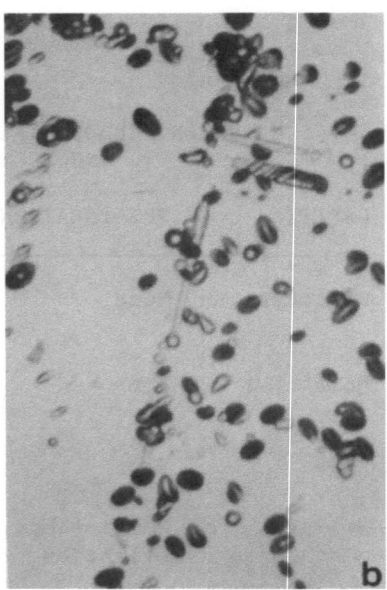

Figure 4. Dislocation etch pits in silicon x 200 (a) (111) silicon, (b) (100) silicon

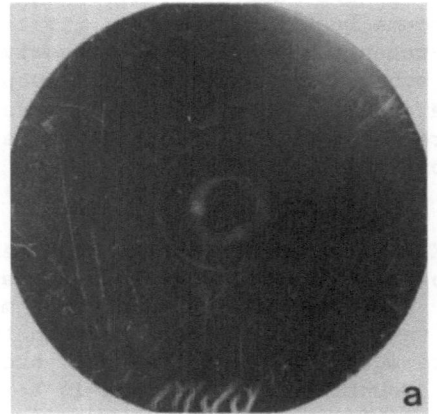 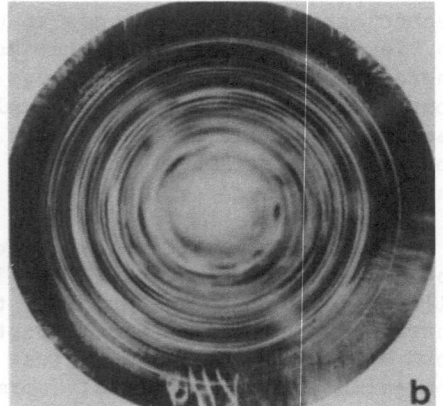

Figure 5. Swirl in etched, oxidized silicon x 1. (a) low swirl, (b) high swirl.

INDUSTRIAL IMPLICATIONS OF CRYSTAL QUALITY

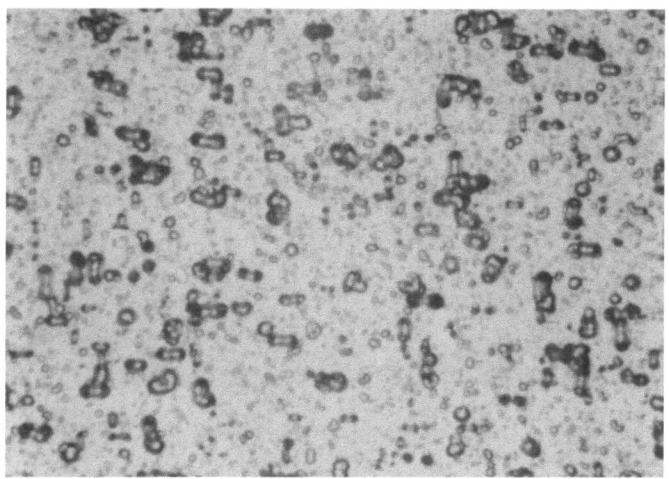

Figure 6. Haze in etched silicon x 220.

Consider the varieties of defects known or suspected to be deleterious to yields - dislocations, precipitation, swirl, haze, etc. Dislocations are easily discerned in single crystal X-ray topographs, or Lang topographs, Fig. 7. Of course, since most crystals contain zero dislocation density, and dislocations are relatively easily measured in an absolute sense by a variety of etching techniques, the use of Lang topography for measuring dislocation density is not of high priority.

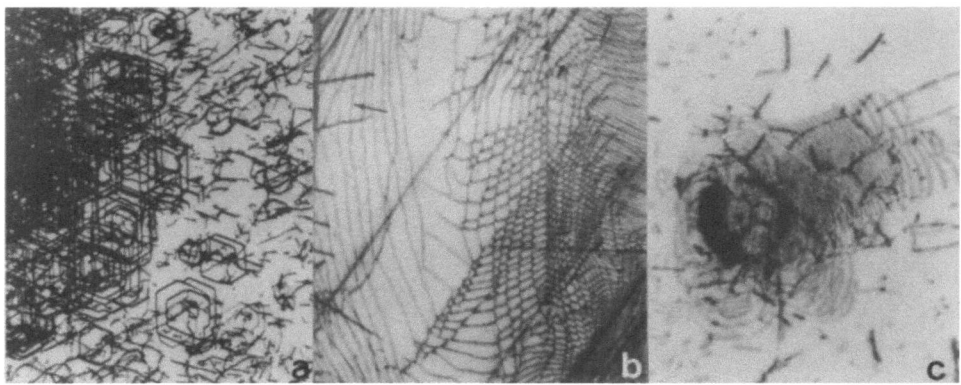

Figure 7. $2\bar{2}0$ transmission X-ray topographs showing dislocations in silicon, (a) as-received Czochralski silicon wafer, (b) annealed silicon web dendrite, (c) diffusion generated dislocations.

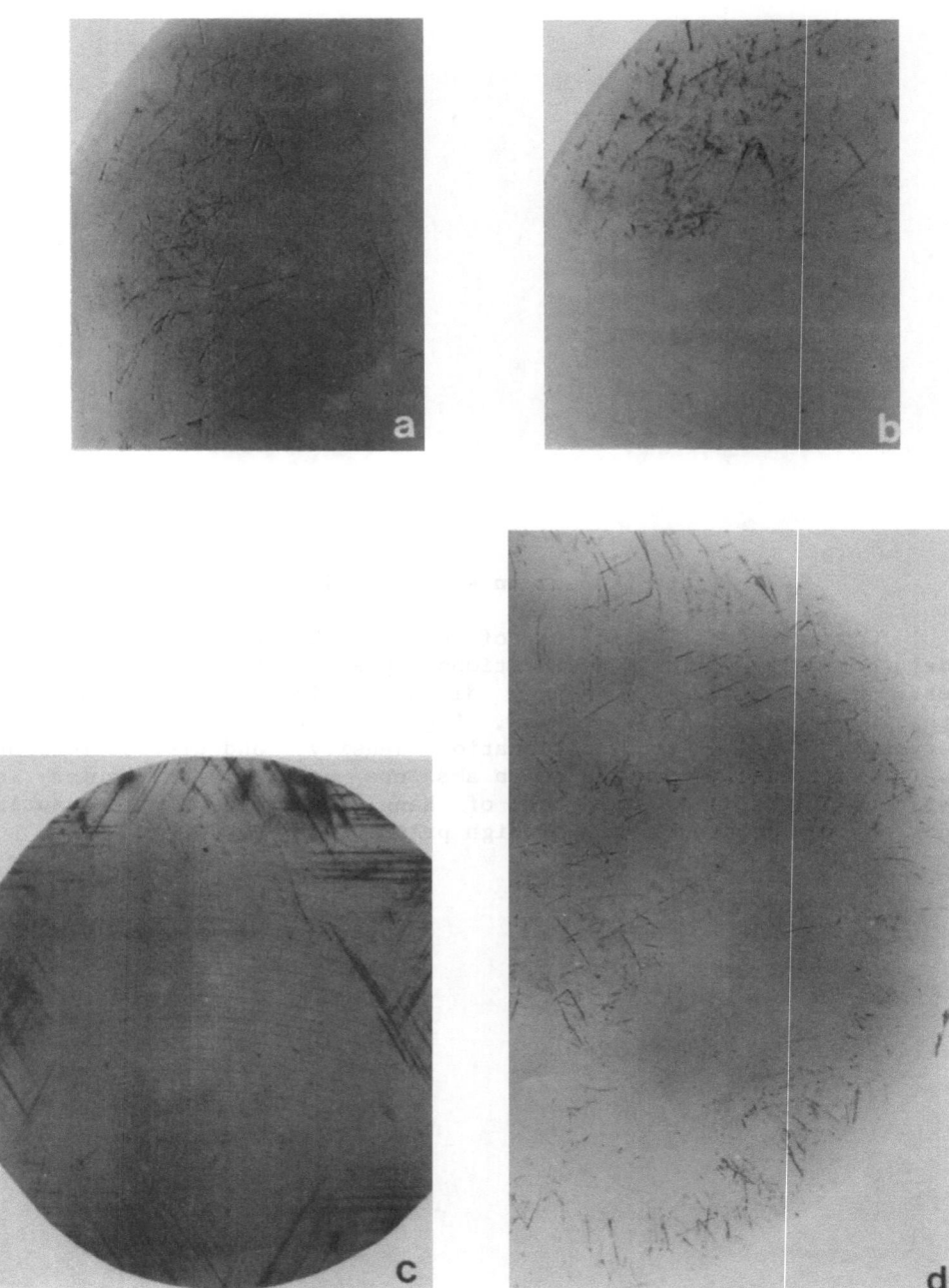

Figure 8. $2\bar{2}0$ transmission topographs showing surface damage in silicon (a) prior to etch x2.7 (b) after etch x2.7 (c) saw damage x6.5 (d) handling damage x3.5.

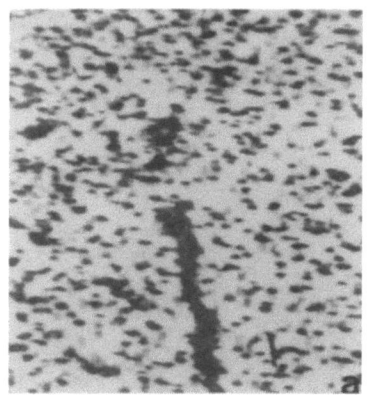 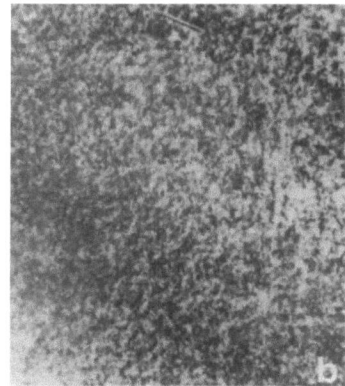

Figure 9. Precipitation in $2\bar{2}0$ topographs x 15, (a) Si, (b) GaAs

Much more important is surface damage, caused by unintentional handling, or by the polishing or lapping procedures used to produce flat wafers. Fig. 8 shows Lang topographs of silicon wafers before and after an etching step, in which about 10 μm of silicon were removed. Clearly visible in Fig. 8a, but not in the etched portion of Fig. 8b, are black images, due to residual surface damage. Since the nature of this damage is not well defined, simple etching techniques would not show the damage in any quantitative manner. The only measureable way to show such damage is by means of X-ray Lang topography. It is important to note that the nature of the damage is really not of concern - all that is needed for this application is an inspection criteria, and correlation to die yields.

Another defect that is usually considered deleterious to yields is the precipitate. Precipitates are observable using Lang topography and either individual precipitates, Fig. 9a, or clusters of precipitates, Fig. 9b, are resolvable. Precipitates can of course be observed in etched samples, but the etched precipitate artifacts may be confused with a variety of other artifacts. Hence X-ray topography provides a better method to measure precipitate density than do etching techniques. However transmission X-ray topography is not the only X-ray technique available; an even better procedure is to use section topography [6] which not only resolves individual precipitates, but also gives their spatial distribution. Fig. 10a shows a section topograph of a good silicon wafer while Fig. 10b shows a crystal containing precipitates. Pendellosung fringes [7] indicating a high degree of crystal perfection, are seen in Fig. 10a; the fringes are lacking and the precipitates are visible, in Fig. 10b. In terms of randomly distributed defects, section topography provides an enormously powerful method to inspect silicon and other crystals. Again, it should be noted that identification of the precise nature of the defects is not necessary in order for the procedure to be useful.

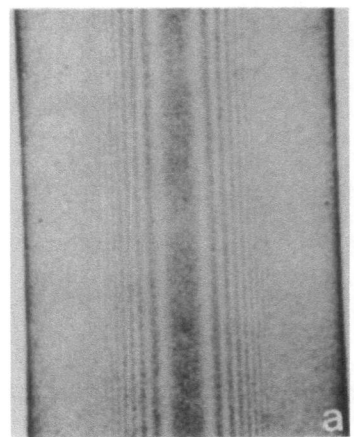 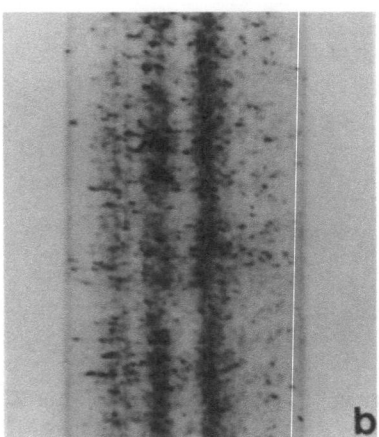

Figure 10. $2\bar{2}0$ section topographs of silicon x16.5 (a) highly perfect silicon, (b) silicon with precipitates.

Of all defects, swirl [8] is probably the most controversial. Swirl is an ambiguous term, often confused with, or substituted for dopant or resistivity striations. In fact, swirl and dopant striations are similar enough to be treated identically so far as incoming inspection is concerned. Meanwhile, disregarding the nature of swirl or striations the following inspection difficulties are encountered:

1. Swirl is not observed by etching silicon wafers directly upon receipt and it is only seen after a multiple stage heat treatment, which tends to decorate the defects contributing to swirl. Fig. 5 showed an example of etched swirl patterns. In some cases, swirl, even if present, may not be observable on crystal surfaces.

2. The lattice parameter shifts that are identified with swirl may be small, with a $\delta d/d$ on the order of 10^{-4} or so, and are difficult to detect with conventional etching techniques or even by means of Lang topography. Since the beam divergence required for Lang topography is on the order of a few minutes of arc, corresponding to a $\delta d/d$ of about 10^{-3}, in order to separate $K\alpha_1$ and $K\alpha_2$ images, only gross cases of swirl can be seen, as shown for silicon, Fig. 11 and for garnet Fig. 12. This is sufficient for some device needs, but is not sensitive for many new small-geometry devices, which may be more sensitive to lower levels of swirl.

INDUSTRIAL IMPLICATIONS OF CRYSTAL QUALITY

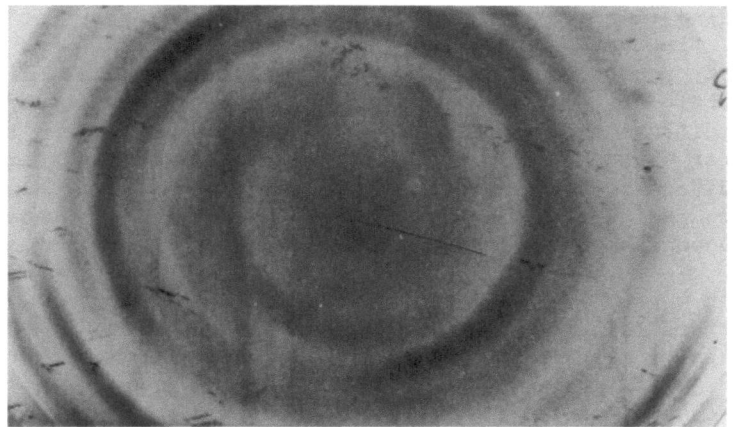

Figure 11. 220 topograph showing swirl in silicon x 3

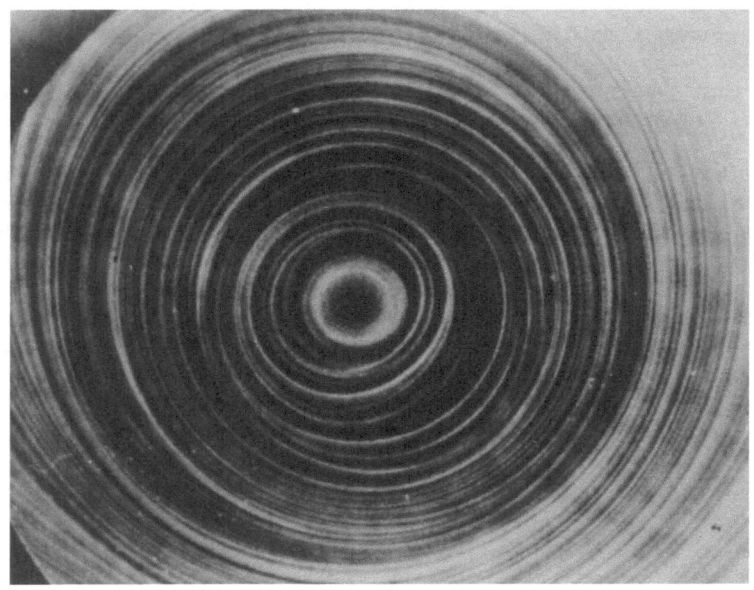

Figure 12. Reflection topograph showing swirl in garnet

In this case, double crystal diffractometry [9] is much more useful. By using high order Bragg reflections and hard radiation, a beam divergence of a fraction of a second of arc is obtainable, corresponding to a $\delta d/d$ of about 10^{-7}. Swirl types of lattice parameter fluctuations can be measured to about 10^{-8}, as shown in the 880 MoKα topograph of a highly perfect silicon wafer, Fig. 13a. The vertical striations are due to the sample wafer (test crystal) while the diffuse horizontal striations are due to the reference wafer (first crystal). The major difficulties associated with double crystal diffractometry, assuming the equipment is available, are in set up time and exposure time; hours of exposure may be required for a single topograph. Fig. 13b and 13c show more examples of swirl in silicon.

The fact that topography is an imaging technique is pertinent, in that maps of strain fields and lattice parameter fluctuations may be made. Fig. 13 is most useful because the extent of lattice parameter change can be quantitatively measured; a change from black images to white images is a 100% change in contrast, and corresponds to $\delta d/d$ of 10^{-7}, and occurs within 1 mm. Even in single crystal topographs, lattice parameter shifts can be mapped. Fig. 14 shows images of phosphorus- and boron-diffused areas in phosphorus-doped silicon. The topograph shown in Fig. 14a was taken from the low end of the Bragg peak, corresponding to the higher d spacing areas (phosphorus diffused), while Fig. 14b shows the high angle end, which corresponds to the low d-spacing boron-diffused area. The differences in contrast between Fig. 14a and 14b are due to lattice parameter changes on the order of 10^{-3}.

Imaging techniques are valuable, but are not absolutely necessary. There are many situations where counting techniques are more useful. Consider the anomalous transmission of X-rays, where the intensity of X-rays transmitted through a perfect crystal is substantially higher than that transmitted through an imperfect crystal of equal thickness. The anomalous transmission or Borrmann effect [10] has been used to look at oxidation-induced precipitation in silicon [11]. The technique is sensitive to small strains, is quick, and requires much less equipment. It would appear that this procedure is the least exploited but perhaps most useful technique that can be used for incoming inspection. In addition to its potential use in incoming inspection, it is valuable for looking at process-induced defects, a topic not considered in this chapter.

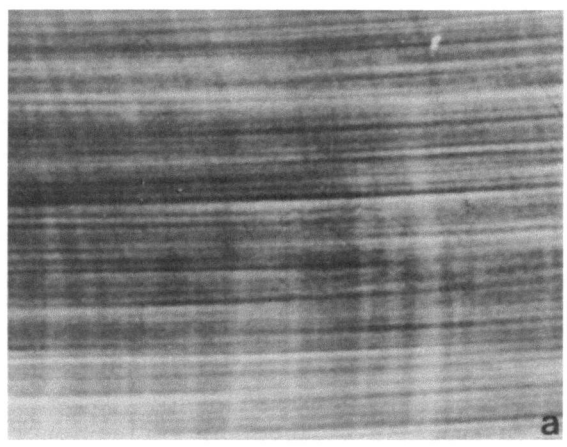
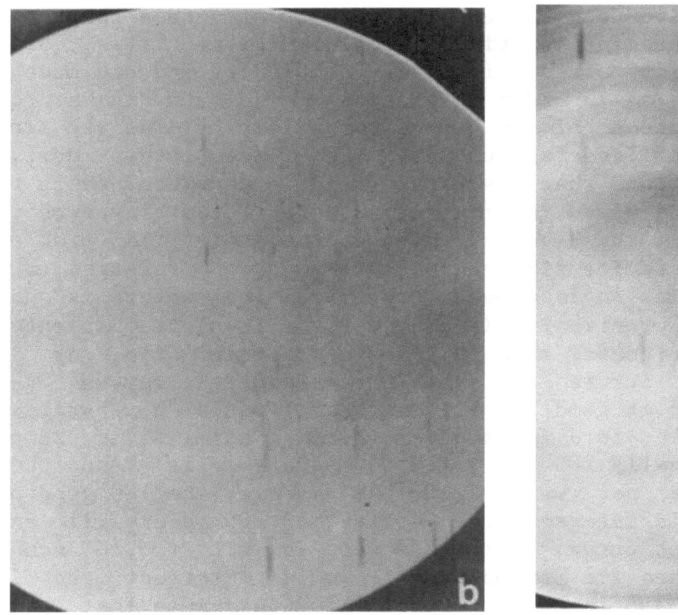
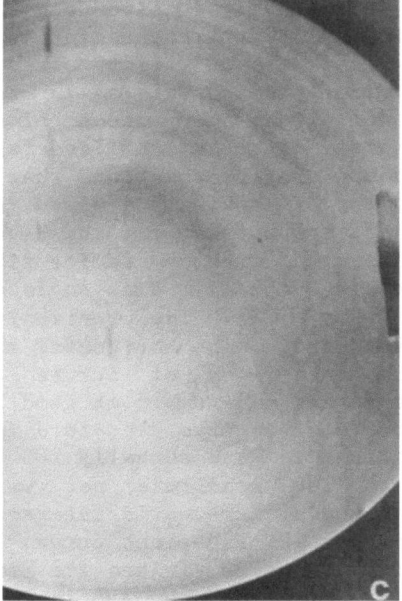

Figure 13. Double crystal X-ray topographs of silicon, showing swirl. (a) $8\bar{8}0$ double crystal reflection topograph showing swirl in silicon, x 5, (b) low swirl, 440, CuKα (c) medium swirl, 440, CuKα

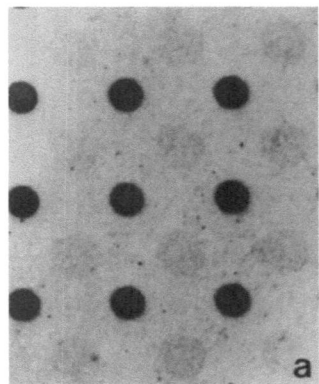
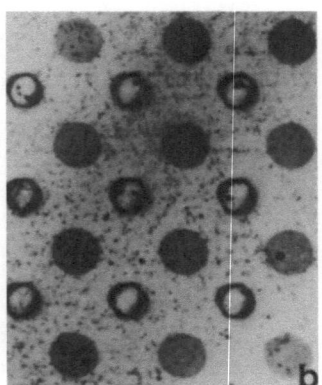

Figure 14. Diffused areas in silicon $4\bar{4}0$ reflection topographs x 20
(a) high angle (low d) (b) low angle (high d)

1.2.2 Yield

In a production environment, yield is of course the bottom line. However, measurement of yield presents difficulties, since devices are of various sizes, of various complexity and are made on variously sized wafers. In order to present yield data in a form that allows comparison of one device to another, yields are often expressed in terms of Isodefect curves. These are curves derived under the presumption that practical yields are determined by the number of masking steps and inherent defect densities involved in the manufacture of the devices, and that the geometrical problems associated with one device size and wafer size can be referred to a standard 100 mil^2 die. An Isodefect curve for 3 inch wafers is shown in Fig. 15a. The vertical axis of the curve is plotted in actual yielded die. The horizontal axis is a measure of die size, or die area. Families of curves are drawn — the Isodefect curves — that represent the number of good 100 mil^2 die that would be available for a given true die yield for a given true die size. As an example if the die were actually 100 mil^2, (10,000 square mils) and there were 100 good die per wafer, on the average, the 100 good die horizontal axis would intersect the vertical 10,000 square mil^2 axis on the 100 Isodefect curve. If the die were 140 mil^2 (19,600 square mils), the 100 yielded die per wafer axis would intersect the 140 mil^2 axis on the 350 Isodefect curve. This would mean that if the die were 100 mil^2, there would be 350 good die per wafer. The higher the Isodefect curve, the better the yield. The top curve represents the maximum number of die theoretically available on a given wafer. Fig. 15b shows an Isodefect curve for 100 mm wafers. The vertical axis is shifted by a factor of 1.8, to give a yield comparable to 3 inch wafers (the area ratio between 3 inch and 100 mm wafers is 1.8).

INDUSTRIAL IMPLICATIONS OF CRYSTAL QUALITY

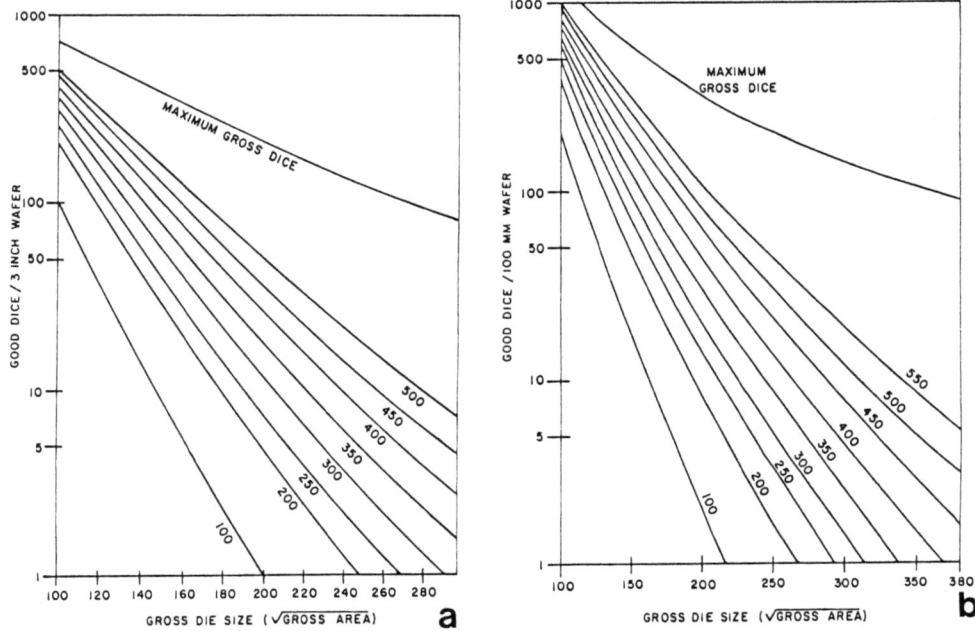

Figure 15. Isodefect curves (a) 3 inch (b) 100 mm

Now, the obvious question is, what do Isodefect curves have to do with X-ray diffraction? The answer comes when actual device yields are plotted on Isodefect curves, and are compared to other, similarly processed devices, and to theoretical maximum yield device yields. The theoretical maximum possible die yield is based on the assumption that all die on a wafer function properly. This is represented by the upper Isodefect curve. That maximum yields are not observed is due in general to defects in the silicon, either as-grown or process induced. By understanding the nature of the distribution of defects, the yield of a given device of a given size can be calculated. This yield can be referenced to the standard 100 mil^2 die, thus allowing the Isodefect curves to be drawn. Added processing steps induce more defects, which lowers the Isodefect level by a fixed (empirical) amount. Hence, a given process can be characterized by its Isodefect level, which is a measure of the quality of the process. Yields less than the expected value for a given process are due to causes beyond the known defect levels.

Typical causes for lower yields are poor resolution in masking, for example due to misalignment of various layers, poor thin-film processes, over- or under- etching, defects in the substrate material, etc. Many of the causes for yield loss are reasonably easy to assess; partial functionality of devices, device parameter shifts, visual misalignment, etc. provide valuable clues. On the other hand, defect effects are quite difficult to measure. The defects are almost always invisible during optical inspection of a device; in general, the devices must be carefully dissected before the defects become visible. This is a difficult and tedious procedure, that is often unsuccessful.

Figure 16. $2\bar{2}0$ transmission topograph of a dynamic 1k RAM. (after metal and oxide strip).

X-ray topography, on the other hand, provides a method to detect such defects, which then can be analyzed to determine why various processes or wafers do not give projected Isodefect levels. An example of this is given in Fig. 16, a transmission X-ray topograph of a 1K RAM. A number of tests showed that wafers exhibiting few dark images in the memory array averaged higher Isodefect curves than wafers with lots of black defect images. The topograph shown in Fig. 16 was taken after stripping off all surface films. Hence, the many defect images are due to either process-induced problems, or to as-grown or incoming wafer defects, and not to elastic strains due to surface films. It is thus possible simply to evaluate some X-ray topographs of wafers, to count defects and to correlate the contribution of defects to Isodefect curves.

INDUSTRIAL IMPLICATIONS OF CRYSTAL QUALITY

This eliminates a large problem in evaluation of yield loss causes, and indeed aids in process improvement. Both transmission and reflection topography can be used. An example of a similar situation is shown in Fig. 17. Here, topographs of a stripped and unstripped CCD (charge coupled device) imaging device are shown. In addition to the expected strains induced by diffusion, or other processing, defects can be observed. The yield of the devices has been related to the number of defects; contributions to Isodefect levels from masking, processing, and defects can be quantified and separated from defect effects.

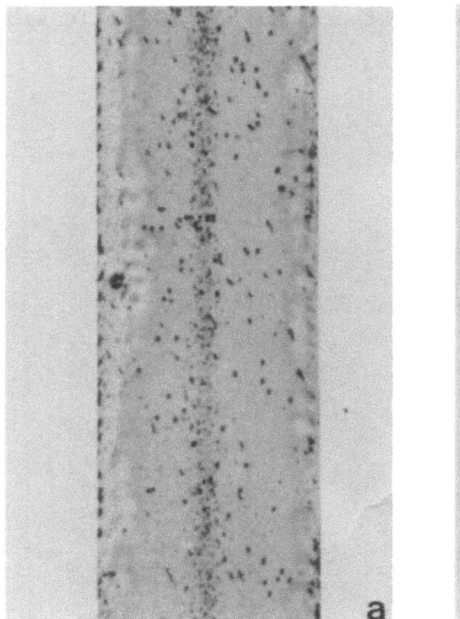
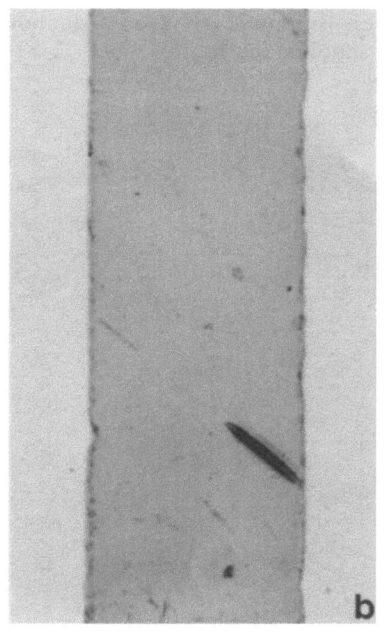

Figure 17. Transmission X-ray topographs of an imaging CCD device (a) defective device, (b) good device

In addition to individual defects that may lead to yield losses, there are some categories of defects that are distributed throughout the bulk of the crystal. Fig. 18a shows an etched silicon wafer that exhibited a very low Isodefect level and Fig. 18b shows an etched wafer with a high yield. It is seen that the phenomenon described earlier as swirl is evident in the low-yield wafer. Many defects contribute to yield loss; swirl appears to be one of these.

Isodefect analysis is common, almost universal, in Fab production areas. X-ray topography is rare, to non-existent. This is due in large part to a lack of attention rather than to a lack of usefulness as X-ray topography is regarded by Fab production engineers as an exotic technique not applicable to production needs.

This in turn is attributable to the lack of attention paid to production needs by X-ray investigations and most X-ray topography analyses have been devoted to explanations of contrast mechanisms, dynamical theory and descriptions of individual defects. Very little has been done to relate yields to topographic images, without regard for the physics of image formation or to specific defect effects. Yield analysis is difficult, but it is the most important factor considered in a production area. Any procedure that helps a Fab engineer to understand and improve yields will be used, provided the procedures are geared to production rather than research use. The fact that X-ray techniques are not used is a reflection of the research nature of X-ray methods, rather than a lack of feasibility of the technology.

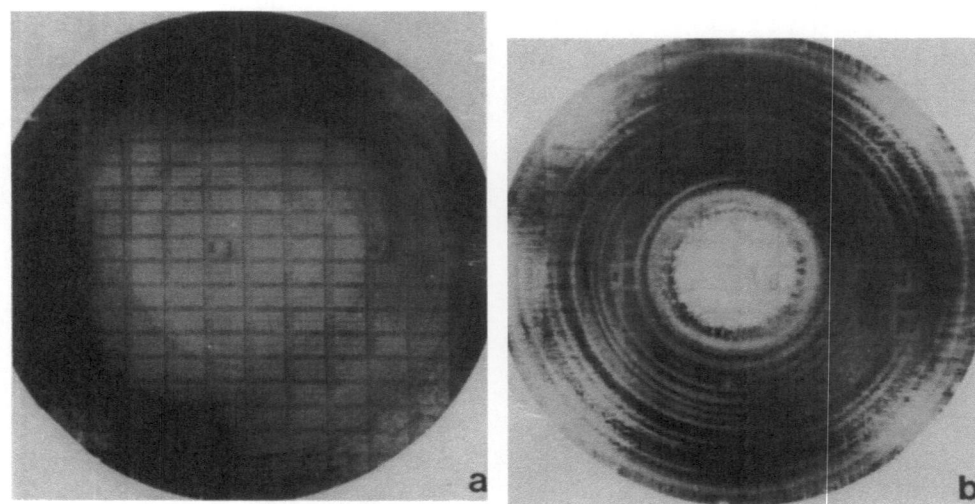

Figure 18. Swirl in silicon wafers (a) high yield, (b) low yield.

1.3 Advances in technology

As has been discussed, X-ray techniques can and should play an important role in characterizing single crystals, in a manufacturing environment. To perform this function the techniques must be non-destructive, and be compatible with current materials, devices and technologies. The particular advances in technology that appear to be of significance and relate to X-ray analysis are the following:

1. Up to 125 mm crystals will be in production by 1981.

2. More perfect bulk crystal properties may be required, to help improve yields, or conversely:

INDUSTRIAL IMPLICATIONS OF CRYSTAL QUALITY

3. Specific types of bulk or surface damage may be intentionally introduced, to act as gettering centres.

4. Wafer distortions, particularly non-symmetrical variations in wafer size, will play an extremely important role in masking technology.

5. Smaller devices, with more dense information storage capacity, where the information is stored in very shallow layers on the device surface is certain for silicon based technology. Such devices will be very sensitive to small defects present in or near the charge storage areas.

6. Higher resistivity silicon substrates will be required for many advanced devices.

7. Highly perfect sapphire and garnet substrates will be required for SOS and GGG technology.

8. Much more attention will be paid to insuring that incoming materials meet quality specifications. Quicker procedures will be useful.

Each of these topics is discussed below.

1. According to Fig. 1, 125 mm silicon, sapphire and garnet wafers will be in production in the early 1980's. Any X-ray technique that is used to analyze materials in a production atmosphere must be able to cope with such large wafers. At present neither commercial single crystal cameras nor double crystal cameras are so constructed. Even X-ray generator tables are small compared to the sizes needed for large size cameras. X-ray techniques must be able to handle the coming materials - at present they are barely able to deal with existing wafer sizes, and there is little evidence to suggest that larger wafers may be handled.

2. Bulk crystal properties probably will tend to become more perfect, and many devices will require very small gradients in properties such as lattice parameter, defect structure, precipitate density, etc., in order to operate properly. Etching techniques will not suffice for material characterization; the methods are too insensitive, and too subject to inspector interpretation. Anomalous transmission techniques will probably be necessary to measure crystal quality; double crystal techniques, although technically suitable, will not be commercially available for a number of years due to a lack of equipment.

3. There is evidence that some devices can be improved if gettering techniques are used to eliminate residual contamination. While many devices use phosphorus gettering techniques, the use of

backside surface damage and possibly buried bulk damage for
gettering will receive more attention. Damaged materials are
difficult to evaluate by etching techniques. X-ray section
topography may provide the answer. The controversy about the effects
of damage on device properties will eventually be resolved, and
X-ray measurements may well be the best technique to help to do so.

4. One of the major problems facing device users is wafer
distortion. This comes about because of the need for extremely high
registration accuracy in new devices, with tolerances of only
fractions of a micron between adjacent layers. Wafer distortion,
especially non-systematic distortion, will be disastrous. The causes
of wafer distortion are not really understood, especially if bulk
distortion (for example, due to local lattice parameter changes) is
present in addition to warpage distortion. It appears that double
crystal diffraction methods will be needed to measure wafer
distortion, for a 100 mm wafer (= 10^9A), a 10^{-5} $\delta d/d$ due to a
process variation will give 1 μm distortion over the 100 mm, and
this low level is barely detectable with Lang techniques. Lattice
distortion maps such as shown in the double crystal contour
topography in Fig. 19 may provide the best method to visualize
distortion. GGG and sapphire will present similar problems, in the
near future. Unfortunately, as pointed out earlier, suitable large
size double crystal diffractometers are not routinely available.

5. As was shown in Fig. 16, X-ray topographs of devices can be
used to inspect for defective bits. However, the device illustrated
in Fig. 16 was a 1K RAM. Present devices are about the same die
size, but contain 16 K bits, and 64 K devices have been sampled.
Since a single defective bit can kill a whole device, it is
absolutely certain that smaller and more dispersed defects will be
of importance in affecting yields. In view of the spatial resolution
of X-ray topography of about 1 μm, it is not clear how these defects
will be resolved; as the images are becoming rather confused, even
at this stage in development.

6. Many new device technologies will require high resistivity
silicon, 50 to 100 Ωcm being likely. In view of the limitations of
Czochralski (CZ) technology to provide high resistivity silicon,
float zone (FZ) silicon will be used more frequently. This presents
a severe problem to process technology, in that FZ silicon has a
tendency to warp more during processing than does CZ silicon, for
identical processes. The cause for this is not understood, and is a
subject of much concern. The need for FZ silicon approaches the
seriousness of the wafer distortion problem, so far as Fab
technology is concerned; warpage is a very serious problem.

7. Sapphire technology has been around the corner, for 15
years. At this time, it does appear that true silicon on sapphire
volume production is - around the corner. Since modern devices are

orders or magnitude more complex than the devices originally conceived, it would appear that the silicon films into which the devices are constructed must be of higher perfection. Since the films are epitaxial in nature, they inherit their structural perfection from the sapphire substrates. Hence the need for more pefect sapphire substrates. X-ray topography will probably be the most reliable method used to inspect substrates, provided cameras of suitable size are available. Similar arguments apply to GGG.

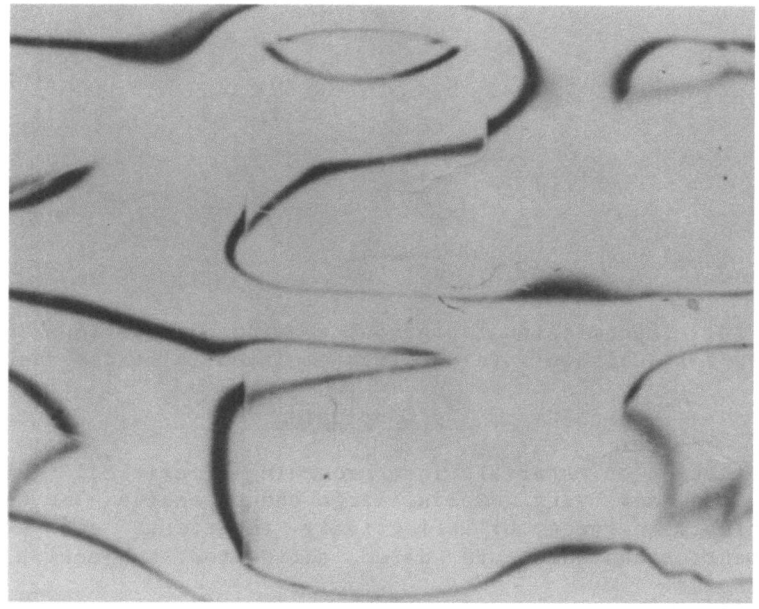

Figure 19. Double crystal contour topograph of a distorted silicon wafer (courtesy M. Hart).

8. Silicon, sapphire, garnet, etc. wafers are increasing in size and in cost. Devices are getting larger and more complex. A simple analysis of Isodefect levels shows that even for a well understood process, i.e. one that maintains a particular high Isodefect curve, the larger size die will mean there are significantly fewer good die per wafer. For example, on a 400 Isodefect curve, on a 3 inch wafer, there are predicted to be 15 good 220 mil^2 die. If the die sizes goes to 260 mils, there will be only 6 good die. At an assumed cost of $100/wafer, the die cost will increase from $6.50 to $15.00. Hence, it is of paramount importance that the die yield not be increased by the presence of additional defects, and control of incoming quality will become more important than it now is. All possible steps will be taken to insure that whatever defect level is specified, will be accepted. Due to some of the considerations discussed above, the characterization of these

materials will be complex and expensive, and all possible tools will be needed. X-rays provide the best methods to handle these problems. Some of the advances in taking rapid topographs will be useful. Fig. 20 shows a video display of an X-ray topograph, compared to its film equivalent. The high resolution of the video image shows that very quick images can be used for some inspections.

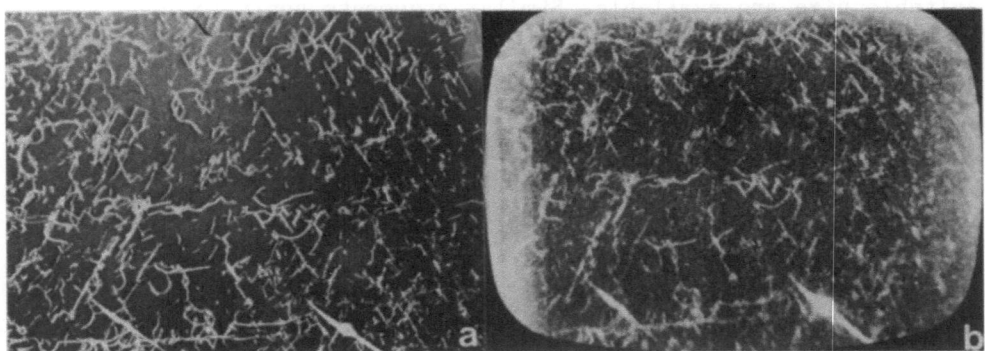

Figure 20. Transmission topographs showing dislocations in silicon (courtesy J-I. Chikawa). (a) nuclear emulsion (b) video display.

1.4 Conclusion

Evaluation of materials is approaching a critical phase. The devices are now very complex, large and expensive. The wafers are large, highly perfect or selectively imperfect, and expensive. Measurement techniques are dated, antiquated, subjective, complex and expensive.

Because of the need for evaluation techniques, there may be renewed emphasis for X-ray methods to be used for non-destructive quantitative measurements. However, this emphasis will not succeed in providing useful techniques, unless attention is paid to making the techniques compatible with production needs. This includes making cameras compatible with large diameter wafers, making specialized cameras to reduce equipment costs, and especially bringing X-ray technology to the Fab engineer level. This latter topic is perhaps the most difficult to implement. Fab engineers are preoccupied with yields, processes, devices; they are uninterested in X-ray theory, defect analysis, equipment design and other classical problems which are of interest to X-ray technologists. Consequently, X-ray technology will have to be brought to the level of the Fab engineer, in order to convince him that he will be helped in his quest for improved yields. This will be a task for the X-ray technologist rather than for the Fab engineer, who is constantly bombarded with new methods for doing everything possible in the Fab area. He has to be convinced that X-ray methods will help him.

INDUSTRIAL IMPLICATIONS OF CRYSTAL QUALITY

The X-ray methods available to help incoming inspection, yield analysis and problem solving are able to cope with such needs. Lang topography, section topography, double crystal diffraction, anomolous transmission all have their uses. The techniques are available, the defects are detectable, the needs are real. All that is required is to join these three topics, to the benefit of device technology. This, rather than X-ray physics, is the need for the 1980's, if the advantages of X-ray technology are to be fully utilized.

REFERENCES

1. A.D. Milne, "The Technical Importance of Growth Defects", Chapter 2 this volume

2. B.K. Tanner (1976) X-Ray Diffraction Topography, Pergamon Press, Oxford

3. E.S. Meieran (1970) Siemens Review 37 Special Issue "X-ray and Electron Microscopy News"

4. A.R. Lang (1970) in Modern Diffraction and Imaging Techniques (ed. S. Amelinckx et al) North Holland Press p.407

5. A. Duncan (1952) Quality Control and Industrial Statistics, Richard D. Irwin, Inc. Chicago

6. A.R. Lang (1958) J. Appl. Phys. 29 597; (1959) J. Appl. Phys. 30 1748

7. M.Hart (1963) Dynamic X-ray Diffraction in the Strain Fields of Individual Dislocations Ph.D. Thesis, Bristol University. See also selected bibliography in Refs 2 and 3

8. International Symposia on Silicon Material and Technology sponsored by the Electrochemical Society 1969, 1973, 1977

9. U. Bonse & M. Hart (1965) Appl Phys Letters 7 238 See also ref 3

10. G. Borrmann (1941) Z. Phys. 42 157 See also ref 3

11. J. Patel & B. Batterman (1963) J. Appl. Phys. 34 2716

CHAPTER 2

THE TECHNICAL IMPORTANCE OF GROWTH DEFECTS

A.D. MILNE

2.1 Introduction

The technical importance of defects has already been considered from the viewpoint of the overall manufacturing process. In this chapter a different approach will be taken and the subject considered from the point of view of the device operation.

Many studies have been undertaken using a variety of techniques and it is well known that it is extremely difficult to make a good correlation between device operation and growth defects. This has been to a large extent due to the fact that device yield is primarily determined by factors such as dust particles, alignment tolerances during masking, pin holes in photoresist layers etc. which obscure the effect of material imperfection. Indeed, with standard large scale integrated (LSI) circuits particularly using MOS technology it is often not until the wafers buckle like potato crisps during processing that the material quality is considered as a problem. However, the semiconductor companies have been steadily improving their production techniques, not only to obtain the obvious commercial advantages of higher yield but also to enable them to produce larger and denser circuits using tighter design rules. This has begun to expose stability problems caused by grown-in stress in the wafers and in the more sensitive devices it has been clearly demonstrated that discrete defects are affecting device operation. This is particularly true for the analogue devices which are normally designed so as to maximise the signal to noise ratio rather than being designed to fixed operating margins as are digital devices. Although, arguably only a small sector of the semiconductor market at present, the drive towards greater and greater systems integration in a single chip, will make mixed digital and analogue LSI chips more common in the future than the few available today and hence greater attention will have to be paid to material specifications.

One design technique within MOS which clearly illustrates the effect of growth defects on device operation is that of charge coupling. More commonly considered as using a separate technology, charge coupled devices (CCD) have been developed as memories, analogue signal processing elements and solid state imagers. With the development of MOS technology, in particular the double layer polysilicon gate process, the identity of the separate CCD technology has been lost and charge manipulation is becoming an accepted technique within the design of analogue MOS structures. It is perhaps worth pointing out that although the use of CCDs for large digital memories received a great deal of publicity it has been almost totally superseded by the improvement in RAM design and apart from some special situations CCDs are not now used for digital signal processing.

The importance of growth defects to device operation is not limited to silicon based devices although they comprise by far the largest sector of the electronics component market. An interesting newcomer which is not unrelated in operation to the CCD and is projected to have a market of £2000M in 1990 is the magnetic bubble domain memory. Its development has been particularly rapid from the perfection of the crystal growth techniques in the early 70s to a fully engineered device with 1M bits of storage in 1979. The fabrication process is relatively simple compared to that of silicon devices but the device operation is strongly dependent on the basic material's structural quality. The problem of residual defects is overcome in an interesting if pragmatic way as will be discussed later.

The use of the optical properties of crystals for engineering purposes is a fast developing field and there are examples which could be considered where structural defects limit the performances of a system. Optical modulators, for instance, for use in communications systems make use of electro optic crystals such as lithium niobate which are cut into blocks about 4 cm long and 6 x 6 mm in section. The performance as measured by the extinction ratio of those crystals varies dramatically with the subgrain and dislocation structure so that the overall specification of the systems can be limited by the perfection of the crystals used.

For the third example, however, a more commonly used device will be considered; the surface acoustic wave (SAW) device. The SAW device has played an important part in modern radar systems for a number of years and is now a standard component in consumer TV sets as an intermediate frequency filter. Based on the use of piezo-electric materials, the devices do not generally place too stringent a requirement on crystal perfection. However, the range of available materials which are suitable is small and they do not offer a wide spectrum of frequencies of operation. There is considerable commercial and military interest in the development of piezoelectric crystals other than quartz and lithium niobate which are available with adequate perfection in order to develop systems operating at higher frequencies and larger bandwidths.

In this chapter the effect of growth defects on operation of the three types of devices, charge coupled devices, magnetic bubble domain memories and surface acoustic wave devices introduced above will be discussed.

2.2 Charge-coupled devices

The basis of charge coupled devices (CCD) is the Metal- Oxide Semiconductor (MOS) capacitor. In n-channel technology this is formed by putting a metal plate on the silicon dioxide which is grown on a p-type silicon wafer, Fig.1. The structure is directly analagous to a two plate capacitor with one of the plates replaced by a layer of silicon. Of course, the characteristics of the MOS structure are rather different from those of metal plate capacitors.

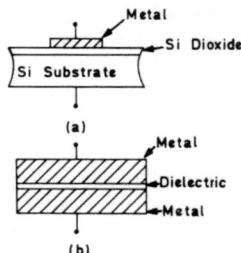

Figure 1. (a) MOS Capacitor, (b) metal plate capacitor

If a positive voltage is applied to the metal plate of such an MOS capacitor, the holes in the silicon will be repelled. The region of silicon under the plate will become depleted and, because of the fixed ionised acceptor atoms, a space charge region will be established. As the voltage is increased, electrons will be attracted to the plate and the interface region between the oxide and the silicon, just in the silicon, will become inverted; that is, a layer of opposite polarity to that of the bulk of the substrate, the inversion layer, will have been established. As the electrons required to produce inversion come from thermally generated electron hole pairs, the process does not take place very rapidly which is fortunate for it is on this fact that CCDs rely for their operation.

If a small negative voltage is applied for some time to the MOS capacitor, the mobile holes in the silicon will be attracted to the surface which will become accumulated. The application then of a large positive voltage to the capacitor will drive the holes away from the interface into the bulk of the silicon and will create an inversion layer as described above. However, for a period immediately after the application of the positive voltage, inversion will not occur and the device is said to be in a state of deep depletion. The positive charge in the metal electrode is balanced by the ionised acceptors in the silicon and in order to preserve neutrality, the width of the depletion layer varies according to the applied voltage. As the voltage is increased so the depletion region becomes larger. This can be thought of as the creation within the silicon of a potential well whose depth is a function of the applied voltage. Such a potential well when suitably isolated can be used to hold a quantity of charge which is proportional to a signal amplitude.

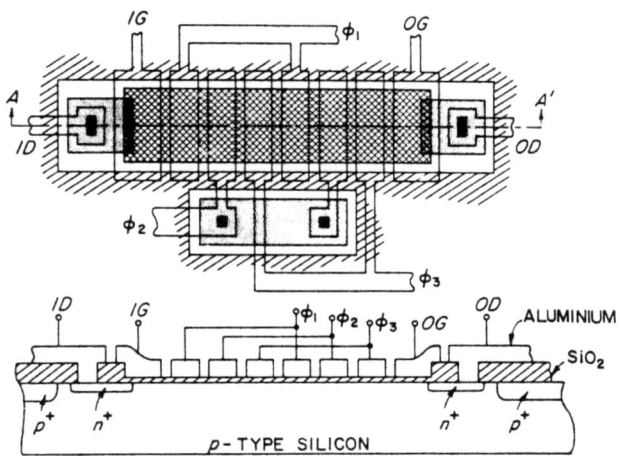

Figure 2. A two bit CCD with 3 phase clocking

Various structures have been developed to manipulate the packets of charge to perform complex electronic functions. The construction of a CCD involves producing a number of the basic capacitors very close together so that by appropriately varying voltages applied to clock electrodes the charge can be moved from one capacitor to the next. A simple example of how this is done is given in Fig. 2. This shows a simple two bit linear CCD with six coupled capacitors connected in pairs to clock lines. The voltages applied to the clock lines create the potential wells under the electrodes with the appropriate timing to control the movement of charge. In this example input and output circuitry is in the form of diodes with gate electrodes by means of which the signal is injected into the device and the charge packet after traversing the six stages is detected at the output.

A simple diffusion contains the charge in the area beneath the gate electrodes. The operation of the device is easily seen with the help of Figs. 3 and 4 which show the applied signals and the shape of the resultant potential wells. At t=0 the voltage on the input diode is high while the gate (IG) is held low, preventing any of the electrons from the source flowing into the device. If the gate is set to some intermediate voltage and the electrodes to a higher voltage, (with and remaining low) a potential well will be formed. When the input diode is pulsed low, charge is allowed to flow into the well under the first ϕ_1 electrode and spill back over the potential barrier under IG as the input diode returns to a high level. The amount of charge contained in the potential well is governed by the difference between the two levels of IG and ϕ_1. To move the packet of charge along the device a voltage is applied to ϕ_2 and ϕ_1 is returned to a low level and the charge spills into the newly created potential well. Similarly with ϕ_3 until the charge packet is output at the output diode OD.

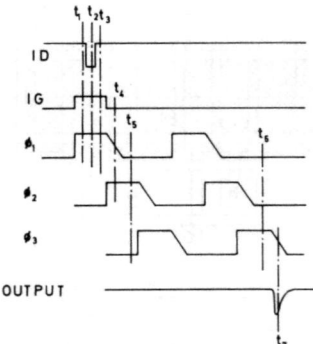

Figure 3. Timing diagram for clock wave forms

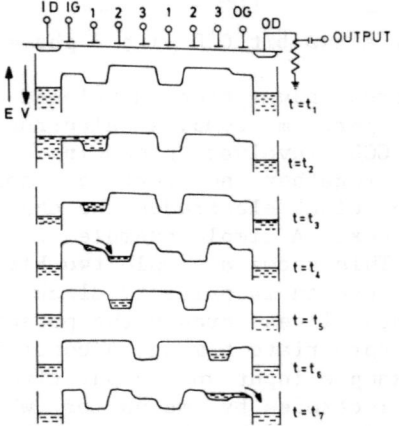

Figure 4. Potential distribution within the silicon

This process of charge transfer will only work satisfactorily if there are no potential barriers between the capacitors and this requires that the electrodes are very close together.

Many schemes have been tried with a number finding favour for particular specialist applications. The recent perfection of double layer polysilicon process which is used for most microprocessors and memories produces very good results for most devices outside the imaging field and has had the beneficial effect of bringing CCDs into the mainstream of MOS processing. With this development it is generally recognised that CCDs are not a technology but simply a design technique which can be used by MOS designers. It is important to recognise that CCDs are analogue devices as the charge packet created can be linearly related over a wide range to the input signal amplitude. This feature, combined with their clocked method of operation makes them attractive from the point of view of power consumption and size for a wide range of signal processing applications.

Perhaps more importantly, they are almost ideally suited to solid state imaging applications where charge is created by the incidence of light on an array of silicon photodiodes. The amount of charge created is proportional to the intensity of the illumination and it is essential if large photo sensitive arrays are to be made, to be able to manipulate these analogue quantities with a minimum use of silicon area and power consumption. Analogue signal processing using charge coupled techniques gives just this possibility.

2.2.1 Applications

Imaging can be performed in two ways. Either with a single line of photo sensors which can be scanned mechanically at right angles to the axes of the device to build up a spatial image, or with a stationary two dimensional image array. The line imagers are most frequently designed with a single line of photodiodes with a CCD shift register along each side. This allows a larger number of image points ($\sim$2000) to be constructed in one device as the photodiodes can be made on 13μm pitch and the CCD registers, which tend to be bulkier, can be limited to half the number of imaging points each. This is an important consideration as will be seen later.

With two dimensional sensors, the topology of the device is more constraining if a dense array of image points is to be achieved. The two most common methods of construction use either frame transfer or line transfer to read out the information. In the former after a period of integration (say 30 ms) the whole frame is rapidly read into a second identical structure known as a frame store which is shielded from the illumination. The information can be read out of the frame store at leisure during the next integration period. In line transfer each line of information is read out individually.

One obvious system use for the area devices is in TV cameras to make them more easily portable and a few companies have built fully compatible colour TV cameras which are similar in size to a single lens reflex camera. These devices have typically 496 x 475 elements and measure 2 x 1.5 cm in area and are rather difficult to make in any quantity.

It is important to appreciate also the sensitivity of solid state imaging devices which allows them to be used for very low light level applications. With suitable cooling (-40°C) and using special amplification techniques images can be formed with input levels of 6×10^{-6} W m^{-2} which is equivalent to only 25 electrons in a single pixel. This type of sensitivity is only useful if sources of noise and spurious signals can be kept to a correspondingly low level and this puts stringent conditions on the defect level. Hence both the processing techniques and raw material quality have to be of the highest order.

The basic signal processing functions which can be performed easily using CCDs are

i) Simple analogue delay

ii) Multiplexing

iii) Transversal and recursive filtering

iv) Correlation

and a number of devices have been developed for these purposes. One of the more recent is the 256 point analogue correlator which was designed at the Wolfson Microelectronics Institute. The chip is about 5mm x 4mm and contains a variety of linear MOS and CCD structures which are very densely packed. With this and the other building blocks complex systems can be constructed. For instance once important application area is sonar signal retrieval in the presence of high reverberation noise with the use of correlation techniques. Also using a recursive filter, a canceller for stationary targets in a MTI radar can be made. Fourier transformations can be carried out using the so called chirp Z transform (CZT) algorithm with an accuracy sufficient to satisfy a number of spectral analysis applications in speech, video, Doppler and sonar processing.

All of these applications require the analogue data in the CCD to transfer through a large number of stages without significant corruption. Typically a filter length of one to two thousand stages would be used for a moderate resolution system which would imply that the charge packets traverse some 50 mm of silicon during the signal processing operation. The demands on the transfer efficiency are therefore quite stringent and consequently any defects which degrade it are a severe embarrassment.

Equally important are the defects which cause extra charges to be generated locally at a rate greater than that of the background. These local "hot spots" are particularly important in imaging devices where, unlike most of the signal processing devices the charge from all the stages is not integrated at the output but the information from each pixel retains its individual identity. These effects are considered in more detail in the following section.

2.2.2 Charge transfer inefficiency

The amount of charge which is left behind when a packet is transferred from one stage to another is known as the inefficiency per stage (E) so that the overall transfer inefficiency of a device is given by ηE where η is the number of stages. The effect of the transfer inefficiency on the operation of a device is to smear the charge which resides in one potential well into other wells hence corrupting both the original and surrounding packets (Fig. 5). For values of $\eta E < 0.6$ the effect is not too serious and usable devices can be made up to this value. For long delays and high frequency operation as low a value as possible is desirable.

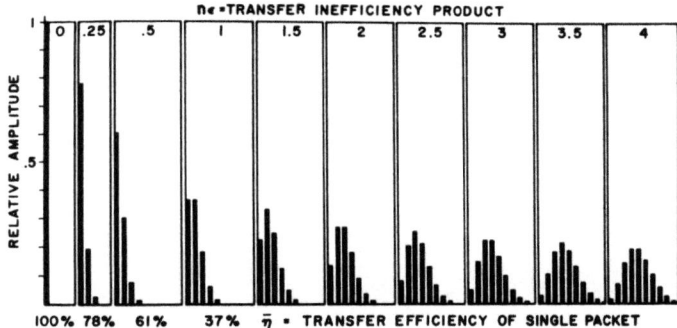

Figure 5. Effect of transfer inefficiency on signal amplitude

There are many factors which contribute to transfer inefficiency but one of the most significant is charge trapping. As a packet of charge encounters trapping sites they are filled by some of the electrons of the packet almost instantaneously. However, the traps relax with a wide range of time constants which can be as long as a few seconds so that charge is redistributed into trailing charge packets. The majority of the traps lie at the silicon – silicon dioxide interface and result from the oxidation process during device manufacture. There are however also bulk traps which are associated with defects and impurities. In simple CCDs the charge packets intersect the interface traps since the charge resides at or close to the interface. These so called surface channel CCDs suffer severe transfer inefficiencies and even with the best starting material and good processing, values only of $E \sim 10^{-3}$ can be achieved. With the use of a background bias charge which helps keep the traps saturated and is known as a "fat zero", devices with 2000 stages have been fabricated. To overcome the interface trapping problem and for other reasons associated with operating speeds, buried channel devices in which the charge moves through the bulk of the silicon several microns below the surface have been constructed. The transfer inefficiency in these devices is much less as the density of trapping sites is lower in good quality material and values of $E=5 \times 10^{-5}$ have been recorded in the literature. In order to achieve this figure however, very high perfection material is required and careful processing is obligatory

2.2.3 Dark current

As CCDs are volatile storage elements the stored information will be continuously corrupted by leakage and the thermal generation of additional charge. These effects are however in general not spatially uniform and signals at different locations in a device are corrupted at different rates. An example is shown in Fig. 6 in which the charge in a device has been allowed to integrate (i.e. without clock operation) for a period of 0.25 sec. and then read out rapidly. Several serious dark current spikes can be seen above the general background white noise. The origin of dark current is attributed to certain impurities in the bulk material. If uniformly distributed they could in principle be removed by a suitable gettering process. However, the impurities coalesce onto dislocations and stacking faults and produce a fixed pattern of noise.

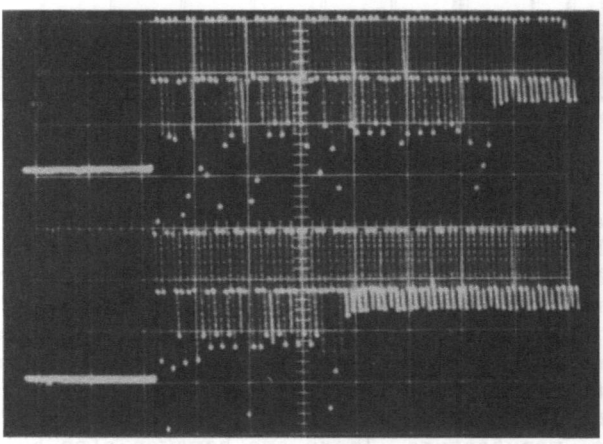

Figure 6. Variation of dark current under stop clock operation Integration time 0.25 secs., Amplitude 1V cm^{-1}

The general effect of high dark current is to put a minimum frequency at which the device can be operated before the background noise becomes significant. As it is the total time for which a charge packet is in the silicon that matters it is not only the lower frequency which is effected but also the overall length of the device. With good quality material and careful processing, current densities as low as 5 nA cm^{-2} have been achieved at room temperature but there is a wide variation from device to device and even within the same device. For instance it is common to have local spikes with densities of several hundred nAcm^{-2}.

The more dramatic effect of dark current spikes is particularly important in imaging devices as they produce white spots in the image which is most objectionable. The effect is often enhanced by charge spilling over into adjacent picture elements causing what is known as blooming. Special design techniques however can be used to overcome this effect to some extent.

2.3 Magnetic bubble memories

The technology of bubble domain memories is based on the properties of thin sheets of uniaxial ferromagnetic single crystals. In such sheets, which have the easy direction of magnetisation normal to the plane of the sheet, magnetic domains are formed which have their directions of magnetisation alternately up and down so that under zero applied field, no net magnetisation arises (Fig. 7a). As a uniform bias field is applied the domains with magnetisation in the same direction as the field grow at the expense of others and at a certain stage the strip domains break up into discrete entities, which are the desired bubble domains. Once created, as long as the applied bias field is maintained, the bubble domains are quite stable and it is possible to create large fields of identical domains which mutually repel each other and form an array (Fig.7b and c). The domains are cylindrical in shape due to

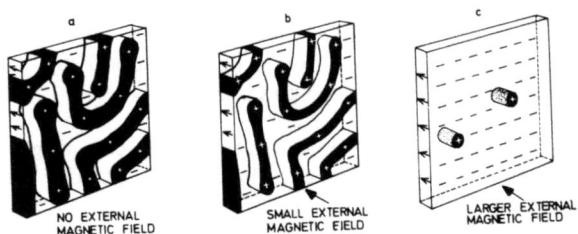

Figure 7. Magnetic Bubble Domain formation

the magnetic surface tension of the wall and extend right through the crystal sheet. They are highly mobile and because the direction is normal to the plane of movement, there is no preferred direction within the plane so that propagation and manipulation can occur in two dimensions. This has immense advantages for designing devices as will be seen later. As the velocity of every bubble is in principle identical for a given sample it is easy to see how digital shift register type memory devices can be created. A binary number can be represented by the presence of a bubble domain at prescribed points within the device and with a suitable technique the bubble can be moved to a detector which reads out the information. Likewise to record information bubbles can be created corresponding to the 1 s, and 0 s of the data.

2.3.1 The manufacture of bubble memories

The principles and device applications of magnetic bubbles were first described by Bobeck of Bell Telephone Laboratories in 1967 and since that time the material problems encountered in making fully engineered devices have been solved until today commercially available 1M bit memories are available and much larger devices, up to 256M bits are projected for the quite near future. Although the manufacturing process is much simpler than for semiconductor memories, (only one or two masking steps are required) there is a corresponding increase in the material quality requirement.

The first devices were built using thin sheets of material which were cut from flux grown orthoferrite crystals. The problem with orthoferrites was that although devices could be readily produced the domains turned out to be over 100 μm in diameter and they were therefore not suitable for high density memories. However, the potential of the technology had been demonstrated and other materials were investigated. The garnets proved to be both suitable and easy to produce but interestingly the magnetic anisotropy in these materials is growth induced and as this was not easy to control in bulk samples, epitaxy techniques, familiar from the production of Yttrium Iron Garnet (YIG) were adopted. Both chemical vapour deposition (CVD) and liquid phase epitaxy (LPE) have been explored but the latter is now the most popular and widely used method. This results from the difficulty of preparing uniform material by the CVD process which suffers the inherent disadvantage that the anisotropy is induced by stress caused by lattice mismatch between the substrate and the epitaxial film.

In the LPE technique the composition of the magnetic garnet is carefully controlled to ensure that the lattice parameter of the film matches that of the substrate which is invariably Gadolinium Gallium Garnet (GGG). To produce the correct magnetic parameters in addition to the precise lattice constant involves a complicated mixture of rare earth elements such as in $(YSmLuCa)_3 (FeGa)_5 O_{12}$, which is a popular recent recipe.

The positions of the bubble in the film and the means by which they can be moved are defined by a permalloy structure evaporated or sputtered onto the garnet layer. The so called T bar and chevron patterns have been most commonly used for propagation. With these patterns an in-plane rotating magnetic field will magnetise the permalloy in such a way that bubbles will move along the pattern (Fig. 8).

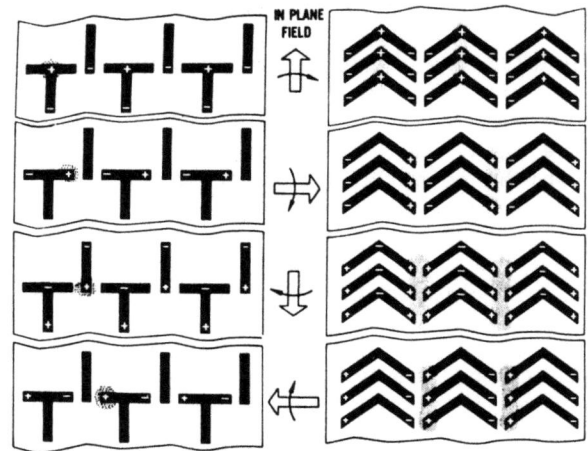

Figure 8. Bubble propagation schemes using permalloy (a) Tbar (b) Chevron configurations

2.3.2 Device architecture

All bubble devices are based on shift register type operation but in order to reduce the access time and to increase the yield of acceptable devices several refinements over having one long register have been devised. The most common is the major-minor loop organisation in which the storage is performed in a series of minor loops which can be connected to a major loop for input and output (Fig.9). A page of data is written serially into the major loop via

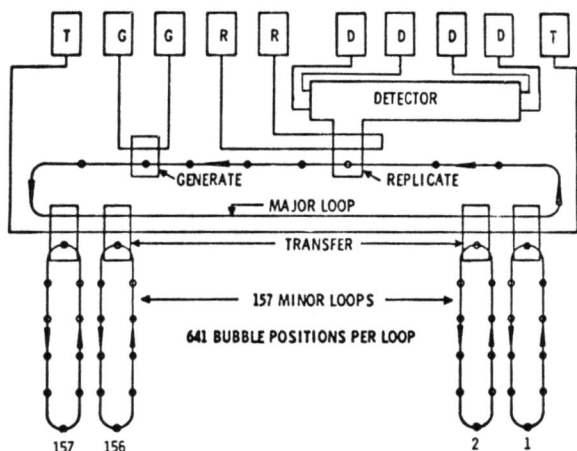

Figure 9. Major-Minor loop architecture of a 92K bit memory

the generator and is transferred in parallel to the minor loops by means of the gates. To read the data it has to be transferred back onto the major loop, replicated into the detector and returned to the minor loops. Improvements on this sequence can be made by modifying slightly the architecture to produce a faster cycle time but the principle is the same.

In a device such as the 92K bit memory produced by Texas Instruments, which uses the simple organisation, there are a total of 157 minor loops each consisting of 641 bubble positions. This approach allows the important concept of redundancy to be introduced. It will be realised that the theoretical storage capacity of the TI chip is 100,637 bits but in order to enhance the yield of good devices, 13 are allowed to be defective. The actual faults are soon determined by the test procedure and their positions can be coded into the memory itself so that the user is unaware of which locations are not used. In their 1M bit device, Intel have extended this concept further by reserving a number of minor loops for error correcting codes which are user definable.

2.3.3 Bubble memory perfomance

The key to the successful fabrication of large bubble memory devices is the production of well controlled magnetic garnet layers which are essentially free from defects. Both the dynamic and static properties of the bubbles are determined by the properties of the layers and a subtle balance exists between the bubbles' velocity, diameter and stability. These parameters are related to the material composition, its thickness and the way in which its parameters vary with the temperature and magnetic field.

The operating margins of the device are a measure of the parameter variations which can occur before the device will malfunction.

As might be expected crystal defects have a catastrophic effect on the operation of devices. Since the early work using much less perfect materials than are currently available it has been known that dislocations can pin bubbles most effectively and that variations in lattice parameter influence the velocity of propagation under a given field. In this latter context the exact mechanism determining the anisotropy in the LPE films is not understood. Many types of defect can occur in GGG substrates from small stacking faults to large helical dislocations several hundreds of microns in diameter. Any defects which intersect the surface of the substrate replicate into the epitaxial layer so that high quality "defect free" substrates are required. With the present day technology which uses wafer diameters of 2 - 3" and bubble diameters of approximately 2.0 μm, defect counts of $0.3/cm^2$ can be achieved. Economics requires that larger wafers have to be used to reduce the

cost still further. (The cost per bit of memory has already fallen
by a factor of 100 in the last 5 years.) The trend also however is
to smaller bubbles where the sensitivity to defects will be enhanced
and even more uniform materials will be required. New ways of
propagating the bubble known as contiguous discs, allows smaller
bubbles to be manipulated without the reducing feature size and
stacking multiple levels of memory are further ways of increasing
the storage of a single device. In parallel with these developments
new compositions of films are being produced which do not require
separate biasing magnets and are therefore simpler to package. In
all these developements careful control of material qualities is
required. The market for bubble memories as a replacement for discs
and tape in small (microprocessor) systems is enormous and the
quality control required for the present and proposed devices as the
volume increases is formidable.

2.4 Surface acoustic wave devices

The basic operation of a SAW can be seen from Fig. 10 which
represents schematically the layout of practically all such
devices.

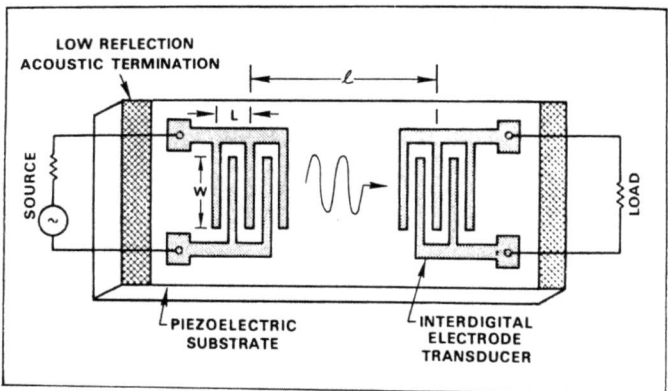

Figure 10. Schematic View of a SAW delay line

There are three major parts, the acoustic medium, the transducers
and the impedance matching circuitry. For our purposes the latter
can be ignored and the transducers considered very briefly. The
acoustic medium usually referred to as the substrate is normally a
single piece of piezoelectric material which for most operational
devices will be either quartz or lithium niobate. When an electric
signal is applied to the transducer, which is in the form of an
interdigitated set of electrodes, an acoustic wave is generated
which travels in both directions along the substrate. The wave
travels with relatively low loss and can be converted back into an
electromagnetic wave by means of a second similar transducer. The

use of this somewhat peculiar procedure is two-fold. As the velocity of the acoustic wave is typically 10^3 m sec^{-1}, delays in signals can be generated much more conveniently than using purely electromagnetic means. More importantly, perhaps, as the signal is travelling along the surface of the material it can be accessed and a variety of signal processing functions can be performed.

As with the charge-coupled devices the SAW devices are analogue signal processing devices but from the geometrical limits imposed on the transducers at high frequencies and the physical length of practical crystals at low frequencies, they are useful in the frequency range from a few to 2000 MHZ.

An important property of the devices is that there is a one to one relationship between the transducer geometry and its impulse response, the overlap of the fingers determining the amplitude and the position of the fingers determining the phase (Fig. 11). The fact that the phase and amplitude of a signal can be determined separately by the geometry of the transducers is also important.

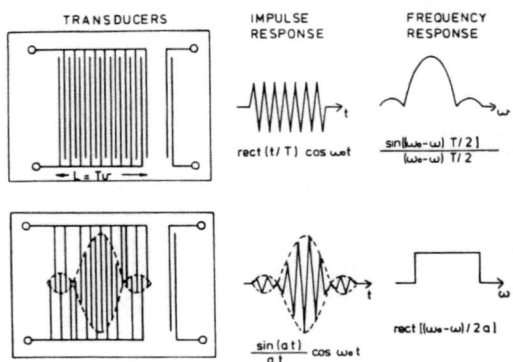

Figure 11. The principles of SAW filtering

The uses to which SAW devices can be put are summarised in Table 1.

The simple delay lines provide delays of up to 20 μsec with a bandwidth of 50 MHz but tend to be used in applications requiring somewhat shorter delays. Filtering is perhaps the most important application of SAW devices as very complicated impulse responses can be implemented simply by varying the position and overlap of the fingers of the transducers. Some devices, such as those required for television are quite small and are comparable to the size of silicon integrated circuits. Others, particularly the pulse compression filters for radar can be as large as 25cm in length. A variety of spatial signal processors such as Fourier Transformers, convolvers and correlators are possible and these are essentially the same in operation as the transversal filter and need not be

TECHNICAL IMPORTANCE OF GROWTH DEFECTS

discussed in detail.

Table 1

Function	Type
Delay line	Fixed or variable
Filter	Fixed band pass or matched
Fourier Transformer	
Convolver	
Correlator	
Signal Generator	
Oscillator	

An interesting use of the basic delay element is in the SAW oscillator. In these devices the output is fed back, with suitable amplification to the input. A comb of frequencies which satisfy the condition that the phase shift round the loop is $2\pi N$ is thereby generated. By suitably choosing the transducer geometry one frequency can be selected and effective stable oscillations in the frequency range 100-200 MHz can be made. This range is somewhat higher than that available with bulk wave devices but also has the advantage that, according to the design, stability can be traded against an ability to modulate the frequency over a range of a few percent.

2.4.1 Crystal properties

The problem of defects in the context of SAW devices is rather different to that of CCD's and magnetic bubble domain devices. The acoustic wave is in general insensitive to small defects and it is not usual to specify the perfection of the substrate other than by the quality of the surface. The normal requirement of this being that it is "optically" flat so that the very fine aluminium patterns can be replicated using photoresist techniques. As long as the substrate is a single crystal it can be used for SAW devices and the major problem is one of finding large enough pieces of single crystal at an economic price.

The key crystal parameters for SAW devices are the acoustic velocity, the coupling coefficient and the temperature coefficient of velocity. Of the commonly used materials quartz has good temperature stability, which makes it popular for large devices and for military operating conditions, but a low coupling coefficient while lithium niobate is the converse with good coupling of electromagnetic radiation but poor temperature stability (Fig. 12). Neither is ideal but no other material has been grown with sufficient quality in the sizes and quantities required to offer a serious alternative.

Material	Crystal Cut and Orientation	Piezo-electric Coupling Constant k^2 (%)	Acoustic Velocity v (m/s)	Delay Per Unit Length (μs/cm)	Delay Time Temperature Coefficient (ppm/ C)
$LiNbO_3$	YZ	114.5	3488	2.87	91
$Bi_{12}GeO_{20}$	(111),(011)	1.7	1680	5.95	128
ZnO	basal	1.0	2700	3.70	40
$LiTaO_3$	YZ	0.74	3200	3.12	37
Quartz	YX	0.23	3170	3.15	-22
Quartz	ST,X	0.16	3150	3.17	0

Figure 12. Physical constants of typical SAW materials

As the device design is being improved and more demanding are specifications there is concern about the quality of the substrates. Subgrain boundaries and groups of dislocations scatter the primary wave into either other unwanted surface waves or into bulk waves hence increasing the signal to noise ratio. Also with oscillators, although the stability is in principle very good, in practice a long term drift is observed which is known as ageing, The mechanism for this is not yet fully understood although it is certainly related to the condition of the surface. Finally as with other devices which are made in large volumes, the uniformity of crystal parameters over large wafers is essential. For the commercial and professional markets, many SAW delay lines and oscillators are being made on 2" or 3" wafers. It is essential that the properties of all devices are identical and this apart from requiring good manufacturing techniques calls for greater control over the basic substrate quality.

BIBLIOGRAPHY

(1) <u>Charge Coupled Devices</u>

Charge Transfer Devices, C.H. Sequin and M.F. Tompsett, Academic Press Inc., New York, (1975)

Proceedings of International Conference on CCDs, Learned Information, New York, (1975)

Charge Coupled Devices and Systems, Ed. Howes and Morgan, Wiley, (1979)

TECHNICAL IMPORTANCE OF GROWTH DEFECTS

(2) <u>Magnetic Bubble Domain Devices</u>

A.H. Bobeck, Bell System Tech. Journal <u>46</u>, 1901, (1967)

Magnetic Bubbles, T.H. O'Dell, MacMillan Press Ltd., London, (1974)

(3) <u>Surface Acoustic Wave Devices</u>

IEE Conference Publication 144, The Impact of New Technologies on Signal Processing, Ed. P.M. Grant, (1976)

IEE Reprint Series 2. Surface Acoustic Wave Passive Interdigitated Devices, Ed. D.P. Morgan, (1976).

CHAPTER 3

DEFECTS AND THEIR DETECTABILITY IN MELT-GROWN CRYSTALS

D.T.J. HURLE AND B. COCKAYNE

3.1 Introduction

In general the topography, and often the type, of defects to be found in a given crystal are characteristic of the method by which it has been grown. Indeed, one can infer much about the growth conditions from a study of defect structure. Broadly speaking, defects are introduced either during the liquid – solid phase change itself or during the subsequent change in temperature or pressure as the crystal is taken to ambient conditions.

Defects can be classified as either chemical or structural inhomogeneities. These categories are often inter-related because one can give rise to the other, e.g. the stress produced by a chemical inhomogeneity can be relieved by plastic flow or a dislocation may attract a Cottrell-type atmosphere thereby producing a chemical inhomogeneity. In the case of growth from the melt the following important mechanisms of defect generation can be identified:

1. Solute striations parallel to the growth interface caused by fluctuations in the growth process arising from either imposed asymmetric rotation or by non-steady convective motion in the melt.

2. Anomalous segregation and strain produced by the presence of facets on the growth surface.

3. Microsegregation at a cellular or dendritic interface formed in the presence of a zone of thermal or constitutional supercooling.

4. The generation of arrays of dislocations by plastic flow or climb during cooling of the crystal from the growth temperature to room temperature.

5. The generation of precipitates and point-defect aggregates during cooling.

6. The entrapment of particulate material and gas bubbles during growth.

To ascertain the value of X-ray techniques in studying these defects we must first enquire about their scale.

Strain and anomalous segregation at facets has the spatial dimensions of the facets themselves and these can be on a scale comparable to that of the whole crystal so there is no problem of spatial resolution. Modern X-ray techniques (double crystal techniques) are capable of measuring extremely small strains and hence the presence of growth facets can be revealed very readily using X-rays.

Microsegregation [1] due to either non-steady convective motion in the melt or to cellular or dendritic growth commonly occurs on a scale governed by either mass or heat transfer processes, ie. with a scale length of D/v or K/v where D and K are respectively the Fick's law diffusion coefficient and the thermal diffusivity and v is the crystallization velocity. For typical values of the latter quantity the scale lengths are very roughly of the order of 10^{-2} and 1 cm respectively, so that, again, the spatial resolution of X-ray techniques is, in general, adequate. The control of dislocation density during growth from the melt of modern electronic materials is now such that dislocation densities are generally in the range 0 to 10^5 cm^{-2} and therefore X-ray techniques are appropriate to image them. However, the detailed characteristics of such dislocations and the micro-aggregates of point defects formed on cooling the crystal from the growth temperature require techniques offering higher spatial resolution, such as transmission electron microscopy, for their study.

In this paper we describe the causes and illustrate the occurrence of the above categories of defects in single crystals of a variety of materials. We confine ourselves to single-phase materials; consideration of eutectic solidification would require an equal amount of space. For a review of the latter topic the reader is referred to the article by Hogan, Kraft and Lemkey [26]

Chemical etching is a powerful method for revealing defects, particularly in semiconductor crystals [2] where the local rate of chemical attack can be made to be dependent on the local density of ionisable impurities as well as on the dislocation strain fields, so that both chemical and structural defects can be revealed by optical microscopic observation of etched surfaces. In the case of oxide crystals, in addition to the study of etched surfaces in reflection, transmission optical microscopy is also a useful tool [3]. Indeed optical techniques, having been available for much longer, and requiring less expensive equipment, have been more commonly used than have X-ray techniques, and the majority of the illustrations in this paper will be of optical rather than of X-ray images.

3.2 Solute striations

Single crystals are seldom required in their pure state for commercial use, the desired properties usually being obtained by careful control of the concentration and distribution of deliberately added solute. Thus one dopes semiconductor crystals with donor or accepter solutes to obtain the required electrical behaviour and one dopes oxide crystals for use as laser host lattices with specific rare earth ions to obtain the required lasing action. The unequal segregation of solute between molten and crystalline phases gives rise to the existence of a solute boundary layer ahead of the growing crystal, whose magnitude is dependent on the growth rate. Any temporal variation in solute concentration in the melt at the interface gives rise to a spatial variation of solute concentration in the crystal. Such variation can arise in two main ways:

1. From a variation in growth rate induced either mechanically (for example by rotating the crystal in Czochralski growth in an asymmetric thermal field) or by thermal variations (e,g, a poorly controlled furnace or by non-steady convection.)

2. From a variation in the flow velocity in the melt which modifies the solute boundary layer thickness.

Such variations are frequently periodic or quasi-periodic and they result in a banded distribution of solute in the crystal (commonly called growth striations or striae), with each band delineating the instantaneous shape of the crystal-melt interface. The amplitude of these concentration fluctuations tends to be in the range of a few to a few tens of per cent but if, as can happen, part of the crystal is periodically melted and re-grown the concentration gradients can become large. Such striations can be readily delineated by etching semiconductor and oxide crystals.

Because of the high temperatures and hence high temperature gradients commonly involved in melt growth, thermal convection in the melt is strong. This serves to minimise the thickness of the concentration boundary layer ahead of the growth interface but also, unfortunately, causes non-steady convection. Thermocouples inserted into high temperature melts reveal marked temperature fluctuations which can be of several tens of degrees centigrade in amplitude [4,5]; in specific cases these can take on a rather regular periodic form. They correspond to fluctuations in the convective flow velocity and give rise to striations by modulating both the growth velocity and the boundary layer thickness [6]. The power spectrum of these temperature fluctuations typically extends from very approximately 10^{-2} to 10 Hz and for normal rates of growth from the melt, give rise to striations spaced from sub-micron to the order of 100 microns apart. Such is the sensitivity of double crystal topographic techniques that the strain fields due to quite low concentration fluctuations can be clearly imaged. A topical and important example is that of carbon in silicon. Fig. 1 shows a topograph of the strain field due to carbon at a level of 8.10^{16}cm^{-3} in a pulled silicon crystal.

By imaging the striation pattern in a plane containing the growth axis one can obtain a profile of the shape of the crystal-melt interface at each stage of the growth. This is a valuable aid to understanding how the heat transport varies as growth proceeds, and shows whether or not facets develop on the growth interface. This is illustrated in Fig. 2 which is an optical micrograph of a Cr-doped spinel crystal in which the interface shape has been modified by changing the crystal rotation rate in discrete steps from 50 to 100 to 150 RPM. The changes in strain are evident from the X-ray transmission topograph of Fig. 3.

Figure 1. Diffraction topograph of a slice from a Czochralski silicon crystal showing strain distribution due to micro-segregation of a carbon solute at a concentration of $8 \times 10^{16} \text{cm}^{-3}$. Mag. x 0.8 (Courtesy of M.J. Hill).

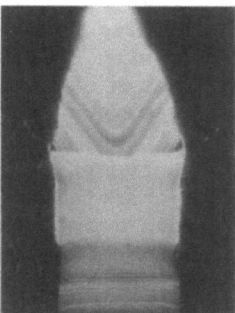

Figure 2. Optical micrograph of longitudinal section through Czochralski Cr-doped spinel crystal. Striations reveal interface shape, changed markedly by changing the crystal rotation rate. Upper region 50 rpm; central, 100 rpm and lower, 150 rpm. Mag. x 1.1.

If the striations are periodic with known period then one can infer how the crystal growth rate varies by measuring variations in the striation spacing. An intriguing recent example of striations due to non-steady convective motion is that of the formation of a helical pattern of segregation of solute or of colloidal particles in a cylindrical ice single crystal grown by the Bridgman technique [7]. The helical pattern has been shown to be due to the rotation about a vertical axis of a convective roll cell.

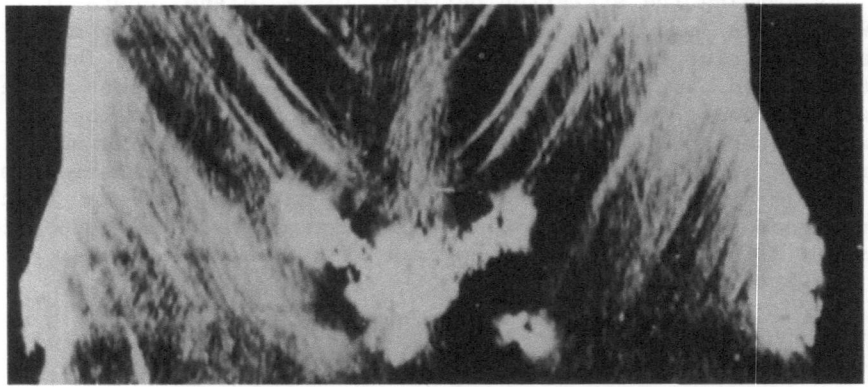

Figure 3. X-ray transmission topograph of the section shown in Fig. 2 revealing dislocations and the pattern of the strain field. Mag. x 5.6.

3.3 Effects of facet formation

Except for materials with a very low entropy of fusion (such as metals), most materials can develop faceted low index surfaces during growth from the melt. They occur wherever the freezing point isotherm is convex towards the melt and parallel to a low-index faceting surface. The development of a facet is attributed to the greater difficulty of nucleating new growth on crystal planes which are atomically smooth. According to an analysis by Jackson [8] materials with $\alpha > 2$ crystallise with smooth surfaces whereas those with $\alpha \leq 2$ crystallise with rough surfaces ($\alpha = \Delta S \eta / R \nu$ where S is the entropy of fusion, R is the gas constant and η/ν is a bonding parameter dependent upon the surface orientation). Most semiconductors and insulators have at least one orientation where $\alpha > 2$ and would therefore be expected to form interface facets. The nucleation difficulty on a facet means that such a region requires a greater supercooling for growth to be initiated. Consequently, a facet extends in width until the maximum supercooling on the facet can sustain nucleation at a rate commensurate with the superimposed growth rate. Note that the size of the facet will, for a given material, depend on the curvature of the isotherm and on the axial temperature gradient. Note also that the facet cannot develop on concave growth surfaces.

The important feature of facets for our purpose is that segregation at them is anomalous, [9] whereas, away from facets, segregation during growth is believed to be nearly at its equilibrium value. On a facet, marked non-equilibrium effects are present. Mullin, [10] in reviewing facets in semiconductors, has pointed out that the region of surface adjacent to a facet approximates to an atomically rough surface although it can be considered as a series of microscopic planes with the same orientation as the facet but separated by steps of atomic height. Nucleation at the edges of such steps should be considerably more favourable from an energetic viewpoint than two-dimensional nucleation on the smooth facet. If the average number of steps available for nucleation on the facet is N_f and their average lateral growth velocity is V_f, with corresponding parameters of N_e and V_e for the edges of the facet, then because faceted and non-faceted regions must grow at the same overall rate $N_f V_f = N_e V_e$. Since N_f is small and very much less than N_e, then V_f must be very much greater than V_e. Thus steps generated at the facet must move with a much higher lateral velocity than the steps at the facet edges and other non-faceted regions of the interface. This rapid lateral velocity of growth on the facets provides the most likely explanation of the differences in impurity incorporation which occur between faceted and non-faceted regions because, with a step velocity of sufficient magnitude, the surface absorbed layer does not have time to equilibrate with the advancing layer and the incorporation of an absorbed impurity can then be markedly different from the off-facet regions where the equilibrium solidus composition is incorporated into the solid.

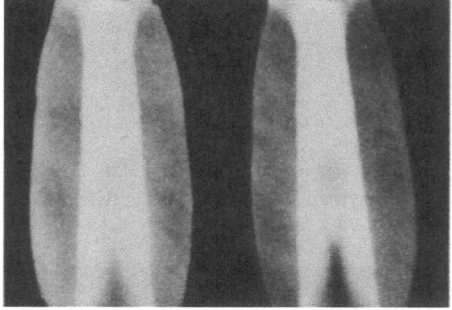

Figure 4. Autoradiograph of the two halves of a longitudinally sectioned <111> axis crystal of radio Te^{128} doped InSb showing a central facetted core of enhanced Te concentration. Note the striations caused by slow crystal rotation. Mag. x 1.0 (Courtesy of J.B. Mullin).

In the practical aspect of producing structurally and chemically uniform crystals, facets are a significant and undesirable defect. The difference in impurity incorporation on the facet with respect to surrounding regions can lead to very non-uniform doping, particularly in semiconductors. This is most strikingly illustrated by the case of tellurium segregation in indium antimonide [9]. The segregation coefficient (the ratio of the solute concentration in the solid to that in the melt at the interface) is approximately 0.5 away from the facet but on the facet rises to a value of ~4. Fig. 4 shows an autoradiograph of a longitudinal section of an InSb crystal grown from the melt by the Czochralski technique containing a small amount of radioactive Te. The crystal was grown in the <111> direction and there was a large {111} facet in the centre of the crystal-melt interface which produced the central Te-rich core seen in the autoradiograph. Similar effects have been observed on {100} type facets in indium antimonide and the same general pattern has been shown to exist in a substantial number of the elements and compounds which have the diamond cubic or zinc blende structure [11] e.g. Ge, Si, InSb, GaAs, GaSb, InP.

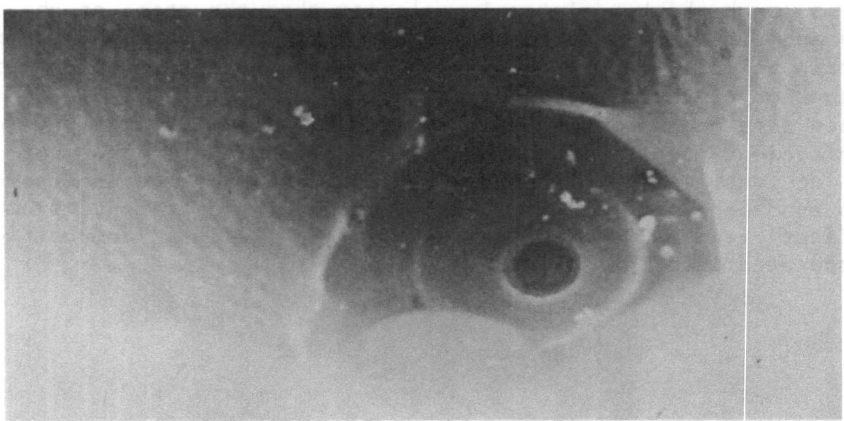

Figure 5. Three {211} growth facets in the central region of the interface of $Y_3Al_5O_{12}$ crystal rapidly withdrawn from the melt. Mag. x 33.

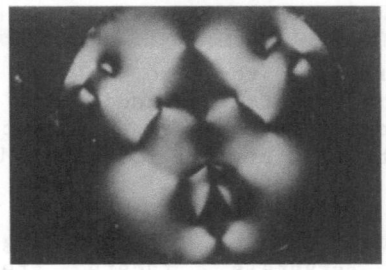

Figure 6. Optically delineated strain pattern due to the three {211} facets shown in Fig. 5. Mag. x 35.

DEFECTS IN MELT-GROWN CRYSTALS

Most oxides have orientations where the α-factor is much greater than 2 and faceting is, in consequence, a common phenomenon and an important defect to avoid in the refractory compounds [12] which have now found a wide range of uses as typified by solid state lasers, insulating substrates for semiconductors, substrates for magnetic materials, surface-wave and bulk-wave delay lines, and non-linear optical and acoustic-optical devices. Differences in the incorporation of an added impurity have been demonstrated, e.g. for Nd in $Y_3Al_5O_{12}$ (yttrium garnet) [3] but the effects so far reported are generally smaller than in semiconductors. In oxides, the most important aspect of facet formation is the strain associated with the facet. The existence of this strain is well recognised and demonstrated (Figs. 5 & 6) although the source of the strain has not been positively identified in many instances.

Most of the efforts to attempt to understand the phenomenon have been concerned with garnets in comparative studies between $Y_3Al_5O_{12}$ and $Gd_3Ga_5O_{12}$ which exhibit both {110} and {211} type facets [13]. The mean lattice parameter of the faceted regions is slightly larger than that of the unfaceted regions ($\delta a/a \sim 1.3 \times 10^{-4}$) so the facets produce a local dilation of the lattice within the bulk crystal. Three principal sources of the strain have been proposed (a) changes in cation ratio between facet and matrix (b) impurity segregation and (c) oxygen segregation. Impurity segregation appears unlikely because of the very high purity of the materials used (e.g. all impurities present in YAG are at levels <1ppm). Changes in cation ratio are possible but the effect should be greater in Gd/Ga than in Y/Al because of the much wider homogeneity range which $Gd_3Ga_5O_{12}$ displays. The lattice spacing data so far available suggests that the magnitude of the effect is the same in both materials. It has been suggested [14] that facet strain due to oxygen segregation is consistent with the entrapment of oxygen vacancies on the facets due to the high lateral growth rates involved in facet formation and the low diffusion constant of oxygen with respect to the cations present. This view is supported by luminescence studies which show that faceted regions contain higher concentrations of electron traps than the matrix.

The presence of facets can modify the morphology of some defects and promote the formation of others. For instance, the growth striations which delineate the shape of the solid/liquid interface at the freezing isotherm appear linear in cross section rather than the curved section typical of non-planar interfaces (Fig. 7). The increased difficulty of nucleation on facets can also restrict the development of structures arising from constitutional supercooling to non-faceted regions (see below). The strain associated with facets is usually accommodated elastically but in some instances, the stress developed exceeds the yield point of the material leading to dislocation generation and low-angle boundary formation in the regions adjacent to the facets (Fig. 8).

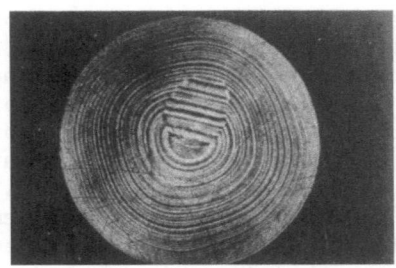

Figure 7. Presence of a (001) facet in a Czochralski crystal of $Ba_2NaNb_5O_{15}$ shown by etching to reveal growth striations. Mag. x 2.

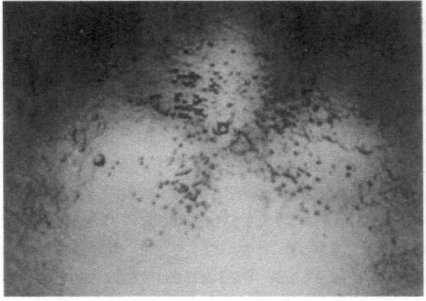

Figure 8. Etched section showing generation of dislocations around a faceted region in a crystal of $MgAl_2O_4$. Mag. x52.

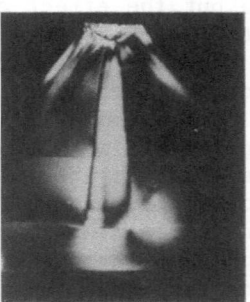

Figure 9. Longitudinal section through a <111> axis crystal of $Gd_3Ga_5O_{12}$ showing strain associated with dislocations propagating toward the crystal edges in the presence of a central facet. Where a planar interface is induced near the bottom of the section (by increasing the rotation rate), the crystal becomes facet-free in addition to being dislocation-free. Mag. x1.7.

Facets can also restrict the formation of dislocations and a faceted interface, being convex towards the melt, forces dislocations which are formed in non-faceted regions to grow out towards the crystal edges. Once this stage has been reached, the production of a planar solid/liquid interface, achieved by altering the rate at which the crystal rotates, can lead to crystals which are both facet- and dislocation-free [15]. This has proved of particular importance in garnets such as $Gd_3Ga_5O_{12}$ which are required in a highly perfect state for magnetic bubble substrates (Fig. 9).

3.4 Cellular and dendritic structures

3.4.1 Introduction

Cellular structures can arise as the result of the occurrence of the condition known as constitutional supercooling [16] which we briefly describe. When solute is rejected at the crystal-melt interface it lowers the freezing point of the melt so that the solute and freezing-point distributions in the melt are as shown in Fig. 10. For a steep positive temperature gradient in the melt (line 2) the actual temperature in the melt is everywhere greater than the equilibrium freezing point. However for a less steep, but still positive, temperature gradient (line 1) there exists a region (shaded) where the actual temperature of the melt is below its equilibrium freezing point and that region of the melt is said to be constitutionally supercooled. The significance of this is that in the presence of such a supercooled region a planar crystal-melt interface becomes unstable and evolves to a cellular form. Consider the chance formation of a projection on the interface. The tip of that projection, finding itself in a region of increasing supercooling, will grow more rapidly than neighbouring regions and, in so doing, will segregate solute laterally thus lowering the freezing point of neighbouring regions and suppressing their growth.

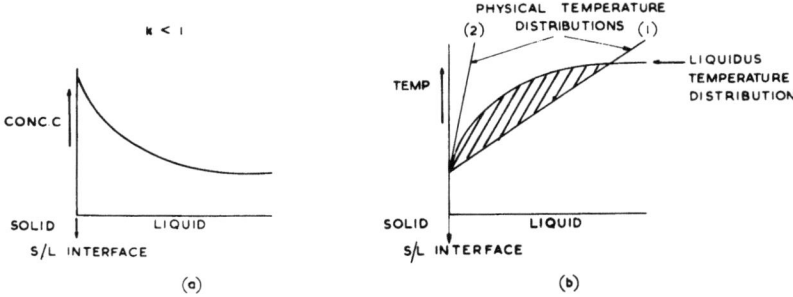

Figure 10. (a) Concentration distribution in the melt ahead of the solid-liquid interface for a growing crystal with a solute with segregation coefficient k <1. (b) Liquidus temperature distribution in the melt corresponding to the solute distribution in (a) together with two possible actual temperature distributions (labelled 1 and 2). For the shallow temperature gradient (1) a zone of constitutional supercooling (hatched) exists in the melt.

This lateral segregation gives each perturbation a "field of influence" of scale D/v so that the interface can be expected to break down into a close packed array of such projections having wavelength of D/v. Roughly speaking this is indeed what happens as can be seen on the interface of a tin single crystal from which the melt has been rapidly decanted (Fig. 11).

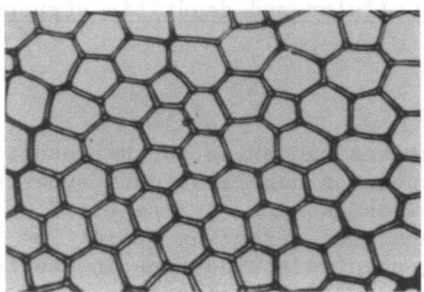

Figure 11. Cellular structure on the rapidly-decanted interface of a Pb-doped Sn single crystal grown under conditions of constitutional supercooling. Mag. x 40.

3.4.2 Morphology

The above behaviour is encountered with non-faceting materials such as metals. With faceting materials, the morphologies obtained are more varied [11, 17, 18]. With these materials an initial perturbation in the shape of the interface will grow in amplitude until it becomes tangential to a low-index faceting plane whereupon a nucleation difficulty will exist, the growth rate in this direction will slow and a facet will develop. Hence the perturbed interface will develop into an array of inclined micro-facets. This is shown schematically in Fig. 12. The morphology of the cellular structure therefore depends on the orientation of the major facets with respect to the growth direction. This is shown in Fig.13 which is of the decanted interfaces of two identically grown crystals of gallium-doped germanium, one grown on a [110] and the other on a [100] axis. If the crystal is grown with a free surface - as in the Czochralski technique - then the onset of the cellular structure is visible on that free surface as a set of "corrugations" corresponding to the intersection of the cell boundaries with the surface (Fig. 14).

Thus far we have focussed attention largely on a single solute present during the growth of an elemental material. Constitutional supercooling can occur during the growth of a compound from a non-congruent melt, [19] and indeed it is often a serious problem which is difficult to avoid, requiring extremely slow and steady growth rates. In general it precludes one from obtaining good quality single crystals of peritectically-melting compounds by growth from the melt. Growth from non-stoichiometric melts can be

DEFECTS IN MELT-GROWN CRYSTALS

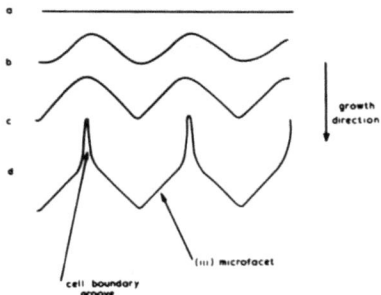

Figure 12. Schematic representation of the development of a cellular structure in a material which develops {111} facets. (a) Planar interface. (b) Initial perturbation of interface shape. (c) Development of {111} micro-facets when the perturbed interface becomes tangential to {111} planes. (d) Deep cell boundary grooves in the fully developed cellular structure.

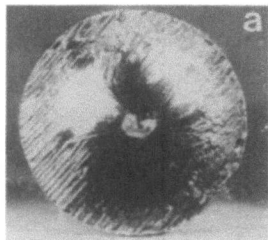
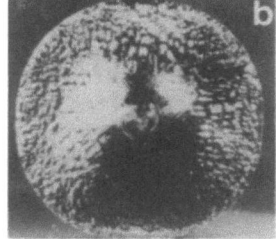

Figure 13. Decanted cellular interfaces of Ga-doped Ge single crystals. Note the large droplet of melt retained in the centre during decanting. (a) <110> axis showing cell structure composed predominantly of pairs of {111} planes inclined at $35°$ to the growth direction. (b) <100> axis showing cell structure composed of four {111} planes inclined at $45°$ to the growth direction. Mag. x 1.5.

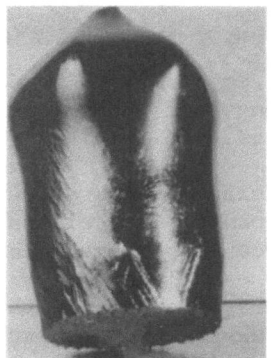

Figure 14. <311> axis crystal of Ga-doped Ge showing "corrugations" which are an external manifestation of cellular structure. Near the bottom, the crystal has multiply twinned. Mag. x 1.2.

illustrated by the growth of indium antimonide from an indium-rich melt. Excess indium, rejected at the growing crystal melt interface, gives rise to constitutional supercooling and the concomitant cellular structure as shown in Fig. 15. The decanted (100) interface is similar to the germanium one shown in Fig. 13. However a very small amount of radio-tellurium was added to the melt. This was preferentially incorporated onto the micro- facets which developed (because of the 'facet effect') and is revealed on the autoradiograph shown in Fig. 15 proving that the cellular structure is indeed composed of small (111) facets.

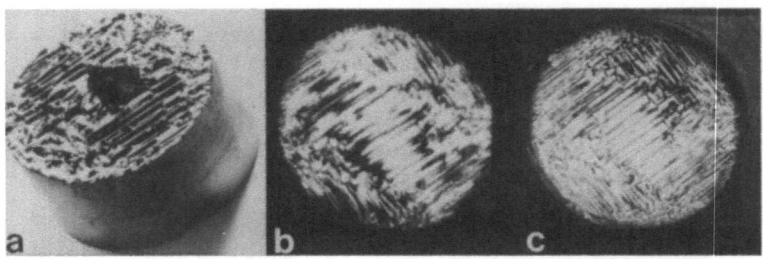

Figure 15. Cellular structure in a <100> axis crystal of InSb obtained by growing from an In-rich melt. (a) Decanted interface. (b) Autoradiograph showing preferential incorporation of radio Te on the {111} micro-facets. (c) Etched micrograph of the section shown in (b) Mag. x 0.85.

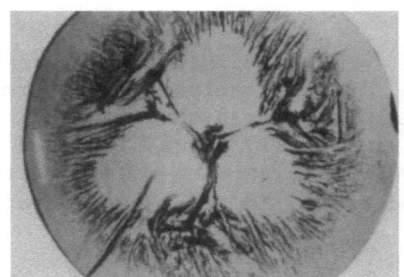

Figure 16. <111> axis crystal of $Nd_3Ga_5O_{12}$ having a cellular structure except on the parts of the interface which were composed of three large {211} facets. Mag. x 3.

Given that the cellular structure is composed of micro-faceted segments then a marked change in cellular morphology is to be expected when the macroscopic plane of the interface coincides with a faceting plane. In this case large areas of facet develop which are stable against cellular breakdown with the cell structure being confined to the neighbouring non-faceted areas of the crystal. This is well illustrated in Fig. 16 which shows the cellular interface of a <111> axis crystal of $Nd_3Ga_5O_{12}$ with the central region dominated by three large {211} facets surrounded by regions of cellular structure.

DEFECTS IN MELT-GROWN CRYSTALS

Materials which form facets do not grow dendritically in the way that metals do because of the nucleation barrier which exists to propagation in certain low-index directions. However the common diamond-cubic and zinc-blende structure semiconductors exhibit a curious form of pseudo-dendritic growth in which a lath-like structure with {111} surfaces and containing two or more closely spaced twin planes can propagate rapidly in the plane of the lath because the presence of two or more twin planes gives rise to a self-perpetuating re-entrancy thereby avoiding the nucleation problem [20]. In the presence of large amounts of constitutional supercooling such laminar-twinned structures develop and the growth takes on a pseudo-dendritic form.

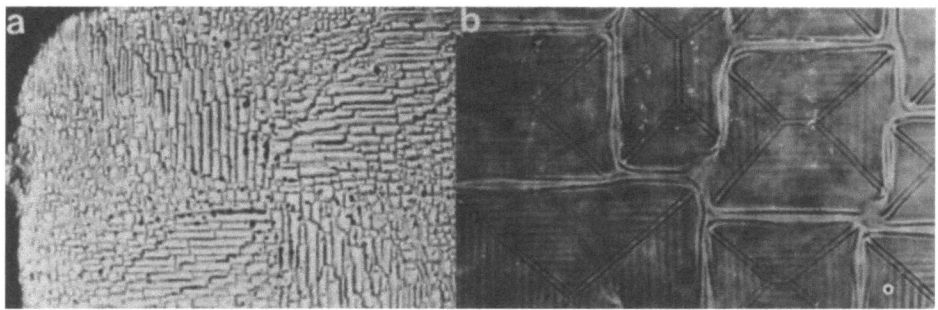

Figure 17. Etched section through a <100> axis crystal of Ga-doped Ge showing pattern of cellular structure, (a) macro-view of whole section, (b) enlargement showing structure of individual cells. The square or rectangular wavy boundaries delineate cell boundaries. The double lines within each cell are the traces of intersection of pairs of {111} planes which form the cell faces. (The doubling is an optical effect caused by slight de-focussing). Growth striae are visible within each cell. Mag. (a) x 30 (b) x 48.

Up to this point we have confined ourselves largely to a description of the external manifestations of cellular structure. If now we examine crystal sections by etching/optical microscopy and X-ray topography a wealth of fascinating detail can be revealed, [11,17] principally because we can expose the interface morphology by delineating the growth striae. This is well illustrated in Fig. 17 which is of a (100) section, normal to the growth axis of a gallium-doped germanium crystal. The growth interface consists of an array of square and rectangular-based pyramids having (111) faces. The regions separating neighbouring cells are the solute-rich cell boundaries. The growth striae delineate the trace of the four {111} planes in the (100) etched section at successive rotations of the crystal during growth. The straight lines, parallel and diagonal to the cell boundaries are the traces in the surface of the ridges formed by the intersection of adjacent pairs of {111} facets of an individual cell.

By studying longitudinal sections containing the growth axis we can learn more of the evolution of the cellular structure. Fig. 18 is of a <110> longitudinal section of a <110> axis gallium-doped germanium crystal. Growth striae delineate the profile of the cellular morphology composed of pairs of {111} planes inclined at 35° to the plane normal to the growth axis. Note the deep cell boundary grooves.

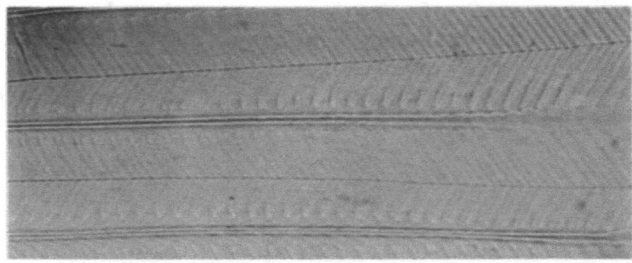

Figure 18. Etched longitudinal (110) section of a <110> axis Ga-doped Ge crystal. Growth striae delineate the profile of the solid-liquid interface. Note the deep cell boundary grooves. Mag x 48.

3.4.3 Chemical inhomogeneities

The narrow channels between the cellular projections (the cell boundary grooves) become progressively richer in solute and become deeper as growth is impeded due to the solute enrichment. Ultimately as the region at the root of the cell boundary cools a phase boundary with a second solid phase will be reached and a precipitate will be formed. Alternatively, if the freezing range is very wide, the deep channels become unstable as the cell boundary concentration increases and spheroidisation occurs leaving liquid droplets entrapped within the crystal. These droplets experiencing the imposed temperature gradient in the crystal dissolve on the hotter side and solidify on the colder side, i.e. they "climb" after the main crystal melt interface by the process known as temperature-gradient zone-melting (TGZM), [21,22], leaving behind highly solute-rich material in the form of "solute trails". This can be seen in Fig. 19. When the crystal is finally cooled to room temperature the droplets solidify to give second-phase inclusions. The concentration of solute in the solute trails can rise to values which are sufficiently high for them to be revealed by absorption topography. Fig. 20, due to M. Hart, is an absorption topograph of a [110] axis indium-doped germanium crystal showing indium-rich solute trails terminating in indium droplets.

The forms of the cell boundaries and of the solute trails can be very variable. Firstly the form depends on the orientation of the crystal with respect to the growth axis. If facets are symmetrically disposed to the growth direction then the cell boundaries will be aligned parallel to the growth direction and the migrating droplets will be confined to the cell boundary regions. However, if the facets are asymmetrically inclined to the growth axis the cell boundaries will also be inclined to the growth axis but the liquid droplets, which climb up the temperature gradient, will move parallel to the growth axis and will therefore not be confined to the cell boundaries. The second factor of importance is the thermodynamics of the system. If the solute-melt system has only a small freezing range then little migration of liquid droplets will be possible, as a solidus will be reached quickly in the cooling phase. However if, as in the case of gallium-germanium illustrated in the previous figures, there is a very extended liquid range then extensive droplet migration can occur.

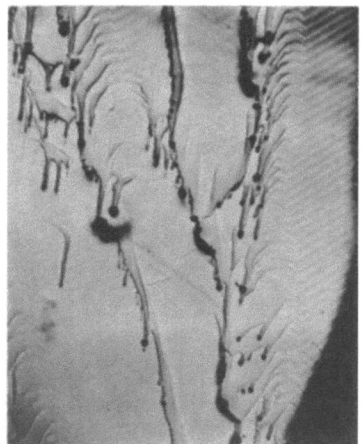

Figure 19. Longitudinal (110) section of a <110> axis Ga-doped Ge crystal showing solute trails due to the motion of trapped liquid droplets. Mag. x 48.

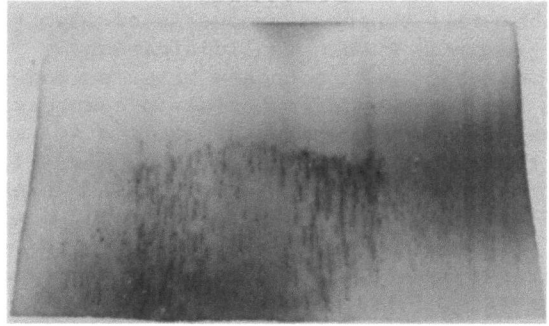

Figure 20. X-ray absorption topograph of a longitudinal section of a <110> axis In-doped Ge crystal showing In-rich solute trails terminating in In droplets. Mag. x 3.2 (Courtesy of M. Hart and A.R. Lang)

If the solute raises the freezing point (i.e. has a segregation coefficient greater than unity) then the solute will be preferentially segregated to the <u>peaks</u> of the cells and the cell boundaries will not be very deep. The resulting pattern of strain is likely to be very different in this case.

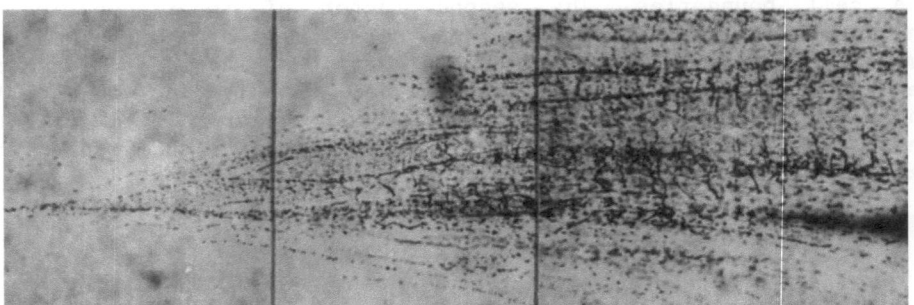

Figure 21. Longitudinal section showing tubular voids in a "c" axis single crystal of $CaWO_4$ grown with a cellular structure. Mag x 9.5.

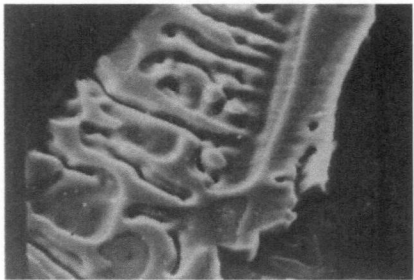

Figure 22. Dendritic solidifications within a cell structure void in a crystal of $Y_3Al_5O_{12}$. Mag x 400.

Finally the sign of the volume change on solidification of the material in the cell boundary groove is of importance. The common semiconductors expand on solidification, and therefore solidification of the cell boundary material and of the migrating droplets places the surrounding material in compression. If, on the other hand, as is more common, there is volume contraction on solidification then freezing of the cell boundary liquid can give rise to long tubular voids in the crystal as is exemplified by Fig. 21, which is of a calcium tungstate crystal. In some instances, the impurity rich liquid becomes trapped within the void and forms a dendritic structure upon solidification as in the case of yttrium aluminium garnet (Fig. 22).

3.4.4 Structural inhomogeneities

We have seen that very gross chemical inhomogeneities can be incorporated into crystals in the presence of constitutional supercooling. Not surprisingly the stresses which they generate can, at the high temperatures, exceed the flow stress and dislocation arrays can be generated in the crystal. Such arrays can be readily revealed by etching techniques or by X-ray topography. Germanium is a particularly attractive material to work with in that by etching non-(111) surfaces, patterns of micro-segregation can be revealed, as shown above, whereas by etching (111) surfaces dislocation etch pits are generated.

Dislocation arrays form both in the cell boundaries and around the solute trails and their solidified liquid droplets. Their form can be seen in an obliquely cut (111) section of a <110> gallium-doped germanium crystal. Both cell peaks and cell boundaries are delineated and massive dislocation arrays are seen to be formed as the cell structure develops (Fig. 23).

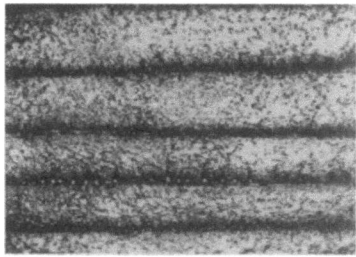

Figure 23. Etched (111) surface of a <110> axis Ga-doped Ge crystal showing dense walls of dislocations formed in the cell boundaries. Note delineation of the cell peaks and occurrence of solute trails in part of one of the cell boundaries (bottom left). Mag. x 24.

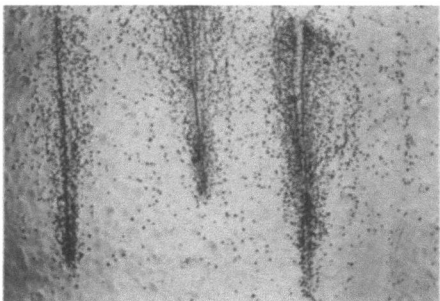

Figure 24. Etched longitudinal {111} section of a <110> axis crystal of Ga-doped Ge showing dislocations associated with solute trails. Mag. x 34.

Dislocation arrays associated with solute trails can take a variety of forms. If the solute has a segregation coefficient which is less than unity but not very much less than unity then solidification can be complete before the liquid droplet concentration reaches a value corresponding to the formation of another solid phase (for example by eutectic reaction). In this case the solute trail merely peters out as shown in Fig. 24. If however a second-phase precipitate forms then an octohedral pattern of slip is evident (Fig. 25). For the case when the solute has a segregation coefficient greater than unity and is therefore segregated preferentially to the peaks of the cells, the dislocation structure is markedly different with the dislocation tending to lie in double rows in the vicinity of the cell peaks. This is shown in Fig. 26 which is of a <110> oriented Si-doped Ge crystal.

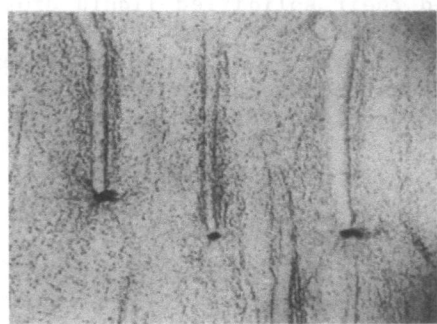

Figure 25. As Figure 24 but with solute trails terminating in second phase precipitates. Mag. x 34.

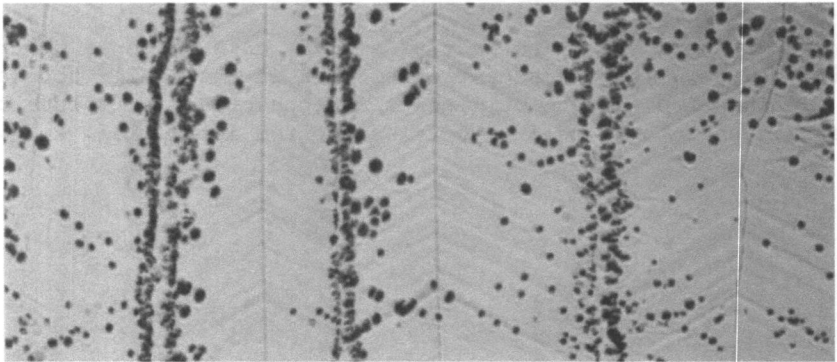

Figure 26. Etched longitudinal {111} section of a <110> axis Si-doped Ge crystal. Striae delineate the cellular structure and double rows of dislocation etch pits occur in the vicinity of the cell peaks. Mag x 86.

The production of these arrays of slip dislocations represents a dramatic deterioration of the structural perfection of the crystal as can be seen in Fig. 27 which is of the longitudinal section of a [110] germanium-doped gallium crystal. A diffraction topograph of a neighbouring region (Fig. 28) shows the similarity of the information obtained by the etching and topographic techniques. Note that prior to the onset of the cellular structure the crystal was essentially free from dislocations.

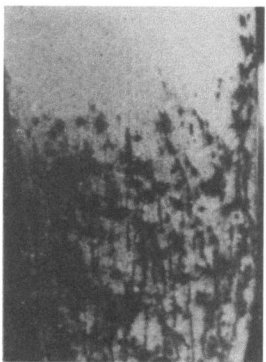

Figure 27. Etched longitudinal {111} section of a <110> axis Ga-doped Ge crystal showing dislocation arrays associated with cell boundaries and solute trails. The near vertical lines in the upper portion of the micrograph delineate the cell boundaries. Mag. x2.5.

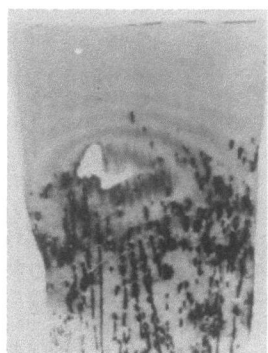

Figure 28. X-ray diffraction topograph of neighbouring section of the crystal shown in Fig. 27. Mag. x2.5. Courtesy of M. Hart and A.R. Lang.

3.5 Dislocations generated during cooling of the crystal

The radial temperature gradients which are produced when a crystal is pulled from the melt into the cooler ambient region above the melt generate hoop stresses [24]. If such stresses exceed the yield stress of the material plastic flow occurs and dislocations are generated. If the dislocations remain in slip bands, characteristic distributions can occur such as the star-like pattern [25] of {111} <110> slip system produced on (111) sections of <111> axis crystals in the group IV and III-V compounds (Fig. 29). Similar distributions can occur in oxides and other inorganic compounds. Fig. 30 shows dislocations lying in slip bands in calcium fluoride.

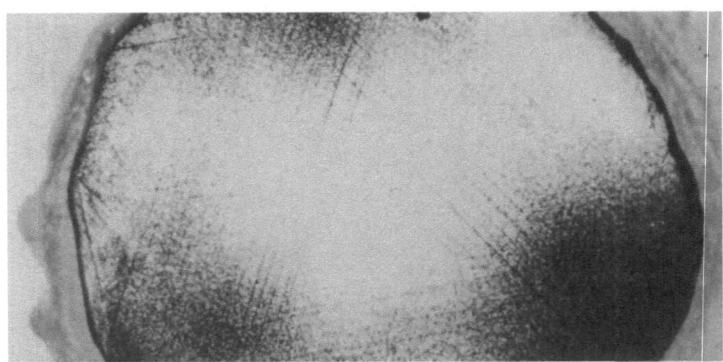

Figure 29. Dislocation-etched cross-section of a <111> axis crystal of GaAs showing octahedral pattern of slip. Mag. x 3.0.

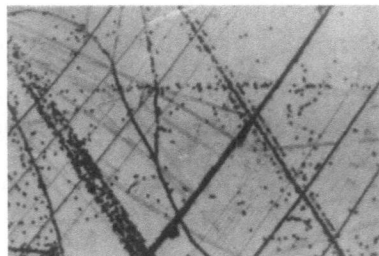

Figure 30. Slip bands revealed by etching a Czochralski crystal of CaF_2. Mag. x35.

In metals, the flow stress is relatively small compared to its value for semiconductors and it is difficult to reduce the thermal stresses during melt growth of single crystals to values which are low enough to avoid dislocation generation and multiplication. However this has been achieved in small single crystals of copper grown by pulling [27]. Production of dislocation-free silicon and germanium single crystals is relatively straightforward and material for commercial device production is commonly specified as 'zero-d'. When the vacancy concentration is high or the vacancies present remain mobile, polygonised structures develop due to climb of the dislocation out of their slip plane. This is a common occurrence in oxides although fully polygonised structures are rare and the more usual situation is to find random networks of low-angle boundaries with a substantial number of dislocations exhibiting no apparent connection with either a polygonised boundary or a slip band (Fig. 31); in other words, the structure observed is not in its lowest energy state and further modification can therefore be achieved by subsequent annealing.

Melt-grown metal single crystals frequently contain a mosaic substructure with low-angle boundaries aligned roughly parallel to the direction of growth [28]. (Fig. 32). The sub-grain boundary misorientations can be as large as several degrees. The structure is referred to as lineage or more confusingly as 'striations'. The origin of the sub-structure is not fully resolved but polygonisation is probably involved [29].

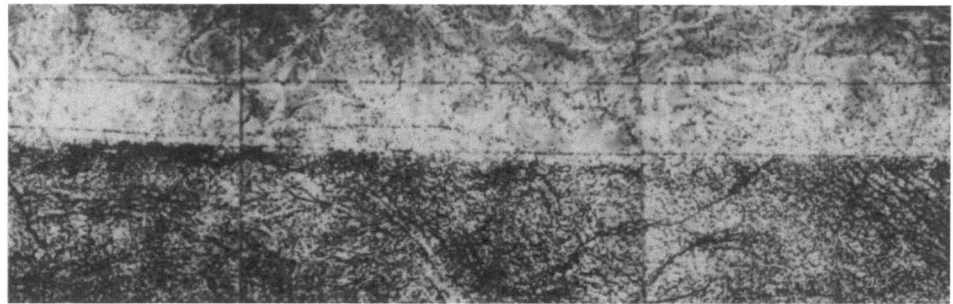

Figure 31. Partially polygonised dislocation network in an as-grown crystal of $CaWO_4$ revealed by etching. Mag. x 40.

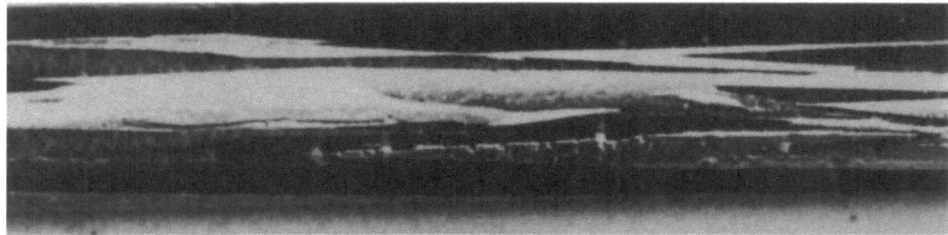

Figure 32. Lineage sub-structure revealed by etching a Bridgman-grown Bi single crystal. Sub-boundary misorientations are of the order of $1°$. Mag x 3.5.

Many inorganic compounds, such as the mixed oxides which are now in device use have complex crystal structures and any dislocations which might form, must have a large Burgers vector and high associated energy. Consequently, in these materials, typified by yttrium aluminium garnet, dislocations are not easily generated by slip and in the absence of other defects the crystals can be produced in a dislocation-free form. If the hoop stress does exceed the yield stress in such materials, brittle fracture ensues and the crystals crack.

The behaviour of dislocations in semiconductors can be strongly influenced by doping. Thus Seki et al [30] have demonstrated that at relatively high doping levels a marked reduction occurs in the dislocation density in as-grown crystals of GaAs and InP and this is ascribed to the pinning of dislocations by the increased bonding between impurity and the lattice. Somewhat different effects appear to occur in Ge. In heavily doped 'p'-type material (Ga doped), dislocation densities are relatively very low with any dislocations tending to remain on their slip planes [11]. In contrast, with n^+ material (As doped), high dislocation densities are encountered with a marked tendency to polygonisation. A possible explanation for this is that the vacancy in Ge behaves as an acceptor so that the vacancy concentration is strongly suppressed in p^+ and enhanced in n^+ material [33].

Dislocations can also be generated by the precipitation of point defects such as vacancies or interstitial atoms or the precipitation of a second phase. The form and distribution of such dislocations is dependent upon a variety of factors including cooling rate. Thus, in dislocation-free as-grown Czochralski crystals of Si, a striated distribution of perfect dislocation loops, believed to be due to the condensation of self-interstitials, can sometimes be found [31].

For sparingly soluble solutes the solidus curve is retrograde: that is, the solubility exhibits a maximum as the temperature is lowered. This is the situation with all the common semiconducting materials and attempts to dope up to near the solubility limit can often result in precipitation from a supersaturated solid solution at lower temperature. Precipitation of oxygen and carbon in Si [31] and of Te in GaAs [32] are typical examples. The defects often occur on a very fine scale and etch with different characteristics than the dislocations generated by slip and usually require the attributes of electron microscopy for meaningful identification. However, complex oxides such as $Gd_3Ga_5O_{12}$ provide notable exceptions as the strain fields associated with closed loops, normally occurring at large precipitates, and helices, which can propagate throughout the length of a crystal can be analysed both optically and by X-ray topography (Fig. 32 and 33). Helices also form in other complex oxides such as the faceted spiral dislocations sometimes observed in $Ba_2NaNb_5O_{15}$ (Fig. 34).

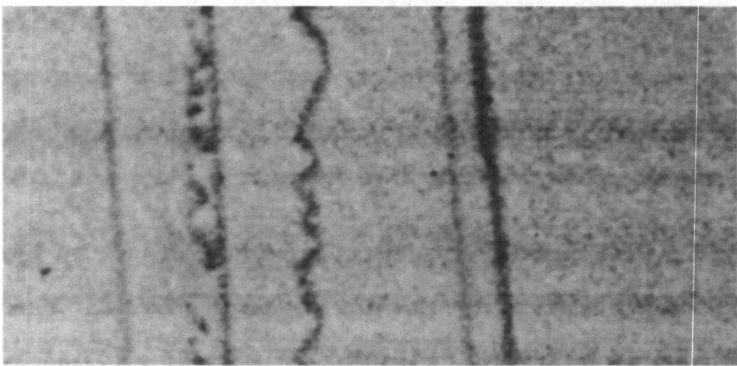

Figure 33. X-ray transmission topograph through a longitudinal section of a $Gd_3Ga_5O_{12}$ crystal showing dislocation strain fields (some helical in character) propagating down the crystal. Mag. x40.

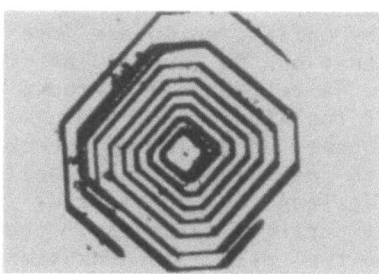

Figure 34. Etched polygonal spiral dislocation in a melt-grown $Ba_2NaNb_5O_{15}$ crystal. Mag. x240.

3.6 Particulate matter and voids

It is well established that, above a critical growth rate, particulate matter in a melt can become embedded in a growing crystal. With proper handling and purification of charge material this is not commonly a problem but it can occur where reaction between melt and crucible produces particles of a new phase. An example of this is the reaction of gallium oxide with an iridium crucible to form particles of iridium oxide which can become trapped in the crystal [34]. A second mechanism can occur whereby evaporation of the crucible material followed by re-condensation in the cooler upper regions of a crystal puller causes fine particles to fall into the melt. This can occur for example during the growth of sapphire crystals using iridium crucibles as shown in Fig. 35. In some cases the stresses due to the differential expansion of particle and matrix can give rise to plastic flow.

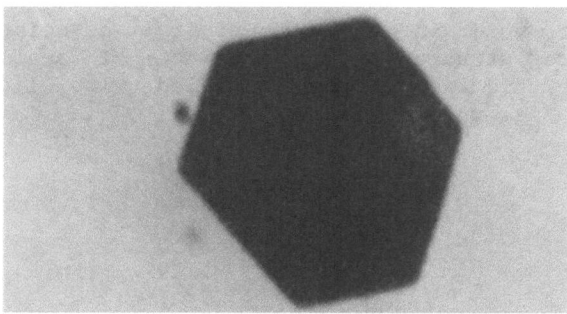

Figure 35. Iridium platelet in a sapphire single crystal grown by the floating-zone technique. Mag. x820.

Voids or cavities occur in many crystals and can vary in size from millimetre to micron dimensions. They are observed only on rare occasions in semiconductors and then usually when a volatile component is present. Thus they can be caused by an imbalance of phosphorus pressure in materials such as InP. In the inorganic compounds exemplified by oxides, voids are more prevalent [23]. As described in section 4, they can be caused by the entrapment of impurity-rich liquid in a cell-boundary groove or solute trail by subsequent contraction of the liquid on freezing. In this case the impurity can be added dopant, an excess of one or other constituent or gaseous material which segregates upon solidification. Other modes of void formation can be considered. For instance, the segegration of gaseous impurities directly, the capture of gas bubbles displaced from the melt, condensation of vacancies and solidification with a limited supply of liquid akin to the piping effect observed in metal castings. Where the void contains solidified liquid of sufficiently different composition, strain can be generated in the surrounding lattice by differential expansion with the matrix. Thus voids can help to obscure the strain fields of other defects. Fig. 36 shows the type of void structure in $CaWO_4$ associated with constitutional supercooling in which the dendrites projecting into the void are indicative of the presence of liquid at some stage. Structureless voids can also form in addition and an example is shown for the fluoride scheelite, $LiRF_4$ (R = Y, Er,Tm,Ho mixture), in Fig. 37.

Figure 36. Tubular void in a $CaWO_4$ crystal associated with a cellular interface structure. The presence of small dendrites indicates that the void was initially filled with liquid before becoming drained. Mag. x370.

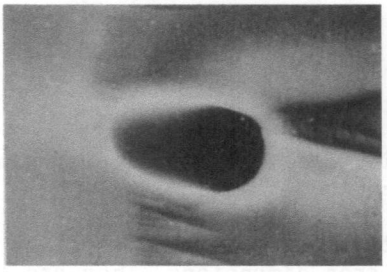

Figure 37. Featureless void in a $LiRF_4$ crystal with no evidence for the existence of trapped liquid. Mag. x1850.

DEFECTS IN MELT-GROWN CRYSTALS

3.7 Summary and conclusions

We have seen that there are several important basic mechanisms by which defects are generated during crystal growth from the melt. These are facet development, the occurrence of constitutional supercooling, non-steady growth induced either mechanically or by non-steady convective processes in the melt, the entrapment of particulate matter and gas bubbles, the relief of thermal and compositional stresses and the precipitation of solutes or native point defects. The manifestations of these effects are diverse and are material specific. Thus the form of the cellular structure is strongly dependent on crystal structure in faceting materials and the resulting pattern of microsegregation depends on the thermodynamic properties of the solute-crystal system. In turn, the structural inhomogeneities depend fundamentally on the basic mechanical properties of the crystal. Hence the interpretation of any particular defect structure demands careful consideration of the likely manifestations of the basic defect-generating mechanisms in a material with given crystallographic, thermodynamic and mechanical properties. Of particular value in such interpretation is the determination of the morphology of the crystal-melt interface at successive stages of growth by the delineation of growth striae.

No one diagnostic technique will be likely to be adequate to reveal the full defect structure of any real crystal. Success comes from the judicious use of a spectrum of techniques as has been illustrated above. X-ray techniques figure prominently in this armoury and are doing so increasingly with the developments in topographic techniques and in techniques for precision lattice parameter determination.

REFERENCES

1. J.R. Carruthers and A.F. Witt (1975) in Crystal Growth and Characterisation, (ed. R. Ueda and J.B. Mullin) North Holland, p 107
2. W. Kern (1978). R.C.A. Review $\underline{39}$, 278
3. B. Cockayne (1968). J. Crystal Growth $\underline{3/4}$, 60
4. E. Jakeman and D.T.J. Hurle (1972) Reviews of Physics in Technology $\underline{3}$, 3
5. B. Cockayne and M.P. Gates (1967). J. Mater. Sci.$\underline{2}$, 118
6. D.T.J. Hurle, E. Jakeman and E.R. Pike (1968). J. Crystal Growth. $\underline{3/4}$, 633
7. M. A. Azouni (1977). J. Crystal Growth $\underline{42}$, 405
8. K.A. Jackson (1958) in Growth and Perfection of Crystals, Wiley. N.Y., p 319
9. K.F. Hulme and J.B. Mullin (1959). Phil. Mag. $\underline{4}$, 1286
10. J.B. Mullin (1962) in Preparation and Properties of III-V Compounds (ed. R.K. Willardson and H.L. Goering) Reinhold, p 365

11. W. Bardsley, J.B. Mullin and D.T.J. Hurle. (1968) in The Solidification of Metals, Iron and Steel Inst. Publication No. 110, p 93
12. B. Cockayne, M. Chesswas and D.B. Gasson (1968) J.Mater. Sci. 4, 450
13. B. Cockayne, J. M. Roslington and A. W. Vere (1973) J. Mater. Sci. 8, 382
14. W.T. Stacy (1974) J. Crystal Growth 24/25, 137
15. B. Cockayne, B. Lent and J.M. Roslington (1976) J. Mater. Sci. 11, 259
16. J.W. Rutter and B. Chalmers (1953) Canad. J. Phys. 31, 15
17. W. Bardsley, J.S. Boulton and D.T.J. Hurle (1962) Solid State Electron. 5, 395
18. W. Bardsley, B. Cockayne, G.W. Green and D.T.J. Hurle (1963) Solid State Electron 6, 389
19. D.T.J.Hurle, O. Jones and J.B.Mullin (1961) Solid State Electron. 3 317
20. J.W. Faust and H.F. John (1961) J. Electrochem. Soc. 108, 860
21. W.G. Pfann (1966) Zone Melting, J. Wiley, N.Y.
22. A. B. Chase and W.R. Wilcox (1966) J. Amer. Ceram. Soc. 49, 460
23. B. Cockayne (1977) J. Crystal Growth 42, 413
24. P.P. Penning (1957) Philips Res. Reports 13, 79
25. N.A. Ardonin, S.S. Vakhrameev, M.G. Milvidskii, V.B. Osvenskii, B.A. Sakharov, V.A. Smirnov and F. Shchelkin (1972) Bull. Acad. Sci. USSR Phys Series 36, 502
26. L.M. Hogan, R.W. Kraft and F.D. Lemkey (1971) in Advance in Materials Research, Wiley - Interscience.
27. C.H. Sworn and T.E. Brown (1972) J. Crystal Growth 15, 195
28. D.T.J. Hurle (1962) Mechanisms of growth of metal single crystals from the melt, Progress in Materials Science 10, 79
29. T. Kuroda and A. Ookawa (1974) J. Crystal Growth 24, 403
30. Y. Seki, H. Watanabe and J. Matsuii (1978) J. Appl. Phys 49, 822
31. A.J.R de Kock (1979) Proc Royal Microscopical Soc 14, 2
32. P.W. Hutchinson and P. S. Dobson (1974) Phil. Mag 30, 65
33. J.R. Patel, R.F. Tramposch and A.R. Chandhuri (1961) Met. Soc. Conf 12, Interscience N.Y. p.45
34. C.D. Brandle, D.C. Miller and J. W. Nielsen (1972) J. Crystal Growth 12, 195.

CHAPTER 4

DEFECTS AND THEIR DETECTABILITY

J.R. PATEL

4.1 Introduction

The Lang X-ray topographic method [1] has played a central role in the characterization of defects in crystals both in the as grown and processed state. Following its introduction and widespread use a large body of information has been gathered on the morphology and topology of defects. The main emphasis has been on line and planar defects and precipitates. In what follows we will confine our attention to the three major topics outlined below. (1) Problems associated with the interpretation of line and planar defects in crystals. (2) Impurity clustering or precipitation and its effect on X-ray anomalous transmission and X-ray diffuse scattering. (3) Point defects in which we discuss epitaxial crystals and recently developed techniques using X-ray standing waves for lattice location of impurity atoms.

4.2 Planar and line defects in crystals

4.2.1 X-ray topography of stacking faults in crystals

(a) Structure

It has been shown that there exists an anomaly between section and traverse topographs of large stacking faults in silicon [2]. In Fig. 1(a) and (b) [3] the section and traverse topographs of an extrinsic fault in silicon show that the first light fringe FH on the section shows up as a dark fringe on the traverse topograph. Computer simulations using the theory of Authier [4] whch takes into account absorption and interference between newly created and old wavefields, are shown for comparison in Fig. 1(c) and (d). The simulation appears to confirm the observed anomaly. Note that there is good qualitative agreement between the fringes observed in the section topographs in Fig. 1(a) and (c). The origin of this anomaly appeared to be due to the excess intensity at H in the section topograph of the asymmetric fault. Since the traverse pattern is an integration of the section along FH, the excess intensity at H overrides the first light fringe to yield a dark fringe FH on the traverse pattern Fig. 1(b). Early simulations also indicated that this was a geometrical effect since symmetrical stacking faults, (i.e. those in which the traces of the fault on both entrance and exit surfaces were parallel to the traverse direction), showed no evidence of such an anomaly. Recent experiments on such symmetric faults [5] are shown in Fig. 2(a) and (b). Note that in Fig. 2(a) the hour glass shape section pattern is symmetric as compared to Fig. 1(a). The anomaly in the experimental pattern however persists and the first white fringe on the section shows up as a dark fringe on the traverse pattern. Computer simulation of this case shows on the other hand a normal behaviour in that the first fringe on the section , Fig. 2(c), shows up as a light fringe, Fig. 2(d), on the traverse pattern. It is evident from Fig. 2(a) that the heavy darkening along IH, which is the trace of the intersection of the primary beam with the fault plane, shows that the observed intensity is always larger than the computed intensity Fig. 2(c).

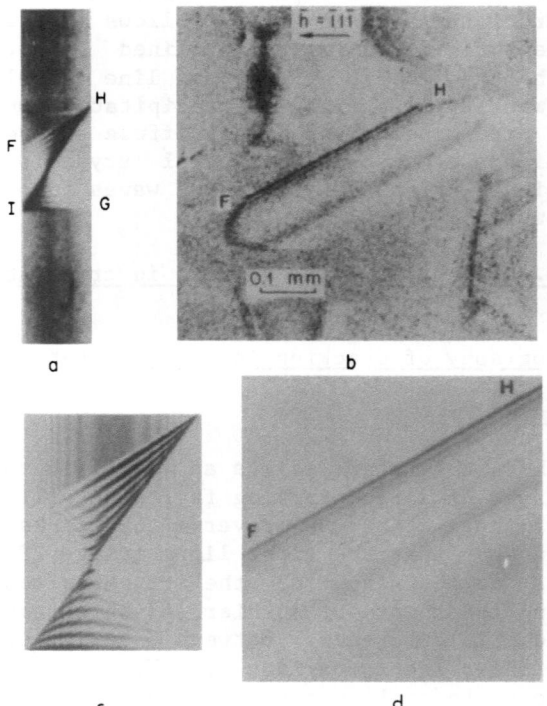

Figure 1. (a) Section topograph of asymmetric fault in silicon CuKα₁ radiation h = 1̄11, (b) traverse topograph of fault in (a), (c) simulation of section topograph shown in (a), (d) simulation of traverse topograph shown in (b).

DEFECTS AND THEIR DETECTABILITY

Computer simulations by Epelboin [6] using Takagi's method for calculating the intensity at the exit surface from a crystal containing a stacking fault rather than the closed form solutions using the stationary phase method developed by Kato et al. [7] (Authier and Sauvage [8]; Authier [4]) show differences in intensity distribution along IH. Calculations for the particular experimental conditions in this study are underway and should provide some definite answers on the origin of this anomaly. In any case this study emphasizes the oft repeated dictum on the importance of section topographs in analyzing the nature of defects.

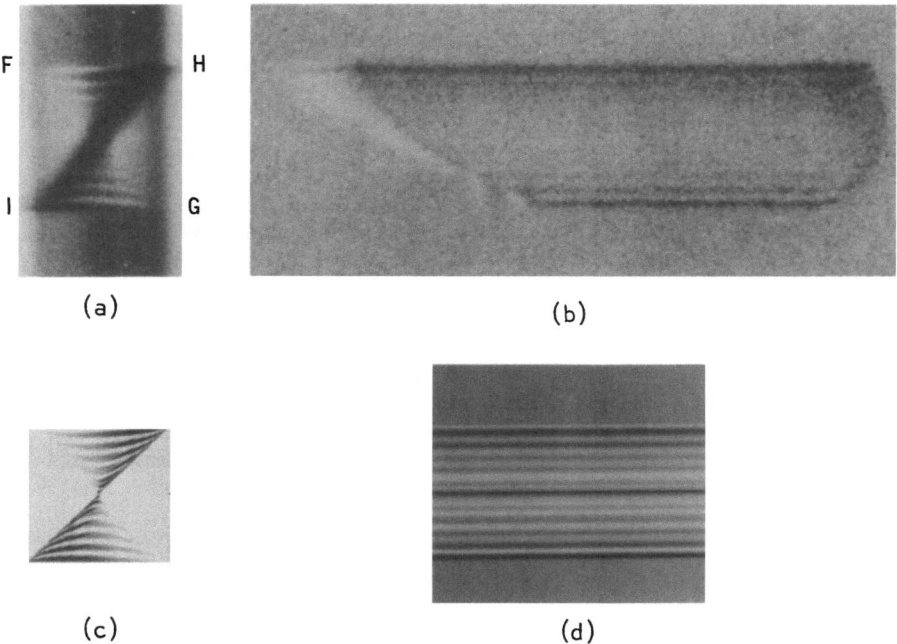

Figure 2. (a) Section topograph of symmetric fault in silicon CuKα_1, radiation h = 111, (b) traverse of section in (a), (c) simulation of section topograph shown in (a), (d) simulation of traverse topograph shown in (b).

(b) Electrical Effects

The large faults described above are formed by the precipitation of oxygen in silicon at temperatures in the range 1000-1200°C. A detailed knowledge of the structure and nature of such defects is crucial to understanding their electrical behaviour. We have verified a model [9] of the growth of such faults in silicon, based on TEM observations that the fault is always associated with a precipitate or precipitate colonies [10]. The model involves (a) the diffusion of oxygen to a precipitate (presumably SiO_2), (b) a slow step which involves the removal of silicon from the precipitate/matrix interface in order to accommodate the large volume change accompanying the reaction $Si + 2O \rightarrow SiO_2$, and (c) a diffusion away of the excess silicon to the fault edges causing it to grow. Both the temperature and time dependence of the observed growth kinetics obey this model in detail. Hence there is strong evidence for believing that except for a small region of the order of microns at the centre of the fault the rest of the fault area is extrinsic in nature. Careful experiments on heating silicon over a range of temperatures and evaluation by junction capacitance methods lead us to believe that the introduction of any contaminating electrically active impurities has been kept to below detectability limits ($10^{11} - 10^{12}$ cm^{-3}). The electrical nature of the faults has been investigated by Kimerling, Leamy and Patel [11] using charge collection scanning electron microscopy. Fig 3(a) and (b) are charge collection scanning micrographs of a stacking fault at 278 and 82 K respectively. The fault has a central precipitate which shows up as an active recombination centre (white contrast). Both the fault plane and its bounding partial show contrast at 82 K. Fig. 4 shows the variation of fault plane contrast with temperature. The data were derived from line scans across the defect and relate the normalized difference in collected current when the beam is far from the defect I_B and in the defect region I_D. The same behaviour is observed for planes inclined or parallel to the surface. The fault in Fig. 3(a and b) is located just below the space charge region of the barrier, about 1 μm as determined by observing the fault image with varying incident beam voltage.

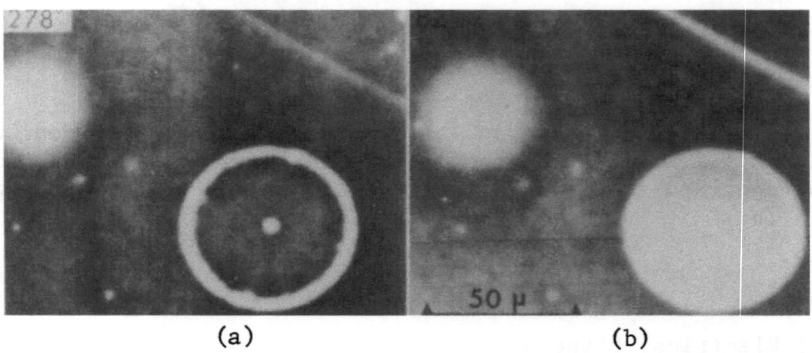

(a)   (b)

Figure 3. (a) Charge collection scanning electron microscopy image of a stacking fault in silicon at 278 K, (b) same fault as in (a) showing strong electrical activity (light region) in the fault plane at 82 K.

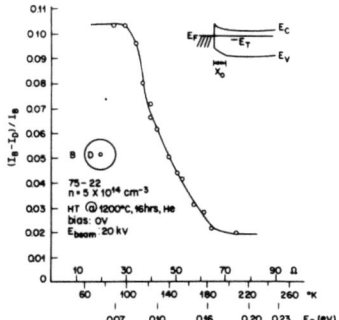

Figure 4. Quantitative variation in fault plane contrast with temperature. I_B is the background collected current far from the fault region and I_D the current measured in the stacking fault plane.

Thus the occupation of fault defect states responsible for the electrical activity is determined by the Fermi level in the neutral base material. The transition in electrical activity shown in Fig. 4 occurs as the Fermi level moves with temperature and changes the defect state occupation. The local minority carrier lifetime at the fault plane decreases as fault states become occupied by electrons and thus become prepared to act as recombination centres for holes. The transition in electrical activity occurs at higher temperature for samples with higher doping, consistent with the movement of the Fermi level towards E_c (conduction band) with doping. As shown in Fig. 4 the position of the Fermi level at the transition indicates an electronic state for the fault located at $E_c - 0.1$ eV. Because the extrinsic fault in silicon is characterized by second neighbour disorder rather than "dangling bonds" the strong electrical activity is somewhat surprising. Theoretical calculations, which are still in a rather crude state, show only minor charge redistribution around an unrelaxed stacking fault defect in silicon.

4.2.2 Dislocations in quartz

X-ray topographic experiments on crystals of quartz with low dislocation densities by Barns et al. [12] revealed certain problems with determining the burgers vector unambiguously. In Fig. 5 we show an X-ray topograph of a crystal cut on the X-face; the direction of c is shown. The majority of these dislocations show no diffraction contrast for $\underline{h}$ = 00.3 indicating that the dislocations are not screw type. The most likely directions for the Burgers vector are the $\underline{a}$ directions of type 11.0. If the dislocations are edge type they ought to vanish for suitably chosen 10.0 type reflections or show reduced contrast if the $\underline{h \cdot b x l}$ term is considered. The results show neither vanishing nor reduced contrast. It was not possible therefore to assign an unique Burgers vector to the dislocations. The only alternative was to take section topographs of

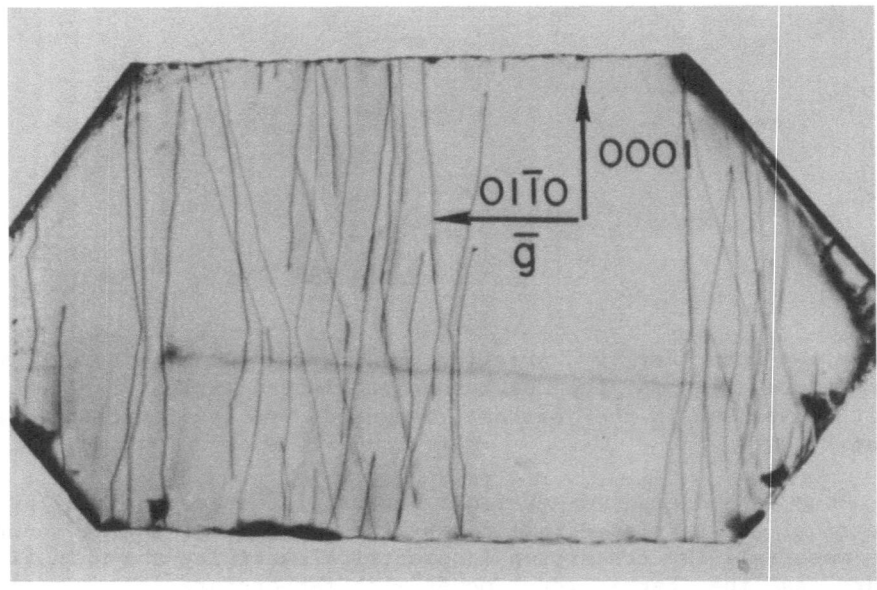

Figure 5. X-ray topograph of an X-cut ($2\bar{1}.0$) crystal showing the c direction. AgKα_1 radiation.

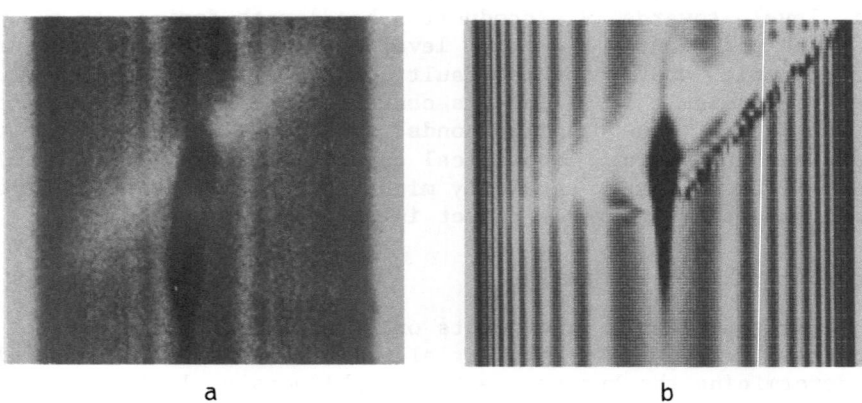

a   b

Figure 6. (a) Section topograph $\underline{h}$ = 10.1 of dislocations in Fig. (5). (b) Computer simulation of a dislocation with the geometry shown in Fig. 5, using anisotropic elasticity and Takagi's theory (from [12]).

the dislocations and attempt to simulate the observed section using Takagi's equations and anisotropic elastic conditions. This has been done by Epelboin [13] and in Fig. 6(a) and (b) we show a comparison between the observed image and the computed one. The reflection used was 10.1 in order to obtain a reasonable section pattern of the dislocation. The pattern agrees well with one of the computed $\underline{a}$ directions as the Burgers vector. The two other $\underline{a}$ directions gave a quite different section pattern. This demonstrates the usefulness of the simulation method in those cases where the usual methods for determining Burgers vectors fail.

4.3 Defects due to clustering or precipitation

4.3.1 Anomalous X-ray transmission

In highly perfect crystals the anomalous transmission method is a powerful technique for detecting the early stages of impurity clustering [14] In anomalous transmission two plane waves of X-ray intensity are coupled in the crystal, one in the primary and the other in the diffracted direction beam. Their interaction gives rise to a standing wave field which for certain reflections produces nodes in the intensity, coincident with the atomic planes. Under these conditions there is a very large reduction in the intensity of photoelectric absorption and a corresponding increase in the amount of transmitted X-rays. The field intensity is given by:

$$E_o = 2E^2 (1 \pm P \cos 2\pi \underline{h} \cdot \underline{r}).$$

where $\underline{E}_o$ is the electric field intensity, P is the polarization factor, $\underline{K}_o + \underline{h} = \underline{K}_h$ (Bragg's law) and $\underline{E}_o \cdot \underline{E}_h = PE_o^2$ where $\underline{E}_o$ and $\underline{E}_h$ are the amplitudes in primary and diffracted beam directions. We see that planes of constant intensity occur when $\underline{h} \cdot \underline{r}$ is constant. This condition implies that the constant intensity planes are parallel to the diffracting planes and are spaced $d = |\underline{h}|^{-1}$ apart. Any source of imperfection which displaces atoms away from the geometric atomic planes reduces the transmitted intensity.

In our case where there is a clustering of impurity atoms, we expect an effect due to a static displacement of atoms giving rise to strains in the perfect host lattice. In Fig. 7(a) we show the reduction of X-ray intensity following heating at $1000^\circ C$ for dislocation-free silicon. The corresponding infrared absorption α at 9 µm, which is a measure of the dissolved oxygen in the crystal, is shown in Fig. 7(b). It is clear that the X-ray intensity measurements detect changes long before any changes in the infrared. In fact the curve A-2 was supposed to be free of oxygen and showed no absorption at 9 µm due to oxygen. Curves B and C are for increasing oxygen concentration as indicated by the values of α. On redissolving the precipitated oxygen at $1350^\circ C$, the initial perfection of the crystal is restored. Subsequent measurements [15] on currently available floating zone crystals with and without carbon show no effects of heat treatment at $1000^\circ C$ in Fig. 8, which is consistent with the idea that small amounts of oxygen are responsible for the effects observed in Fig. 7(a) curve A-2. The defects that are responsible for the decrease in anomalous transmission can be revealed by X-ray topography. In Fig. 9 we show section patterns at various time intervals of a crystal whose anomalous transmitted intensity is shown in Fig. (2) of reference [15]. One hour at 1000°C is sufficient to reveal the strain field of the defects (dark spots). These defects are too small to be fully resolved by X-ray topography. We are undoubtedly seeing only the largest defects of some unknown size distribution.

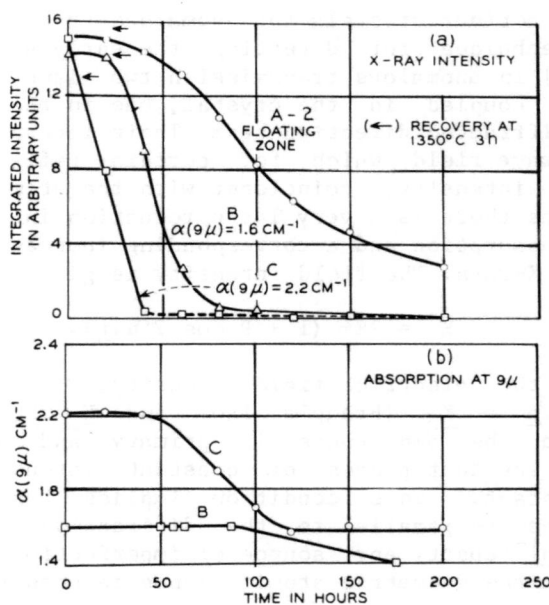

Figure 7. (a) Intensity of X-ray anomalous transmission MoKα1, 220, following heat treatment at 1000°C. Curve A-2 is a floating crystal, B and C are for crystals with increasing oxygen concentration. (b) Infrared absorption at 9 μm corresponding to curves B and C as a function of heat treatment at 1000°C for various times.

Defects of a much smaller size distribution can be produced by a heat treatment at a low temperature, 650° C for 10 days, [16]. X-ray anomalous transmission experiments on such crystals show that the transmitted intensity is not as sensitive to clustering. In this instance the defect size is very much smaller than the Pendellosung length and the average strain field from the defect cluster was about 50Å which is very much smaller than the Pendellosung length of about 17 μm. The results showed that the anomalous X-ray intensity decreased only by 40% while the infrared absorption at 9 μm decreased by a factor of 4 indicating almost complete precipitation of oxygen at this temperature.

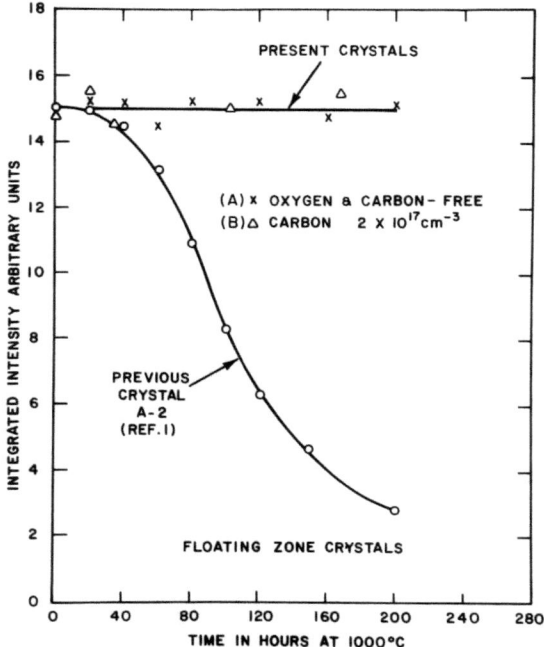

Figure 8. Heat treatment effects at 1000°C on floating zone crystals compared to the earlier crystal A-2.

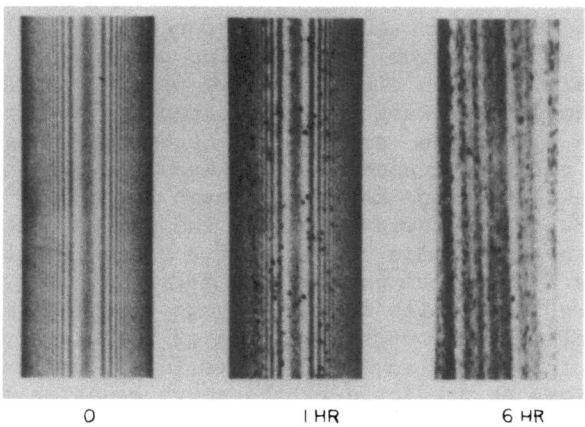

Figure 9. X-ray section topographs of heat treated silicon. $\underline{h}$ = 220 after 0, 1 and 6 hr at 1000°C AgKα_1.

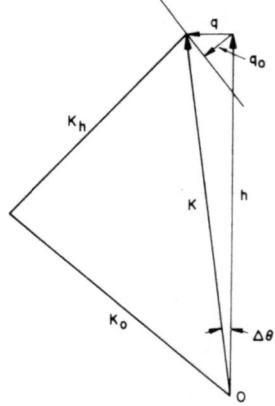

Figure 10. Ewald sphere construction showing the contribution due to various q for a given $\underline{K}$ in the vicinity of a reciprocal lattice point $\underline{h}$.

4.3.2 X-ray diffuse scattering

When the defects are very small information of the mean size of the defect may be obtained from X-ray diffuse scattering in the vicinity of Bragg reflections as described below. From the theory of X-ray diffuse scattering [17,18], one can show that the symmetrical part of the diffused scattered intensity in the region of small q where $q < 1/R_o$ is given by

$$I_h^s(q_o) \sim A \ln|e^{\frac{1}{2}}/q_o R_o|.$$

where $\underline{q}$ is the deviation from the Bragg position and is defined as $\underline{K} = \underline{q} + \underline{h}$, see Fig.10, $q_o = |\underline{h}|\Delta\Theta \cos\Theta_o$ and R_o is the mean cluster radius; the constant A involves the size and number of clusters as well as the diffraction parameters. Thus a plot of $I_h^s q_o$ vs $\ln q$ should be linear and at the intercept where $I^s(q_o) = 0$, $q_o = e^{\frac{1}{2}}/R_o$. Thus we can from the angular dependence of diffuse scattering obtain the mean cluster radius size R_o. Since the defect clusters are likely to be large the scattering measurements were made [19] with a double crystal spectrometer using a 333 $CuK\alpha_1$ reflection. The results of rocking curve measurements near the tails of the Bragg peak before and after treatment for 5 hrs. at 975°C are shown in Fig. 11. The half widths of the two curves before and after treatment were the same, with the peak of the heat treated specimen being lower. The symmetrical part of the X-ray intensity is $I_h^s(q_o) = [I(+q_o) + I(-q_o)]/2$ which is the average of the intensity change measured at equal q_o or $\Delta\Theta$ on either side of the Bragg peak. It is shown plotted in Fig. 12 as $I_s(q_o)/I_o$ vs $\ln \Delta\Theta$, the deviation from the Bragg position. As predicted by the above equation the curve is linear at small $q_o < 1/R_o$ with an intercept $\Delta\Theta \sim 37"$ which gives a mean value of $R_o = 2300$. Defects in this size range were revealed by TEM after sectioning the specimen, see reference [19] Fig. 10.

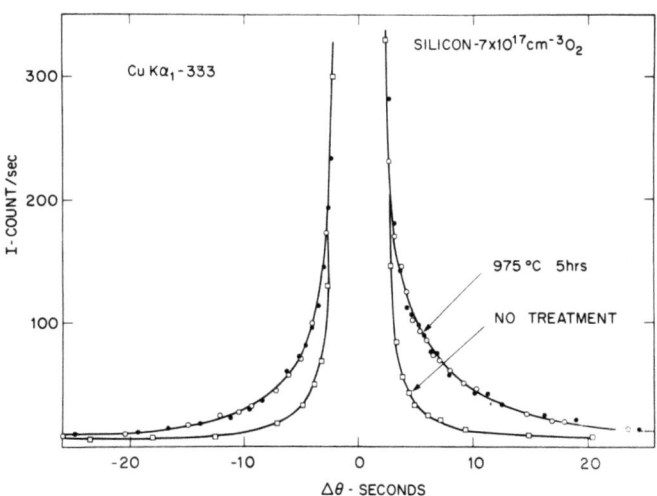

Figure 11. Integral diffuse scattered intensity measured near $\underline{h}$ = 333: (□) before heat treatment (○) after heat treatment at 975 C for 5 hr in hydrogen..

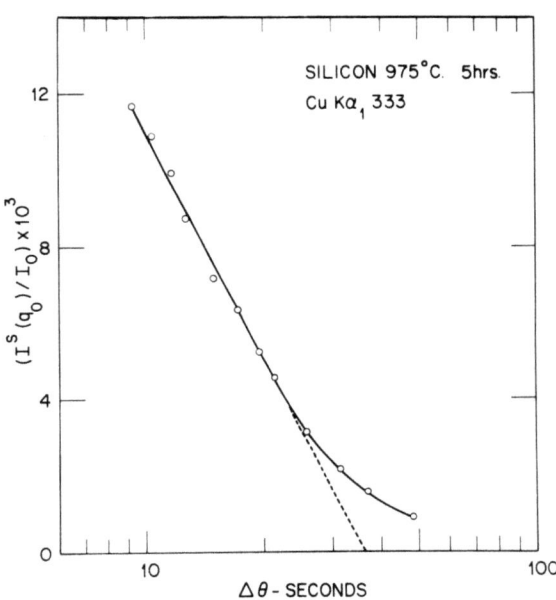

Figure 12. Integrated diffuse scattered intensity vs $\ln_{\Delta\theta}$; intercept gives the value R_0 = 2300

J.R. PATEL

4.4 Point defects

4.4.1 Perfection of epitaxial layers

There is a great deal of effort currently expended in the growth of high quality epitaxial layers on crystal substrates. Methods of growing such layers by liquid phase or molecular beam epitaxy have been highly developed. Since the properties of the epitaxial layer are usually controlled by different alloy compositions the relative misfit between the substrate and layer is important. In particular because of its significance in laser devices the system GaAs–$Ga_xAl_{1-x}As$ has been intensively studied. Lattice parameter measurements of epitaxial layers on substrates have provided useful information on the mismatch problem. Unfortunately any direct measurement of the lattice parameter change between substrate and layer has to take into account elastic strains due to mismatch between the layer and substrate. There are two ways in which this may be accomplished. (1) The actual elastic distortion is measured by measuring the curvature of the specimen and performing a suitable correction to the measured lattice parameter change. (2) An analysis is done using anistropic elasticity and a correction factor calculated for the observed lattice parameter change by X-rays. In the first approach Estop, Izrael and Sauvage [20] have measured the relative lattice parameter change between a GaAs substrate with epitaxial layers of various compositions of $Ga_{1-x}Al_xAs$. The relative lattice parameter change measured normal to the epitaxial layer is

$$\left(\frac{\Delta a_s^\ell}{a}\right)_s = \left(\frac{\Delta a_s^\ell}{a}\right)_o + \frac{\Delta a_\ell}{a} - \frac{\Delta a_s}{a}$$

where $(\Delta a_s^\ell/a)_o$ is the relative lattice parameter change between layer and substrate in the unstrained condition and $(\Delta a_\ell/a)$, $(\Delta a_s/a)$ are the strains normal to the reflecting planes in the layer and substrate respectively. The latter two quantities are calculated from the measured curvature of the specimen, knowing the thickness of the layer and substrate and the elastic constants. Estop et al. [20] have computed $(\Delta a_s^\ell/a)_o$ over a range of compositions of $Ga_{1-x}Al_xAs$. They find that Vegard's law is obeyed over the whole range of compositions and the curve extrapolates to the lattice parameter of AlAs measured independently Fig.13. These findings agree with data previously published by Rozgonyi et al. [21] and resolves the controversy between this previous paper and the work of Druzhinina et al. [22], who measured only the $(\Delta a_s^\ell/a)_s$ values in the strained condition.

More recently Hornstra and Bartels [23] have proposed a scheme for analyzing the distortion in an epitaxial layer using anisotropic elasticity. Their analysis yields various quantities which, depending on the crystal geometry, must be multiplied by the measured $(\Delta a^\ell/a)_s$ in order to obtain $(\Delta a^\ell/a)_o$. For instance for $Ga_{1-x}Al_xAs$ grown on a 100 GaAs substrate $(\Delta a_s^\ell/a)_o = 0.525x(\Delta a_s^\ell/a)_s$. The factors are 0.644 and 0.683 for 110 and 111 substrates respectively. Bartels and Nijman [24] have used the calculated $(\Delta a^\ell/a)_o$ measurements for the unstrained layer and substrate to obtain the Al concentration from the data of Estop et al. [20] Fig. 13, and find excellent agreement (within 1% Al) with the Al content obtained by photoluminescence or electron microprobe measurements. These authors have extended their measurements to complex double heterostructure junctions and report the observation of Pendellosung fringes in the Bragg geometry.

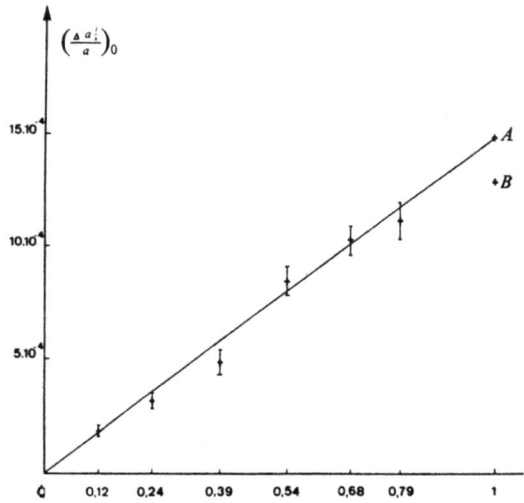

Figure 13. measured relative lattice parameter change between GaAs and $Ga_{1-x}Al_xAs$ corrected for strain versus x, from [19].

Observations of Pendellosung fringes in deliberately thinned perfect crystals of silicon were first reported by Batterman and Hildebrandt [25] who have outlined the conditions necessary for the observation of the fringes. Briefly the dynamical theory predicts such oscillations for a plane wave incident on a thin crystal in which two tie points on the same branch of the dispersion surface interact and produce interference effects in the diffracted intensity. The nature and the origin of these fringes is treated in some detail by Hart in Chapter 9. Slight distortions in the crystal will affect fringe contrast and the observation of a clear set of well resolved fringes is a good indication of the perfection of an epitaxial layer. A good example of such observations in epitaxial garnet films is shown in Fig. 14 by Stacey and Jannsen [26]. The epitaxial layer is a complex rare earth garnet on a substrate of

$Gd_3Ga_5O_{12}$. For this layer thickness of 1.6 μm fringes were observed only on the low angle side of the peak corresponding to branch 1 of the dispersion surface where absorption is low. Fringes were observed on the high angle side of the peak from tie points excited on branch 2 of the dispersion surface only when the thickness was less than 1 μm. In thicker layers the absorption of branch 2 is too high to give clearly resolved fringes. The layer thickness can be calculated from expressions of the reflection coefficient of a thin crystal. Neglecting absorption the thickness t is related to the fringe separation Δθ by

$$t = \lambda \gamma_H / \Delta\theta \sin 2\theta_B$$

where γ_H is the direction cosine of the reflected beam direction and the normal to the epitaxial layer. Layer thickness measured from Pendellosung fringe separation agrees remarkably well with thickness determined from direct measurements.

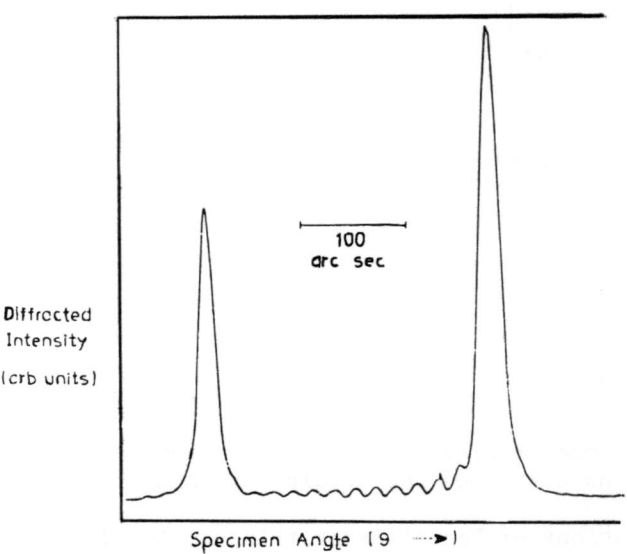

Figure 14. Rocking curve displaying Pendellosung phenomena, substrate peak of $Gd_3Ga_5O_{12}$ is on the left, layer peak on the right t = 1.6 μm, after [26].

4.4.2 Impurity lattice location studies

In section 4.3.1 we discussed the standing wave pattern in a crystal and saw that planes of constant field intensity occur when $\mathbf{h} \cdot \mathbf{r}$ is a constant. The photoelectric absorption of an atom is proportional to the electrical field intensity present at the atom. When the nodal planes of intensity coincide with the atomic planes, absorption much smaller than normal absorption can occur. If however the antinodal planes are at the atom sites an anomalously high absorption occurs. In the Bragg reflection geometry we can easily see from considering the dispersion surface Fig. 15 that on the low angle side of the reflection, branch 1 of the dispersion surface is excited and the intensity and direction of the energy flow at the atom is that of the incident beam. It is evident that on the low angle side of the reflection only tie points on the heavy portion of the dispersion curve on branch 1 will be excited. As we approach the reflecting region indicated by $P_2 P_3$ a weak diffracted beam is excited and its interaction with the primary beam reduces the intensity at the atoms. The direction of energy flow which is normal to the dispersion surface approaches parallelism with the Bragg planes. As the tie point moves up on branch 1 to the diameter points, the intensity at the atoms is reduced and the photoelectric absorption coefficient becomes very much less than its normal value in the crystal. Near the diameter points where there is total reflection it approaches a value of the order of $\mu_0(1-P\varepsilon)$ where ε can be very nearly unity for full reflections. As the Bragg angle sweeps through the not quite totally reflecting range because of absorption, the X-ray intensity increases and at the end of the range where the tie point is on branch 2 the linear absorption has changed from very nearly zero to $\mu_0(1+P\varepsilon)$ or twice the normal value. Thus when the tie point is on branch 2 higher than normal absorption takes place at the atoms. With more energy being absorbed at the atoms, less is diffracted and the diffracted curve is asymmetric with less diffracted intensity on the high angle side.

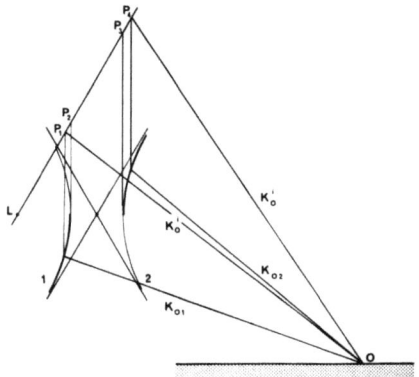

Figure 15. Dispersion surface construction for Bragg geometry showing the excitation of the tie point in the crystal for given incident directions K_o^1. The region $P_2 P_3$ corresponds to the region of total reflection.

In an elegant experiment Batterman [27] has used the fluorescence excited at Ge atomic planes to probe the electric field intensity at the atoms. In the experimental arrangement shown schematically in Fig. 16, MoKα_1 radiation was used on a dislocation free Ge crystal. As the crystal is rotated through its Bragg angle the GeKα fluorescence is measured in a detector placed normal to the reflecting planes and the field at the atoms can essentially be determined. The experimental result is shown in Fig. 17 [28]. The upper curve is the reflection curve and shows an asymmetry which is due to absorption. The fluorescence curve shown below it can be correlated to the structure of the electric fields. Far off the reflection curve the fluorescence is given by the solid line I_F. As the incident beam direction approaches the region of diffraction from the low angle side, corresponding to a tie point on branch 1, intensity is channelled away from the atoms and the primary beam penetrates more deeply. The self-absorption of the deeply generated X-ray fluorescence in getting out of the crystal causes a drop in the observed fluorescence in the low angle region of the reflection curve. Correspondingly on the high angle side with antinodes coincident with the atomic planes the same amount of fluorescence is generated but now very near the surface, because of the high fields at the atoms and shallow penetration of the X-ray beam. Consequently very little self absorption occurs and produces a measured fluorescence greater than that observed when all the energy is absorbed with the crystal well off its diffracting position. This enhancement is evident in Fig. 17. In the region of total reflection, strong extinction effects come into play and both the Compton and temperature diffuse scattering background intensities do not remain constant, but vary quite drastically. Fluorescence intensities from this region, unless corrected for the background, do not agree with theory, and as indicated by Batterman, reference 27, Fig. 5, it is mainly the intensities in the tails that can be accounted for theoretically.

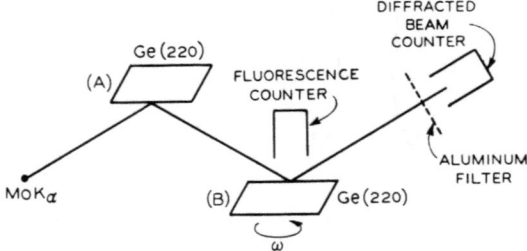

Figure 16. Schematic of parallel crystal arrangement for measuring the fluorescence of Ge excited by the standing wave fields after [27].

Extending these ideas to the determination of the interstitial or substitutional nature of impurity atoms Batterman [29] was able to show that in highly doped dislocation free silicon containing 5×10^{19} cm^{-3} As, a signal due to As fluorescence could be detected. Analysis of the results with reflections from various crystallographic directions led to the conclusion that arsenic was located in substitutional sites.

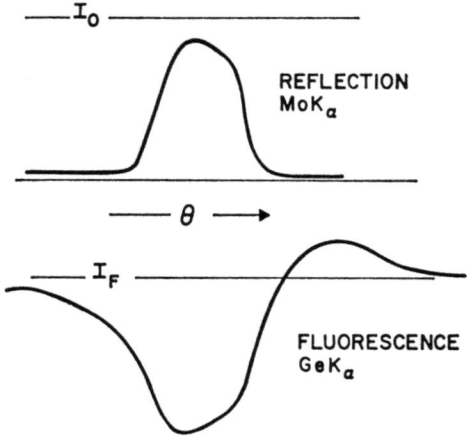

Figure 17. Fluorescence measured during rotation through a Bragg peak [28].

In a subsequent paper Golovchenko, Batterman and Brown [30] pointed out that in the angular region where there is most control over the motion of the wavefields (i.e. in the almost total reflection region), the angular yield of the arsenic fluorescence is dominated by primary extinction, i.e. the reduction in X-ray intensity with depth in the crystal due to enhanced scattering by the atoms of the solid. The diminution of X-ray fluorescence at low angles and its enhancement at high angles (Fig. 17) result from primary extinction which diminishes the number of arsenic atoms that the incident X-rays are able to penetrate and cause fluorescence. Since for the doped crystal, arsenic is distributed uniformly in depth throughout the sample, this effect is independent of the actual location of the arsenic atoms and serves to mask interesting detail due to motion of the internal wave fields during Bragg reflection. It also places severe restrictions on the sensitivity with which the position of the impurity atoms can be determined. Golovchenko et al. [30] propose a clever way out of this difficulty by preparing a crystal in which all the impurities are located at a very small depth (less than an extinction distance) below the surface. The effects observed are then all due to the angular variation of the wavelength in the crystal and effects due to extinction are practically eliminated. Another technical improvement in their scheme was to use a lithium drifted silicon X-ray detector of much higher energy resolution than the scintillation detector previously used. The results of their experiments on an As diffused silicon crystal are shown in Fig. 18. The left curve shows the reflection curve from silicon and the right curve the arsenic fluorescence yield; the solid line is the theoretical expected yield. The agreement between theory and experiment is reasonably satisfactory but discrepancies remain originating probably from crystal strain due to high As concentration at the surface and problems related to the distribution of As after the short diffusion cycle. If the As atoms were all situated between the atomic planes their fluorescence yield would be given by the dashed curve shown in the inset of Fig. 18(b)

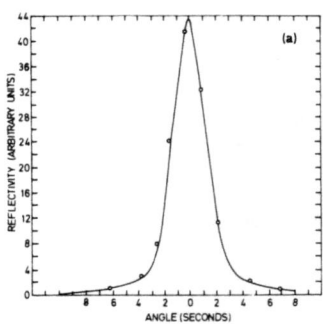 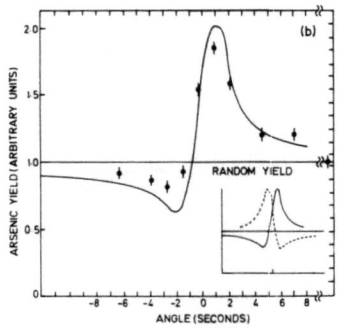

Figure 18. (a) Reflection curve for a silicon crystal parallel arrangement. (b) As fluorescence yield (solid curve theory). Dashed curve in inset shows the yield curve if all the As was between the silicon reflecting planes; after [30].

Substantial improvements in both the analysis and instrumentation were subsequently reported by Andersen, Golovchenko and Mair [31]. Let us assume as previously that the fluorescence yield of an impurity at any point $\underline{r}$ is proportional to the standing wave intensity. Then for the region of total reflection in the symmetrical case, we can write for $-1 < \eta < +1$:

$$I = 1 + \left|\frac{E_h}{E_o}\right|^2 + 2P \left|\frac{E_h}{E_o}\right| \cos(\phi + 2\pi\underline{h}\cdot\underline{r})$$

where $E_h/E_o = \eta \pm \sqrt{\eta^2 - 1}$, η is a dimensionless parameter proportional to $\Delta\theta$ the deviation from the Bragg position and ϕ is a phase factor since η is complex in the total reflection region. These relations will be discussed in some detail in chapter 9 on Dynamical Theory. The quantity $\underline{h}\cdot\underline{r}$ is the displacement of the atom from the reflecting planes and is given by $\delta d/d$ where δd is the displacement. Corresponding relations apply to the case where $\eta > 1$ and $\eta < -1$. Calculations of the standing wave field intensity as a function of η and for various $\delta d/d$ are shown in Fig. 19(a) and (b) [32]. These curves are similar to those for the case shown in James [33] and the reader is referred to the paper of James for details. The curves shown in Fig. 19(a) and (b) take into account absorption and are for a point just below the crystal surface. The assumption being that one may ignore extinction effects because the impurities are all located near the surface at a depth much smaller than the Pendellosung distance. The astonishing sensitivity of the yield curves to the position of the impurity is impressive: differences of 0.1 of the interatomic distance alter the yield curves in a recognisable fashion. If one can achieve high resolution in the angular counts and good counting statistics then the authors [31] claim that position resolution of impurities of the order 1% of the interplanar spacing can be achieved. To this end the authors have used a highly collimated incident beam obtained from an asymmetrically cut silicon crystal and devised a system to obtain high stability over long counting times. The system consists of a piezoelectric drive on the sample crystal servomechanically controlled from the Bragg reflection detector. The stability that can be obtained with a similar device has been demonstrated by Miller et al [34] Fig. 20(a) and (b). The figure shows the almost perfect stability that is achieved when the correlator is switched on as compared to the drift observed in Fig. 20(b). With these improvements Andersen et al [31] demonstrated in a rather remarkable experiment the variation of the fluorescence yield with impurity location in the standing wave field. To keep impurities in the surface region, the crystal was implanted with 60 keV arsenic ions. The total dose was 10^{15}/cm^2, approximately a monolayer (the range for this energy is about 400Å). The cold implanted specimen showed essentially a random distribution of As atoms, Fig. 21. After annealing for 20 min. at 750°C the Rutherford backscattering spectrum indicates that most of the As is in substitutional sites. The standing wave pattern after annealing is shown in Fig. 22(a). Comparing this to the cold implanted specimen, we see a quite different yield curve corresponding very nearly to the curve for $\delta d/d = 0$ in Fig. 19(a). A more sophisticated analysis of the fluorescence yield shows ($\delta d/d$) = .048 or very nearly zero. The solid line in Fig. 22(a) is the fitted theoretical fluorescence curve.

In order to move the thin 400Å surface layer containing As with respect to the standing wave field, a damaged layer behind the As was created by implanting N atoms at 60 keV and 500°C. The damage peak due to N occurred at 2000Å with straggling to 1000Å as determined by backscattering. this is well below the As layer. A dose of 10^{15} N creates sufficient displacement due to damage to shift the yield curve to the position shown in Fig. 22(b). The solid curve was fitted to the yield curve for $(\delta d/d) = 0.254$ and compares well with the expected curve in Fig. 19(a). Increasing the dose to 2×10^{15} N/cm^2 shifts the yield curve further to the left Fig. 22(c). Finally after a dose of 5×10^{15} N/cm^2 the curve returns very nearly to its original position in Fig. 22(d). In this procedure the implanted layer has been shifted upwards one interatomic layer distance a_{220}.

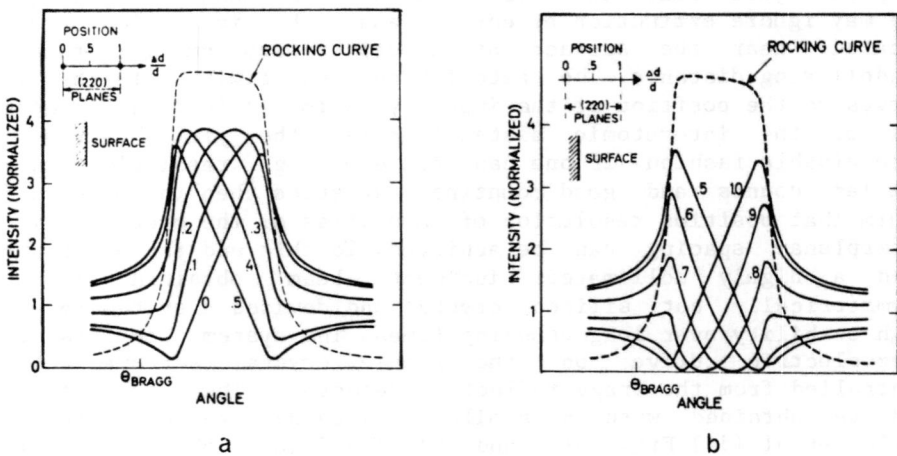

Figure 19. (a) Theoretical reflectivity and intensity of the standing wave field for several atomic positions between 220 planes ($\delta d/d$) = 0-0.5, (b) same as (a) ($\delta d/d$) = 0.5-1.0; after [32].

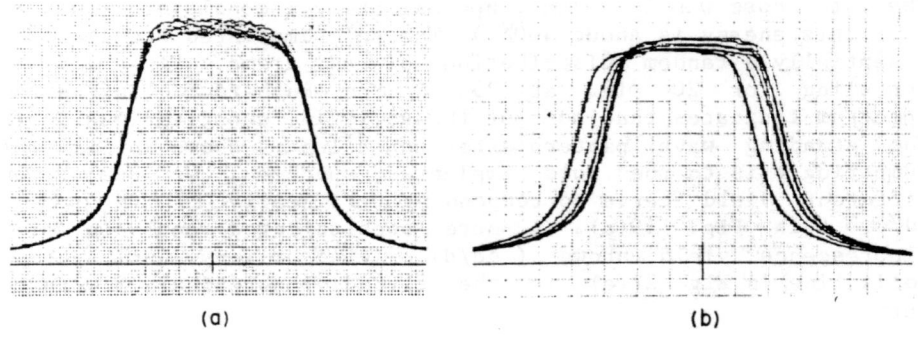

Figure 20. (a) A family of silicon 220 rocking curves taken at 10 minute intervals over 1 1/2 hrs. (b) Shows the PMT anode current versus demanded position signal over the same time period, thereby displaying the correcting action of the complete servo system; after [34].

The lattice expansion determined from fitting the yield curves is linear as a function of dose, Fig. 23, which shows the lattice expansion, as determined from the curves for fluorescence yield, Fig. 22(a)-(d), plotted against the nitrogen dose. The expansion is sensibly linear with dose. Since both the implanted region and the damaged region due to N is very much smaller than the extinction distance the standing wave field is determined by the underlying wavefield and the increasing N dose merely serves to move the As

DEFECTS AND THEIR DETECTABILITY

implanted layer through this fixed standing wave field. These experiments raise many exciting prospects. One that has been pointed out by the authors is the possibility of locating an adsorbed surface layer of atoms. The standing wave field does not terminate at the surface but exists outside the crystal because of overlap between incident and diffracted waves. preliminary results of experiments on adsorbed Br layers on a silicon surface show a well resolved fluorescence signal [35].

Additionally, TDS profiles [36] and photoelectron yields during Bragg reflection have been investigated [37, 38].

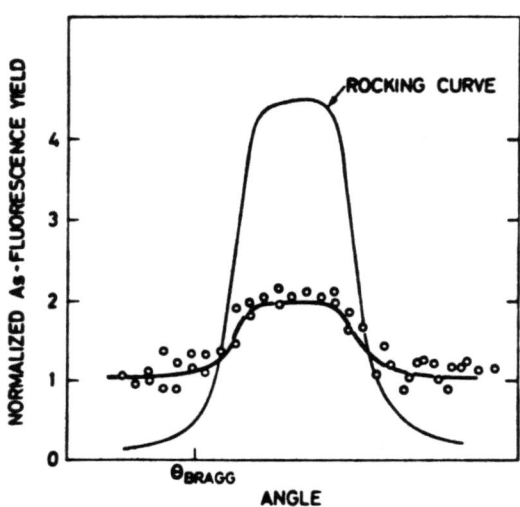

Figure 21. Fluorescence yield from a cold implanted crystal at 60 keV. Heavy line indicates fluorescence yield from randomly distributed As atoms; after [32].

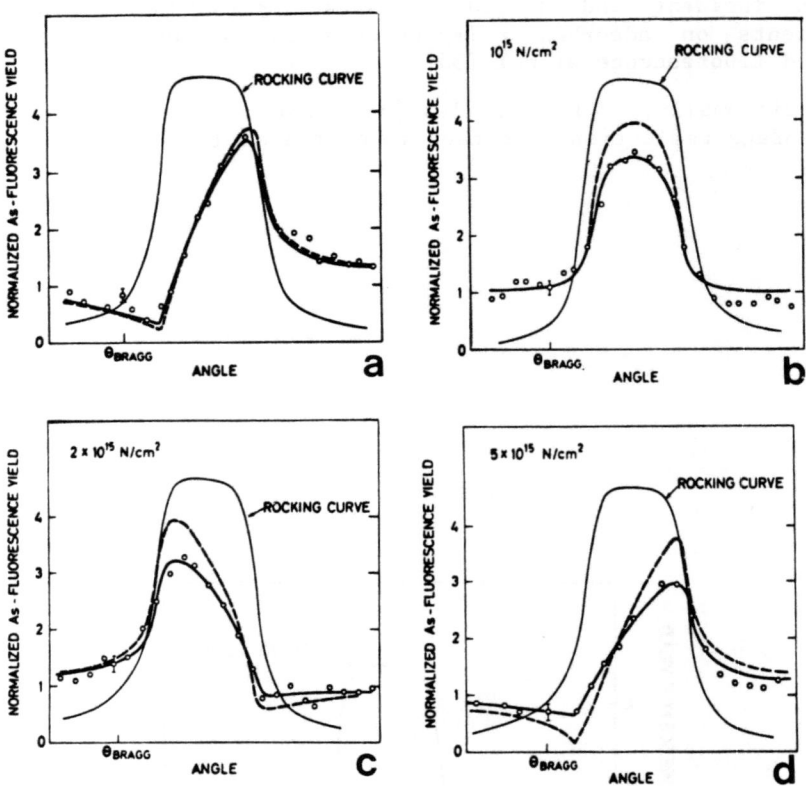

Figure 22. (a) Fluorescence yield from the same crystal as that in Fig. 20 annealed 20 min. at 750^0 C. Solid line fitted for $(\delta d/d)$ = 0.048. (b) Fluorescence yield from crystal shown in (a) after a N dose of 10^{15} /cm^3, fitted for $(\delta d/d)$ = 0.254. (c) Fluorescence yield from crystal shown in (b) after a dose of 2×10^{15} N/cm^2, fitted for $\delta d/d)$ = 0.371. (d) Fluorescence yield from crystal shown in (c) after an implant of 5×10^{15} N/cm^2, fitted for $(\delta d/d)$ = 0.039; after [32].

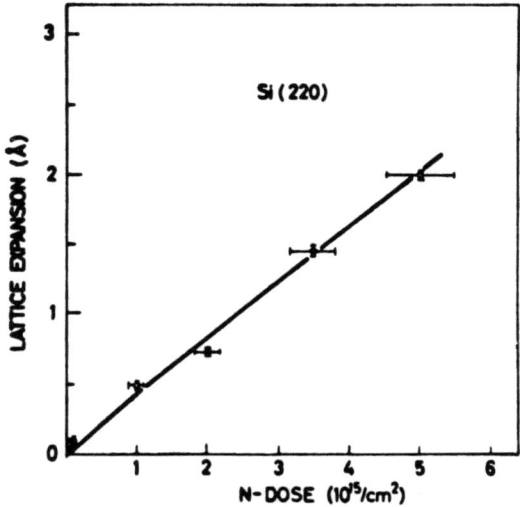

Figure 23. Lattice expansion determined from Fig. 22(a)-(d) plotted as a function of implanted N dose.

REFERENCES

1. A.R. Lang (1958), J. Appl. Phys. <u>29</u>, 597: See also (1970) in Modern Diffraction Techniques in Materials Science (ed. S. Amelinckx et al.), North Holland p. 407.
2. A. Authier and J.R. Patel (1975), Phys. Stat. Sol. <u>27</u>, 213.
3. B.C. Wonsiewicz and J.R. Patel (1976), J. Appl. Phys. <u>47</u>, 1837.
4. A. Authier (1968), Phys. Stat. Sol. <u>27</u>, 77.
5. J.R. Patel (1979), Acta Cryst. <u>A35</u>, 21.
6. A. Authier, private communication.
7. N. Kato, K. Usami and T. Katagawa (1967), Adv. X-ray Anal. <u>10</u>, 46.
8. A. Authier and M. Sauvage (1966), J. Phys. (Paris) <u>27</u>, C3-137.
9. J.R. Patel, K.A. Jackson and H. Reiss (1977), J. Appl. Phys. <u>48</u>, 5279.
10 D.M. Maher, A. Staudinger and J.R. Patel (1975) J. Appl. Phys. <u>47</u>, 3813.
11. L.C. Kimerling, H.J. Leamy and J.R. Patel (1977), Appl. Phys. Lett. <u>30</u>, 217.

12. R.L. Barns, P.E. Freeland, E.D. Kolb, R.A. Laudise and J.R. Patel (1978), J. Cryst. Growth 43, 676.
13. Y. Epelboin and J.R. Patel, to be published.
14. J.R. Patel and B.W. Batterman (1963), J. Appl. Phys. 34, 2716.
15. J.R. Patel (1973), J. Appl. Phys. 44, 3903.
16. P.E. Freeland (1980) J. Electrochem. Soc. 127, 754.
17. P.H. Dederichs (1971), Phys. Rev. B4, 1041.
18. M.A. Krivoglaz (1969), Theory of X-ray and Thermal Neutron Scattering by Real Crystals, Plenum, New York.
19. J.R. Patel (1975), J. Appl. Crystl. 8, 186.
20. E. Estop, A. Izrael and M. Sauvage (1976), Acta Cryst. A32, 627.
21. G.A. Rozgonyi, P.A. Petroff and M.B. Panish (1974) J. Cryst.Growth 27, 106.
22. L.V. Druzhinina, V.T. Bublik, L.M. Dolginov, P.G. Eliseev, M.P. Kerbelev, V.B. Osvenski, I.Z. Pinsker and M.G. Shumskii (1975), Sov. Phys. Tech. Phys. 7, 935.
23. J. Hornstra and W.J. Bartels (1978), J. Cryst. Growth, 44, 513.
24. W.J. Bartels and W. Nijman (1978), J. Cryst. Growth 44, 518.
25. B.W. Batterman and G. Hildebrandt (1968), Acta Cryst. A24, 150.
26. W.T. Stacy and M.M. Janssen (1974), J. Cryst. Growth 27, 282.
27. B.W. Batterman (1964), Phys. Rev. 133, A759.
28. B.W. Batterman and H. Cole (1964), Rev. Mod. Phys. 36, 681.
29. B.W. Batterman (1969), Phys. Rev. Lett. 22, 703.
30. J.A. Golovchenko, B.W. Batterman and W.L. Brown (1974), Phys. Rev. B10, 4239.
31. S.K. Andersen, J.A. Golovchenko and G. Mair (1976), Phys. Rev. Lett. 37, 1141.
32. S.K. Andersen (1977), Application of Standing Wavefield in Impurity Lattice Location Studies, Licentiate Thesis Univ. of Aarhus
33. R.W. James (1963), Adv. in Physics, 15, Academic Press, New York, p. 127.
34. G.L. Miller, R.A. Boie, R.L. Cowan, J.A. Golovchenko, R.W. Kerr and D.A.H. Robinson (1979), Rev. Sci. Inst. 50, 1062.
35. J.A. Golovchenko (1980), Phys. Rev. Lett.
36. S. Annaka, S. Kikuta and K. Kohra (1966) J.Phys.Soc. Japan, 21, 1559.
37. V.N. Schchemelev, M.V. Kruglov and V.P. Pronin (1971), Soc. Phys. Sol. State, 12, 2005.
38. T. Takahashi and S. Kikuta (1979), J. Phys. Soc. Japan, 46, 1608.

CHAPTER 5

DEFECT GENERATION IN METAL CRYSTALS

G. CHAMPIER

5.1 Introduction

There is no comprehensive study setting down explicit relations between the growth conditions of a metal crystal and its crystalline perfection. From studies published so far we can infer at best tendencies that qualitatively indicate the influence of a given parameter of the growth method on the perfection of the crystal obtained. Difficulty arises from different factors: the number of parameters quantitatively defining a growth process is very high. Those parameters sometimes are ill defined and they are dependent on one another. Besides, within the growth process must be included the initial phase of crystal generation, generally at high temperature, as well as the phase of cooling to room temperature as crystalline perfection is generally assessed at such temperature.

X-ray transmission topography, direct image by the Lang method [1,2] or dynamic image by the Borrmann method [3,4] constitutes a choice method for the study of crystal perfection. However, we will keep in mind that the imperfections in size less than one micron cannot be detected and that the method applies to crystals of dislocation density less than 10^4 cm^{-2}. Moreover, the notion of perfection is related to the volume of the sample under observation; this can be either a small crystal or a small part of some large one, with maybe two dimensions inferior or equal to one millimeter, or one large crystal, two dimensions of which can reach several millimeters. In all cases X-ray topography can involve only limited thickness of the metal.

In situ examination of the imperfections formed during the initial phase of crystal generation has only just started; it can only be achieved by synchrotron radiation owing to the shortening of time of exposure and the number of the results available to this day is still very limited. The imperfections observed at room temperature may either be produced by the evolution of the imperfections already existing in the generation phase, or they may come out during the cooling phase. It is difficult to avoid the stress field due to possible contact with the crucible or to the temperature gradients which develop in various parts of the crystal. Since the yield stress of metals is comparatively low, and so much the lower as the temperature is higher, slip and multiplication of dislocations can be expected to occur. Moreover, those crystals generally have a low dislocation density and the vacancies do not find an adequate number of sinks during cooling; they will stay in the lattice and beyond a given supersaturation they produce climb of the dislocations. This supersaturation effect can be enhanced if the crystal reacts with the environment, during oxidation for instance. Furthermore, in some cases, the preparation of the sample requires cutting and polishing of the crystal. If they are not carried out very carefully, those operations can bring more imperfections in the lattice. Thus one must note the importance of the cooling phase and preparation of the sample in connection with the structure and density of the defects in metal crystals at room temperature.

There are several ways of stating the results obtained so far concerning the relations between growth conditions and crystal perfection, either by crystalline structure or growth method. The latter classification seems to be more relevant and it is the one we have chosen taking into account the initial state of the material.

5.2 Growth from liquid state

Two main methods allow metal crystals to be prepared from the liquid state. The first one is the Bridgman method using a horizontal or vertical crucible, with or without a single crystal seed, and the second one is the Czochralski method. Our purpose is to review for each method the works carried out with a view to link the growth conditions to the crystal perfection.

5.2.1 Pridgman metbod

Several studies have been published on the X-ray topographic characterization of metal crystals obtained by the Bridgman method. However, only the study by Young and Savage [5] on copper concerns various growth conditions.

Copper crystals

Young and Savage [5] take into account the following parameters which are likely to affect the crystal perfection:

- presence of single crystal seed and its orientation or absence of seed; growth direction;

- growth rate;

-rigidity of the crucible;

DEFECT GENERATION IN METAL CRYSTALS

- existence of a radial temperature gradient.

The first three parameters can be considered as exerting direct influence on the generation phase and the last two as tending to intervene rather during the cooling phase.

The crystals are cylindrical, 25mm in diameter and 150 mm long. When the crystal grows without seed, the dislocation density is the lower as the growth direction is farther from the [111] zone. For instance, if this direction is [123], high density of dislocations and sub-boundaries are observed. If, on the contrary, the growth direction is [111], the dislocation density is lower and decreases when passing from the initial part to the final one. Growth direction [111] also leads to a low dislocation density. If a single crystal seed is used and if the growth direction is [100] or [111], the crystal perfection is not as good as it would be for the same growth directions, if no seed were present. The dislocation density is higher ($\geqslant 10^5$ cm^{-2}) and sub-boundaries are formed. The crystals obtained from a melt at $1180°C$ have a lower dislocation density than those obtained from a melt at $1125°C$. The growth rate, from 25 to 50 mm/H, does not affect the crystal perfection. A thinning down of the graphite walls of the crucible from 9.5 to 2.4 mm causes the dislocation density to be divided by ten. Lastly, using a Mullite tube as a loose sleeve around the crucible lowers by a factor of 2 to 5 the dislocation density.

The dislocations that have occurred during the growth probably result from slip and multiplication under the stresses developed either by the crucible, by the longitudinal temperature gradient or by possible fluctuations in the impurity concentration. Besides, if the growth direction is in the [111] zone there is a set of slip planes parallel to it and dislocations can develop slipping along the whole length of the crystal. In the opposite case, the slip planes are inclined to the growth direction, and one may infer that, beyond a given length of growth the initially present dislocations will emerge on the surface of the crystal causing a decrease in the dislocation density.

The dislocation density of these crystals remaining at least equal to 10^4 cm^{-2} under the best conditions and the topographs obtained being hardly exploitable, Young and Savage annealed them at $1075°$ C in vacuum, 10^{-6} Torr, or in a hydrogen atmosphere. When the initial dislocation density is more then 10^5 cm^{-2} annealing results in polygonization without any decrease in the dislocations number. When the density is less than 10^5 cm^{-2}, it is then possible to divide by a factor 100 the dislocation number at the end of an adequate time, namely several weeks. In the best cases the dislocation density was about 10^2 cm^{-2}. The dislocations generally are of mixed character and do not lie on {111} planes or along simple crystallographic directions. Merlini and Young [6] conclude that climb is very important in establishing the dislocation configuration (Fig. 1).

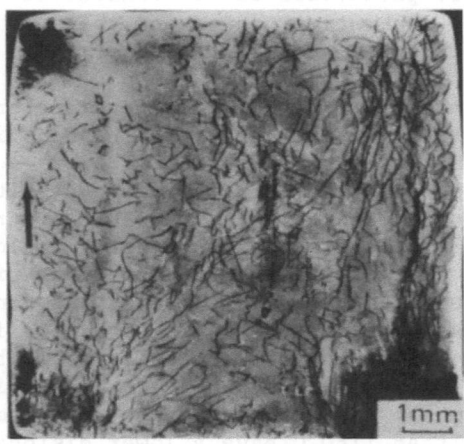

Figure 1. X-ray topograph 111 of a copper lamella, thickness 0.45 mm [6]

Other metals

Crystals from other metals have been prepared by the Bridgman method and characterized by X-ray topography, in some cases, after prolonged annealing. The growth conditions mentioned by the authors are on the whole those which led to the best perfection of the crystal.

Lang and Polcarova [7] studied by X-ray topography thin plates of iron-3.5% silicon spark-cut from a crystal prepared by the Bridgman method, annealed at $1000°C$ in dry hydrogen for 10 hours, then slowly cooled. The defects introduced by cutting were eliminated by electrolytic polishing. The topograph in Fig. 2 provides an example of the quality of those crystals.

Badrick and Puttick [8] have prepared 5x5x120 mm cadmium crystals. They cut slices and annealed them in argon atmosphere to decrease the dislocation density down to 10^4 cm^{-4}. They noted a high sensitivity of crystal perfection to cooling rate after anneal, the density of the dislocations and sub-boundaries being higher as the cooling rate is higher. On the topographs they observed dislocation lines, and also a large number of loops lying in basal plane and showing various contrast. The high density of dislocations and sub-boundaries prevented the authors from assessing with any certainty the Burgers vectors of loops.

Vale and Smallman [9,10] have prepared magnesium crystals by the Bridgman method in a purified argon atmosphere. They have mainly studied the stability of the dislocation configuration during high temperature oxidation treatment.

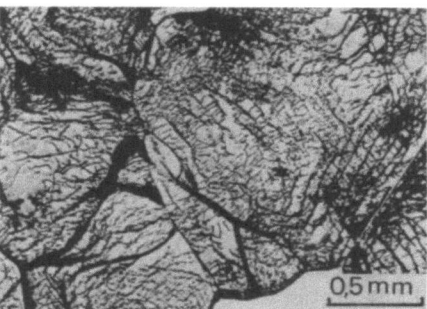

Figure 2. X-ray topograph 002 of an iron-3.5% silicon plate; the crystal is divided in subgrains inside which the dislocation density is low [7].

Preparation of small crystals

In order to reduce strains caused by the crucible and the temperature gradient, on the one hand, and on the other hand, to obtain directly samples likely to be examined by X-ray topography after simple polishing, G'Sell, Champier and Iwasaki [11] have proposed a growth method of small platelike crystals. This method is related to the Bridgman method without use of a seed. The crucible is machined in pure graphite and it simply consists of two plates held apart at a steady interval by two wedges, from 0.3 to 0.9 mm thick. The samples are rolled in order to become slightly thinner than the wedges, they are then put between the plates, heated to a temperature slightly higher than their melting point in an argon atmosphere and eventually slowly cooled down to room temperature, 5-6° C/hr (Fig. 3). That method has been applied to magnesium, zinc and cadmium. In the first case, it has led to crystals of poor quality, a result accounted for by the insufficient purity of magnesium used, 99.98% (Fig. 4). In the last two cases, the crystals are of high perfection (Fig. 5 and 6) and the dislocation density is low enough to make an accurate analysis of the configuration and determine Burgers vectors with precision. The dimensions of the samples so obtained are 10 x 4 x t mm with t between 0.1 to 0.6 mm; the final thickness being achieved by chemical polishing. This method has been applied to indium by Blind, George and Champier [12]. The perfection of the crystals depends on the cooling rate after solidification. When it is more than 5°C/hr, the crystal shows a rather high density of sub-boundaries.

After cooling at 3° C/hr crystals contain little or no sub-boundaries (Fig. 7). Annealing for several days, possibly several weeks, at room temperature improves the perfection of the crystals.

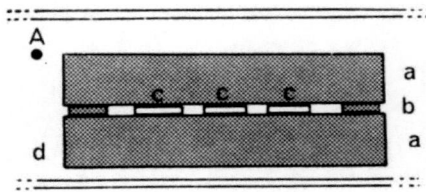

Figure 3. Graphite crucible for the growth of flat small crystal, a and b pieces of graphite, c crystal samples, d quartz tube, A argon.

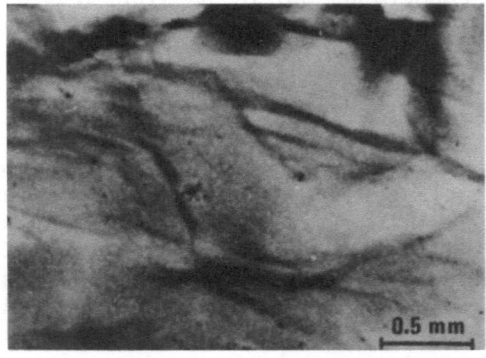

Figure 4. X-ray topograph $10\bar{1}0$ of a magnesium crystal [11].

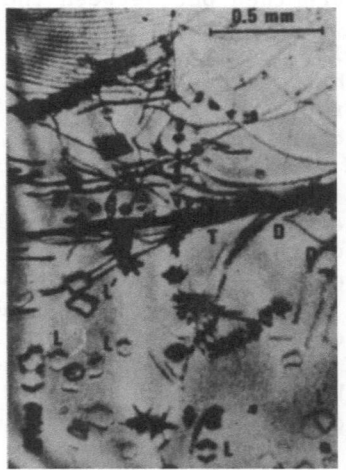

Figure 5. X-ray topograph $11\bar{2}0$ of a cadmium crystal, dislocations lines $\underline{a}$ loops $\underline{c}$ and $\underline{c} + \underline{a}$ [11].

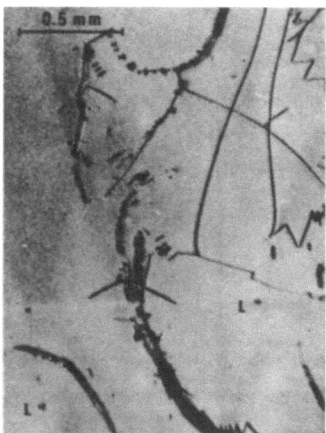

Figure 6. X-ray topograph 11$\bar{2}$0 of a zinc crystal. Dislocations lines and tangles a small loops c and c + a [11]

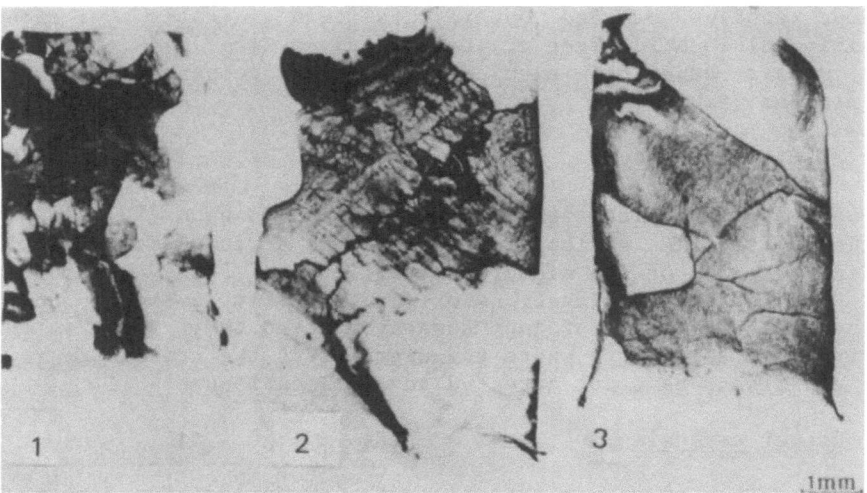

Figure 7. X-ray topograph 111 of indium crystals obtained with different cooling rate 30°C/hr; 10°C/hr; 3°C/hr

McFarlane and Elbaum [13] have used a similar method to prepare gallium crystals. They melt partially a single crystal plate and then they cool slowly the new crystal. With high purity gallium they obtain crystals wih a low dislocation density, 30 cm^{-2}. With less pure gallium, crystals show a "banding structure" with striations parallel to the solidification interface, which have been accounted for by fluctuations in the impurity concentration.

5.2.2 Czochralski method

The various parameters stating the conditions of Czochralski growth are very numerous:

- Seed: diameter, crystalline perfection, single or poly crystal, orientation, rotation rate;

- Crucible: diameter, height, rotation rate, after-heaters;

- Neck: length, diameter;

- Crystal: diameter, length, pulling rate, orientation, tail

- Atmosphere, dust

The measures concerned are directly accessible but other parameters could be considered e.g. radial temperature gradients, longitudinal temperature gradient, movement of the liquid in the melt etc. The various studies carried out in this field generally have aimed at obtaining dislocation-free crystals of given size and orientation. Five metals have been examined: aluminium, copper, silver, nickel and niobium (table 1).

Aluminium crystals

The values of the parameters used by Howe and Elbaum [14] are shown in Table 1. The dislocation density is a rising function of the pulling rate and of the diameter of the crystal. The authors state that the crystal dislocations originate mostly in the dislocations of the seed and in the vacancy supersaturation that appears during cooling. In some parts where the diameter of the crystal is reduced Howe and Elbaum have not detected any dislocations.

Copper crystals

The preparation of copper crystals with low dislocation density has been dealt with in four studies by Fehmer and Uelhoff [15], Sworn and Brown [16] and Tanner, [17] Kuriyama, Early and Burdette [18] and more recently Buckley-Golder [19]. The values of the parameters stating the growth conditions are shown in Table 1. The small crystals (SB, B-G) are examined directly by X-ray topography; the other ones (FU, KEB) are cut with a chemical saw, the slices are polished to eliminate the defects introduced during the cutting and to ensure suitable thickness.

DEFECT GENERATION IN METAL CRYSTALS

Table 1

Metallic crystals grown by the Czochralski method

	GD	θ rpm	V mm/H	Crucible	Susceptor	After heater	Atmos. Torr	ϕ_n mm	L_n mm	ϕ_n mm	L_c mm
Al HE	110 111	ni	24-300	ni	ni	ni	He	ni	ni	<1	10
Cu FU	100 110	25	60-600	graph	ni	ni	5.10^{-6} Ar-H	0.5-1.5	ni	5-10	100
Cu SB	100 110 321	3-4	33 66	graph	graph	with and without	Ar	0.3-0.5	ni	1-3	10-50
Cu KEB	uvw 111 100	0.4-12.6	7.8-60	graph	ni	with	vacuum	0.5-4	ni	15-30	60
Cu B-G	123	0	8-30	graph vitre carb.	graph molyb.	without	2.10^{-6} Ar 95% H 5%	<0.3	3-5	<1.2	20
Ag T	110	3-4	8.5-30	graph	graph	with	Ar	<0.3	sev.	1	10
Ni KBB	Poly 111 110 100	0.1-4	24-60	Alum.	-	with	Ar	<1	ni	20-30	120
Nb N	Poly	2-3	9-300	pedes.	without	without	10^{-5}	ni	ni	0.3-1.2	30-50

HE   Howe and Elbaum (1961)
FU   Fehmer and Uelhoff (1972)
SB   Sworn and Brown (1972)
KEB   Kuriyama, Early and Burdette (1974)
B-G   Buckley-Golder (1977)
T   Tanner (1973)
KBB   Kuriyama, Boettinger and Burdette (1978)
N   Naramoto (1978)

ni   no indicated
uvw   direction at about 20° from both <111> and <110>
GD   growth direction
θ   rotation speed of the crystal
V   pulling rate
ϕ_n neck diameter
L_n neck length
ϕ_c crystal diameter
L_c crystal length

Regarding large diameters, the authors (FU, KEB) agree upon admitting that growth direction [100] is optimum the other directions studied <110>, <111> and 20° from both <111> and <110> give a higher dislocation density and sub-boundaries. As for smaller crystals (BB) the orientations studied (<100>, <110> and <123>) have failed to lead to different results. The perfection of the seed does not seem to be a decisive factor so long as its dislocation density is less than $10^5/cm^2$ (FU); beyond this it is hard to obtain copper crystals with a low dislocation density and without sub-boundaries. For large diameters (KEB) more perfect seeds are required. An increase in the rotation speed causes an increase in the dislocation density (KEB); a low rotation speed or even the suppression of the rotation (B-G) seems to be an important condition so long as good crystal quality is concerned. An excess of pulling rate causes sub-boundaries to come out (FU).

The material the crucible is made of and possibly that of the susceptor can give rise to small particles, as occurs with graphite, and those can get into the melt or settle on the surface of the new crystal and cause dislocations likely to impair the crystal perfection (FU, BB, B-G). This defect can be remedied by using a vitreous carbon crucible and a molybdenum susceptor (B-G).

The presence of afterheaters allows a reduction of the radial temperature gradient and brings down the number of dislocations in the crystal (SB). In other cases, the presence of afterheaters is a handicap when strict control of the diameter of the crystal must be ensured.

The atmosphere also plays an important part in the production of good quality crystals. Bad vacuum or partial oxygen pressure produces copper oxide particles which, when settling on the surface of the newly formed crystal, cause dislocations (FU, SB, B-G). A hydrogen atmosphere leads to the production of sub-boundaries (FU) while impure argon causes small dislocation loops to be formed. A mixture of argon (95%) and hydrogen (5%) will prevent copper oxide particles from forming (B-G).

All the authors agree upon the need for a neck. Its diameter must be the smaller as the diameter of the crystal is larger (SB). Its length can be a decisive factor in connection with the perfection of the crystal. The longer it is, on the one hand, the easier is for the seed dislocations to reach the surface before the crystal starts growing, and on the other hand, the smaller is the temperature gradient along the crystal (B-G).

In the dislocation-free crystals prepared by Fehmer and Uelhoff, X-ray topographs still show images such as small dashes, about 10 µm long, and black dots. The authors regard these defects as clusters of vacancies or loops caused by the vacancy supersaturation during cooling (Fig. 8a). To support this explanation, they point out that in low dislocation density crystals, the density of those clusters decreases strongly in the neighbourhood of the dislocations which act as sinks (Fig. 8b)

Vacancy supersaturation is not put forward in the case of small crystals. Sworn and Brown ascribe the origin of the dislocations in the crystal to collisions with oxide or graphite particles on the surface and to the existence of stresses caused by the temperature gradients. Buckley-Golder pays special attention to the length of the neck as a factor decreasing the temperature gradient in the crystal; he also points out the necessity of suppressing all vibration sources (pumps, water circulation, RF coils, ...) and of ending the crystal in a tapering tail.

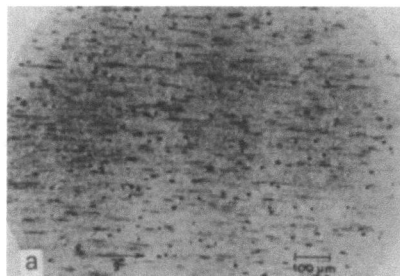 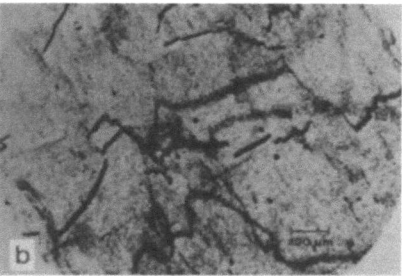

Figure 8. X-ray topograph of copper crystals a) dislocation free with point defect clusters b) with dislocations [15]

Kuriyama, Early and Burdette have proved the presence in low dislocation density crystals of configurations which they account for in terms of Lomer-Cottrell locks. These locks result from interaction between the first dislocations which slip through the crystal and they constitute obstacles to the following dislocations. In the case when the growth direction is <100> eight slip systems are likely to be activated under the stress due to the weight of the crystal and there are several opportunities for Lomer - Cottrell locks formation. The authors thus account for the fact that this growth direction favours crystal production of good perfection.

We cannot avoid vacancy supersaturation in large crystals, especially if the pulling rate is high, nor the stress due to the weight of the sample (except perhaps in space), but it can be said that the dislocation density of a copper crystal prepared by Czochralski method is the lower as:

- the seed is perfect; the growth direction lies along <100> for large crystals;

- mechanical perturbations are more reduced and so are the rotation speed of the seed and vibrations from all sources;

- the pulling rate is not too high;

- the crucible, the heating device and the atmosphere will not produce particles likely to collide with the crystal;

- the neck before crystal growth has a shorter diameter and will be long enough not to allow extensive propagation of the seed dislocations and to decrease the temperature gradient;

- any factor causing temperature gradient is reduced during the growth as well as during the last stage of growth.

Silver crystals

The experimental device used by Tanner [20] to prepare small silver crystals is similar to that used by Sworn and Brown. He has found high sensitivity of crystalline perfection to the pulling rate and an inverse relationship between pulling rate and the maximum diameter of dislocation-free crystal for given neck diameter.

Nickel crystals

Using the same technique as for copper crystals, Kuriyama Boettinger and Burdette [21] have prepared nickel crystals. They have found a marked orientation effect and growth direction <111> is the only one to give good quality crystals, whether starting from a poly- or single crystal seed. They have also found that a low pulling rate improves the crystal perfection. Lastly they have confirmed that, as with copper, the presence of Lomer-Cottrell locks attests to good crystal perfection.

Niobium crystals

The crucible-less pulling method, pedestal method, has been employed by Naramoto [22] for growing niobium single crystals excellent in quality. He has found an orientation effect and only crystals with a growth direction close to <110> can be obtained free of dislocations though the seed be single or polycrystal. Naramoto also notes that in that case the dislocations which travel from the seed to the crystal have a Burgers vector of <110> type and that they disappear quickly reaching the surface of the crystal. Concerning crystals containing dislocations, he points out a difference depending whether the diameter of the crystal is less or more than 0.6 mm. In the first case, the dislocation density is lower and the Burgers vectors are either $a/2$ <111> or a <100> type; in the second case, all Burgers vectors are $a/2$ <111> type and rows of loops and helices are observed (Fig.9). Naramoto ascribes this configuration of dislocations to the climb resulting from a vacancy supersaturation.

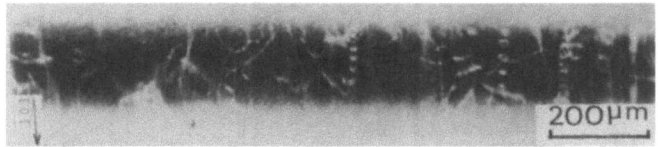

Figure 9. Section topograph $\overline{10}\overline{1}$ of a niobium crystal, helices and loop rows [22]

Evaluation of the temperature gradient along the crystal

The first theoretical evaluations of the temperature gradient in the growing crystal have taken into consideration the radiation heat transfer into an environment at constant temperature [23], or the radiation and gaseous convection heat transfer into an environment at non-constant temperature [24] or more the melt flow during the growth [25]. The study by Buckley-Golder and Humphreys [26] however, is the only one to take into account the geometry of the seed, neck and crystal as well as the heat transfer. The greater are the radiation losses, the stronger is the temperature gradient, which reaches its maximum value at the liquid-solid interface. Buckley-Golder and Humphreys take up the worst conditions of radiation losses in order to study the influence of geometrical parameters: length (Λ) and radius (R) of the seed (index s), of the neck (n) and of the crystal (c). The dimensions of the seed do not affect the temperature distribution along the crystal (Fig. 10a and b). On the other hand, the size of the neck has a marked effect on it: if length Λ_n decreases from 3 to 1 mm, the temperature gradient at the liquid-solid interface increases from 1.307 to 2.0°C/mm. If radius R decreases from 0.15 to 0.075 mm, the gradient decreases from 1.307 to 0.915°C/mm (Fig. 10c and d). With a neck of adequate thinness and length, it is possible to reduce, or even suppress the afterheater, a result experimentally corroborated by Buckley-Golder [19]. The radius of the crystal has a marked effect on the value of the temperature gradient. If radius R_n decreases from 1 to 0.5mm, the gradient increases from 1.307 to 2.869° C/mm (Fig 10f). The gradient is less affected by length Λ_c (Fig. 10e).

Since the temperature gradient produces a stress field and since, moreover, above a certain value of the latter, dislocations slip and multiply, when searching for a crystal of good perfection, one must work under minimal gradient conditions.

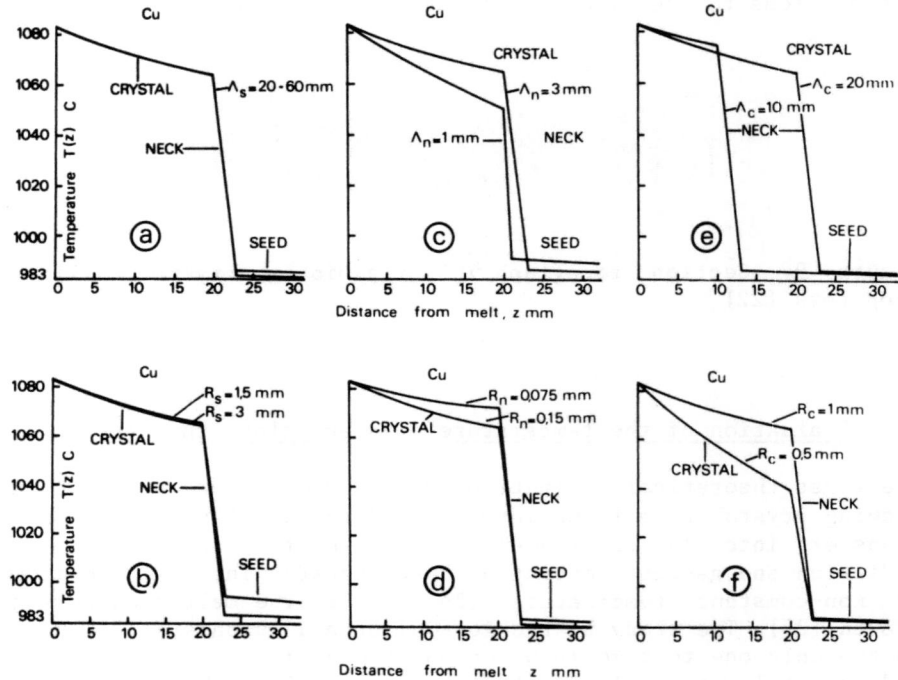

Figure 10. Theoretical study of temperature distributions along the sample of copper versus geometrical parameters, length Λ and radius R of the seed (s), the neck (n) and the crystal (c) [26]

5.3 Growth from the solid state

Metal crystals can be prepared from the solid state by two methods; strain-anneal and phase transformation. Both methods can give plate-shaped crystals to be observed directly by X-ray topography. The first method was largely applied in the case of aluminium and it has even allowed <u>in situ</u> examination in the generation phase. The second one has been applied successfully in the case of titanium.

5.3.1 Strain-anneal method

Our aim is to compare the result obtained by research teams who observed by X-ray topography metal crystals prepared by strain-anneal. We shall then try to state relationships between preparation conditions and crystal perfection. Among preparation conditions we can retain:

- metal purity

DEFECT GENERATION IN METAL CRYSTALS

- size of the sample

- tensile strain or other mode strain

- heating rate, temperature and time of annealing, cooling rate down to room temperature

- atmosphere in which the annealing is carried out

We will report results concerning crystals taken at room temperature just after preparation, but for one single polishing and we will be concerned mainly wtith the case of aluminium. The examination of annealed crystals will be considered in another paragraph.

Aluminium crystals

A large number of papers dealing with the preparation and examination of aluminium crystals has been published [27-37]. The values of the parameters defining growth conditions are stated in Table 2.

Under identical conditions of preparation an improvement in purity entails a decrease in dislocation density (Fig. 11). Moreover, the thinner is the sample, the lower is the dislocation density. Within certain limits the value of tensile strain before annealing does not seem to affect the final perfection of the crystal. Conversely, a strong unidirectional rolling (90% reduction rate) improves the crystalline quality; in comparison with tensile strain, the dislocation density is reduced by a factor 10 (Fig. 12).

The heating rate, the time and the temperature of annealing do not practically affect the final perfection of the crystal. Conversely, the cooling rate is a determinative factor. Nes and Nost have had the cooling rate vary from 2 to 55°C/hr. For metals of given purity they have found an increase in dislocation density ranging from 5×10^2 to 10^4 cm^{-2} though for equal purity metal and a less pure one they have not pointed out any variation in dislocation density in connection with cooling rate. Deguchi et al. report an increase in dislocation density as the cooling rate increases: 1.2×10^3 cm^{-2} for 2°C/hr, 5×10^3 cm^{-2} for 27°C/hr and 10^4 cm^{-2} for 500 C/hr. In their attempts to obtain crystals with a low dislocation density, most authors used relatively low cooling rates.

The atmosphere in which the annealing is carried out affects the dislocation density. With annealing in air taken as a reference, Fremiot et al. find that annealing in 5×10^{-6} Torr vacuum multiplies the dislocation density by about 2. Conversely, in 10^{-6} Torr, Deguchi et al. find the density to be divided by about 2. In 10^{-8} Torr vacuum Gastaldi et al. point out a very marked effect: the

Table 2

Aluminium crystals grown by strain anneal method

	Purity	Thickness mm	Strain	Anneal Heating rate °C/H	Anneal Temp. Time °C H	Anneal Cooling rate °C/H	Atm.	Dislo. density cm^{-2}
LM	5N+ZM	0.02-0.2	t.2%		495 14		Air	10^4-10^5
BE	4N8	0.25	t		450		Air	2-4 x 10^4
ARL	5N+ZM	<1.25	t.1-8%		500 15 550 16		Air	10^4
N	ρ_e 10000	1	t.3%	14	440	4	Air	2-5 x 10^2
NN	ρ_e 10000 less pure	1	t.3%	30	510	2-55	Air	10^2- 10^4 5 x 10^4
FC	4N8+ZM	0.8	t.3.5%		450	5	Air	1.5 x 10^3
FBC	4N8 +5ZM +20ZM	0.75	t.3.5%	25	420 72	5	Air Air Air 5.10^{-6}	>10^4 10^4 1.2 x 10^3 2.3 x 10^3
GCJ GMJ	3N8	0.8	t.1.6% Rolled	20	515 72	3.5	Air	10^4- 10^5 10^3- 10^4
			t.1.6% Rolled	20	515 72	3.5	10^{-8}	10^2- 10^3 10-10^2
GJ	3N8	0.8	Rolled	20	515 72 600 72	3.5	10^{-8}	10-10^2
DKKK	ρ_e 16000	0.5	t.2%		550	500 27 27 2 27-2-27	10^{-6} Air 10^{-6} " "	10^4 10^4 5 x 10^3 1.2 x 10^3 5 x 10^2- 3 x 10^3

LM Lang and Meyrick (1959)
BE Basu and Elbaum (1964)
ARL Authier, Rogers and Lang (1978)
N   Nøst (1965)
NN Nes and Nøst (1966)
FC Frémiot and Champier (1967)
FBC Frémiot, Baudelet and Champier (1968)
GCJ Gastaldi, Grange and Jourdan (1972)
GMJ Gastaldi, Marzo and Jourdan (1973)
GJ Gastaldi and Jourdan (1976)
DKKK Deguchi, Kamigaki, Kashiwaya and Kino (1978)
ZM zone melting pass
ρ_e electrical resistance ratio
t   tensile strain

DEFECT GENERATION IN METAL CRYSTALS

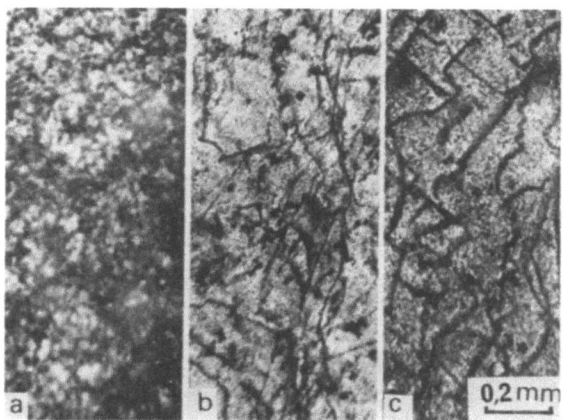

Figure 11. X-ray topographs 111 of crystals of aluminium a) 99,998% b) purified by 5 zone melting passes c) purified by 20 zone melting passes [33]

dislocation density is divided by a factor 100 whatever the mode of strain (Fig. 12). Annealing conditions, more especially atmosphere and cooling rate, strongly affect the configuration of the dislocations being examined at room temperature.

Gastaldi and Jourdan [38,39] have used synchrotron radiation in order to investigate *in situ* by X-ray topography some configurations which develop just at the recrystallization process. After a 3% tensile strain and during annealing at 300-350°C, they have followed the migration of a boundary of a growing grain and noticed the presence of a dislocation normally ending on the boundary. Moreover, that dislocation moves within the recrystallized grain and partially vanishes at the surface and the remaining part of the line aligns as segments which probably coincide with simple crystallographic directions. The authors have also followed the surrounding of an obstacle by a mobile boundary and observed the emission of straight dislocations and segmented helices when both parts of the boundary join again beyond the obstacle. Furthermore Gastaldi and Jourdan have followed the process of recrystallization of two grains on either side of a twin boundary. Such studies are expected to provide data concerning the configuration of dislocations observed in growing crystals.

Dislocation shapes observed at room temperature by most authors (cylindrical or prismatic helices, rows of loops, single more or less complex loops, curved or straight lines) suggest that vacancy supersaturation has taken place during cooling and dislocations have climbed. That supersaturation originates from the fact that the low number of vacancy sinks (dislocations, boundaries, surfaces) is unable to absorb, as the temperature decreases, vacancies in excess as regards the thermodynamic equilibrium. If the cooling rate increases, we may expect a higher vacancy supersaturation and therefore a higher dislocation density. The kind of atmosphere in which annealing is carried out affects the vacancy supersaturation as it generates a surface state which conditions the efficiency of the surface as a vacancy sink. High vacuum prevents the aluminium sample from oxidizing and enables the surface to act as an efficient sink. Thus the supersaturation is lower and the crystalline perfection of the sample consequently improves.

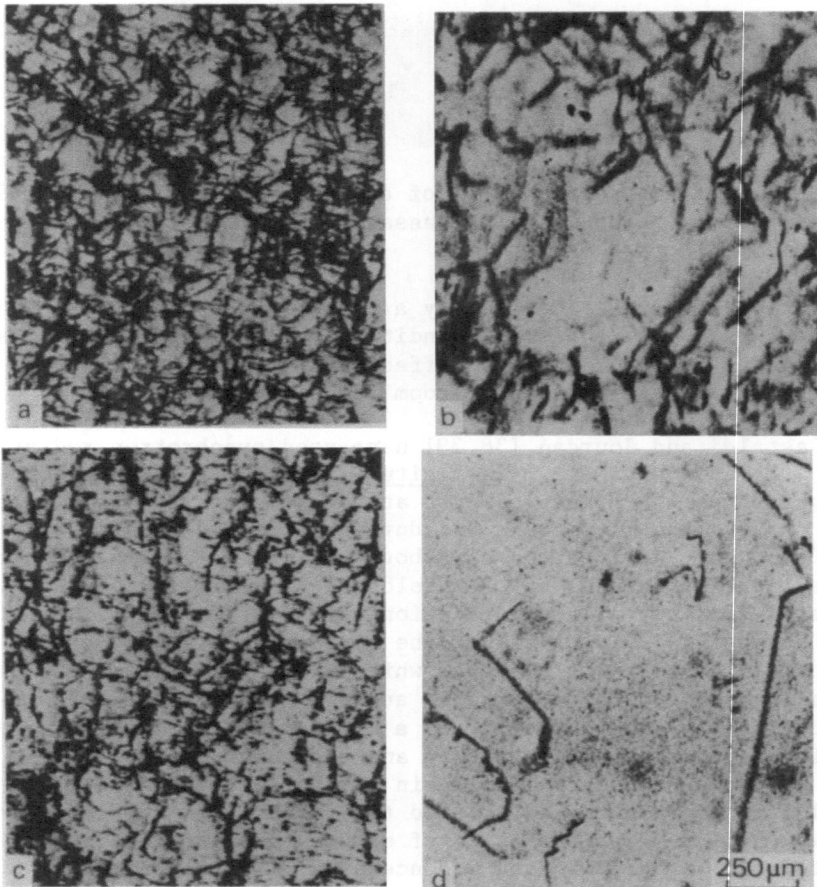

Figure 12. X-ray topographs 111 of aluminium crystals, grown by strain-anneal (a) in air (b) in vacuum and by secondary recrystallization in air (c) (d) and in vacuum [35]

A higher perfection connected with large strain as a result of rolling is explained by the existence of a strong texture and relatively easy transition from textured grains to recrystallized ones. The migration of the boundary is easily achieved without it being necessary to create a large number of dislocations. Consequently, immediately after recrystallization, a low dislocation density is regained.

Iron and iron-silicon crystals

Kadeckova and Saleeb [40] and Kadeckova and Bradler [41] have prepared by strain-anneal method some iron and dilute iron- silicon crystals and they have studied them by X-ray topography (Fig.13). Observations made during recrystallization have shown a higher dislocation density away from the recrystallization front than near it. Hence the authors conclude that the dislocations of the recrystallized grain are essentially caused by stresses resulting from the temperature gradient. Dislocation density is lower in alloys than in pure iron and this might be due to a higher critical shear stress in alloys than in pure iron.

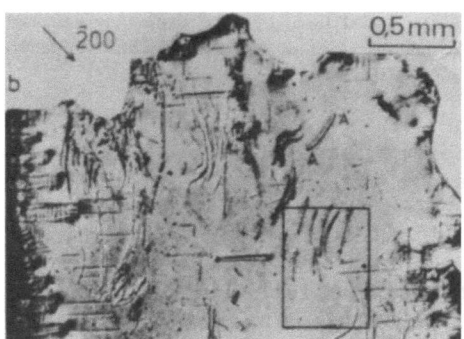 

Figure 13. X-ray topographs of iron-0,5% silicon and iron-0,9% silicon crystals [41]

MacCormack and Tanner [42] have followed in situ by X-ray topography, with synchrotron radiation, incipient recrystallization of a Fe-3.5% Si sample. This study has just started and will provide data on the state of the dislocation configuration at recrystallization.

Tin crystals

Brummer and Alex [43] have prepared tin crystals of good crystalline quality by the strain-anneal method. The samples are cut from strips rolled and annealed at 100° C. They are compressed under 3 MPa stress, then annealed at 150°C for 4 weeks and cooled down slowly (5 C/hr). In the parts wihich had been thinned down by electrolytic polishing the authors have noted areas perfect enough to enable precise determination of Burgers vectors of dislocations.

Copper crystals

Minari, Pichaud and Capella [44] have prepared copper crystals by a method similar to the previous one. Large size grains of a very good crystalline perfection, 10 - 100 dislocation lines per cm^2, could be obtained (Fig.14). When annealing is carried out directly without strain compression taking place, the perfection is lower and some sub-boundaries will show up.

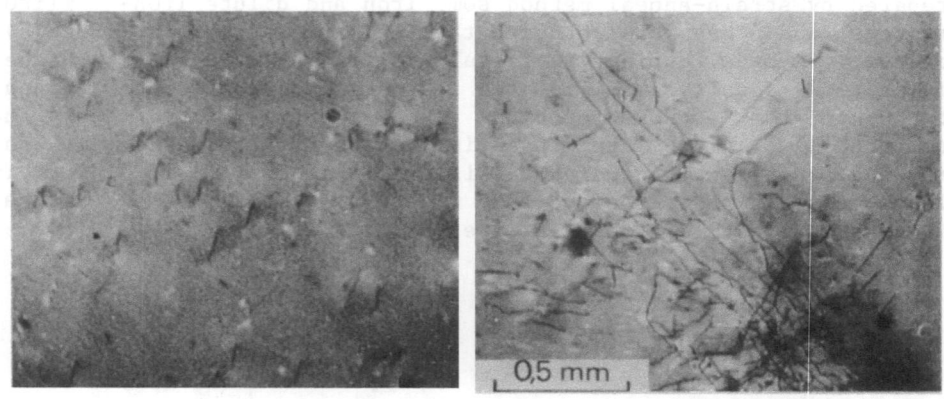

Figure 14. X-ray topographs of copper crystals grown by strain anneal and secondary recrystallization [44]

Molybdenum crystals

Becker and Pegel [45] have strongly rolled a molybdenum crystal along its <110> axis and found out after annealing at 2000°C for one hour in 10^{-5} Torr vacuum that the sample was perfect enough to be studied by X-ray topography. They have observed long screw dislocations, helices and rows of loops and they ascribe those configurations to the vacancy supersaturation taking place during cooling.

Crystals of solid solution of tin in silver

Pichaud, Minari and Bernardini [46] have prepared crystals from a solid solution of tin in silver. They merely rolled strips and annealed them for 72 hours in 10^{-6} Torr vacuum at 20° C below the solidus temperature. Topographs show a maximum perfection (10^3 lines per cm^2) for a tin concentration between 0.5 and 3 at %. The authors assign that effect to the high values of the interaction energies dislocation-solute atom and vacancy-solute atom in the silver matrix. The vacancy supersaturation in the core of the dislocations would make their elimination through climbing easier.

DEFECT GENERATION IN METAL CRYSTALS

5.3.2 Phase transformation method

The method has been successfully applied to the case of titanium by Jourdan, et al. [47, 48]. From titanium of various degrees in purity they rolled strips 200 microns thick, annealed them in 5×10^{-10} Torr high vacuum up to a temperature slightly higher than the transformation $\alpha \rightarrow \beta$ temperature, then down to a slightly lower temperature (the cycle can be repeated several times) for 8 days and finally slowly cooled down, 3º C/hr. The use of several cycles around the transformation temperature permits larger grains to be obtained. Dislocation density is between 10^2 and 10^4 cm^{-2}. Purity hardly affects the crystalline perfection of the sample. The dislocations observed by X-ray topography have $\underline{a}$ and $\underline{a} + \underline{c}$ as Burgers vectors. They sometimes show as prismatic loops emitted by a stress concentration around a particle (Fig.15).

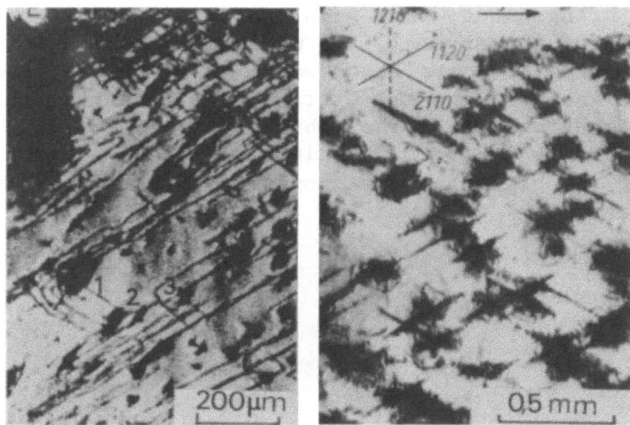

Figure 15. X-ray topographs of titanium crystals [48]

5.4 Growth from gas state and whiskers

We will study both growth methods simultaneously in so far as they lead to crystals one dimension at least of which is particularly small.

5.4.1 Growth from gas state

This growth method has been used by Michell and Ogilvie [49] in the cases of cadmium and zinc and by Michell and Smith [50] in the cases of cadmium and magnesium. Crystals are platelets about 5 microns thick prepared by the Coleman and Sears method [51] according to a mechanism of sublimation-condensation in a gas flow. X-ray topographs taken a short time after removal from the furnace show that the crystals are dislocation-free or contain only very few dislocations (Fig.16). The nature of the gas flow (neon, argon, xenon) does not affect the morphology and the perfection of crystals.

5.4.2 Whiskers

Characterization of whiskers by X-ray topography has been effected in the case of copper [52], iron [53.54], copper-silver alloy [55] and copper-zinc alloy [56]. Copper whiskers are prepared by the Brenner method [57] at 700°C in a hydrogen flow, starting from copper iodide. There are three possible orientations; axis <001> and square cross section with faces {110}; axis <111> and hexagonal cross section with faces {110}; axis <110> and rectangular cross section with faces {100} and {110}. The lateral dimensions vary from 10 to 300 microns, the length from 1 to 30 mm. Micrographical examination brings out the existence of surface inclusions, most probably copper iodide. X-ray topography examination brings out some dislocations and also some tangles limiting parts of the crystal, the misorientation of which can reach one minute of arc. No correlation seems to be made between surface defects and internal ones. Nittono and Nagakura [52] have not detected dislocations along the whisker axis in any case. Slow cooling after annealing at 700 - 1000°C in high vacuum improves the perfection of the crystal, the content of surface and internal defects decreases.

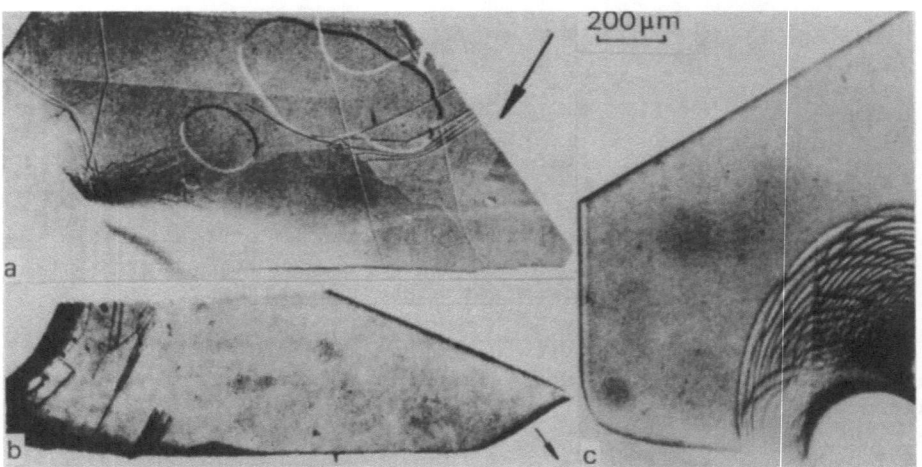

Figure 16. X-ray topographs of platelets of a) cadmium b) zinc c) magnesium, dislocations were brought in accidently in the course of the fixation of the sample. [49, 50]

Iron whiskers are also prepared by the Brenner method at 720 -750°C in a hydrogen flow by reduction of ferrous chloride, then annealed for several hours at 850°C in vacuum. They have the same orientations and morphologies as copper whiskers. Using X-ray topography Bojarski and Surowiec [54] have found that whiskers had no axial dislocation, that some were perfect and some were not and contained parts with high dislocation density as well as perfect parts, that those dislocations were screw type, inclined to the axis and running in {110} planes and sometimes in {112} planes (Fig.17). According to the authors, these dislocations are emitted during cooling or annealing from sources located near the edges of the crystals as a result of stresses due to the accumulation of impurities.

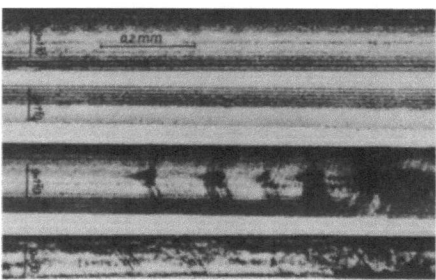

Figure 17. X-ray topographs 110 of iron whiskers with various degrees of perfection [54].

Thus it turns out that platelets and whiskers obtained from pure metal are perfect at the time of their growth and that the dislocations observed by X-ray topography are a result of the further treatments the crystal was submitted to.

Copper-silver and copper-zinc alloy whiskers show on X-ray topographs striations parallel to the axis of the crystal, the latter consisting of longitudinal subgrains slightly misorientated in relation to one another. Moreover Nittono et al. [55,56] found out dislocations located in {111} planes inclined on the axis. The perfection of alloy whiskers is far less than that of pure metal whiskers.

5.5 Stability of dislocation configurations at room temperature

Once the crystal is prepared, one may question whether the dislocation configuration observed will remain stable at room temperature for a given position of the crystal, in a given environment. In the cases of aluminium, copper, nickel, iron, iron-silicon, niobium, molybdenum and magnesium crystals, keeping slices or crystals exposed to air at room temperature in any position, and even for a very long time, will not modify the dislocation configuration in any way.

This is not true in the cases of cadmium and zinc [49,50,58-60], of gallium [13] or of indium [12]. Depending on cases, the dislocation density may remain constant, with a rearrangement of the configuration taking place, though it may also decrease or increase. The dislocation density may remain constant while some isolated dislocations change direction or even move keeping the same direction. This happens to be the case for indium.

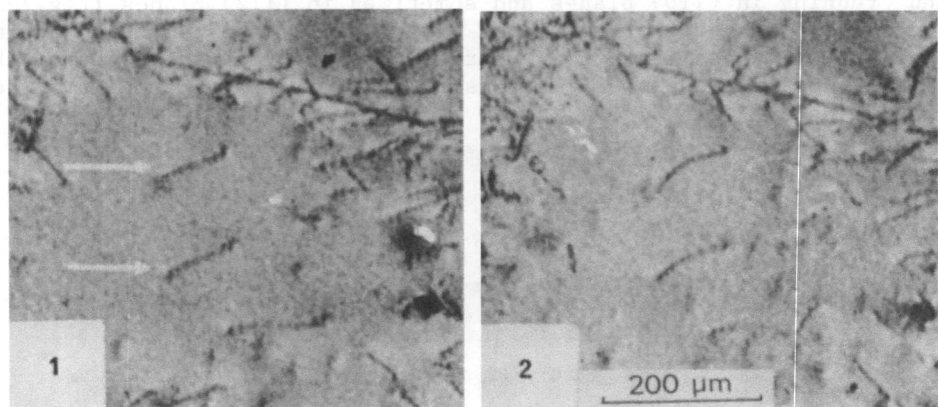

Figure 18. X-ray topographs $\bar{1}11$ of an indium crystal before and just after a rotation of the crystal for one hour [12].

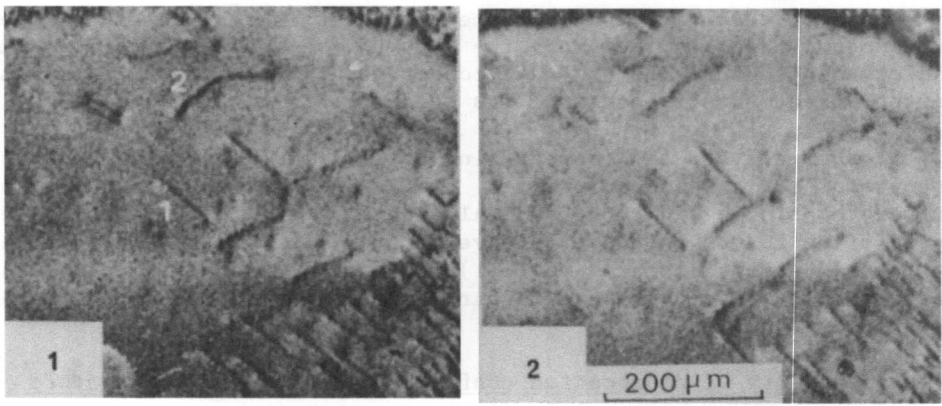

Figure 19. X-ray topograph $\bar{1}11$ of an indium crystal; displacement of dislocation 1 after 70 hours [12]

Both topographs in Fig.18 were taken respectively before and just after the crystal was turned over 90° for one hour around the normal to its largest faces placed horizontally. Both lines with reference arrows have changed direction and after a certain length of time both take again the initial direction as in the first topograph. This direction corresponds to the minimum energy of the dislocation segment when it crosses the crystal running on its slip plane. Still in the case of indium a dislocation segment may slip without losing its direction. Both topographs in Fig.19 have been taken at 70 hours interval without any manipulation of the crystal. Line marked 1 has shifted with respect to line 2. Modifying the crystal position and observing the sense of the shift show that the dislocation slips under the stress due to the weight of the sample; the specific mass of indium is relatively high and room temperature constitutes an important fraction of the melting point (68%), so the critical shear stress for the slip of a dislocation must be very low. Indium also provides an instance of a case of decrease in dislocation density. The topographs in Fig.20 were taken at three week intervals without any manipulation of the crystal. A steady decrease in dislocation density from 10^5 down to less than 10^2 cm^{-2} is observed and it is possible to infer that the initial configuration of dislocations has been recovered.

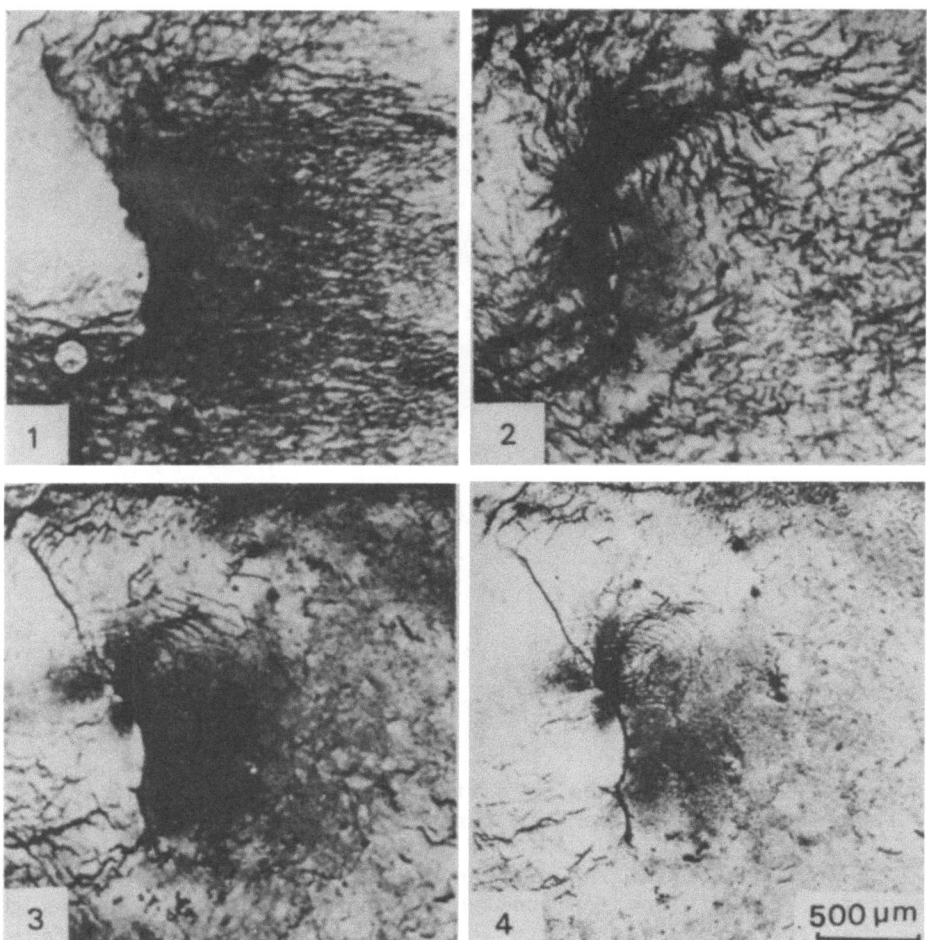

Figure 20. X-ray topographs 022 of an indium crystal, change in dislocation configuration after 3(2), 6(3) and 9(4) weeks [12]

In the case of cadmium, maintaining it in air at room temperature results in a fast increase of the dislocation density, growth of dipoles by climbing of superjogs up to the production of tangles, growth of $\underline{c}$ hexagonal loops becoming more and more irregular, growth of $\underline{c}$ or $\underline{c} + \underline{a}$ intricate loops and spirals (Fig.21). In the case of zinc, phenomena of the same type, with far lower rates of evolution, have been observed (Fig.22). All authors agree upon explaining that evolution as coming from the oxidation mechanism of these metals. As the oxide layer is growing, the metal atom diffuses through the layer leaving a vacancy behind, that in its turn diffuses into the lattice of the metal resulting in a vacancy supersaturation which causes the climb of the jogs and of the prismatic loops. This model is proved by the fact that evolution stops as soon as the crystal is put in vacuum or in inert gas atmosphere.

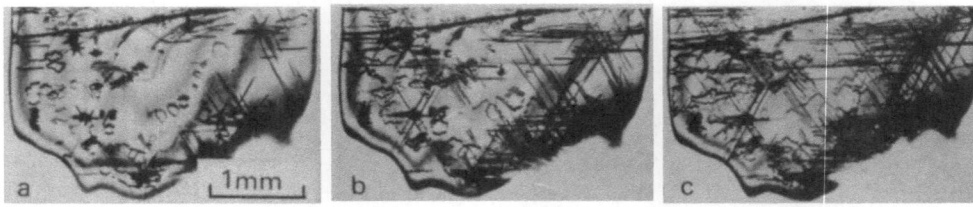

Figure 21. X-ray topographs of a cadmium crystal (a) 5, (b) 19 and (c) 34 days after removal from the furnace [62]

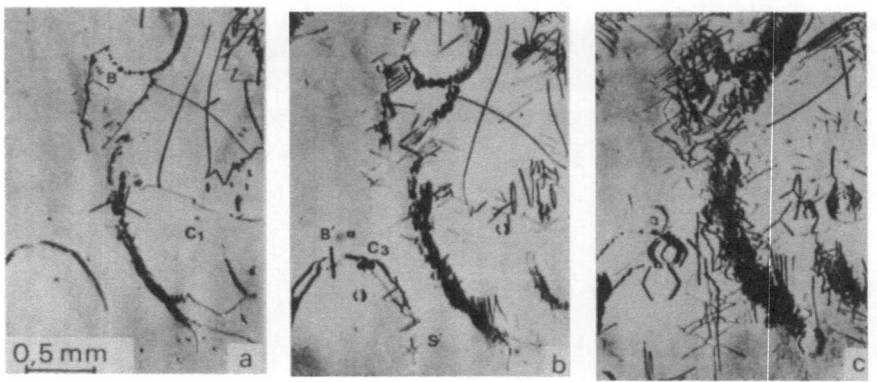

Figure 22. X-ray topographs of a zinc crystal (a) 14 (b) 74 and (c) 186 days after removal from the furnace [62]

In the case of gallium, in the purest crystals, without banding structure, the dislocation configuration does not change during the time they are maintained in air at room temperature. It is different with crystals of lower purity which contain banding structure where straight screw dislocations climb and become irregular helices, loops grow and lead up to complex configurations. McFarlane et al.

[13] relate the origin of the dislocation climb to the spatial fluctuations in the impurity content. It may be also considered that the impurities make gallium more likely to oxidize and that a mechanism identical to that of cadmium may be suggested. This would require checking that maintenance in vacuum or inert atmosphere stops any evolution of the dislocation configuration. In the case of tin crystals obtained by the Bridgman method Fiedler and Lang [61] observed in some samples the growth of edge dislocations parallel to the specimen surface.

5.6 Stability of dislocation configurations at high temperature

It is questionable whether the dislocation configurations observed at room temperature will stay constant when the crystal is heated, maintained at high temperature, then cooled down. Those observations may be carried out *in situ* provided an oven is erected on the X-ray topography camera and the evolution is not too fast. They may also be carried out at room temperature after an anneal, the various parameters of which are made to vary: heating rate, time and temperature of annealing cooling rate and finally atmosphere. We will report the results obtained in both cases and show how cooling may affect the dislocation configurations.

5.6.1 Annealing effect: observation at room temperature

All the authors have observed that one anneal generally results in improving the crystalline perfection of a sample provided the dislocation density is not too high ($\leqslant 10^4$ cm^{-2}) and that the dislocations do not form a Frank network. Some authors have suggested successive annealing or even cyclic annealing between two temperatures. As far as improvement of crystals quality is concerned there are few systematic studies and we will merely report the results in connection with aluminium [63, 64] G'Sell et al. [64] have submitted 0.45mm thick aluminium crystals with a dislocation density of 10^3 cm^{-2} to the following heat treatment. 10^{-3} Torr vacuum, heating up to the temperature T_s (300, 400, 500 and 600°C) at the rate of 30°C/hr, keeping at this temperature ten hours, then cooling back to room temperature at a rate ranging about 10, 100, 1000 and 10,000°C/hr. The four topographs in Fig. 23 show the configurations obtained for T_s = 300°C. At low cooling rate curved dislocations, at intermediary rate curved dislocations and rows of loops and eventually at high rate a continuous background can be seen. Observations with T_s = 400°C and 500°C are of the same type. The evolution of the various geometrical parameters versus T_s temperature and cooling rate are shown in Fig. 24. The authors suggested that curved dislocations are initial dislocations in the crystal at T_s. To account for rows of loops, they suppose the existence of microprecipitates which might be alumina or intermetallic compound particles. From a critical value of the vacancy supersaturation during cooling, dislocation segments at

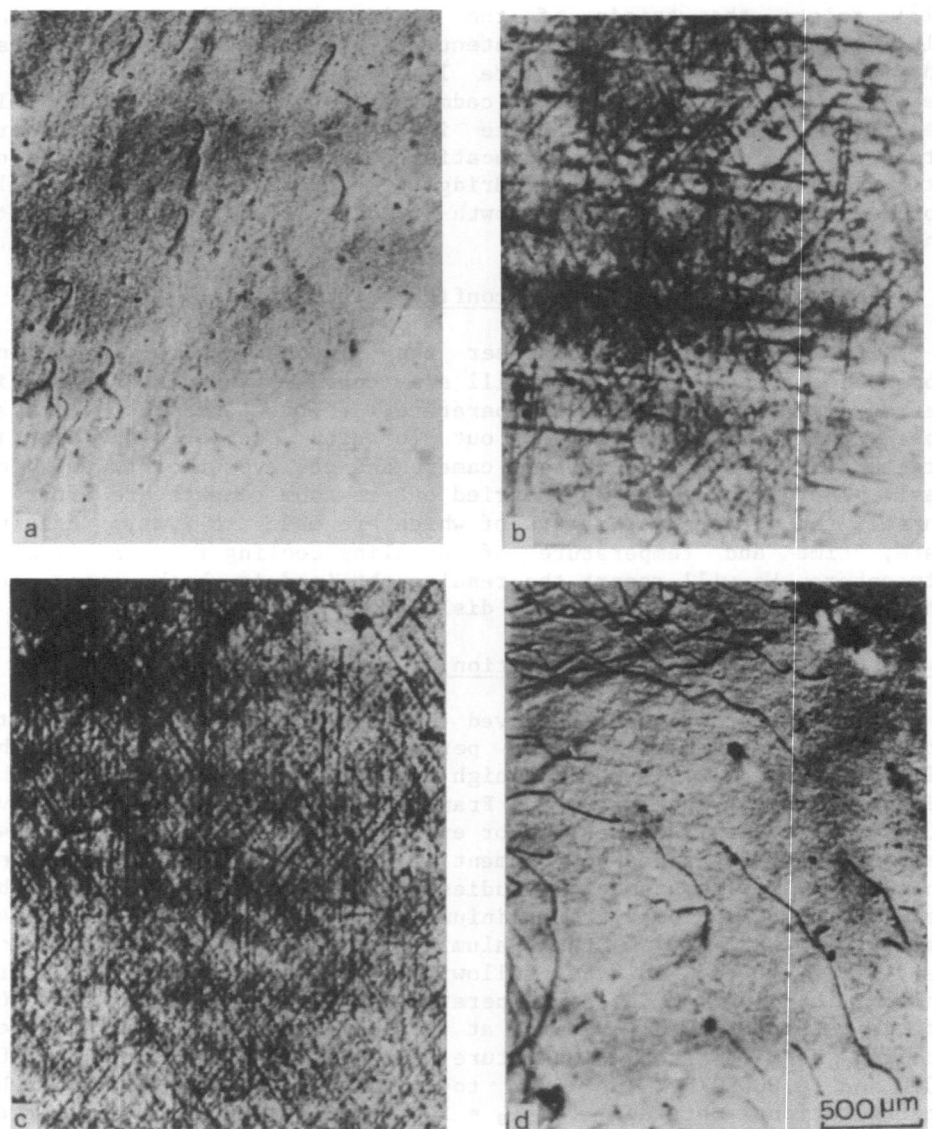

Figure 23. X-ray topographs 111 of an aluminium crystal cooled from 300°C with a cooling rate a) 10°C/hr, b) 150°C/hr, c) 3800°C/hr and d) 22,000°C/hr [63]

interface microprecipitate-matrix could climb and act as sources emitting dislocation loops in rows according to the mechanism described by Bardeen and Herring [65]. An estimate of the supersaturation at the time when loops begin to appear on the topographs gives a value 1.1. This implies dislocation segments of 0.3 microns in length [66]. The variations of density and diameter of the loops can be explained as arising from the competition between the nucleation mechanism and the growing one versus the supersaturation and the mobility of vacancies.

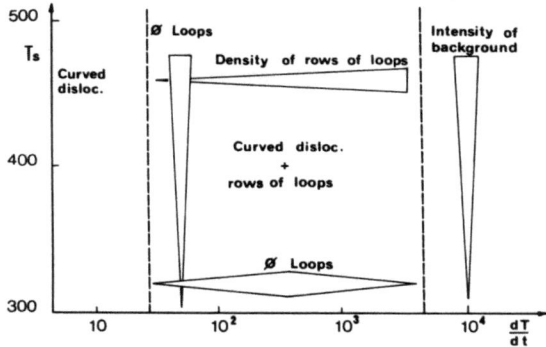

Figure 24. Effects of the annealing temperature and the cooling rate on the geometrical parameters of the dislocation configuration [63]

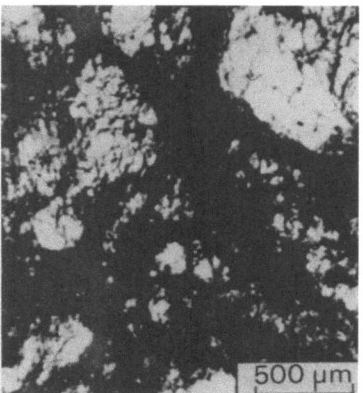

Figure 25. X-ray topographs 111 of an aluminium crystal cooled from 600°C with a cooling rate 10°C/hr [63]

Crystals cooled down from $T_s = 600°C$ contain dislocation walls bounding parts in which dislocation density increases with the cooling rate (Fig. 25). Above 520°C the surface of aluminium is covered with a layer of crystallized oxide and it cannot be considered as a perfect vacancy sink any longer. Rows of loops are developing and interacting, eventually resulting in a highly tangled configuration of dislocations. These above results show that annealing does not necessarily improve the perfection of a crystal and consequently the most important factor remains the cooling rate.

5.6.2 Observation at high temperature

When carried out at steady high temperature, experiment shows that the dislocation density decreases. Nost and Sorensen [67] using section topographs, have found out for aluminium the law of variation of dislocation density against time t: $(a+bt)^{-1}$, a and b are two constants depending upon the initial dislocation density and the temperature. The topographs in Fig. 26 show the evolution of the dislocation configuration in an aluminium crystal maintained at 300 °C [68]. Moreover Baudelet and Champier [69,70] have found that at 350°C dislocations tend to orientate along <110> directions.

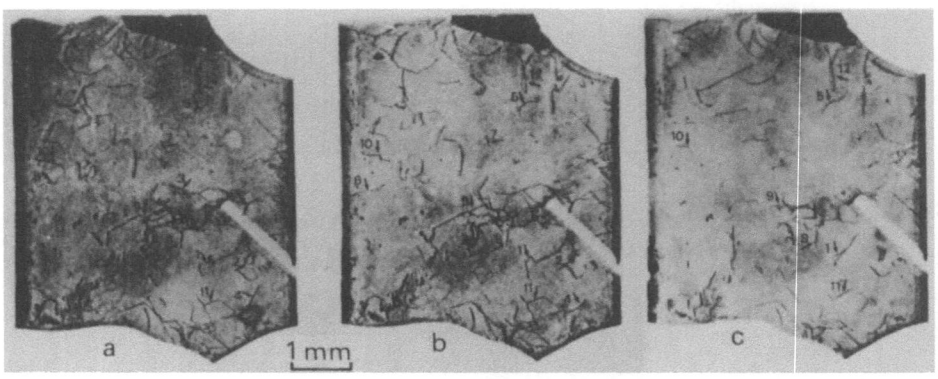

Figure 26. Evolution of the dislocation configuration in an aluminium crystal at 300°C versus time at a) t = 0, b) t = 4H and c) t = 24 H [68]

Further observations at high temperatures also are relevant to aluminium. The authors have observed *in situ* the evolution of a dislocation configuration when temperature varies between two limits [66,71-4]. Nost et al. have studied from section topographs variation of the dislocation density against the temperature (or time) for given heating or cooling rates and they have found relatively intricate variations. Baudelet et al. have followed on projection topographs the evolution of a dislocation configuration from the temperature θ_i (270-350°C) to the temperature θ_f (20-350°C) at a cooling rate 400° C/hr. Starting from a crystal with a dislocation density 100 ± 100cm^{-2}. they have defined three areas in the diagram temperature difference ($\theta_i - \theta_f$) -initial temperature (θ_i) (Fig. 27). In the first area (small temperature difference) dislocations slip and climb; in the second area (intermediate temperature difference) rows of loops appear during the maintenance at the temperature θ_f ; in the third area (large temperature difference and high initial temperature) rows of loops appear in the course of cooling. The row density and the loop radius depend on the temperatures, θ_i and θ_f, and on the time of maintenance at θ_f. It is worth noticing that in the same crystal submitted to various thermal cycles, rows of loops always appear in the same positions in the crystal (Fig. 28). This result supports the assumption on the existence of microprecipitates and of a source mechanism by climb of the interface dislocations for the rows of loops.

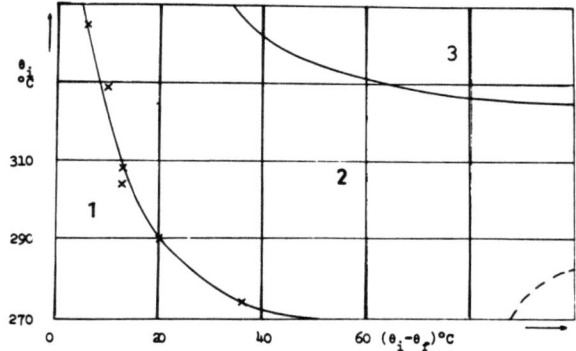

Figure 27. Temperature domains allowing one to classify the experimental results during the cooling of a crystal of aluminium between θ_i and θ_f [66]

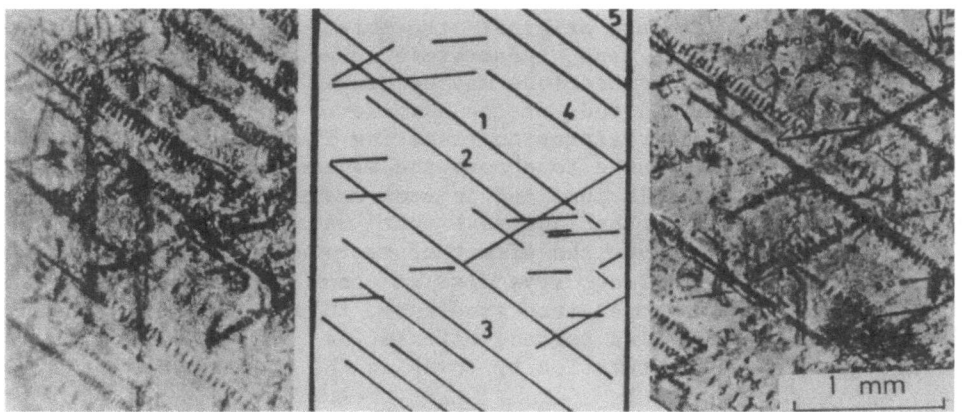

Figure 28. X-ray topographs $\bar{1}\bar{1}1$ of an aluminium crystal after cooling a) from 345 to 212°C and c) from 345 to 20°C [66]

A second instance of the evolution of the dislocation configurations in the course of a rise in temperature is shown in Fig. 29 in the case of cadmium [60]. On successive topographs, one observes progressive shrinkage of loops until complete vanishing, rotation of spirals towards a reduction in length until disappearance and at last nucleation, growth and vanishing of new loops. It is pointed out that those new loops do not have the same contrast as the initial ones. The initial loops had been generated by vacancy absorption due to the vacancy supersaturation occurring in the course of cooling; the latest loops are generated by vacancy emission due to the vacancy undersaturation occurring in the course of heating.

Vale and Smallman [9] have followed by X-ray topography together with *in situ* weight gain measurements the high temperature (550°C) oxidation of magnesium crystals. During the initial stage the oxide layer remains coherent and loops appear and grow with increasing time of exposure in air. During the last stage, when decohesion of the metal-oxide interface becomes marked, the surface acts as a vacancy sink and continued annealing causes the loops to shrink and eventually disappear.

5.7 Conclusion

By now it is clear that we have only scanty data at our disposal concerning the defects which develop with the generation of a metal crystal whereas all data available concerning the evolution of a dislocation configuration attest to the importance of the effect of the cooling rate on the crystal perfection.

In the course of the cooling phase, various sources of slip and climb of dislocations are activated. Slip is caused by stresses resulting from various origins: crucible, vibrations or other mechanical influences and temperature gradient, mainly in large crystals. Moreover, dislocations may be locked by reaction one with another or be eliminated at the surface of the crystal. Climb is caused by vacancy supersaturation resulting from the scant amount of vacancy sinks being unable to absorb the vacancies in excess of the thermodynamical equilibrium and in some cases metal reactions with the environment can further increase that vacancy supersaturation. It is possible to reduce the effect of certain factors directly. In order to obtain perfect crystals at room temperature, it is necessary to avoid temperature gradients, to promote elimination of dislocations by selecting correct orientation and, above all, to reduce vacancy supersaturation by lowering the cooling rate.

X-ray topography has been shown to be a choice method for studying those phenomena as far as it permits detection of defects in the crystal bulk and *in situ* observation of their evolution against temperature and reactions with the environment. New prospects opened by synchrotron radiation suggest that, in the near future, data concerning the defects which occur in the initial growth phase will be obtained with regard to the various growth methods.

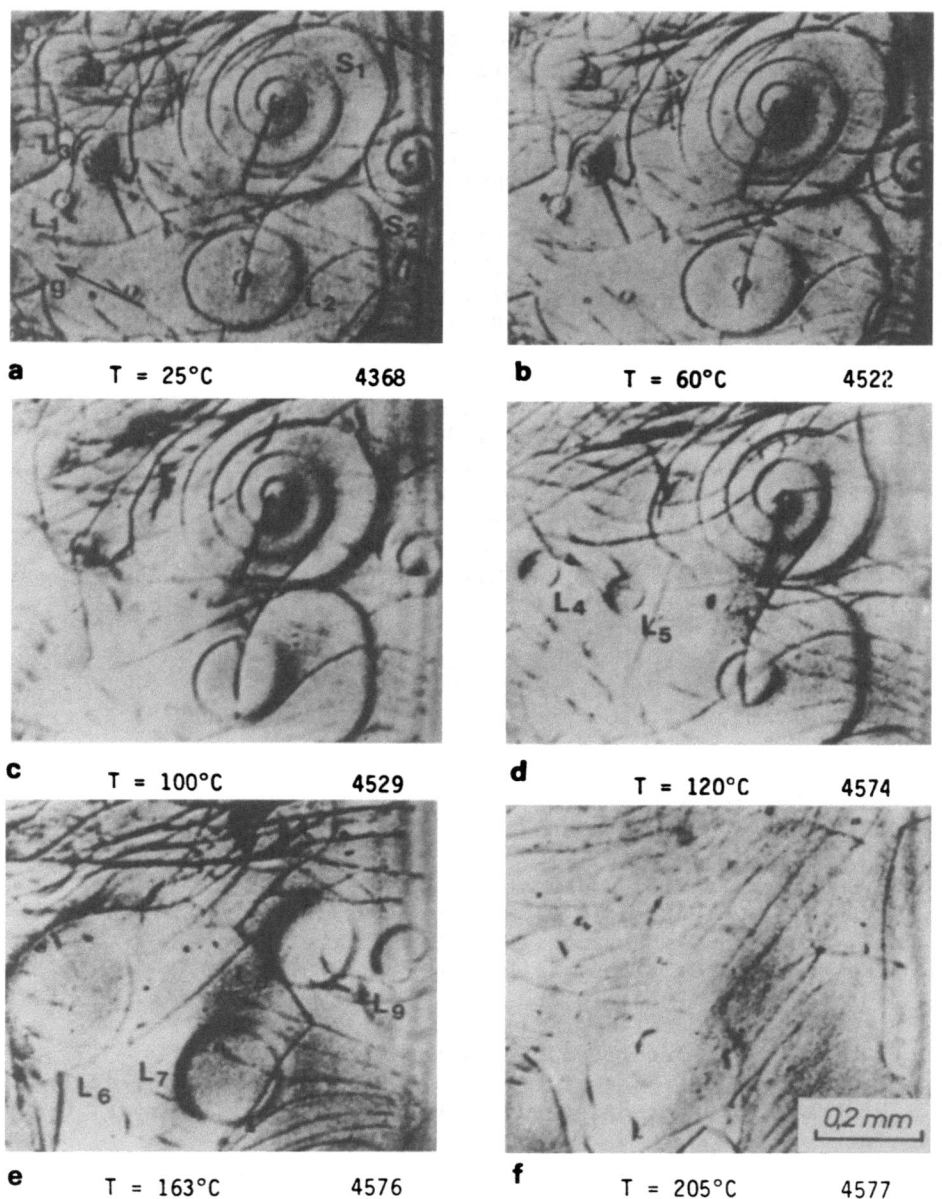

Figure 29. X-ray topographs 11$\bar{2}$0 showing the evolution of dislocations in a cadmium crystal during heating (heating rate 10° C/hr) at different temperatures [60]

REFERENCES

1. A.R. Lang (1958) J. Appl. Phys 29, 597
2. A.R. Lang (1959) Acta Cryst. 12, 249
3. G. Borrmann (1941) Phys. Z 42, 157
4. G. Borrmann (1950) Phys Z 127, 297
5. F.W. Young and J.R. Savage (1964) J. Appl. Phys. 35, 1917
6. A. Merlini and F.W. Young (1966) J. Phys. Colloque Suppl. C3 27, c3-319
7. A.R. Lang and M. Polcarova (1965) Proc. Roy. Soc. A285, 297
8. A.S.T. Badrick and K.E. Puttick (1971) Phil. Mag. 23, 585
9. R. Vale and R.E. Smallman (1977) Phil. Mag. 36, 209
10. R. Vale and R.E. Smallman (1978) Cryst. Lattice Defects 7, 177
11. C. G'Sell, G. Champier and Y.J. Iwasaki (1974) J. Cryst. Growth 24/25, 527
12. J.M. Blind, A. George and G. Champier (1979) Acta Met. 27, 471
13. S.H. McFarlane and C. Elbaum (1967) J. Appl. Phys. 38, 2024
14. S. Howe and C. Elbaum (1961) J. Appl. Phys. 33, 1227
15. H. Fehmer and W. Uelhoff (1972) J. Cryst. Growth 13/14, 257
16. C.H. Sworn and T.E. Brown (1972) J. Cryst. Growth 15, 195
17. B.K. Tanner (1972) J. Cryst. Growth 16, 86
18. M. Kuriyama, J.G. Early and H.E. Burdette (1974) J. Appl. Cryst. 7, 535; (1974) Proc of AIAA 12th Aerospace Sciences Meeting no. 74-204
19. I. Buckley-Golder (1977) J. Cryst. Growth 40, 189
20. B.K. Tanner, Z. Naturforsch (1973) A28, 676
21. M. Kuriyama, W.J. Boettinger and H.E. Burdette (1978) J. Cryst. Growth 43, 287
22. H. Naramoto (1978) J. Cryst. Growth 44, 475
23. V.H.S. Kuo and W.R. Wilcox (1972) J. Cryst. Growth 12, 191
24. T. Arizumi and N. Kobayashi (1972) J. Cryst. Growth 13/14, 615
25. N. Kobayashi (1978) J. Cryst. Growth 43, 357
26. I. Buckley-Golder and C.J. Humphries (1979) Phil. Mag. A39, 41
27. A.R. Lang and G. Meyrick (1959) Phil. Mag. 4, 878
28. B.K. Basu and C. Elbaum (1964) Phil. Mag. 9, 533
29. A. Authier, C.B. Rogers and A.R. Lang (1965) Phil Mag. 12, 547
30. B. Nost (1965) Phil. Mag. 11, 183
31. E. Nes and B. Nost (1966) Phil. Mag. 13, 855
32. M. Fremiot and G. Champier (1967) C.R. Acad. Sci. 265, 1331
33. M. Fremiot, B. Baudelet and G. Champier (1968) J. Cryst. Growth 3/4, 711
34. J. Gastaldi, G. Grange and C. Jourdan (1972) C.R. Acad. Sci. 274, 186
35. J. Gastaldi, P. Marzo and C. Jourdan (1973) J. Cryst. Growth 18, 77
36. J. Gastaldi and C. Jourdan (1976) J. Cryst. Growth 35, 17; (1977) J. Less. Com. Met. 56, 141
37. Y. Deguchi, N. Kamigaki, K. Kashiwaya and T. Kino (1978) Jap. J.

Appl. Phys. 17, 611
38. J. Gastaldi and C. Jourdan (1978) Phys. Stat. Sol. 49a, 529
39. J. Gastaldi and C. Jourdan (1979) Phys. Stat. Sol. 52a, 139
40. S. Kadeckova and K.Z. Saleeb (1975) J. Cryst. Growth 30, 335
41. S. Kadeckova and J. Bradler (1978) J. Cryst. Growth 43, 301
42. I.B. MacCormack and B.K. Tanner (1978) J. Appl. Cryst. 11, 40
43. O. Brummer and V. Alex (1970) Phys. Stat. Sol. (a) 3, 193
44. F. Minari, B. Pichaud and L. Capella (1975) Phil. Mag. 31, 275
45. C. Becker and B. Pegel (1969) Phys. Stat. Sol. 32, 443
46. B. Pichaud, F. Minari and J. Bernardini (1978) J. Cryst. Growth 43, 273
47. C. Jourdan, D. Rome-Talbot and J. Gastaldi (1972) Phil. Mag. 26, 1053
48. C. Jourdan and J. Gastaldi (1977) Phys. Stat. Sol. (a) 43, 425; (1979) Scripta Met. 13, 55
49. D. Michell and G.J. Ogilvie (1966) Phys. Stat. Sol. 15, 83
50. D. Michell and A.P. Smith (1968) Phys. Stat. Sol. 27, 291
51. R.V. Coleman and G.W. Sears (1957) Acta Met. 5, 131
52. O. Nittono and S. Nagakura (1969) Jap. J. Appl. Phys. 8, 1180
53. W. Hagen and H.H. Mende (1977) Z. Metallk. 68, 550; H. Galinski and H.H. Mende (1975) Phys. Stat. Sol. (a) 27, 35
54. Z. Bojarski and M. Surowiec (1979) J. Cryst. Growth 46, 43
55. O. Nittono, N. Onodera and S. Nagakura (1970) Jap. J. Appl. Phys. 9, 328
56. O. Nittono, N. Totsuka, K. Hamamura and S. Nagakura (1976) J. Cryst. Growth 36, 41
57. S.S. Brenner (1956) Acta Met. 4, 62
58. C. G'Sell and G. Champier (1973) Mat. Sci. Eng. 12, 203
59. C. G'Sell and G. Champier (1975) Phil. Mag. 32, 283
60. C. G'Sell and G. Champier (1976) Phil. Mag. 34, 733
61. R. Fiedler and A.R. Lang (1972) J. Mater. Sci. 7, 531
62. C. G'Sell (1977) These Institut National Polytechnique de Lorraine, Nancy
63. C. G'Sell (1971) These de Troisieme Cycle, Institut National Polytechnique de Lorraine, Nancy
64. C. G'Sell, B. Baudelet and G. Champier (1972) J. Cryst. Growth 13/14, 252
65. J. Bardeen and C. Herring (1952) Imperfections in nearly perfect crystals (Eds. W. Shockley, J. Hollomon and R. Maurer) Wiley, New York, p 261
66. B. Baudelet (1970) These Institut National Polytechnique de Lorraine, Nancy
67. B. Nost and G. Sorensen (1966) Phil. Mag. 13, 1075
68. R. Argemi (1976) These Institut National Polytechnique de Lorraine, Nancy
69. B. Baudelet and G. Champier (1969) J. Physique 30, 999
70. B. Baudelet and G. Champier (1969) C.R. Acad. Sci. 268, 1194
71. B. Nost, G. Sorensen and E. Nes (1967) J. Cryst. Growth 1, 149
72. O. Lohne and B. Nost (1967) Phil. Mag. 16, 341
73. B. Baudelet, C.G'Sell and G. Champier (1969) Quantitive Relation

behaviour properties and microstructure, Proc. Int. Conf. Haifa p 97
74. B. Baudelet and G. Champier (1973) Cryst. Lattice Defects 4, 95

CHAPTER 6

DEFECTS IN NON-METAL CRYSTALS

H. KLAPPER

6.1 Introduction

The group of non-metal crystals comprises a vast variety of materials with extremely different properties. Their mechanical strength ranges from the high hardness of diamond or corundum to the wax-like softness of organic compounds. They are grown by diverse methods from the melt, from the vapour phase and from solution. The growth temperatures vary from values above $2000°$ C (e.g. for high-melting oxides) to values close to or even below room temperature for melts of organic compounds and solutions in aqueous or organic solvents. In hydrothermal growth (quartz, zinc oxide) pressures of several kbars are applied.

This wide range of mechanical properties and of growth conditions results in a great variety of crystal defects and in a widely varying degree of crystal perfection. Nevertheless, many features of the generation and geometrical arrangement of defects are common to extremely different crystal species. Thus, crystals grown on planar faces under various conditions of growth from solution, supercooled melt and vapour phase exhibit the same typical appearance of their grown-in defects. It is the purpose of this paper to stress these common aspects which permit prediction of the nature of the defects and their distribution in the crystals. Only those imperfections are dealt with which can be visualized by X-ray methods.

According to their formation, the defects will be treated in the following two categories:

(a) defects generated during growth

(b) defects generated subsequent to growth

In the first group (sections 2 - 4) those defects are described which are independent of the properties of the specific material but typical for the growth method applied. In the second group (section 6.5), changes of grown-in defects and generation of new defects during cooling of the crystal or during specimen preparation are treated. For additional information the review articles by Lang [1,2] and Authier [3,4] are recommended.

6.2 Growth defects: crystals with planar growth faces

6.2.1 General features of grown-in dislocations

Crystals growing from solution, the vapour phase and from supercooled melt exhibit, as a rule, planar (habit) growth faces. The crystals can be divided into different growth sectors, i.e., into regions grown on different habit faces and thus having different but sharply-defined growth directions (Fig. 1a). The growth sectors are separated by growth-sector boundaries which represent the trajectory of the crystal edge between two neighbouring faces during growth.

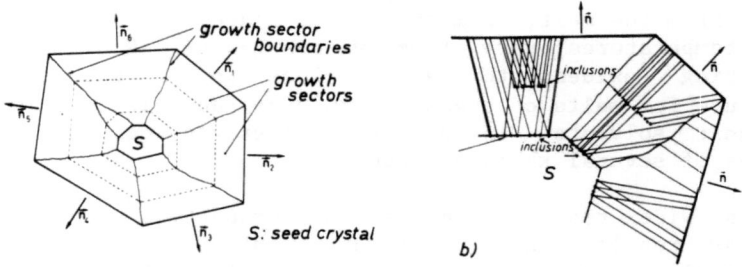

Figure 1. (a) Division of a crystal into growth sectors. $\underline{n}_i$ = growth directions, ------ contours of the crystal at different stages of growth. (b) Typical geometry of grown-in dislocations. The preferred directions of dislocation lines within one growth sector result from different Burgers vectors.

The grown in dislocations originate from inclusions which frequently occur at the surface of the seed crystal or are produced by growth accidents (section 6.2.2). The geometry and distribution of these dislocations within the crystal is strongly related to the growth on planar faces and to the different growth sectors. As a rule, the dislocations follow straight lines with rather sharply defined directions. These preferred directions (see section 6.2.3) which in most cases lie within an angle of ±15° to the growth normal $\underline{n}$, depend on the growth direction and on the Burgers vector involved. The dependence on the growth direction becomes strikingly apparent when the dislocation line passes through a growth sector boundary, i.e., when the dislocation outcrop shifts from one face to the other. According to the abrupt change of growth direction, the dislocation line undergoes a sharp bend and proceeds along a new direction which is characteristic of the new growth sector.

This typical arrangement of grown-in dislocations is schematically shown in Fig. 1b and is demonstrated for some solution- and melt- grown crystals in Figs. 2 - 4. It has been observed in many other crystals too, for example in triglycine sulfate [5,6], KDP [7-9], ammonium hydrogen oxalate hemihydrate [10], benzil [11], thiourea [12], natural quartz [13], hydrothermal zinc oxide [14], flux-grown yttrium iron garnet [15], flux-grown potassium nickel fluoride [16] and vapour-grown silicon carbide [17,18].

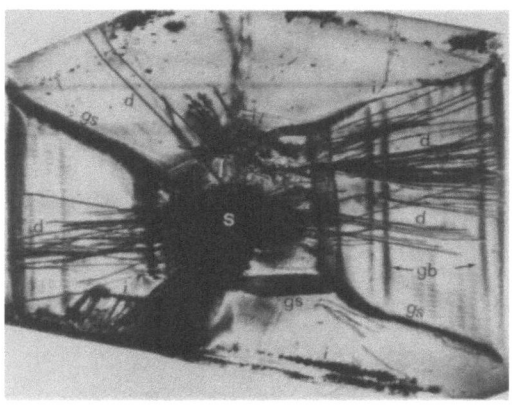

Figure 2. Grown-in dislocations d, growth sector boundaries gs and growth bands gb in triclinic $NH_4H_3(C_2O_4)_2 \cdot 2H_2O$ (an ammonium hydrogen oxalate) grown from aqueous solution. Plate 1.2 mm thick, horizontal dimension 24 mm. X-ray topograph ($MoK\alpha_1$). S = part of the seed crystal, i = inclusions. Crystal grown by H. Kuppers.

(a)   (b)

Figure 3. X-ray topographs ($CuK\alpha_1$) of plates cut from crystals grown in ~1°C undercooled melts [36]. (a) 1.4 mm (001) benzophenone plate, vertical edge 27 mm. Melt temperature $T_{gr} \simeq 47°C$. (b) 1.2 mm (0001) benzil plate, height 32 mm. $T_{gr} \simeq 95°C$. S = seed crystal, gs = growth sector boundary; spotted contrast results from surface damage.

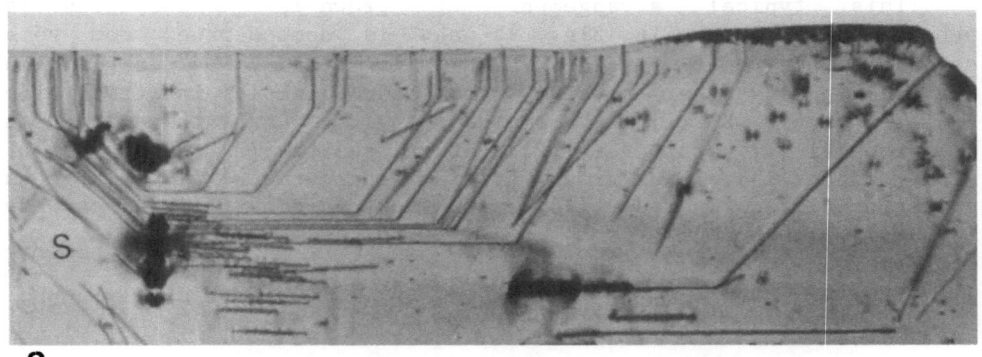

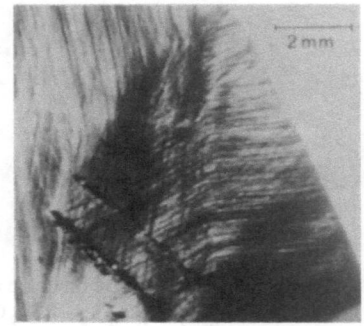

Figure 4. Bending of grown-in dislocation lines in growth sector boundaries. (a) Salol grown at 41°C from the undercooled melt. X-ray topograph (CuKα_1). The growth sector boundaries are invisible in the reflection used. They are recognisable by the bends in the dislocation line. Some dislocations continue from the seed S into the crystal. (4.5 x 15 mm). (b) $NH_4H_3(C_2O_4)_2 \cdot 2H_2O$, grown from aqueous solution (H. Küppers). Most of the dislocations inside the triangular growth sector originate from two rows of inclusions and show two different preferred directions (two different Burgers vectors). By enhanced growth due to the newly generated dislocations the corresponding growth face has vanished. X-ray topograph (MoKα_1).

6.2.2 Origin of dislocations. Formation of inclusions.

For topological reasons dislocation lines cannot begin or end in the interior of a perfect crystal. During crystal growth they originate from three sources:

(a) they continue from dislocations present in the seed crystal

(b) they are nucleated at the interface of seed and crystal

(c) they arise from defects due to growth accidents.

The nucleating defects in (b) and (c) are inclusions of foreign particles or of solvent. When the crystal lattice is "closed" behind such inclusions, closure mistakes, giving rise to dislocations, frequently occur. The dislocation lines proceed with the growing crystal unless they end at other inclusions produced by growth accidents.

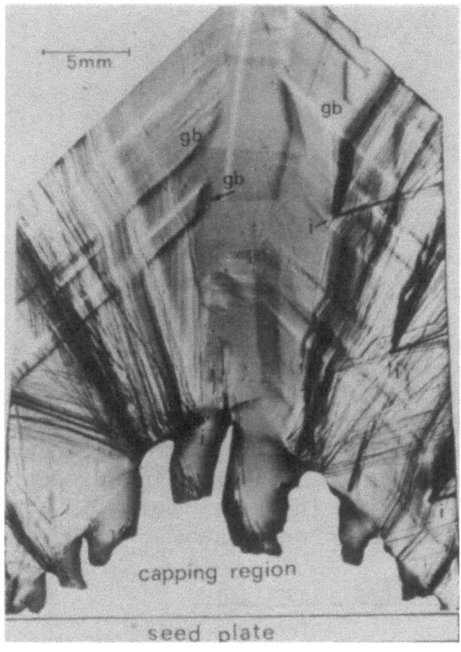

Figure 5. (100) plate of KDP. Thickness 1.5 mm. X-ray topograph AgKα Growth bands gb, inclusions i. The greater part of the capping region broke off during polishing and etching. Crystal grown by Van der Linden [35].

Thus, the generation of dislocations is strongly correlated with the formation of inclusions. At the interface of seed and crystal, inclusions arise from foreign particles or from mechanical damage such as scratches. Inclusions may also be due to the improper shape of the seed. If the seed is not bounded by habit faces, facets of habit faces are formed in the early stages of growth. The regions between these facets are micro-facetted and have re-entrant edges or corners. In these regions solvent is easily entrapped, giving rise to the so-called veils. Clear crystal growth begins after the habit faces have met and formed an edge. This process of capping is well-known in the growth of KDP on (001) seed plates. (For details see Jansen-van Rosmalen et al., [19]). A plate of KDP cut through the capping region is shown in Fig. 5.

During growth inclusions result from changes in growth conditions which induce very rapid growth or a period of dissolution or melting followed by rapid growth. Redissolution and remelting leads to a rounded crystal and to a "capping" process which favours the formation of inclusions. In Fig. 6 the generation of inclusions and dislocations by slight remelting and fast growth of salol (undercooled melt) is demonstrated.

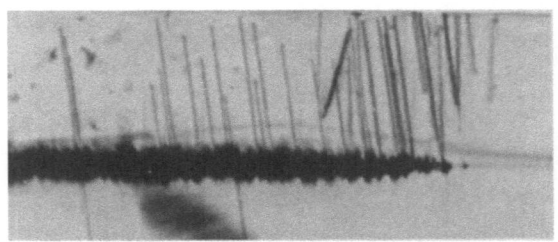

Figure 6. Generation of inclusions and dislocations by slight back-melting and fast regrowth of salol (undercooled melt) [36]. (11 x 5 mm).

In crystals grown from solution, liquid inclusions also arise from local variation of supersaturation due to the solvent flow around the crystal. This has been studied in detail on KDP by Janssen-van Rosmalen et al. [19, 20]. A dependence of inclusion and dislocation formation on the solvent-flow direction is observed by Lefaucheux et al. [21] in hydrothermally grown calcite, and by Gits-Leon et al. [22] in potash alum.

6.2.3 Theory of preferred directions

The preferred directions of grown-in dislocation lines, as described in 2.1, can be explained by two different but equivalent approaches.

(a) Minimum-energy approach [10, 23, 24]

A dislocation line emerging from a growth face proceeds into the newly-grown layer and increases the free energy of this layer by the amount of its own energy. The tendency to minimize the free energy leads to the requirement that the energy of the dislocation within the layer should be a minimum. Thus, a straight dislocation line is favoured because any bending would increase both its energy per unit length and its whole length within the layer. The total energy of a straight dislocation line in the layer is (Fig. 7)

$$E = E(\underline{b}, \underline{\hat{\ell}}, c_{ij}) \cdot d/\cos\alpha$$

($\underline{b}$ = Burgers vector, $\underline{\hat{\ell}}$ = unit vector of dislocation-line direction, $\alpha = \alpha(\underline{\ell},\underline{n})$ = angle between $\underline{\hat{\ell}}$ and the growth direction, d = thickness of the layer). E ($\underline{b},\underline{\ell},c_{ij}$) is the elastic energy per unit length of the dislocation line which depends also on the elastic constants of the crystal. Here only the elastic energy is considered because the entropy contribution to the free energy of a dislocation can be neglected (Cottrell [34]). Since the direction of minimum energy does not depend on d, we set d = 1 and regard the energy W per unit growth length:

$$W(\underline{b},\underline{\ell},\underline{n},c_{ij}) = E(\underline{b},\underline{\ell},c_{ij})/\cos\alpha$$

Now the preferred direction $\underline{\ell}_0$ is given by the condition

$$W(\underline{b},\underline{\ell}_0,\underline{n},c_{ij}) = \text{Minimum}.$$

It is clear that $\underline{\ell}_0 = \underline{\ell}_0(\underline{b},\underline{n},c_{ij})$ defined in this way depends on the Burgers vector $\underline{b}$ and the growth direction $\underline{n}$ and, thus, accounts for the geometrical features of grown-in dislocations shown in Fig. 1b.

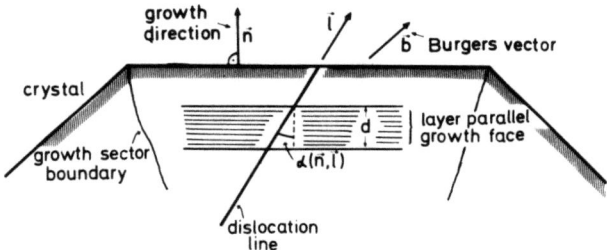

Figure 7. Energy of a dislocation within a layer parallel to the growth face (see section 6.2.3)

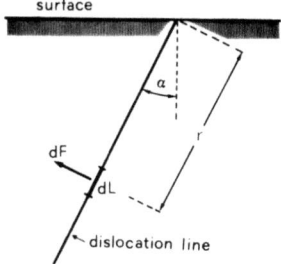

Figure 8. Force dF exerted by the surface on a dislocation line element dL (see section 6.2.3)

<u>(b) Zero-force approach [24]</u>

A dislocation-line element dL close to the crystal surface receives a force dF which is inversely proportional to the distance r of the line element from the surface and depends on the angle between the line element and the surface normal (Fig. 8). After Lothe [25] this force is given by

$$dF = -\frac{1}{r} \cdot f(\alpha) \cdot dL$$

with
$$f(\alpha) = E \cdot \tan\alpha + \frac{\partial E}{\partial \alpha}$$

Here E is again the elastic energy per unit length of the dislocation line in the interior of the crystal. At an angle α_0 for which $f(\alpha_0) = 0$ the force dF upon the line element is zero in any depth below the surface. A dislocation line of this orientation, therefore is free of forces. It is reasonable to assume that during crystal growth a dislocation follows this orientation of zero force.

Both approaches are equivalent and lead to the same preferred orientations of grown-in dislocation lines. This is easily proved by setting $\partial W/\partial \alpha = 0$, which results in the above condition $f(\alpha) = 0$ for the direction of minimum W. In the following section the minimum energy concept is applied.

<u>Verification of the theory</u>

In order to verify the minimum-energy approach of preferred directions the dependence of $W = E/\cos\alpha$ on the dislocation-line direction ℓ has to be calculated. Some problems arise in the calculation of the energy E which has to be split into two parts:

$$E = E_a + E_c$$

E_a is the elastic energy of the long-range strains which surround the dislocation line and which can be described by linear elasticity theory. E_c is the energy of the dislocation core in which the strains are so high that linear elasticity theory cannot be applied. The core energy E_c also contains contributions of strongly-deformed or broken bonds in the centre of the core region. Since the structure of the dislocation core is not known (except for very simple crystal structures), a calculation of E_c and its dependence on ℓ is not possible. In general, however, the core energy is one order of magnitude smaller than E_a [26, 27] and can be neglected without introducing serious mistakes. The energy of the long-range strain field is given [28 - 30] by

$$E_a = Kb^2 \ln(R/r_0)/4\pi,$$

with b the modulus of the Burgers vector, R the outer and r the inner cut-off radius, and $K = K(\underline{e}, \underline{\ell}, c_{ij})$ the so called energy factor, which depends on the direction $\underline{e}$ of the Burgers vector, the line direction $\underline{\ell}$ and the elastic constants c_{ij}. The inner cut-off radius r_0 is identical with the core radius. (For the meaning of R see Hirth & Lothe [30], page 62.)

It is now assumed that for a given Burgers vector the factor $b^2 \ln(R/r_0)/4\pi$ is constant with respect to the line direction $\underline{\ell}$. Thus the preferred direction $\underline{\ell}$ can be determined from the simplified condition

$$K(\underline{e}, \underline{\ell}, c_{ij})/\cos\alpha = \text{Minimum}.$$

The energy factor K can be calculated using the linear anisotropic elastic theory of straight dislocations developed by Eshelby et al. [29] and described in detail by Hirth & Lothe [30]. In the general case of elastic anisotropy K cannot be expressed analytically, and the calculations have to be done numerically with the aid of a computer. Fig. 9 presents the calculated variation of $K \sim E$ of a dislocation with Burgers vector [100] in the planes (100), (010) and (001) of orthorhombic $NH_4HC_2O_4 \cdot 1/2H_2O$, a crystal of high elastic anisotropy. The pure screw orientation has the lowest energy E, as is usually the case. Diagram (a) contains only directions of pure edge character. In the case of elastic isotropy this plot would be a circle.

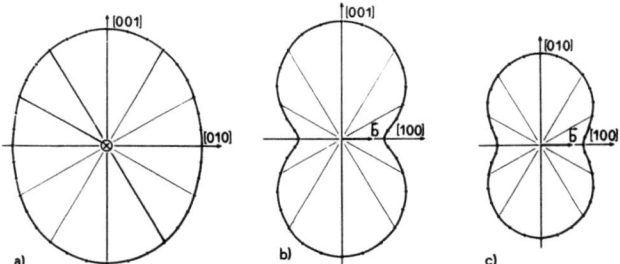

Figure 9. Variation (polar coordinates) of energy factor K of a dislocation with $\underline{b}$ = [100] in planes (100), (010) and (001) of orthorhombic ammonium hydrogen oxalate hemihydrate $NH_4HC_2O_4 \cdot 1/2H_2O$.

Fig. 10a shows, as an example, the calculated variation of the energy W per unit growth length for four different Burgers vectors in the (011) growth sector of KDP [9,24]. The directions of minimum energy are indicated by arrows. In the topograph Fig. 10b grown-in dislocations corresponding to curves 1, 2 and 3 are present. For these dislocations the calculated directions deviate from the observed ones by less than $4°$. Curve 4 ($\underline{b}$ = [011]) shows a very flat minimum, so that the preferred direction is not sharply defined. Correspondingly, the observed directions spread over an angular range of $10°$. The observed direction, however, deviates by about $20°$ from the calculated direction. Due to the flat minimum of the energy curve the value of W along the observed direction is only 2% higher than the minimum value of W along the calculated direction. Thus, a small variation of the (neglected) core energy could be responsible for the difference between observed and calculated directions.

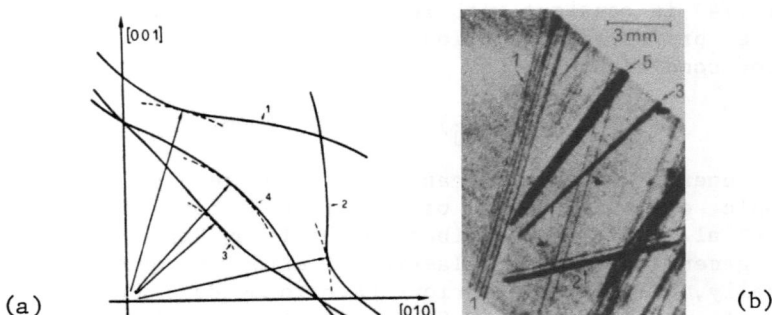

Figure 10. (a) variation of $W \sim K/\cos\alpha$ (polar coordinates) for dislocations with $\underline{b}$ = [001] (curve 1), [010] (2), [011](3) and [0$\bar{1}$1] (4) in growth sector (011) of KDP [9,24]. The preferred directions are marked by arrows. The dashed lines are part of circles with the minimum value of $K/\cos\alpha$ as radius. (b) Grown-in dislocations with Burgers vectors corresponding to the curves 1,2,3 of Fig. 10a. The dislocations 5 have $\underline{b}$ = 1/2 [111]. (X-ray topograph made by Y.M.Fishman [7,8]).

Preferred directions have been calculated for many crystals grown from aqueous solutions and organic solvents [9, 10, 23, 24, 31, 32], for hydrothermal zinc oxide [24] and for flux-grown $KNiF_3$ and $KCoF_3$ [16]. In the majority of cases a satisfactory or even excellent agreement of observed and calculated directions is found.

There exist, however, grown-in dislocations which do not fit into the above concept. They are mainly

(a) dislocations following irregularly curved paths,

(b) straight dislocations aligned along low-indexed crystallographic directions which deviate strongly from the calculated orientations.

Curved dislocations can be due to movement after crystal growth (see section 6.5). During growth, deviations from the straight path result from long-range stresses in the neighbourhood of inclusions or in growth bands. These stresses produce forces upon the dislocation line, shifting the zero-force direction (see 2.3.2.) to an orientation different from that of the unstressed crystal. Other causes may be a varying impurity decoration of the dislocation line during growth, or macroscopically high growth steps which sweep over the growth face and influence the path of grown-in dislocations

[45]. The change of dislocation line direction by growth bands in KDP has been described by Fishman [8] and is strikingly apparent from Fig. 5.

The tendency of grown-in dislocation lines to follow low-indexed directions has been observed, for example, in ammonium hydrogen oxalate [10], lithium formate hydrate [32], and aluminium iodate [33]. It corresponds to the alignment of dislocations, during slip, along low-indexed lattice directions and is due to the fact that a crystal is not an elastic continuum, but has a discrete submicroscopic structure. It is reasonable to interpret this tendency in terms of the core energy and to assume that in the above crystals certain dislocations have a relatively high core energy with a sharp minimum along a low-indexed direction. This direction may be favoured, even though the consideration of long-range strain energy alone leads to another preferred direction. It should be noted, however, that there is a problem in attributing a definite direction to a dislocation line which is inclined with respect to a low-indexed lattice row. This is macroscopically possible but on a submicroscopic scale this dislocation may have a kinked course with line elements parallel and perpendicular to the lattice row.

Due to the factor $1/\cos\alpha$ in the expression of W the preferred directions ℓ_0 lie in most cases close to the growth direction $\underline{n}$. Deviations of ℓ_0 from $\underline{n}$ by more than 15° are rare. This results in a specific distribution of grown-in dislocations. If no growth accidents have occurred (i.e. no inclusions have been formed during growth), only the sections extending from the seed toward the centre of the growth face along the growth directions contain dislocations. The regions in the neighbourhood of the sector boundaries are free of dislocations. This is schematically illustrated in Fig. 11 and apparent from the topographs Fig. 2 and 3a.

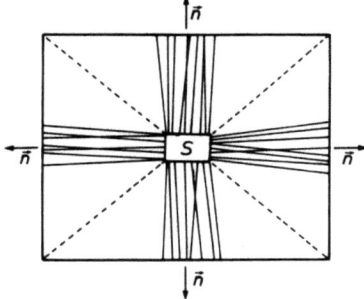

Figure 11. Distribution of grown-in dislocations in a crystal grown on planar faces. S = Seed crystal. Dashed lines: growth sector boundaries.

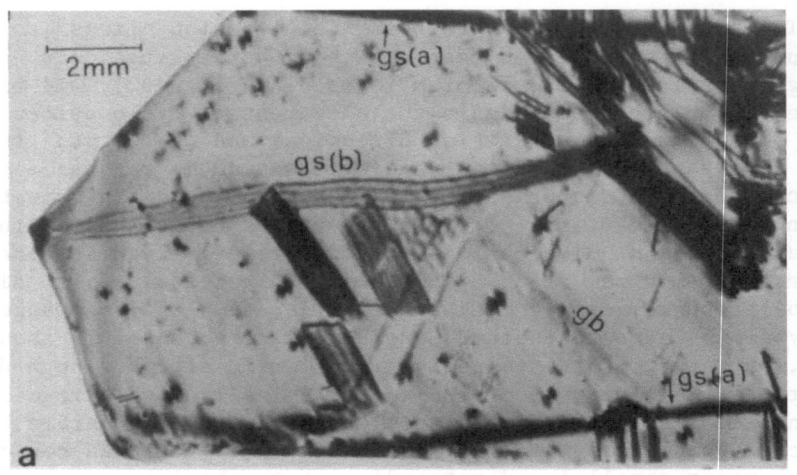

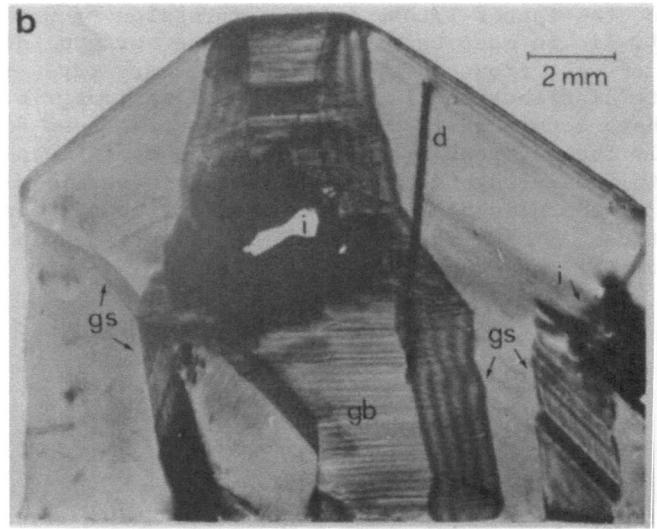

Figure 12. Growth bands gb and growth sector boundaries gs. (a) Plate of salol grown from undercooled melt. Growth sector boundaries gs(a) show kinematical, boundaries gs(b) dynamical fringe contrast (b) Plate of benzil grown from solution in xylene. The growth sector boundaries show mainly fringe contrast. i = inclusions of solution, d = dislocation line.

6.2.4 Growth bands and growth sector boundaries

Growth bands

Growth bands arise from local variations of the impurity content. They are created by fluctuations in the growth conditions, such as changes of temperature, cooling rate, pressure, convection of the solution or the melt. The varying impurity concentration causes local changes of the lattice parameters which give rise to X-ray topographic contrast. The impurities may be

(a) contaminants of the solvent or of the melt

(b) solvent components incorporated

(c) deviations from stoichiometry.

The regions of different impurity concentration form layers coinciding with the instantaneous growth front. They are faults which appear on X-ray topographs by direct image or dynamical fringe contrast as "bands" normal to the growth direction. These growth bands may occur isolated or in dense sequence (striations). Examples are shown in Figs. 2, 5 and 12. The following features of growth bands are noteworthy:

Usually, the X-ray topographic contrast vanishes in reflections with diffraction vector $\underline{g}$ parallel to the growth band. This is also observed for faulted growth sector boundaries, if $\underline{g}$ is parallel to the fault surface. It indicates that the displacements in the distorted regions or the fault vectors are mainly normal to the faulted layer (see [11, 53, 54]).

The occurence and "intensity" of growth bands depends on the growth sector. This selective incorporation of impurities on different growth faces is apparent in Figs. 2 and 12.

Pronounced growth bands are found in almost all flux-grown crystals. This results from the increased incorporation of solvent components due to the relatively high growth temperature.

Among the great number of crystals showing growth bands, the following are mentioned: KDP [7, 8, 37 - 39], natural quartz [13, 40], benzil (see Fig. 12b, [11]), topaz [41], beryl [42], flux-grown iron yttrium garnet [15], and potassium nickel fluoride [16, 43]. Refs. [39] ([KDP) and [40] (quartz variety amethyst) prove the correlation of impurity content with growth bands.

Growth sector boundaries

In many crystals growth sector boundaries are accompanied by fault surfaces. On projection topographs they appear by kinematical contrast due to lattice distortions or by dynamical fringe contrast similar to that of stacking faults. This is shown in Figs. 2, 3a and 12. In the topograph of salol, Fig. 12a, both kinds are present. By applying section topography to the fringe-contrast boundaries, the following cases can be distinguished [11, 52, 53]:

(a) The lattices on both sides of the boundary are exactly parallel (i.e. coincidence of the reflection curves of both sectors).

(b) The lattices have different interplanar spacings or are tilted with respect to one another (partial or no overlapping of the reflection curves).

In the first case the boundary is assumed to consist of a thin layer of impurities. The effective misorientation in the second case results from differences in impurity content (i.e. in interplanar spacings) of the two growth sectors. The difference in the lattice parameters may also produce a distorted transition layer which gives rise to direct image contrast. Such contrast appears frequently at the outcrops of the boundary in the specimen surface because here, due to stress relaxation, stronger distortions are present (see [11, 47]). This feature is also observed in the contrast of isolated growth bands.

Detailed studies of growth sector boundaries have been performed by Yoshimura & Kohra [46] and Parpia [47 - 49] on synthetic quartz and by Lutsau & Fishman [50] on KDP. The change of lattice parameters in different growth sectors of solution-grown NaCl has been investigated by Ikeno et al. [51].

Actually, growth bands and faulted growth sector boundaries are present in all crystals. Even if they are not detected by normal X-ray projection topography, they can be revealed by more sensitive methods like the double-crystal technique.

6.3 Growth defects: crystals grown on curved interfaces

6.3.1 General remarks

Curved growth faces appear in almost all methods of melt growth (except for undercooled melts) and in special techniques of condensation from the vapour phase (e.g. the temperature oscillation method of Scholz [55, 56]). In these cases the growth front is determined by the freezing or condensation isotherm. There are, however, no systematic studies of vapour-grown crystals, so that the following treatment concerns melt growth only. For the investigation of the true grown-in defect arrangements in melt-grown crystals the

DEFECTS IN NON-METAL CRYSTALS

following facts have to be considered:

(a) In some techniques like the Verneuille method, growth takes place under high thermal stresses which give rise to very poor crystal quality. An investigation of isolated grown-in defects in such crystals usually is not possible. Other methods, like the Czochralski or the floating zone technique, allow growth under low thermal gradients. Applying these methods, crystals of high perfection are obtained, permitting the study of isolated grown-in defects and their arrangements.

(b) Most important melt-grown crystals have high melting temperatures. Thus, the cooling-down to room temperature may introduce a drastic change of grown-in defects. In addition, new defects may be created. The distinction between grown-in and post-growth defects is treated in section 6.5.

There are not many high-temperature grown crystals preserving the true defect arrangement created during growth. An excellent example is Czochralski-grown gadolinium gallium garnet (GGG, melting point T_m = 1750° C, [57]). GGG is highly important as a substrate material for magnetic bubble devices and is, therefore, extensively studied with respect to its defects [58 - 65]. Interesting examples of low-melting Czochralski crystals are benzophenone (T_m = 48°C) and benzil (96 C) [66].

6.3.2 Main growth features

In Fig. 13 the shape of a typical Czochralski-grown crystal, the traces of the crystal-melt interfaces at different stages of growth and the typical arrangement of grown-in defects are shown. During the first period after seeding, the crystal grows with a conically increasing diameter on a strongly-curved growth front until the desired diameter is reached ("crystal cone"). At this point the growth conditions are adjusted in such a way that growth proceeds with constant crystal diameter on a slightly convex interface. The grown-in defects exhibit features very similar to those in crystals grown on planar faces. The most important imperfections which are briefly reviewed here are striations, strains due to facet growth, inclusions and dislocations. These defects can be seen on the photograph Fig. 14 (Schmidt [67]) of a GGG slice in polarized light, where they become visible by stress birefringence.

6.3.3 Striations

Striations correspond to growth bands in crystals grown on planar faces. They follow the shape of the instantaneous crystal-melt interface and present a clear picture of the growth process and the growth conditons. Like growth bands, striations are

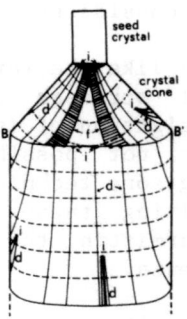

Figure 13. Typical shape of a Czochralski-grown crystal and arrangement of grown-in defects. i = inclusions; d = dislocations f = facet growth regions. The dashed lines indicate the shape of the growth front at different stages of growth. B-B´ is the boundary between the crystal cone and the region of constant diameter growth.

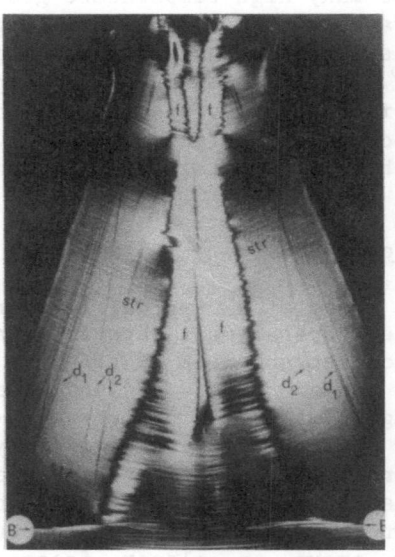

Figure 14. Plate cut from the cone crystal of a Czochralski-grown gallium gadolinium garnet (GGG) between crossed polarisers. [111] pulling direction vertical. Optical contrast of defects due to birefringence. f = facet growth regions; d_1 and d_2 dislocation lines with different Burgers vectors; str = striations. B-B´: see Fig 13 (after Schmidt & Weiss [67]).

layers of varying impurity content or, in doped crystals, of varying dopant concentration. They also arise from fluctuations in the growth conditions. In addition to the parameters mentioned above (section 6.2.4), the following cause of striations is important in melt growth. The temperature distribution in the melt deviates in practically all cases from rotational symmetry. Consequently, within one cycle of crystal rotation partial back-melting and growth occurs. Thus, the rotation causes periodic striations the frequency of which is correlated with the rotation rate of the cyrstal. As an example, in Figs. 14 - 16 striations in GGG are shown. They are due to deviations from stoichiometry. An extensive treatment of the various kinds of striations is given by Carruthers and Witt [68].

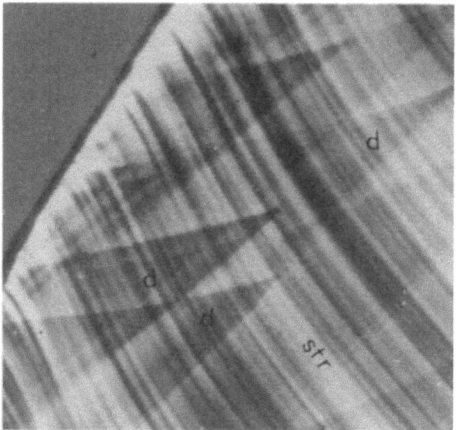

Figure 15. Section (2.5 x 2.5 mm) of a plate cut from the cone of a Czochralski GGG (crossed polarisers). Dislocation bundles d fan out from small Ir inclusions. str = striations. (Courtesy D. Mateika)

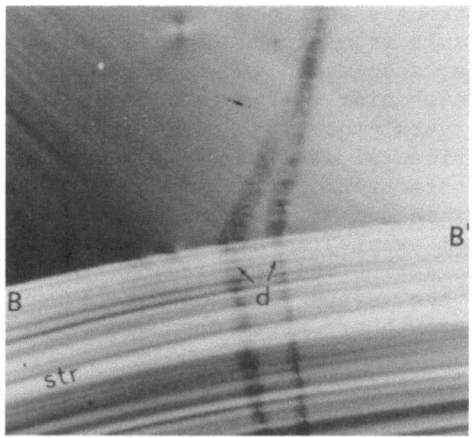

Figure 16. Section (2.5 x 2.5 mm) of a GGG plate between crossed polarisers. At B-B' the growth front orientation changes abruptly. Dislocation lines d are bent at the intersection with B-B'. The dislocations d form helices, which are assumed to arise from "straight" dislocations by climb due to absorption of interstitials [76]. (Courtesy D. Mateika).

6.3.4 Facets

Facet formation occurs in those areas of the crystal-melt interface where the growth front is parallel to morphologically important crystal faces, e.g. parallel to {211} and {110} in GGG [57, 59, 60] or parallel to {111} in germanium [69]. The facets lie "behind" the normal freezing isotherm. Thus, the crystal grows on facets from the undercooled melt. Due to the different growth modes the impurity concentrations and the composition of the crystal in the facet sectors differ considerably from those of the non-facet regions. This results in changes of lattice parameters and macroscopic strains which are detrimental for technical applications.

Facet regions are always present in the cone (Fig. 13) since the interface has a strong curvature in this part of the crystal. In the cylindrical crystal, which is used for applications, the formation of facets can be avoided by the choice of a suitable pulling direction and by a slightly convex interface. In Fig. 14 facet regions in the cone of Czochralski-grown GGG are shown.

6.3.5 Inclusions

In principle, the inclusions in melt-grown crystals are of the same kind as those mentioned in 2.2. Among foreign particles incorporated into the crystal small grains of crucible material are common. Fig. 15 shows some iridium inclusions acting as dislocation sources in a GGG crystal grown in an iridium crucible. Inclusions of misoriented and strongly strained, or of compositionally different, matrix material result from strong changes in the growth conditions. They are, as a rule, associated with a period of very fast growth which leads to a roughening of the interface by micro-facetting . A change of the crystal-rotation rate by 10% may be sufficient to generate these detrimental defects [70] which are the origin of dislocations. Imperfections of this kind frequently occur at the seed-crystal interface and at the boundary between the cone and the cylindrical crystal where, in order to attain a constant diameter, the conditions are appreciably changed. These defects can usually be avoided when the steps of seeding-in and of conversion to constant-diameter growth are performed carefully.

6.3.6 Grown-in dislocations

The main features of grown-in dislocations in melt-grown crystals are described as follows.

(a) The dislocations originate from dislocations already present in the seed and from inclusions and growth defects mentioned in 6.3.5.

(b) The dislocation lines follow curved trajectories, which are normal or nearly perpendicular to the local growth front.

(c) Dislocation lines with different Burgers vectors form different angles with the local direction of the interface and, thus, proceed along different paths (Schmidt [67], see dislocations d_1 and d_2 in Fig. 14).

(d) When the local orientation of the interface is abruptly changed, the dislocation line undergoes a sharp bend (Fig 16). This generally occurs in the boundary (B - B' in Fig. 13) between the cone and the cylindrical crystal and corresponds to the refraction of dislocation lines in growth sector boundaries.

The following consequence of item (b) is noteworthy. The trajectories and the distribution of grown-in dislocation lines depend strongly on the shape of the interface. An interface which is concave toward the melt focusses the dislocation lines into the centre of the crystal boule. If there is a convex growth front, they diverge and grow out of the crystal at the side faces of the boule (Fig. 13). This behaviour can be used to eliminate grown-in dislocations by growing the crystal for some time on a strongly convex interface.

Items (b) - (d) indicate that the path of grown-in dislocation lines follows the minimum-energy theory of section 6.2.3. This has been investigated by Schmidt & Weiss [67] who calculated the trajectories of minimal energy per unit growth length for dislocations with Burgers vectors [100], [110] and [111] in Czochralski-grown GGG, taking into account the curvature of the growth front. The calculations are very extensive even for a computer because the dislocation path has to be traced stepwise in small increments. The results show an excellent agreement of calculated and observed dislocation paths, demonstrating the validity of the minimum-energy theory. Moreover the behaviour of dislocations with different Burgers vectors is shown. For pulling direction [111] and a spherically convex interface (radius 20 mm) dislocations with $\underline{b}$ = [110] and [111] are focussed into the (110) plane whereas dislocations with $\underline{b}$ = [100] grow out asymmetrically with respect to the [111] boule axis.

6.4 Growth defects: crystals grown under high thermal gradients

Crystals grown under high thermal gradients are generally of very poor quality. Examples are given in Figs. 17 and 18 which show transmission X-ray topographs of a Verneuille sapphire (Fig. 17) and a KCl crystal grown by the Nacken-Kyropoulos method (Fig. 18). Both crystals exhibit a pattern of large blocks with misorientations up to a few degrees. Although a rather divergent primary X-ray beam has been used, a part of the misoriented subgrains are out of reflection condition. Within the blocks strong distortions due to low-angle boundaries and dislocations of high density are present. Single dislocations can hardly be resolved and a distinction between grown-in and post-growth defects is impossible.

Figure 17. AgKα_1 transmission topograph of plate cut from a Verneuille sapphire.

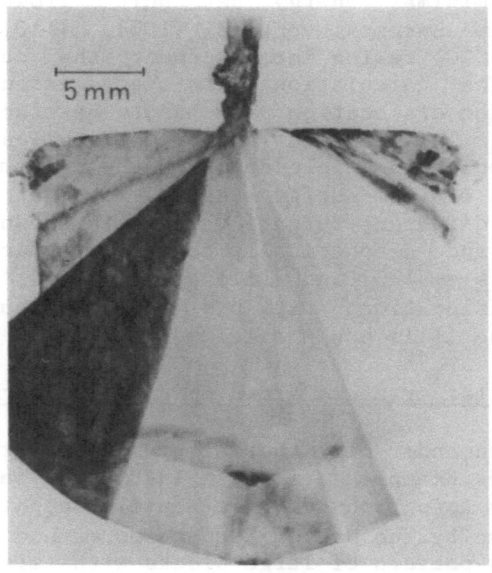

Figure 18. AgKα_1 transmission topograph of plate cut from a KCl crystal grown by Nacken-Kyropoulos method (growth by Dr. Eckstein).

In contrast to the specimen of Fig. 17 the KCl plate (Fig. 18) is cut parallel to the [100] growth direction and shows the development of the subgrains during growth. Growth was initiated by dipping the strongly cooled seed rod into the melt, the temperature of which was 50° C above the melting point. Due to gradually increased cooling of the seed (or gradually lowered temperature of the melt) the freezing isotherm and the growth front proceed into the melt. The nucleation of subgrains mainly occurs at the interface between seed and crystal where the highest thermal gradients are present. Consequently, the subgrains originate from the seed-crystal interface and radiate into the crystal boule as shown in Fig. 18.

6.5 Post-growth defects

6.5.1 Distinction between grown-in and post-growth dislocations

Subsequent to the growth process the configuration of dislocations may be changed by

(a) movement of grown-in dislocations due to glide or climb,

(b) generation of new dislocations due to glide or condensation of vacancies and self-interstitials.

Grown-in and post-growth dislocations can be distinguished in many cases by the following considerations. Let us regard the curved dislocation line L in Fig. 19. It is clear that this geometry of the dislocation line cannot occur when the growth front advances from A-A to A'-A'. A dislocation line cannot proceed against the growth direction as it does along the segment 1. For energetical reasons it is also highly improbable that the dislocation line is parallel to the growth front (segments 2 and 3). This would mean that the whole energy of the (macroscopically long) segment must be supplied within the period of the spreading out of <u>one</u> growth layer across the growth face. Also from another point of view the incorporation of parallel segments is prohibited due to the forces which the surface exerts on dislocation lines. These image forces ([30, 34], see section 6.3.3) are infinitely high in the surface and drive a parallel segment out of the crystal. It should be noted that these considerations hold only for smooth growth faces. If there are macroscopically high growth steps moving across the face, additional forces arise and "sweep" the dislocation line in the direction of step motion. Under certain conditions dislocation segments parallel to the "macroscopic" growth face may be trapped by such growth steps (Duckett & Lang [45]).

Thus the following simple criterion can be formulated for the distinction between grown-in and post-growth dislocations.

Dislocations with segments "reversed" or perpendicular with respect to the growth direction have undergone post-growth movement. This criterion always holds for dislocation loops or half-loops created after growth. It is a sufficient but not a necessary condition to conclude dislocation motion after growth.

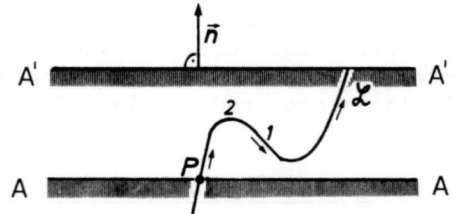

Figure 19. Geometry of a dislocation line which cannot occur during growth.

6.5.2 Dislocation glide

Grown-in dislocations

Frequently grown-in dislocations have been decorated during growth with impurities and, therefore, are rather immobile. Post growth movement of grown-in dislocation lines has been observed in ammonium hydrogen oxalate [10], lithium formate hydrate [32], sodium chlorate [71] and triglycerin sulphate [6]. These dislocation lines originate from inclusions and follow a somewhat irregular path until they end at the growth surface. In many cases they form rows of arcs with pinning points which indicate the presence of obstacles against dislocation motion [32,71]. An example is given in Fig. 20. In many cases the glide is activated by stresses due to inhomogeneous temperature changes (e.g. during cooling after growth).

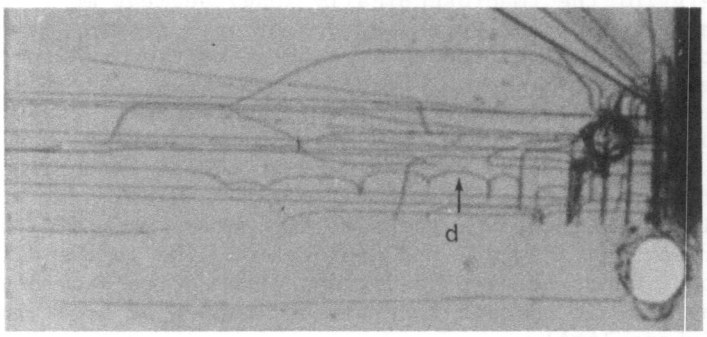

Figure 20. Glide of grown-in dislocations after growth in sodium chlorate. The dislocation line d forms a row of arcs with pinning points (X-ray topograph, $MoK\alpha_1$; 2.6 x 5.5 mm).

DEFECTS IN NON-METAL CRYSTALS

Generation of dislocations

New glide dislocations originate preferably from two sources:

(a) Internal defects which are surrounded by high stresses and associated with a "break" of the crystal lattice (i.e. with some kind of "internal" surface), Such defects are inclusions, grain boundaries, etc.

(b) Mechanical damage in the external surface (e.g. scratches).

In general, dislocations newly created by glide have the shape of half-loops extending from the source. Closed loops centered at inclusions may occur, too [86]. They result from stress concentrations in the neighbourhood of these defects. The stresses may be increased by temperature changes or temperature inhomogeneities. Moreover, dislocation multiplication by sources such as Frank-Read spirals is possible.

Some examples of dislocation generation by plastic deformation are presented in Fig.21. The topograph of Fig.21a shows dislocation half-loops in lithium fumerate tetrahydrate grown from aqueous solution. The half-loops extend from strongly disturbed regions of local dehydration. The plate of benzophenone in Fig.21b contains the platinum wire which held the seed in the solution. The crystal was grown at 40°C from a solution in xylene. Due to the different thermal expansion of platinum and benzophenone stresses developed during cooling to room temperature that exceeded the yield strength and generated numerous glide dislocations. In the plate of sodium chlorate (Fig 21c), in addition to grown-in dislocations, many glide dislocations are present which show a strong tendency to polygonization [71].

Among numerous further examples, dislocation generation by plastic processes has been observed in α- sulphur [72,73], thiourea [12,31], lithium formate hydrate [32], anthracene [74], tetraoxan [86], melt-grown benzophenone [66], vapour- grown silicon carbide [17,18], etc. Since all materials are plastic at temperatures close to the melting point, post - growth dislocation movement and generation occurs, above all, in crystals grown from the melt. This has recently been shown for benzil wich is brittle at room temperature but plastic at temperatures approaching the melting point of about 96°C. Crystals of benzil grown from solution at 40°C never show glide dislocations, whereas crystals grown from the undercooled melt contain numerous dislocation half-loops [75].

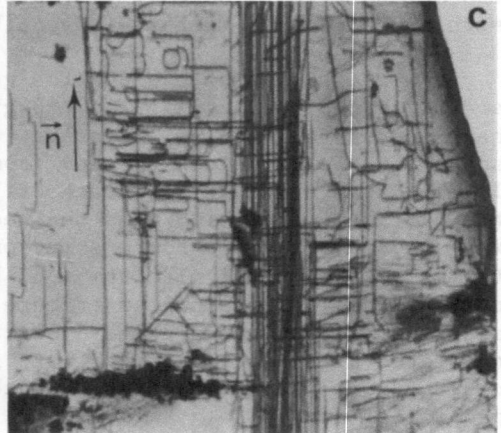

Figure 21. Generation of dislocations by glide (X-ray topographs) (a) lithium fumarate tetrahydrate (5 x 3.3 mm) (b) benzophenone (3.5 x 4.2 mm) (c) sodium chlorate (6 x 6mm) For details, see text.

6.5.3 Dislocation climb

Crystals grown at high temperatures contain point defects (vacancies, self-interstitials, impurity atoms) in high concentration. The configuration of these point defects is thermodynamically stable close to the growth temperature. During cooling after growth the crystal becomes supersaturated with point defects which mainly diffuse to lattice imperfections already present in the crystal. Dislocations act as such sinks for excess point defects. The absorption of vacancies and self-interstitials leads to dislocation climb, whereas the accumulation of impurity atoms results in a decoration of dislocations. The diffusion of point defects is strongly increased by annealing of the crystal. Thus a grown-in dislocation arrangement may be drastically changed by the condensation of point defects. It is assumed that the dislocation helices frequently observed in GG [59, 64, 76, 77] develop from "straight" dislocations with predominant screw character by the absorption of interstitials [76]. A very complicated dislocation configuration which has developed by glide, by climb and by dislocation reactions has been observed in as-grown and annealed Czochralski sapphire [78,79].

6.5.4 Dislocation loops, stacking faults and precipitates

The condensation of point defects may also generate prismatic dislocation loops, stacking faults and precipitates of impurities or dopants. The formation of these defects is pronounced if the crystal is free of dislocations, which may act as sinks for the excess point defects. They are common in dislocation-free semiconductor crystals where they give rise to the so-called "swirl" defects. An extensive review of these crystal imperfections and their deleterious effects on the function of electronic devices is given by de Kock [80,81].Investigations of the formation of stacking faults and the precipitation of oxygen in silicon were performed by Patel [82,83] and by Authier & Patel [84]. A highly interesting fact is presented by Lang in his study of natural diamonds [85]. He observed numerous impurity clusters in the shape of platelets with crystallographic orientations. It is suggested that these precipitates consist of nitrogen and that they have been formed during an annealing period following growth.

REFERENCES

1. A.R. Lang (1978) in: Diffraction and Imaging Techniques in Materials Science (ed. S. Amelinckx, R. Gevers & J. Van Landuyt), North Holland Publ. Co., Amsterdam-New York-Oxford, p.623
2. A.R. Lang (1973) in: Crystal Growth: An Introduction (ed. P. Hartman), North-Holland Publ.Co., Amsterdam- London, p.444
3. A. Authier (1977) in: 1976 Crystal Growth and Materials (ed. E. Kaldis & H.J. Scheel), North-Holland Publ. Co., Amsterdam-New York-Oxford, p 515
4. A. Authier (1972) J. Crystal Growth $\underline{13/14}$, 34
5. V.F. Miuskov, V.P Konstatinova and A.I. Gusev (1966) Soviet Phys.-Cryst. $\underline{13}$, 791
6. A. Izrael, J.F. Petroff, A. Authier and Z. Malek (1972) J. Crystal growth $\underline{16}$, 131
7. V.G. Lutsau, Y. M. Fishman and I.S. Res (1970) Kristall und Technik $\underline{5}$, 445

8. Y.M. Fishman (1972) Soviet Phys.-Cryst. 17, 607
9. H. Klapper, Y.M. Fishman and V.G. Lutsau (1974) Phys. Stat. Sol. (a) 21 115
10. H. Klapper and H. Kuppers (1973) Acta Cryst. A29, 495
11. H. Klapper (1971) J. Crystal Growth 10, 13
12. H. Klapper (1972) J. Crystal Growth 15, 281
13. A.R. Lang (1967) Advan. X-Ray Analysis 10, 91
14. D.F. Croxall, R.C. Ward, C.A. Wallace and R.C. Kell 1974) J. Crystal Growth 22, 117
15. W. Tolksdorf and F. Weiz (1978) in: Crystals for Magnetic Applications (e.d. C.J.M. Rooijmans), Springer-Verlag, Berlin-Heidelberg-New York, p1
16. M. Safa, B.K. Tanner, H. Klapper and B.M. Wanklyn (1977) Phil. Mag 35, 811
17. S. Amelinckx and G. Strumane (1960) J. Appl. Phys. 31, 1359
18. H. Posen and J.A. Bruce (1972) AFCRL Report 72-0225
19. R. Janssen-van Rosmalen, W.H. vander Linden, E. Dobbinga and D. Visser (1978) Kristall und Technik 13, 17
20. R. Janssen-van Rosmalen and P. Benema (1977) J. Crystal Growth 42, 224
21. F. Lefaucheux, M.C. Robert and A. Authier, (1973) J. Crystal Growth 19, 329
22. S. Gits-Leon, F. Lefaucheux and M.C. Robert (1978) J. Crystal Growth 44, 345
23. H. Klapper (1972) Phy. Stat. Sol (a) 14, 99
24. H. Klapper (1975) Habilitationsschrift, Aachen
25. J. Lothe (1967) Physica Norvegica 2, 154
26. H.B. Huntington, J.E. Dickey and R. Thomson (1955) Phys. Rev. 100, 1117
27. A. Maradudin (1959) J. Phys. Solids 9, 1
28. A.J.E. Forman (1955) Acta Metall. 3, 322
29. J.D. Eshelby, W.T. Read and W. Shockley (1953) Acta Metall. 1, 251
30. J.P. Hirth and J. Lothe (1968) Theory of Dislocations, McGraw-Hill, New York
31. H. Klapper (1972) Phys. Stat. Sol. (a) 14, 443
32. H. Klapper (1973) Z. Naturforschung 28a, 614
33. A. Hunsche (1977) Diplomarbeit, Universitat Koln
34. A.H. Cottrell (1953) Dislocations and Plastic Flow in Crystals, Clarendon Press, Oxford, p.39
35. W.J.P. Van Enckevort, R. Janssen-van Rosmalen, H. Klapper and W.H. van der Linden, in preparation
36. H. Klapper, T. Scheffen and S. Szreder, in preparation
37. C. Belouet, E. Dunia and J.P. Petroff (1974) J. Crystal Growth 23 243
38. C. Belouet, M. Monnier and J.C. Verplanke (1975) J. Crystal Growth 29, 109
39. C. Belouet, M. Monnier and R. Crozier (1975) J. Crystal Growth 30, 151

40. B.H. Schlössin and A.R. Lang (1965) Phil. Mag 12, 284
41. C. Giacovazzo, E. Scandale and A. Zarka (1975) J. Appl. Cryst 8, 315
42. E. Scandale, F. Scordari and A. Zarka (1979) J. Crystal Growth 12 70
43. M. Safa, B.K. Tanner, B.J. Garrard and B.M. Wanklyn (1977) J. Crystal Growth 39 243
44. A.R. Lang (1967) in Crystal Growth Suppl. Journal of Physics and Chemistry of solids, Pergamon Press, New York-Oxford, p. 833
45. R.A. Duckett and A.R. Lang (1973) J. Crystal Growth 18, 135
46. J. Yoshimura and K. Kohra (1976) J. Crystal Growth 33 311
47. D.Y. Parpia (1976) Phil. Mag 33, 715
48. D.Y. Parpia (1978) Phil. Mag 37, 375
49. D.Y. Parpia (1978) Phil. Mag. 37, 401
50. Y. M. Fishman and V.G. Lutsau (1970) Phys. Stat. Sol (a) 3, 829
51. S. Ikeno, H. Maruyama and N. Kato (1968) J. Crystal Growth 3/4, 683
52. M. Sauvage and A. Authier (1965) Bull. Soc. Franc. Miner. Crist. 88, 379
53. H. Klapper (1970) Dissertation, Universitat Koln
54. G. Calas and A. Zarka (1973) Bull. Soc. Franc. Milner. Crist. 96, 274
55. H. Scholz (1974) Acta Electron 17, 69
56. I. Beinglass, G. Dishon, A. Holzer and M. Schieber (1977) J. Crystal Growth 42, 166
57. F.J. Bruni (1978) in: Crystals for magnetic Applications (ed. C.J.M. Rooijmans), Springer-Verlag, Berlin-Heidelberg- New York, p.53
58. C.D. Brandle, D.C. Miller and J.W. Nielsen (1972) J. Crystal Growth 12, 1972
59. B. Cockayne, J.M. Roslington and A. W. Vere (1973) J. Materials Science 8, 382
60. H.L. Glass (1972) Mat. Res. Bull. 7 1087
61. D. Becker, E. Zsoldos and A. Weber (1976) Phys. Stat. Sol. 34 519
62. J.W. Matthews, E. Klokholm, V. Sadogapan, T.S. Plaskett and T.S. Mindel (1973) Acta Met. 21, 203
63. H.L. Glass (1973) Mat Res. Bull 8 43
64. W.T. Stacy, J.A. Pistorius and M.M. Janssen (1974) J. Crystal Growth 32, 37
65, Krishan Lal and S. Mader (1976) J.Crystal Growth 32, 357
66. J. Bleay, R.M. Hooper, R.S. Narang and J. N. Sherwood (1978) J. Crystal Growth 43, 589
67. W. Schmidt and R. Weiss (1978) J. Crystal Growth 43, 515
68. J.R. Carruthers and A. F. Witt (1975) in: Crystal Growth and Characterization (ed. R. Ueda & J.I.B. Mullin), North-Holland Publ. Company, Amsterdam, p. 107
69. J.A.M. Dikhoff (1960) Solid State Electron 1, 202
70. D. Mateika, private communication

71. R. Zilber and H. Klapper, in preparation
72. J. Di-Persio, B. Escaig, E.M. Hampton and J.N. Sherwood (1974) Phil. Mag 29, 733
73. E.M. Hampton, R.M. Hooper, B.S. Shah and J.N. Sherwood, J. Di-Persio and B. Escaig (1974) Phil. Mag 29, 743
74. D. Michell, P.M. Robinson and A.P. Smith (1968) Phys. Stat. Sol. 26, K93
75. H. Klapper and T. Scheffen, in preparation
76. K. Takagi, T. Fukagawa and M. Ishii (1976) J. Crystal 36, 185
77. J.W. Matthews, E. Klokholm, T.S. Plaskett and V. Sadgopan (1973) Phys. Stat. Sol. 19, 617
78. J.L. Caslavsky and C.P. Gazzara and R.M. Middleton (1972) Phil. Mag. 25, 35
79. J.L. Caslavsky and C.P. Gazzara (1972) Phil. Mag 26, 961
80. A.J.R. de Kock (1973) Philips Research Rep. Suppl Nr. 1
81. A.J.R. de Kock (1977) in: 1976 Crystal Growth and Materials (ed. E. Kaldis and H.J. Scheel), North-Holland Publ. Com., Amsterdam, p.662
82. J.R. Patel (1972) Bull. Soc. Franc. Miner. Crist. 95, 700
83. J.R. Patel (1972) J. Appl. Phys 44, 3903
84. J.R. Patel and A.Authier (1975) J. Appl. Phys 46, 118
85. A.R. Lang (1977) J.Crystal Growth 42, 625
86. T. Watanabe and K. Izumi (1979) J. Crystal Growth 46, 747

CHAPTER 7

DEFECT VISUALISATION: INDIVIDUAL DEFECTS

A.R. LANG

7.1 Introduction

7.1.1 Experimental background

This chapter will try to provide explanations and advice for the practising X-ray topographer. Much of the material has not previously been set down in print. The topics to be included have been selected with the needs of the inexperienced worker particularly in mind. Consequently, only the simpler techniques will be discussed: the section topograph, the projection topograph, and the scanning reflection topograph. However, X-ray topographers will surely want to read more comprehensive descriptions of the many techniques that have been devised over the years for various purposes, and to find out what results have been achieved using them. To do so, they may consult reviews by Armstrong [1] and Lang [2] (the latter preferably in the revised edition), and the monograph by Tanner [3]. Since X-ray topography is most widely known and practised as a non-destructive method for 'seeing' dislocations in crystals, a simply-worded review [4] spanning the topics of X-ray diffraction contrast from dislocations, other methods of observing dislocations, and the properties of dislocations themselves, may be found helpful.

7.1.2 Theoretical Background

X-ray topographers may consider themselves fortunate because they really need to remember only two formulae (in addition, of course, to Bragg's Law). The first formula is that for the extinction distance, ξ_g, of a reflection whose diffraction vector is $\underline{g}$. The second formula is for the angular range of reflection by a perfect crystal. Some of the many circumstances in which topographers need to apply these formulae in order to design their experiments and interpret their topograph images will be instanced below, in Section 7.3. The reciprocal of the extinction distance is given by

$$\xi_g^{-1} = \lambda |F_a| / \pi V \cos\theta_B, \qquad (1)$$

in which λ is the wavelength of the radiation used, F_a is the structure factor of the reflection in <u>absolute</u> units, V is the volume of the unit cell and θ_B is the Bragg angle. It may help towards remembering this equation to recall that F_a has the dimensions of length, whether it be electrons, X-rays or neutrons that are involved. In the case of X-rays $|F_a|$ is the product of the modulus of the familiar structure amplitude (expressed in electron units) and the classical electron radius, 2.818 fm (1fm = 10^{-15}m). Also in the X-ray case, the right-hand side of (1) must be multiplied by the polarisation factor, C; C = 1 for the σ polarisation state (electric vector perpendicular to the plane of incidence) and C = $|\cos 2\theta|$ for the π polarisation state. (Some implications for defect-imaging consequent upon the different orders of magnitude of X-ray and electron scattering factors, and their different variation with $\sin\theta_B/\lambda$ are discussed in [4].)

In the geometrical situations of symmetrical transmission (symmetrical Laue case) and symmetrical reflection from the specimen surface (symmetrical Bragg case), and under conditions when absorption is negligible or small, the expression for the angular range of reflection is

$$W_0 = 2d\, \xi_g^{-1} \qquad (2)$$

Here d is the interplanar spacing of the Bragg reflection concerned. In the Bragg case W_0 is the width of the flat summit of the 'top-hat' reflection curve, i.e. of the range of total reflection, whereas in the symmetrical Laue case W_0 is the full width at half maximum height of the reflection curve whose intensity, I, as a function of angular departure, $\Delta\theta$, from the position of peak intensity is

$$I = I_{max} \left[1 + (\Delta\theta/W_0)^2 \right]^{-1} \qquad (3)$$

It is characteristic of dynamical diffraction theory that formulae which are simple in the symmetrical diffraction geometries become rather cumbersome under asymmetric conditions. In Laue case experiments it rarely occurs that formulae for the symmetrical case are insufficiently close approximations: in Bragg case experiments the consequences of asymmetry need more frequently to be considered. The serious-minded topographer will doubtless seek more grounding in diffraction theory: good starting points are the reviews [5-7]. The really serious-minded will keep Kato [8] as constant companion for study and reference.

DEFECT VISUALISATION: INDIVIDUAL DEFECTS

7.2 Production of X-ray topographs

7.2.1 Section topographs

Kato has often emphasised the fundamental nature of the diffraction pattern contained in the section topograph image. In the course of the spatial integration that transforms the section topograph image into the projection topograph image some information is unavoidably lost. The section topograph technique is certainly the method of choice for analysis of Pendellosung fringe phenomomena of all sorts, including fault-surface fringes: Authier [9] has presented a useful review of this activity. Moreover, the experimenter turns to the section topograph when the density of defects is too high or the specimen is too thick to permit the resolution of defects individually on a projection topograph. Furthermore, for the most precise determination of the position and orientation of a fault surface, or an inclusion, or a dislocation line, recourse to section topographs is necessary. Here we will just go over three matters of practical importance for the production of section topographs which will give the information required and exhibit acceptable quality. These are (a) the resolution limits imposed by axial divergence and by dispersion, (b) the most appropriate choice of specimen orientation, and (c) the necessary adjustments to secure uniform reflection by the specimen.

Topic (a) has been adequately treated in [2] and [3]: recapitulation here can serve as a reminder of the basic diffraction geometry of the section topograph technique shown in Fig. 1. This diagram is a plan view of the plane of incidence and diffraction, the goniometer axis being perpendicular to the plane of the drawing.

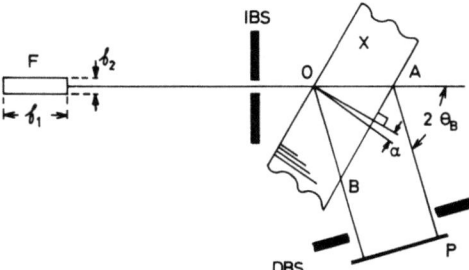

Figure 1. Basic diffraction geometry of section topograph technique. Symbols explained in text. The Bragg planes make angle $90^\circ - \alpha$ with the surfaces of the parallel-sided specimen, X

F represents the X-ray tube focal spot which is assumed to be a well-defined rectangle with major and minor edges of lengths f_1 and f_2 respectively. This rectangle is viewed 'end-on' at a mean 'take-off' angle τ. The slit assembly providing collimation of the incident beam is IBS, the specimen crystal (assumed to be plate-shaped) is X, the diffracted beam (deviated by angle $2\theta_B$ from the incident beam) passes through the diffracted beam slit, DBS, and falls perpendicularly on the photographic plate, P. Denote the distance FX by a, and the distance XP by b. As a consequence of divergence in the plane containing the goniometer axis, the image of a point in the specimen will be drawn out in this plane to a length $(b/a)f_1 \sin\tau$. Note that if the axial dimension of the specimen subtends some degrees at F, quite easily detectable differences in geometrical resolution may exist between parts of the image formed with τ lower and τ higher than the mean take-off angle. Also note that 'axial divergence' is a better term than the more familiar term 'vertical divergence' in view of the fact that a horizontal orientation of the goniometer axis will be favoured when synchrotron radiation sources are used. Typical laboratory values of f_1, f_2, a and τ are 2 mm, 0.2mm, 0.5 m and $3°$. With these values, b must be kept less than 7mm to secure a resolution in the axial plane not worse than $1\frac{1}{2}\mu$m. Such relatively small values of b are easily accommodated if the specimen is not large, and both specimen mount and plate casette are carefully designed

Dispersion introduces streaking of the image in the plane of incidence. Often other factors, largely unavoidable, such as relatively great distance of a particular defect from the X-ray exit surface AB together with a wide 'energy-flow fan', AOB, dominate in producing streaking in this plane; but it is wise to check that the contribution of dispersion is not excessive. For example, suppose $CuK\alpha_1$ radiation is used and θ_B is $45°$. Adopting an acceptable value for the full width at half maximum intensity of α_1 emission line profile, we derive a measure of relative spread of wavelength, $\Delta\lambda/\lambda = 3 \times 10^{-4}$. Then one sees that to keep the image spread due to dispersion less than $1\frac{1}{2}\mu$m b must not exceed 5 mm. Note that if IBS is narrow, say 10 µm wide, the angular range of rays incident on the specimen is effectively limited by the value of (f_2/a), i.e. to 4×10^{-4}. Hence a reduction in f_2 or increase in a, combined with a stable experimental set-up, can produce a useful truncation of the range of wavelengths reflected by the specimen when θ_B is relatively high. However, a recently developed technique [10] provides both a spatially very narrow incident beam for section topography and also a beneficial limitation of the wavelength range effective in the reflection. This is achieved by using an asymmetrically reflecting flat crystal monochromator to function as spatial beam compressor in place of a narrow IBS.

Figs. 2 and 3 relate to the topic (b), and illustrate the choices open to the experimenter when taking section topographs in asymmetric transmission, using reflecting plane inclined at angle α to the normal to the specimen (which is assumed to be parallel-sided, or nearly so). We take α to be positive in Fig. 2 and negative in Fig. 3; and find for the width of the section topograph image [11] the value $t\sec(\theta_B-\alpha)\sin2\theta_B$, for which expression $2t\sin\theta_B$ is often an adequate approximation when $|\alpha|$ is small. It follows that if we want a section topograph in which the incident beam makes a cut through the crystal as near as possible perpendicular to its surfaces we go for the geometry of Fig. 2, and look for reflecting planes which most nearly satisfy the condition $\alpha = \theta_B$, (θ_B taken as positive with the sense of deviation shown in Figs. 2 and 3). On the other hand, if we want to proceed to projection topographs which give a perpendicular view of the plate-shaped specimen we opt for the geometry of Fig. 3, and look for reflecting planes which approach the condition $\alpha = -\theta_B$. When analyzing the geometry of internal defects in crystals, topographers should consider carefully the choice of α, and avoid situations in which the geometry of intersection of OA with the defect concerned is ill-conditioned. For accurate measurements, remember that the magnification of the image in the axial direction is $(b + a)/a$ which may exceed unity by several percent in the case of laboratory sources but is not likely to exceed unity by a significant amount when synchrotron sources are used.

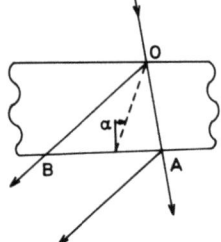

Figure 2. Section topograph geometry with positive angle α

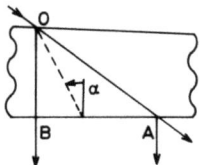

Figure 3. Section topograph geometry with negative angle α

We will discuss topic (c) in two parts, dealing first with the problem of a crystal which is well-aligned, its g-vector lying accurately in the plane normal to the goniometer axis, i.e. the equatorial plane, but which has an overall height H (say 0.1 m) which is not small compared with a typical laboratory value of a, say 0.5 m. The stereographic projection in Fig. 4 refers to this situation. The centre of the projection is the mean incident-beam direction, and the diameter PP' is the equatorial plane of the goniometer. The rectangle S, drawn with interrupted lines, encompasses the directions of all rays that fall upon the crystal:

its width, w, is determined by the range of cross-fibre between the focus width f_2 and the width of IBS (Fig. 1), its height is determined by the specimen height and corresponds to the angle H/a. All incident and diffracted rays must lie on a 'diffraction cone' whose axis is the diffraction vector G. (located in the remote hemisphere) and whose semi-angle is $(\pi/2) - \theta_B$. The shaded rectangle, D, represents the range of directions of all possible diffracted beams. In the sketch, the cone generators corresponding to incident directions are shown all lying comfortably within the rectangle S, and the entire height of the crystal can reflect. However, if f_2 is small and IBS has to be made narrow in order to get good topographic resolution in section topographs, there can arise the situation when the projection of the sagitta corresponding to chord length H of the cone base circle (whose radius is $a.\cos\theta_B$) exceeds w. This condition arises when $(H/2)^2$ exceeds $2aw.\cot\theta_B$. Thus the maximum specimen height that can reflect, H_{max} is proportional to $(aw \cot\theta_B)^{\frac{1}{2}}$. It would appear that high diffraction angles and uniform reflection cannot be simultaneously achieved. However, in practice the natural wavelength spread may rescue the situation, for it endows the diffraction cone with a finite range of angles. This range will be proportional to $(\Delta\lambda/\lambda)\tan\theta_B$, and (as indicated earlier) may exceed w when θ_B is high. Then H_{max} will be almost independent of θ_B, becoming proportional to $[(\Delta\lambda/\lambda)a]^{\frac{1}{2}}$. Here the advantages of a synchrotron radiation source for examining specimens with large H are manifest: it provides both a very large a and an effectively unlimited $\Delta\lambda$.

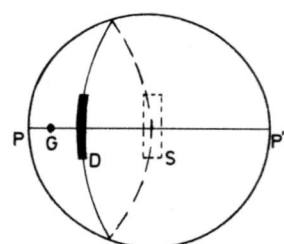

Figure 4. Stereographic projection, case of well-adjusted crystal (see text)

The second part of our discussion of topic (c) concerns the everyday problem of setting the g-vector in the equatorial plane. This operation has to be performed frequently, but not for every topograph, for if an important zone axis is set parallel to the goniometer axis then all planes in that zone will be properly aligned. (This provides a good reason for adopting such a setting, in addition to the benefits it bestows in respect of easier crystallographic interpretability of topograph images.) The stereographic projection in Fig. 5 shows G lying out of the equatorial plane. The median incident ray and G together determine the plane of incidence and diffraction, represented by the diametrical plane QQ . The diffraction cone now falls within only a fraction of the height of S, the diffracted beam is incomplete, and is represented by the shaded band in D. (In practice, there exist two coaxial cones, with openings appropriate to the α_1 and α_2 wavelengths, so two bands may appear in D when the tilt of QQ is relatively high. Indeed from the separation of the bands the tilt may be calculated.) Let γ_i, γ_g and γ_d be the angles made with the equatorial plane by the plane of incidence QQ′, by $\underline{g}$ and by the diffracted beam, respectively. Then $\gamma_g = \gamma_i \cos\theta_B$ and $\gamma_d = \gamma_i \sin 2\theta_B$, from which there follows $\gamma_d = 2\gamma_g \sin\theta_B$. Hence as γ_d is reduced to zero, so is γ_g. Given a goniometer whose detector-carrying arm is topped by a track in the manner of an optical bench, it is an easy matter to withdraw the detector to some distance away from the specimen and then run a screen, whose height is adjustable, back and forth along the track. When the horizontal edge of the screen is at the median beam height, and obscures the same fraction of the diffracted beam whatever its distance from the specimen, one knows that $\gamma_d = 0$, and hence $\gamma_g = 0$. As can be seen from the formulae, the sensitivity of this adjusting process increases with increasing θ_B. A net saving of time may sometimes be achieved by performing the adjustment on a higher order of the reflection used in the topograph. Indeed, for accurate final adjustment of a zone axis parallel to the goniometer axis, the use of higher-order reflections is recommended. The manipulations just described may be remotely controlled for safety reasons, if required.

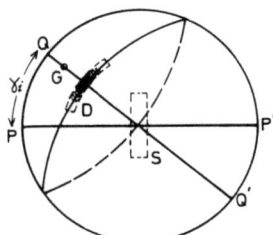

Figure 5. Stereographic projection similar to Fig. 4. Case of maladjusted crystal (see text)

7.2.2 Projection topographs

Proper specimen adjustment for a section topograph is a good preliminary to taking a projection topograph, but in the latter experiment the topographer will probably have to tolerate some loss of resolution because of the need to increase the distance b. This increase may be a geometric necessity when the traverse range is long and θ_B relatively high, in order to avoid collision of either the specimen or the plate cassette with the diffracted beam slits DBS. Two other reasons for increasing b will now be mentioned. Background scatter on a projection topograph will be at least somewhat greater than on a section topograph (in the ratio of the width DBS to the width of the section image, plus some allowance for the greater width of the IBS), and the contrast of the projection topograph is intrinsically lower than that of the section topograph. However, the photographic density due to Compton scatter and fluorescent radiation will be proportional to the solid angle subtended by the emitting regions of the specimen at the point on the plate P. It will fall off faster than linearly with increase in b, so a larger b gives improvement in signal-to-background ratio.

A second need to increase b may arise when unwanted reflected beams get through the DBS and produce streaks across the image. Usually there will be a pair of streaks corresponding to the α_1, and α_2 reflected beams from the unwanted reflection whose g-vector is inclined to the equatorial plane (like G in Fig. 5) and whose diffraction cones by chance fall across S. As the figure shows, the unwanted diffracted beams are directed above (or below) the equatorial plane. Hence an increase in b will always raise (or lower) them, and can be employed to shift them away from regions of interest on the topograph. In practice it usually happens that a small increase in b suffices to throw them outside the aperture of the DBS. Also, a judicious rotation of the specimen about the principal g-vector (i.e. that used for taking the topograph) may be effective in turning the diffraction cones of the unwanted reflection away from S. Accompanying strong accidental reflections from planes whose diffraction vectors are inclined to the equatorialplane are 'Aufhellungen' on the topograph image [12]. These are seen on both section and projection topograph images, and are especially numerous and trouble- some in the case of crystals with large unit cells, such as crystals of organic compounds. At first glance, Aufhellungen look like the undesirable artefacts that arise from dirt or irregularities in the IBS. Rotation about the principal g-vector will shift the former, but not, alas, the latter. In defence of Aufhellungen, recall that interest resides in their interaction with Pendellosung fringes, and that analysis of such interaction led to the breaking of a 50-year old taboo by showing that phases of X-ray reflections could be determined directly by a physical method not involving anomalous dispersion [13].

The limited projection topograph [2, 14] can be a useful method when dealing with relatively thick, X-ray transparent crystals, and has proved its value in correlating dislocation outcrops with etch-pits [15]. However, it must be remembered that to achieve good definition of sampling depth in limited projection topographs, the IBS must be made as narrow as it would be to obtain correspondingly good depth resolution in a section topograph. This condition, combined with the fact that only part of the diffracted beam is received on the limited projection topograph makes exposure times with this technique long compared with those involved in standard projection topograph experiments. For the latter, IBS widths may be increased to the limit imposed by the condition that no significant intensity of α_2 image finds its way on to the α_1 image.

7.2.3 Scanning reflection topographs

With laboratory radiation sources and collimated incident beams, means for traversing the specimen across the incident beam are necessary if substantial specimens are used and freedom to employ large as well as small grazing angles of incidence upon the specimen surface is essential. The choice of scanning direction is less confined than in the transmission case; nor are there overriding reasons for traversing both specimen and photographic plate rigidly together. Here we just describe the two limiting cases, that in which the plate, P, is stationary (Fig. 5), and that in which specimen and plate move together (symbolized by the enclosing rectangle in Fig. 7). The system shown in Fig. 6 permits

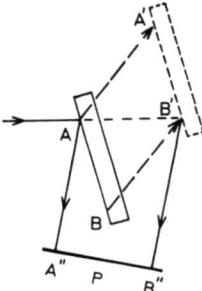

Figure 6. Scanning reflexion topograph technique producing unit horizontal magnification (explained in text)

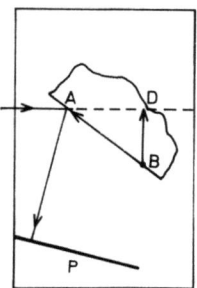

Figure 7. Scanning reflexion topograph technique with rigid translation together of specimen and plate (explained in text)

control of horizontal magnification. Suppose it is desired to scan the range from A to B on the specimen surface. The direction of specimen motion may be chosen, as shown with AB moving to A'B', so that the length A"B" on the image is the same as AB. Note however, that to maintain good topographic resolution when AA' and A"A are not colinear, the incident beam must be kept very narrow. This condition was fulfilled fairly well when the technique was used for mapping growth lineages and their mutual misorientations [16]. The spatial width of the incident beam does not limit topographic resolution with the arrangement shown in Fig. 7. (unless it is so wide that some α_2 or much white radiation is reflected). The travel direction chosen is not likely to fall outside the range between B´A´ and B´D´. The former, being parallel to BA, keeps the diffracted beam stationary in space and permits use of the customary, stationary diffracted beam slit. The latter direction, being perpendicular to the incident beam, is the shortest traverse providing scanning of the length AB on the crystal surface; and when very large specimens are handled, circumstances may compel this choice. For traverses other than parallel to BA, the diffracted beam slit must either be much widened, removed altogether, or itself translated at an appropriate rate. Scanning reflection topographs have been useful in studying abraded crystal surfaces [17] and in correlating X-ray and cathodoluminescence images of individual dislocations [18]. There are many problems in which the experimenter must be prepared to employ several techniques on the same specimen: section topographs, projection topographs, scanning reflection topographs, and 'stills' taken in the last-mentioned arrangement. These stills we term reflection section topographs.

7.2.4 Stereo methods

Take two topographs of a specimen (either by the transmission projection method or by the scanning reflection method) in which the diffracted beams leave the specimen in different directions. Then the two views of the specimen recorded on plates placed perpendicular to the diffracted beam should contain all the information needed to reconstruct the three-dimensional geometry of defects in the crystal, provided that the visibility and type of diffraction contrast exhibited by the defects is not substantially different in the two images. The latter condition generally, rules out the possibility of using Bragg reflections from <u>different</u> planes, hkl and h´k´l´, to form the two images. Under a certain range of absorption conditions (say μt equals a few units) it also makes difficult the derivation of the required geometrical information from a given reflection and its inverse, hkl and $\bar{h}\bar{k}\bar{l}$. In practice, the X-ray topographer, whatever techniques of photogrammetry he may have at his disposal and may wish to use to obtain exact measurements, desires to produce stereopairs of images which he can view comfortably. Then he can enjoy the fascination that direct, three-dimensional visualization affords. The important quantity to get right is the convergence angle. Correct choice of this quantity depends upon the size of the objects whose depth variation is to be resolved, the specimen thickness and the magnification at which the topographs are viewed. Here we will not attempt quantitative analysis of the problem, but will just offer a few hints based on practical experience. The two stereo methods in common use are that of hkl, $\bar{h}\bar{k}\bar{l}$ pairs [19] and that due to Haruta [20] in which the two views of the specimen are obtained with the same Bragg reflection, but with the specimen rotated about the g-vector. The hkl, h̄k̄l̄ method has two disadvantages, firstly that the convergence angle (which is $2\theta_B$) may be either too small or (more likely) too large, and secondly that the so-called 'failure of Friedel's Law' in diffraction contrast may render the left-eye view qualitatively too dissimilar. Haruta's method has more flexibility, but can be inconvenient with large plate-shaped specimens when more than a few degrees rotation is needed. Here are the practical hints. Sensitivity of course is increased by increasing the parallax, but the eyes cannot take in comfortably more than a certain maximum parallax. To select the convergence angle, take a partly transparent, three-dimensional object (a crystal structure model serves excellently). Choose certain near and far objects in the model whose difference in depth corresponds to that of, say, the front and back surfaces of the crystal specimen viewed at the magnification it is desired to use when examining the topograph. Hold the model at a distance from the eyes that gives comfortable three-dimensional viewing between these near and far objects, and measure the angle subtended by the interocular distance at the midpoint between the near and far objects. This angle is an appropriate convergence angle. With luck, it may not differ much from one or more of the $2\theta_B$ values to be used in the topographic survey of the specimen. An alternative, rather more precise procedure is available if one or more topographs are already available. Recognise defects which are near and far (this can be done on the basis of certain features of the images they exhibit, as discussed in Section 7.3) and visualize a suitable (or perhaps a maximum) relative shift between them that would be appropriate in the stereopair. This shift divided by the specimen thickness roughly gives the angle of convergence needed. Following this procedure in the case of the thick (1½ mm) diamond specimen whose projection topograph is shown in Fig. 12, 7° was estimated as a suitable convergence angle for stereo. Since this angle was well below any $2\theta_B$ available, the Haruta method was used. On the other hand, the LiF slice, fields from whose 020, 0̄2̄0 stereopair are shown in Fig. 8 and 9, respectively, was so thin (~50μm) that the convergence angle determined by $2\theta_B$ for $CuK\alpha_1$, 45°, was not excessive, even when the topographs were viewed at magnifications as high at 100. For reflection topographs the hkl, h̄k̄l̄ method is inapplicable; and when using the Haruta method with large Bragg angles it must be remembered that to obtain a convergence angle γ, the rotation about g needed is $\gamma \sec\theta_B$.

Figure 8 (overleaf). Projection topograph of (001) cleavage plate of LiF, chemically polished, approx. 50 μm thick. Reflexion 0̄2̄0, $CuK\alpha_1$ radiation. Arrow, 0.4 mm long, shows direction of g-vector.

Figure 9 (overleaf). Stereo pair topograph with topograph shown in Fig. 8. Inverse reflexion, 020, $CuK\alpha_1$ radiation. Arrow shows direction of g-vector.

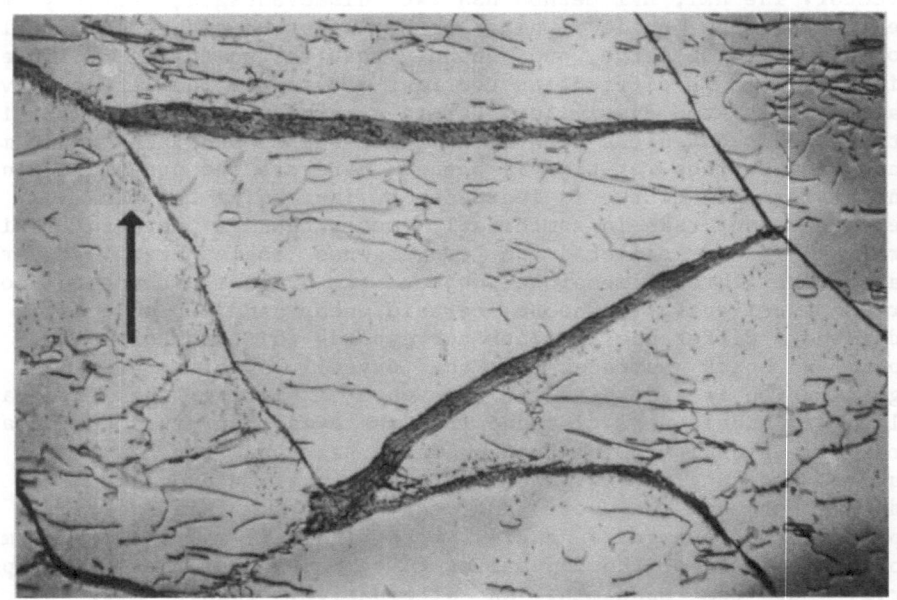

Fig 8

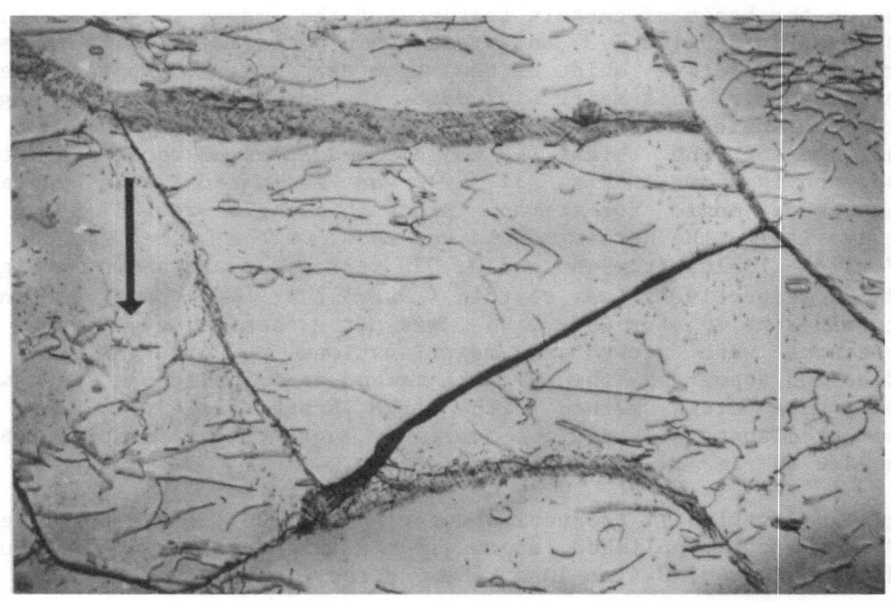

Fig 9

DEFECT VISUALISATION: INDIVIDUAL DEFECTS

7.3 Interpretation of X-ray topographs

7.3.1 Types of defect individually visualizable

We shall refer to the six topographs, Figs. 8-14, to demonstrate the X-ray diffraction contrast images of some important defects which can be seen individually by X-ray topographic methods, and shall commence with a bald list of defects to be considered. These are: individual dislocations, low angle boundaries, precipitates, inclusions, specimen surface damage, stacking faults, growth-sector boundaries, and growth banding. (The last-mentioned feature can also be called impurity zoning, a term used by mineralogists.) In this lecture we will go into details only to the extent necessary for correctly interpreting the feature seen on the topograph image, and for understanding its geometry. Topographs Figs. 8-11 are of a chemically polished thin cleaved plate taken from a region of fairly low dislocation density in a large boule of LiF grown about 20 years ago at the university of Aberdeen. The customary topograph orientation procedure is followed: the topograph is viewed looking towards the X-ray source, and its magnified image is printed to give this same direction of view. Thus, in Figs. 8-11, having indexed the X-ray exit face of the specimen (001), [001] points out of the print towards the observer. The arrow on the prints gives the projections of the g-vector, and its length corresponds to 0.4 mm on the topograph. The 020, 0$\bar{2}$0 pair, Figs. 8 and 9 were taken in symmetrical transmission: hence with $\cos \theta_B = 0.92$, the topograph image of the specimen has a small lateral compression by this factor. Figs. 10 and 11 are transmission topographs of the same area taken in the 1$\bar{1}$1 reflection with CuKα and MoKα_1 respectively. The angle α, $35\frac{1}{4}°$, is negative, according to our convention, and the image is compressed in the direction parallel to the projection of $\underline{g}$ by $\cos(\theta_B - \alpha)$ i.e. by the factors 0.96 and 0.95 for the Cu and Mo cases, respectively.

Fig. 12 is a 1$\bar{1}$1 CuKα_1, projection topograph of a whole natural diamond whose habit was a flattened octahedron. The large pair of faces were indexed (111) and ($\bar{1}\bar{1}\bar{1}$), and the [$\bar{1}$10] axis was set parallel to the goniometer axis. The ($\bar{1}\bar{1}\bar{1}$) face was the X-ray entrance surface for this projection topograph and also for the section topographs Figs. 13 and 14 which used the same reflection, in the same geometry. The angle $\alpha = 19\frac{1}{2}°$ is negative, and is so close to $\theta_B (=21.95°)$ that the projection topograph is a virtually undistorted view of the crystal perpendicular to (111).

Figure 10 (overleaf). Same area of LiF specimen as in Figs. 8 and 9. Reflexion 1$\bar{1}$1 in asymmetric transmission. CuKα_1 radiation. Arrow shows projection of g-vector.
Figure 11 (overleaf). Same reflexion as in Fig. 10, MoKα_1 radiation.

Fig 10

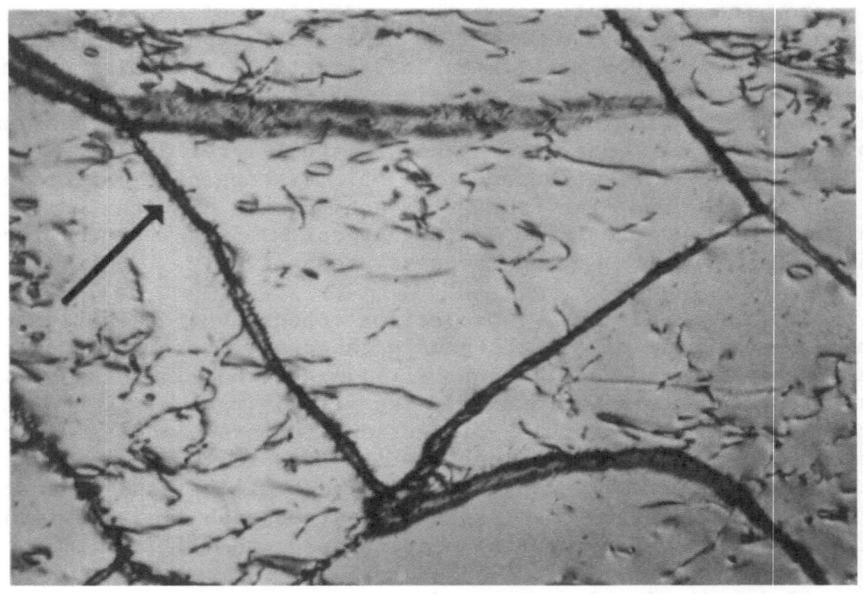

Fig 11

DEFECT VISUALISATION: INDIVIDUAL DEFECTS

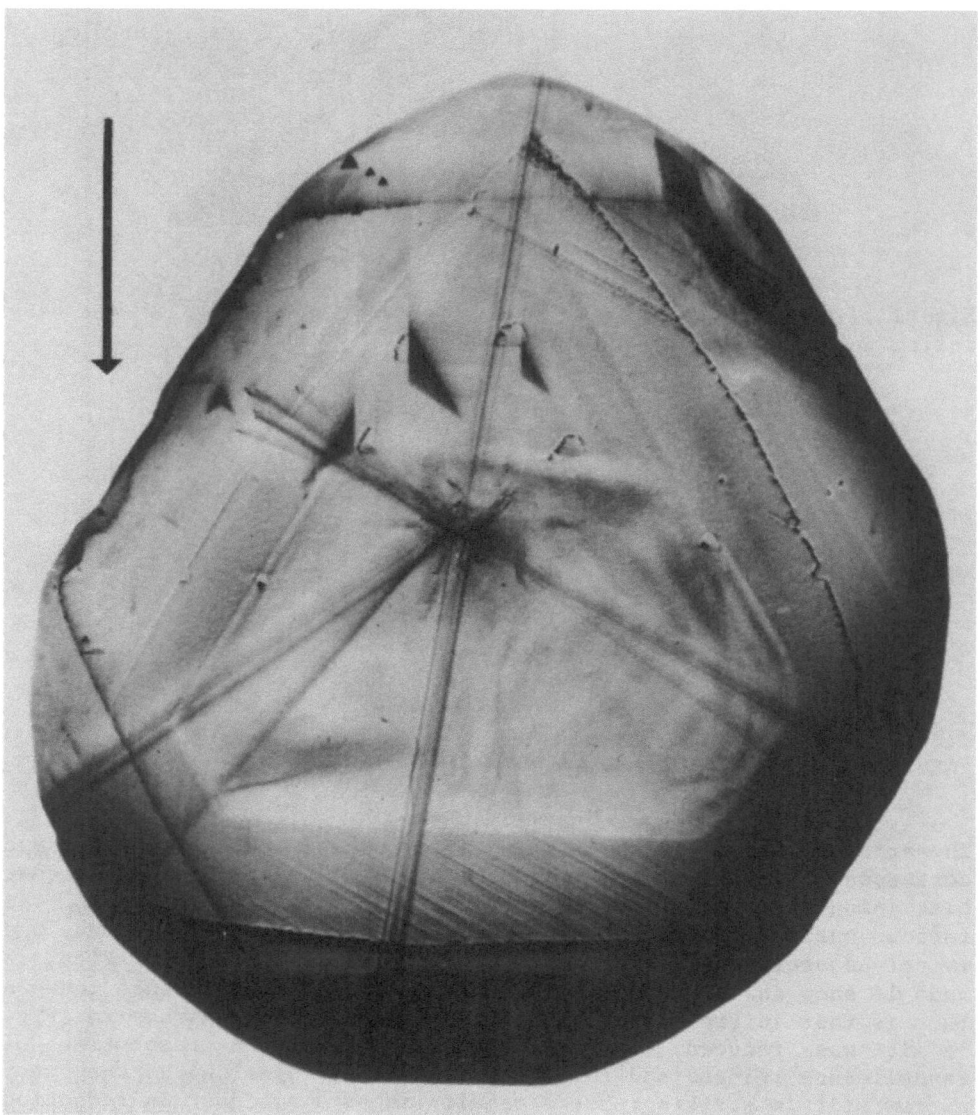

Figure 12. Projection topograph of whole natural diamond. View nearly ⊥ to $(\bar{1}\bar{1}\bar{1})$. $g=11\bar{1}$, $CuK\alpha_1$, arrow (1.0 mm) is projection of g.

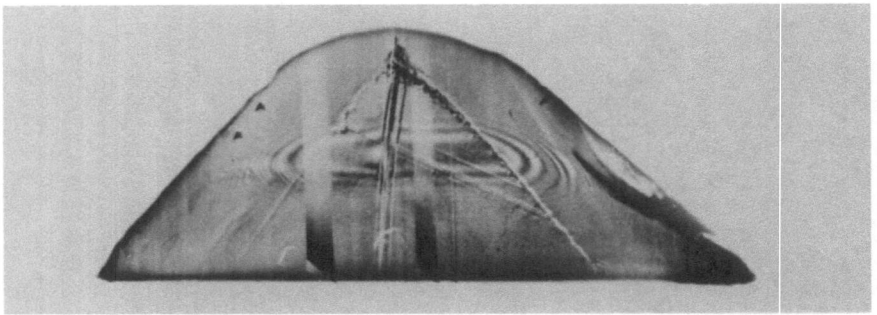

Figure 13. Section topograph through specimen shown in Fig. 12. Same reflexion and radiation. Incident beam slit 25 μm wide.

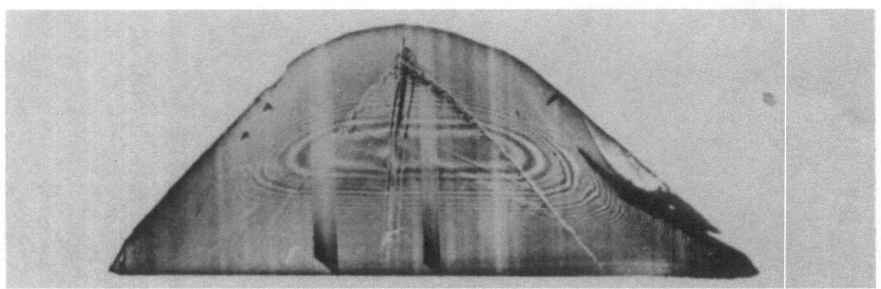

Figure 14. Similar to Fig. 13 but with incident beam slit 7 μm wide

The arrow on Fig. 12 shows the projection of the g-vector, and corresponds to 1 mm on the topograph. The region of intense blackening at one limit of the projection topograph image is the intense surface reflection from the (111) facet of the specimen, and is not an accidental defect of the topograph. The topograph Figs. 13 and 14 show the same section of the crystal. The difference between them is that in Fig. 13 an IBS 25 μm wide was used whereas in Fig. 14 it was reduced to 7μm. The improvement in resolution of Pendellosung fringes and in stacking fault fringes with the very narrow slit is striking; but a penalty is paid not just in reduction of intensity but also in the streaking introduced by irregularities and dirt in the slit which are extremely hard to avoid with such small openings.

7.3.2 Effects of specimen thickness on defect contrast

One cannot clearly observe diffraction contrast (i.e. extinction contrast) from defects such as dislocations when the specimen is less than about 1/4 to 1/3 times ξ_g in thickness: this

DEFECT VISUALISATION: INDIVIDUAL DEFECTS

is concluded from observations on tapering specimens of MgO and of iron- silicon alloy studied with radiations of various wavelengths. One can still observe orientation contrast at lesser thicknesses: sensitivity for detecting the latter will only deteriorate when the crystal is so thin that 'diffraction broadening' according to kinematic diffraction theory is apparent. The diffraction broadening from a crystal of thickness t is given by the old Scherrer formula: $\Delta(2\theta) = K\lambda/t\cos\theta_B$, in which K, the Scherrer constant, is about unity. In the zero-absorption, symmetrical Laue case, the first Pendellosung fringe maximum occurs at t = 0.382 ξ_g and the first minimum at t = $0.878\xi_g$. For further data on intensities and positions of fringe extrema see [2] and [8]. Counting up Pendellosung fringes from the first maximum (or any known fringe order) provides a precise, non-destructive method of determining local specimen thickness that can be strongly recommended to topographers. When counting large numbers of fringes, the periodic polarization fade, or 'beat', [2,8] provides a short cut to finding the fringe order. The beat is well seen in Fig. 14 The strongest positive dislocation contrast is obtained when the specimen thickness corresponds to the first Pendellosung fringe minimum. The consequence of increasing specimen thickness as regards the contrast and resolution of images of defects can be roughly divided into those arising just from the widening of the base of the energy-flow triangle AOB in Fig.1, and those arising from the Borrmann effect; for in principle these are independent. Most attention is usually paid to the Borrmann effect probably because topographers have concentrated so much on crystals with simple structures and whose lattices can be highly perfect except for localized defects such as dislocations and precipitates. The measure of possible importance of the Borrmann effect is the ratio of the Fourier coefficient of order g to that of the zero order of the imaginary part of the structure factor: $\varepsilon_o = F''_g/F''_o$. This can be close to unity for even- ordered allowed reflections in diamond structures, and in all reflections of structures containing one atom per primitive unit cell; but (for example) in certain reflections in alpha quartz, and in II-VI compounds when diffracting particular wavelengths, ε_o may be reduced to a small fraction. Thus we must remember that when topographers refer to 'contrast reversal when μt is large', they really mean 'when $\varepsilon\mu t$ is large' ($\varepsilon = \varepsilon_o \exp(-M)$ where M is the Debye- Waller factor.) However, one may use the phrase 'under Borrmann conditions' to signify the case of large $\varepsilon\mu t$. In Figs. 8-11 the strong positive diffraction contrast expected from the low absorption conditions (μt= 0.16 for CuKα) is evident. The absorption due to fluorine is so much greater than that due to lithium that ε_o is little less in the 111 reflections than it is in the 220 reflections; and for all low-order reflections ε is close to unity. In Figs. 12-14 the value of μt is 2.5 but the images in these topographs show very little manifestation of the Borrmann effect. This is due partly to the lower value of ε_o for odd reflections in the diamond structure (it cannot exceed $1/\sqrt{2}$) and also to its additional reduction (by an

177

unknown amount) arising from unresolved lattice imperfections having dimensions and mutual separations small compared with X-ray extinction distances. What Fig. 12 does show markedly is the loss of definition and contrast of images of defects remote from the X-ray exit surface that arises from diffusion of the image within the energy flow triangle in this thick specimen. Damage on the exit face gives sharp images, that on the entrance face, very diffuse images. The stacking faults (the dark triangles seen in Fig. 12) outcrop on the X-ray entrance surface. Because of their oblique orientation to the plane of incidence, the stacking fault fringes are all smeared out and lost on the projection topograph. Section topographs are needed to reveal them in this diffraction geometry, and they are well seen in Fig. 14 which was taken with the very narrow IBS.

7.3.3 Dislocations and low-angle boundaries

Figs. 8-11 show dislocation configurations typical of melt-grown crystals which have been slowly cooled, whereas Fig. 12 is typical of the configurations found in crystals grown from flux or solution. In Fig. 12 the majority of dislocations radiate (in extraordinarily straight lines) from a central nucleus. This suggests heterogeneous nucleation of the crystal. Topographers may be called upon to answer several questions concerning the crystals they examine, such as: (a) what is the dislocation density, (b) what are the dislocation line directions, (c) what are the Burgers vectors, (d) are the dislocations 'clean' or 'decorated', and (e) what is the origin of the dislocation. We will try and offer guidance on how to set about answering these questions. Firstly, as regards dislocation density, this is usually defined as number of lines crossing unit area, or (rather better) as total line length per unit volume. A topograph such as Fig. 12 shows an example when it is impossible to give any reasonable answer other than 'very low'. Enclose the nucleus by a sphere of radius 100 µm, say. The dislocation density within this sphere is about 500 lines mm^{-2}. Take the external surface of the crystal instead, and the density drops to about 2 lines mm^{-2}. On the other hand, the LiF crystal can be handled more quantitatively: we have to find the dislocation density within sub-grains, and in the low-angle boundaries. Now, from stereopairs or Pendellosung fringe measurements, we know the specimen thickness. A pair of orthogonal cube-plane reflexions, or three of the 111 reflexions should show all Burgers vectors of type $\frac{1}{2}<110>$ as occur in LiF. Then, either manually or by an image analyser, the total line length of individually resolved dislocations in a given volume of specimen can be computed. Next we consider the low-angle boundaries. We need the total area of low-angle boundary per unit volume, and the average dislocation density in a low-angle boundary. In favourable cases, when dislocation content is very low, the individual dislocations in a low-angle boundary can be individually resolved. (The boundary that runs parallel to the long axis of the field shown in Figs. 8 - 11 is a good example.) If the misorientation at a low-angle boundary is more than about 0.1 milliradian, a series of topographs taken with a very narrow slit-collimated beam will give the data from which the distribution of misorientations, and from this the mean misorientation, can be found. With smaller average misorientations, double-crystal topographic methods are needed.

On question (b), concerning line orientations, the X-ray topographic method is a powerful tool. Consider a simple example. In the LiF specimen we can recognise some fairly straight dislocations which run from X-ray exit surface to X-ray entrance surface.(The former outcrop appears sharp, the latter more diffuse, because of the geometrical factors discussed in 3.2.) Co-ordinates of these outcrops relative to a fixed landmark in the crystal can be measured on two or more topographs. Then computation or a purely graphical procedure enables the orientation of the line joining these two outcrops to be found. How accurate can such determination be made? The specimen shown in Fig. 12 provides an opportunity for precise measurement. Some lines run straight for over 2 mm (as far as straightness can be judged within the resolution of the X-ray method). By using section topographs with α positive, measurements of distance of any of these lines from the exit or entrance face can be measured to within 5 - 10 µm. A range of 10 µm on a base length of 2 mm corresponds to an angular range of $\frac{1}{4}°$. However, to translate this into crystallographic measurement of corresponding precision requires that the specimen have facets or edges of equally well defined orientation. In their absence, more elaborate X-ray goniometry is needed to establish reference crystallographic directions within the specimen. This is tedious but is perfectly feasible.

Question (c), on Burgers vector determination, falls outside the scope of the chapter, and question (d) will be considered in section 7.3.5. Question (e) deserves a treatise to itself: all that need be said here is that when a 'grown-in' origin, as exemplified by Fig.12, is not obvious then consideration of anisotropy in distribution of dislocation line directions (as would be found in answering (b)) and of anisotropy in distribution of Burgers vectors (discovered in answering (c)) may provide helpful clues towards solving question (e).

A final question concerns the maximum dislocation density at which individuals may be resolvable. As has been discussed in [2] and elsewhere, the apparent width of the intense diffraction contrast image of a dislocation in the low-absorption case is proportional to the extinction distance (keeping $\underline{g} \cdot \underline{b}$ constant). Extinction distances for the reflections shown in the Fig. 8,9 pair, in Fig. 10 and in Fig. 11 increase in the ratio 1: 1.56: 3.53, approximately. Even from looking only at photographic prints, one can see that, on average, the apparent dislocation image widths increase pro rata. It follows that for highest resolution of dislocations one tries to use short extinction distances. It also follows (from both geometrical and contrast considerations) that a section topograph taken with a narrow beam will give higher resolution than a projection topograph, even with specimens as thin as the LiF sample here described.

7.3.4 Surface damage

We shall interpret the term surface damage broadly, so as to include all sources of inhomogeneous strain at specimen surfaces which attain a magnitude sufficient to produce detectable X-ray diffraction contrast. Thus, besides mechanical damage, we include strains due to a 'skin', partially or completely covering the surface, in which chemical composition differs from the bulk crystal. The types of mechanical damage usually encountered can be classified into gross damage, consisting of random scratches, percussion damage (which produces ring cracks), and cracks generally, on the one hand; and, on the other hand, systematic, fine-scale damage resulting from saw damage or polishing damage (including such damage that has been incompletely removed by etching). This classification is made because in the case of systematic, fine-scale damage, such as that caused by parallel polishing striae, the stresses produced in the underlying material by adjacent striae oppose each other, and lattice curvature on a scale sufficient to produce significant diffraction contrast is not generated within the polished area although the polished surface is in a state of compressional stress. However, at edges of the specimen where two polished surfaces meet, or a polished and unpolished surface meet, the surface stress is incompletely supported, relaxation occurs giving local lattice curvature that produces strong diffraction contrast. Such contrast appears at edges of polished facets on diamonds, even when little or no contrast from polishing is seen on the facets themselves. Similar phenomena are associated with an incomplete, or non-uniform, chemical skin. The skin may arise accidentally, i.e. by local oxidation promoted by ionisation produced by an intense X-ray beam, or it may be deliberate, i.e. the familiar oxide masks on silicon. One must remember that in the case of a uniform skin adhering coherently to the surface, there is no deformation <u>parallel</u> to the surface. Hence no contrast is produced in symmetric Laue case diffraction. There can be, indeed probably will be, a gradient in cell dimensions normal to the surface. This will produce contrast in asymmetric transmission, or in surface reflection. At the edges of the skin, one will expect contrast in Bragg reflections from all planes except that normal to the edge, this exception arising because the elastic situation is one of plane strain in the plane normal to the edge of the skin. From the sense of Friedel's Law departure (which will be observed over quite a wide range of $\varepsilon\mu t$ values, from about 0.5 to 5, say), the sense of lattice curvature at the skin edge can be determined, and hence information as to whether the skin is under tensional or compressional stress. Both simple [17] and the more sophisticated [8] calculations of the strength of diffraction contrast can be used to find the magnitude of the stress.

Returning to gross mechanical damage, in the case of the surface scratch (of which regrettably many examples appear in the X-ray topographic literature) one finds it very hard, if not impossible, to distinguish by X-ray topography whether the deformation has been plastic or by fracture. (In the latter case, the scratch consists of a chain of incomplete ring cracks.) Whether performed by dislocation motion or by crack opening, the dominant displacement of material is a parting away from the median line of the scratch. In reflections in which $\underline{g}$ lies in (or nearly in) the surface, and also makes a large angle with the scratch, a strong double line of contrast will be seen (the two lines equally strong when $\varepsilon\mu t$ is small). When $\underline{g}$ is parallel to the scratch, the diffraction contrast will vanish to the extent that (averaged over distances of a few tenths of ξ_g) the lattice deformation is limited to plane strain in the plane perpendicular to the scratch.

On Figs. 8 - 11 we see no signs of surface damage, on Fig. 12 we see many. In an attempt to remove the severe naturally produced damage from the surfaces of this specimen, the (111) and ($\bar{1}\bar{1}\bar{1}$) faces had been both polished and etched. Polishing striae are seen, in particular near the edge between (111) and (11$\bar{1}$). The edges of the (111) face stand out clearly because they retain the chains of little cracks that nature has produced in the course of burring and chamfering these edges. Note how the projection topograph and section topograph images of these edges can be correlated. The specimen also contains one crack approaching millimetre dimension. This obtrudes in one of the tapering parts of the section topograph image.

7.3.5 Precipitates and inclusions

If a precipitate or inclusion (be it solid, liquid or gaseous) does not deform the matrix then it will be invisible on X-ray topographs unless it be of such size and composition that it shows up by absorption contrast, or of such shape and size as to produce phase contrast. (Note that in X-ray topography we are always working under phase-contrast conditions). For inclusions and precipitates that do deform the matrix there are many similarities with the contrast phenomena observed in electron microscopy. However, the extreme sensitivity to misorientation that occurs in the case of Bragg- diffracting nearly perfect crystals scales up the X-ray diffraction contrast image of a given strainfield by a factor between 10^3 and 10^4 compared with the electron case. It follows that the true size of an inclusion or precipitate will in general be impossible to determine from an X-ray topographic image. However, the topographic context can usefully assist in establishing whether the matrix-deforming body is an inclusion incorporated during growth, or a precipitate formed after growth. If dislocations originate at the body and proceed in a bundle (often gently diverging) towards the local growth face, then it can be assumed that the body is an inclusion, and that the dislocations were generated by lattice-closure errors when it was engulfed in the crystal. But by no means all inclusions generate dislocations. On the other hand, the decoration of dislocations is such a common phenomenon that when (as frequently occurs) we observe diffraction-contrast producing bodies strung out along dislocation lines it is reasonable to call them precipitates. But we must not be dogmatic about this: under certain conditions on a growing crystal face, it is possible for foreign bodies to lodge preferentially at dislocation outcrops. If these bodies are occasionally incorporated into the crystal, then they too will be found strung on dislocations. Another question to be asked is whether a small diffraction-contrast-producing object is an unresolved dislocation loop. Here experience with small prismatic loops in aluminium gives some guidance. At sizes much too small for a loop to be recognised as such (i.e. not giving any semblance of a 'doughnut' shape) these objects followed the visibility rules appropriate to their Burgers vector. But again care in interpretation is needed, for platelet precipitates causing a matrix displacement normal to the platelet plane will give similar visibility behaviour.

The small dots seen in all the LiF topographs and largely confined to the vicinity of the low-angle boundaries are believed to be precipitates. The low-angle boundaries will have condensed out of the dislocation population during slow cooling after growth. At a certain stage precipitation on their dislocation networks occurred. During the precipitation epoch, and also subsequently, the low-angle boundaries migrated, sometimes leaving behind a trail of precipitates, sometimes a sheet of precipitates delineating a previous position of the low-angle boundary.

7.3.6 Fault surfaces and growth bands

To lump together fault surfaces (taken to include stacking faults and growth sector boundaries) with growth bands may at first sight seem strange, but we will see that there is quite close kinship in the way they come to produce diffraction contrast. The classification by Amelinckx of fault surfaces into α boundaries and δ boundaries (and combinations thereof) can usefully be adopted in X-ray topography [2]. Consider one type of α boundary first. Let there be a change of crystal structure on either side of a plane surface, the size, shape and orientation of the unit cell being unaltered on either side of the boundary, i.e. similar Bravais lattices, in similar orientation, exist on either side of the boundary. The boundary will, in general, cause a phase shift δ of F_g

DEFECT VISUALISATION: INDIVIDUAL DEFECTS

upon passing across the boundary (phases being measured in a co-ordinate system common to the whole crystal). This phase shift will cause interbranch scattering and diffraction contrast proportional to $\sin^2(\frac{1}{2}\delta)$ in the case of low absorption. A good example of this type of boundary is the Brazil twin boundary in alpha quartz. Note that δ will depend upon g, and that there will in general be a different set of values of δ for each orientation of twin composition plane, since the bonding structure in the composition plane depends upon its orientation. An α boundary with similar diffraction behaviour will be produced in any structure by the insertion of a thin twin lamella, one whose thickness is so much smaller than ξ_g that its own scattering amplitude is negligible: the action of the lamella is simply to introduce a mutual translation between the matrices on either side of the boundary. The analogy to this defect in growth banding phenomena is the presence at a particular growth horizon of a thin sheet ($<<\xi_g$ in thickness) whose cell dimensions are abnormal in the direction perpendicular to the sheet. Again interbranch scattering is the contrast-producing mechanism. However, in the case where a sheet of anomalous material is intercalated in the matrix, stress relief can occur at surface outcrops of the sheet; and the concomitant lattice curvature produces diffraction contrast which may be dominantly due to intrabranch scattering. Returning to α boundaries arising from twinning of the parallel-lattice type, it must be remembered that F can change in magnitude as well as undergo a phase shift on crossing the boundary. Perhaps the best known case is the Dauphiné twin boundary in alpha quartz. But here we have the complication that the composition surface is generally stepped on a fine scale. Each step facet orientation gives its own characteristic relative translation between the twins, and consequently at each step edge or kink there exists a strainfield equivalent to a stair-rod dislocation.

With the δ boundary we have a mutual misorientation at the boundary, introduced in the process of coherently matching dissimilar-shaped unit cells. The 'effective tilt', τ, for determining the change in satisfaction of the Bragg condition when crossing the boundary is given by $\tau = \tau_0 + (\Delta d/d)\tan\theta_B$ where τ_0 is the lattice rotation and $\Delta d/d$ the relative change in interplaner spacing for the reflection concerned. The X-ray topographic example, par excellence, is the ferromagnetic domain boundary. This has been previously discussed [2] and will be further considered in other chapters. Here we just recapitulate that the type of contrast observed will vary from coherent interbranch scattering when $\tau \leqslant W$ (W being the perfect-crystal angular reflecting range, discussed in Section 7.1.2), to the incoherent superimposition of intensities diffracted from matrices on either side of the boundary when $\tau >> W$. Note that when g and the rotation vector $\underline{\tau}$ are colinear we expect a vanishing of contrast. Thus by varying $\underline{g}$, and varying extinction distance (by changing λ, for example), the direction and magnitude of $\underline{\tau}$ can be investigated. The prevalence in both natural and man-made crystals of growth sector boundaries producing contrast of the δ-boundary type has been one of the more intriguing discoveries made by X-ray topography; and it is an observation of significance to the crystal grower who is concerned with the growth sectorial dependence of quantity and of state of aggregation of impurities. To conclude, two points will be emphasized. Firstly, it may not be at all obvious from observation of one or even a few projection topographs whether a series of bands on the image is a true projection of an internal sequence of growth banding, or a sequence of stress reliefs at surface outcrops of the bands, or a set of Pendellosung-type fringes arising from interbranch scattering at a single interface. To determine the true situation a careful geometrical analysis of images on high-resolution section topographs as well as on projection topographs should be made. Also, extinction distances should be calculated, and the fringe periodicities predicted therefrom compared with observed banding periods, making use of observations at more than one wavelength if possible.

REFERENCES

1. R.W. Armstrong and C.M. Wu (1973) in Microstructural Analysis Tools and Techniques (ed. J.L. McCall), Plenum Press, New York - London, p. 169
2. A.R. Lang (1978) in Diffraction and Imaging Techniques in Materials Science, Volume II: Imaging and Diffraction Techniques; Second, revised edition (ed. S. Amelinckx, R. Gevers, and J. Van Landuyt) North-Holland Publishing Co., Amsterdam - New York - Oxford, p. 623
3. B.K. Tanner (1976) X-ray Diffraction Topography, Pergamon Press, Oxford.
4. A.R. Lang (1973) in Crystal Growth: An Introduction (ed. P. Hartman) North-Holland Publishing Co., Amsterdam - New York - London, p.444
5. B.W. Batterman and H. Cole (1964) Rev. Mod. Phys. $\underline{36}$, 681
6. A. Authier (1970) in Advances in Structure Research by Diffraction Methods, $\underline{10}$, (ed. R. Brill and R. Mason) Pergamon Press, Oxford, p. 1
7. M. Hart (1971) Reports on Prg. in Phys., $\underline{34}$, 435
8. N. Kato (1974) in X-ray Diffraction (ed. L. V. Azaroff) McGraw-Hill Book Co., New York - London, p. 176
9. A. Authier (1977) in X-ray Optics (ed. H-J.Queisser) Springer-Verlag, Berlin-Heidelberg - New York, p. 145
10. Z-H. Mai, S. Mardix and A.R. Lang, (1980) J. Appl.Cryst.$\underline{13}$, 180
11. A.R. Lang (1957) Acta Metall. $\underline{5}$, 358
12. A.R. Lang (1957) Acta Cryst. $\underline{10}$, 252
13. M. Hart and A.R. Lang (1961) Phys. Rev. Lett. $\underline{7}$, 120

14. A.R. Lang (1963) Brit. J. Appl. Phys. $\underline{14}$, 904
15. A.R. Lang (1964) Proc. R. Soc. Lond. $\underline{A278}$, 234
16. A.R. Lang (1957) Acta Cryst. $\underline{10}$, 839
17. F.C. Frank, B.R. Lawn, A.R. Lang and E.M. Wilks (1976) Proc.Roy. Soc. Lond. $\underline{A301}$, 239
18. I. Kiflawi and A.R. Lang (1976) Phil. Mag. $\underline{33}$, 697
19. A.R. Lang (1959) Acta Cryst. $\underline{2}$, 249
20. K. Haruta (1965) J. Appl. Phys. $\underline{36}$, 1789

CHAPTER 8

EXPERIMENTAL TECHNIQUES FOR THE STUDY OF

STATISTICALLY DISTRIBUTED DEFECTS

JOCHEN R. SCHNEIDER

8.1 Introduction

Because phonon energies $\hbar\omega \sim kT \sim 5\times 10^{-4}$ to 0.5 eV cannot be resolved in x-ray scattering experiments until now, inelastic scattering experiments are best performed with neutrons and are not discussed in this lecture. For the application of neutron diffraction techniques to the study of disordered crystals we would like to refer to a recent review article by Schmatz [1], which includes a presentation of recent results from neutron small angle scattering, and diffuse elastic scattering by nonmagnetic crystals as well as scattering by disordered magnetic systems.

The various diffraction techniques for the study of statistically distributed defects can be characterized in reciprocal space. Small angle scattering, which is essentially due to fluctuations in the scattering density, occurs around the origin of the reciprocal lattice. Bragg scattering occurs at the reciprocal lattice points. The shape and the integrated reflecting power of the Bragg peaks are affected by dislocations, grain boundaries or, roughly speaking, by the "mosaic structure" of the sample. In the tails of the Bragg peaks Huang diffuse scattering can be observed which is mainly governed by the displacement field at large distances from the defect. Huang diffuse scattering provides information about the symmetry and the strength of the elastic strain field. The measurement of changes in the lattice parameter and the anomalous transmission are complementary tools in the study of the long range strains due to defects. Point defects in a crystal also give rise to coherent elastic diffuse scattering of x-rays into the region between Bragg reflections, which we shall call "Zwischenreflex scattering". This scattering is sensitive to the atomic configuration in the immediate neighbourhood of the defect and therefore allows the investigation of the distribution of the forces exerted by the defect on its close neighbours.

STUDY OF STATISTICALLY DISTRIBUTED DEFECTS

In this chapter we discuss only the following two topics out of this large field of current research:

The investigation of the diffuse x-ray scattering due to point defects.

The study of the mosaic structure of large single crystals by means of γ-ray diffractometry.

The first subject was discussed recently in a number of review papers and we do not wish to repeat this work here. Therefore we will essentially give a number of references concerning this topic. On the other hand γ-ray diffraction as a tool for the study of defects in crystals is not yet well known and therefore in the following we will concentrate on this subject.

8.2 Investigation of diffuse X-ray scattering due to point defects

The displacement field of point defects decays with r^{-2} and therefore these defects cannot be studied with current topographic methods. However, once dedicated sources of synchrotron radiation are available, "spike topography" as presented by A.R. Lang [2] in this school may allow the investigation of diffuse scattering. The diffuse scattering due to point defects is very weak compared to Bragg scattered intensities and can be safely interpreted in terms of the kinematical theory. The crystals containing the point defects must not be perfect in the sense of dynamical theory but they should show narrow Bragg peaks.

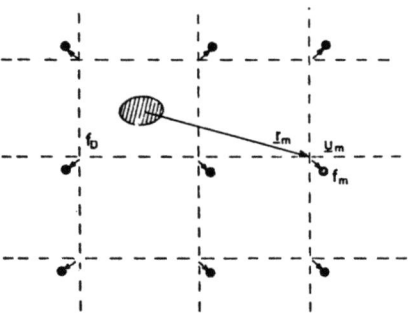

Figure 1. Schematic presentation of the displacement field $\underline{u}(\underline{r})$ caused by an interstitial atom in a cubic matrix.

Fig. 1 shows the displacement field caused by an interstitial atom. $\underline{r}_m$ is the coordinate of the matrix atom m in the undistorted lattice, $\underline{u}_m$ is the displacement of atom m caused by the defect. f_m and f_D are the scattering amplitudes of the matrix atom and the defect respectively. It is useful to introduce a virtual force field $F_i(m)$ which could create the same displacement field $\underline{u}_m$ in the absence of the defect. If these forces are applied on the coordinates $x_i^{(m)}$ of the undistorted lattice one obtains the elastic dipole tensor or double force tensor

$$P_{ij} = \sum_m F_i^{(m)} x_i^{(m)}$$

The sum is over all atoms in the distorted region of the crystal.

The static displacement field $\underline{u}_m$ gives rise to a reduction of the Bragg scattered intensities which can be described by means of a static Debye Waller factor. The defect will also lead to an average expansion of the lattice. For unoriented defects in a cubic lattice the corresponding change of the lattice parameter $\Delta a/a$ is related to the trace of the double force tensor via the relation:

$$\text{trace } P_{ij} = \sum_m \underline{F}^{(m)} \cdot \underline{r}^m = 3 \cdot \frac{\Delta a}{a} (C_{11} + 2C_{12}) \cdot V_a$$

c_{11} and c_{12} are the elastic constants and V_a is the atomic volume of one atom. If $\underline{K} = \underline{k} - \underline{k}_o$ is the momentum transfer in the scattering experiment one obtains the following expression for the intensity of the diffuse scattering due to point defects:

$$I_{diff}(\underline{K}) = n \left| f_D(\underline{K}) + \sum_m f_m(\underline{K}) e^{i\underline{K} \cdot (\underline{r}_m - \underline{u}_m)} - \sum_m f_m(\underline{K}) e^{i\underline{k} \cdot \underline{r}_m} \right|^2$$

$$\phantom{I_{diff}(\underline{K}) = n |} \text{Laue} \text{distortion} \text{Bragg}$$
$$\phantom{I_{diff}(\underline{K}) = n |} \text{scattering} \text{scattering} \text{scattering}$$

where n = number of defects; $f_D = f_m$ (self interstitial); $f_D = -f_m$ (vacancy); $f_D \approx f_I - f_m$ (impurity; f_I = scattering amplitude of the impurity). For small displacements $\underline{u}$:

$$I_{diff}(\underline{K}) = n \left| f_D(\underline{K}) + if_m(\underline{K}) \cdot \underline{K} \cdot \underbrace{\sum_m \underline{u}_m e^{i\underline{K} \cdot \underline{r}_m}}_{\tilde{u}(\underline{K})} + \ldots \right|^2$$

$$I_{diff}(\underline{K}) = n \left| f_D(\underline{K}) + if_m(\underline{K}) \cdot (\underline{K} \cdot \tilde{\underline{u}}(\underline{K})) \right|^2$$

Thus, it is the projection of the scattering vector $\underline{K}$ on the Fourier transform of the displacement field. $\tilde{\underline{u}}(\underline{K})$, which is measured with diffuse scattering experiments, in addition to the defect scattering itself. Close to Bragg peaks $\underline{G} = \underline{K}$ the diffuse scattering is called Huang scattering. Interference between Huang and Laue scattering amplitudes gives rise to an asymmetry of the scattering intensity distribution with respect to the Bragg peak. This asymmetry depends on the sign of f_D and $\underline{\tilde{u}}(\underline{K})$; thus one can tell whether a measured scattering intensity distribution is mainly due to interstitial or vacancy type defects. Away from the Bragg peaks (Zwischenreflex scattering) both scattering amplitudes are of the same order of magnitude. The scattering intensity here mainly depends on the relative phase factor between scattering from the defect and scattering from the displaced neighbouring lattice atoms. Scattering for $\underline{K} \approx \underline{G} = 0$ (small angle scattering) is only observed if a defect or its displacement field is connected with a change of the local electron density. For a discussion of the applications of small angle scattering in materials sciences we refer to a recent review paper by Gerold and Kostorz [3]. For a detailed presentation of the various cross sections involved in the description of x-ray and neutron scattering from distorted crystals we refer to a review article by Schmatz [4].

8.2.1 Huang diffuse scattering

In order to gain a better idea of what one can learn from Huang diffuse scattering experiments, the displacement field $\underline{u}$ of a single point defect and its Fourier transform $\underline{\tilde{u}}$ will be considered in some detail. At some distance from the point defect the displacement falls off as

$$\underline{u}(\underline{r} \to \infty) \simeq \frac{1}{r^2} \cdot \frac{\underline{r}}{r}$$

and yield a Fourier transform centred close to the reciprocal lattice point

$$\underline{\tilde{u}}(\underline{g} \to 0) \simeq \frac{1}{g} \cdot \frac{\underline{g}}{g}$$

The Huang scattered intensity is proportional to

$$I_{HDS} \simeq n \cdot |f_m(\underline{K})|^2 \cdot |\underline{K} \cdot \underline{\tilde{u}}(\underline{K})|^2$$

and therefore we expect zero intensity whenever $\underline{K} \perp \underline{u}$

In the special case of an isotropic displacement field shown in Fig. 2 this happens on a plane $\perp \underline{G}$ through the reciprocal lattice point. Equal scattering intensity is expected on spheres touching the reciprocal lattice point (Huang sphere). Maximum intensity is expected in the direction of $\underline{G}$ where $\underline{K} // \underline{\tilde{u}}$ Similarly to thermal diffuse scattering, the Huang scattering intensity for the assumptions defined above is proportional to the square of the scattering vector $\underline{K}$ and falls of as g^{-2}, where g is the distance from the Bragg peak.

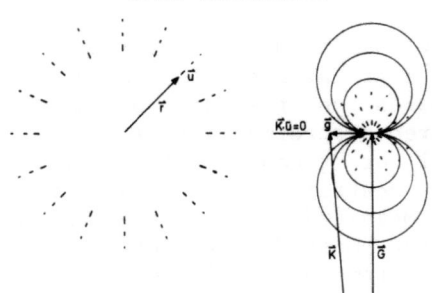

Figure 2. Isotropic displacement field u(r) in real space and its Fourier transform ũ(K) in reciprocal space (from H. Peisl [7])

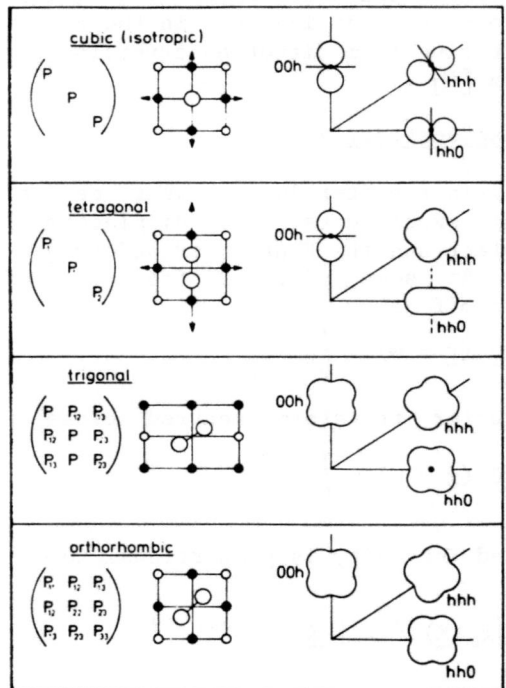

Figure 3. Schematic isointensity contours for various defect symmetries, averaged over the possible defect orientations. P_{ij} is the dipole force tensor (from H. Peisl [7])

For an anisotropic displacement field the Fourier transform also becomes anisotropic. Anisotropic defects can have different orientations in the lattice. For a random distribution of the defect orientations the diffuse scattering intensity has to be averaged [5,6]. Resulting isointensity contours for various defect symmetries averaged over the possible defect orientations are shown in Fig. 3. Measuring the diffuse scattering intensity around some high symmetry reciprocal lattice points in some low index directions immediately gives the symmetry of the displacement field which resembles the symmetry of the defect. For a review on Huang diffuse scattering including results on KCl, KBr, LiF and MgO, one should refer to two papers by Peisl [7, 8].

8.2.2 Zwischenreflex scattering

As we shall see in some detail later for the case of irradiation induced defects in Al, with Huang scattering alone the structure of the defect cannot be determined unambiguously, because it reflects only the symmetry of the long range displacement field caused by the defect. For a further determination of the defect configuration one has to look for differences in the displacements of the atoms close to the defect, which affects the diffuse scattering between the Bragg peaks. In metals this diffuse scattering intensity can be calculated assuming simple force models with forces acting only on nearest neighbours. This is a good approximation, since in spite of the long range of the displacements the forces are highly restricted to the neighbouring atoms only, because in metals the electrons screen out the defect potential very efficiently at large distances.

In the frame of lattice statics the displacement $u_i^{(m)}$ of atom m in the i-direction is calculated via the relation

$$u_i^{(m)} = \sum_{j=1}^{3} \sum_n G_{ij}^{mn} \cdot F_j^{(n)}$$

where G_{ij}^{mn} is the response function which can be calculated from phonon dispersion curves, which have been measured very precisely in many materials with inelastic neutron scattering. For a calculation of the diffuse scattering the Fourier transform of the displacement field $\underline{u}^{(m)}$ is needed, which is obtained from the product of the Fourier transform of the forces and the response function

$$\tilde{u}_i(K) = \sum_j \tilde{G}_{ij}(K) \cdot \tilde{F}_j(\underline{K})$$

Results of a numerical calculation for interstitials in Al are shown in Fig. 4.

There are only a few of these measurements of the Zwischenreflex scattering performed up till now because of the serious background problems demonstrated in Fig. 5. The Compton cross section grows monotonically with increasing scattering vector.

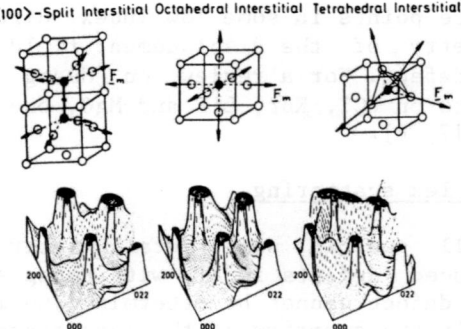

Figure 4. Three possible configurations for self interstitials in Al and the corresponding calculated intensity distribution of the diffuse scattering (from H.G. Haubold [9]).

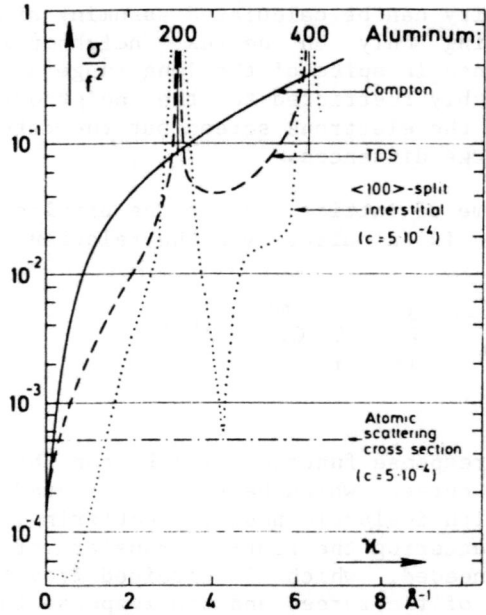

Figure 5. Cross sections for the various scattering processes involved in Zwischenreflex scattering experiments (from H.G. Haubold [9].

Between the Bragg reflections the Compton cross section is about five times larger than the thermal diffuse scattering in the case of aluminium. The magnitude of the complete scattering cross section, which includes the displacement field scattering, varies strongly with the length of the scattering vector. At small scattering vectors the cross section of the diffuse scattering due to the interstitials is very small and nearly cancels the scattering of the interstitial itself. The strongest displacement field scattering occurs in the vicinity of the Bragg reflections. Here the defect scattering is stronger than the background scattering and no severe background problems result. In contrast to this, the defect scattering between the Bragg peaks is typically 20 to 100 times smaller than the background scattering. The defect induced scattering is nevertheless up to about 60 times higher than that for the scattering of the interstitials themselves. This effect is caused by the additional scattering of the displaced lattice atoms surrounding the defect. For a description of Zwischenreflex scattering and especially its use for the structure determination of self interstitials in Al, which will be discussed next, see the review paper by Haubold [9].

8.2.3 Structure determination of self-interstitials in aluminium

The structure determination of point defects in metals is a rather old problem, but has been solved only recently after the appearance of strong x-ray sources and large multidetectors. X-ray methods seem to be most useful in determining the structure of self-interstitials in metals, because other techniques like electron spin resonance and optical methods, which have been very powerful tools for the investigation of point defects in insulators cannot be used in metallic systems.

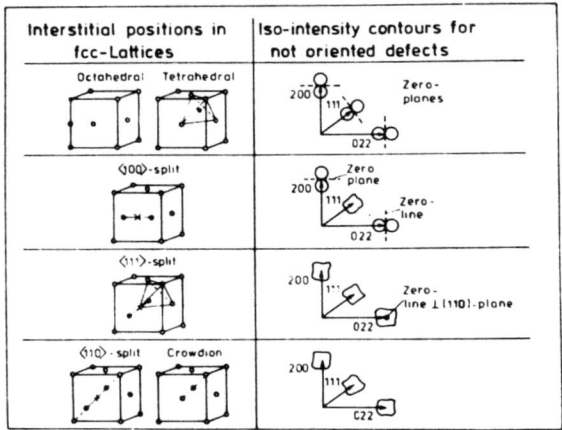

Figure 6. Isointensity contours of the diffuse scattering near the Bragg peaks for different configurations of self-interstitials in a fcc lattice (from H.G. Haubold [9]).

From symmetry considerations six configurations of an interstitial atom in a fcc lattice are possible, as shown in Fig. 6. The calculations of the diffuse scattering were performed by Trinkhaus [5] and Dederichs [6] and the iso-intensity contours for Huang diffuse scattering are also shown in Fig. 6. The measurements of the Huang scattering by Ehrhart and Schilling [10] ruled out the <111> -split, the <110> -split and Crowdion configurations, but only from the investigation of the Zwischenreflex scattering by Haubold [11] could the structure of the self-interstitial in Al due to electron irradiation be determined as the <100> split configuration.

The diffuse scattering experiments, as well as the irradiation of the samples, were performed at 5 K in order to avoid clustering of the point defects. The samples contained some 10^{-4} atomic fraction of interstitials and were about 10 - 100 μm thick, A typical result from the measurements of the Huang scattering around the 400 reflection for scans parallel to <100> and <110>, respectively, is shown in Fig. 7. There is no Huang diffuse scattering in the scan parallel to <110> as expected for the octahedral, tetrahedral and <100> split configuration. Even though the long range displacement fields are similar for these 3 interstitial configurations, the displacement of the matrix atoms close to the defect are radically different as indicated in Fig. 4, which leads to different cross sections for the Zwischenreflex scattering as shown in Fig. 8. The contribution of the vacancy, which is produced in equal number with the interstitial during the irradiation, to the distortion scattering was included in the cross section calculation. Because of the smaller displacements around a vacancy, its scattering is only important at small scattering angles. The experiment clearly favours the <100> -split interstitial configuration. By fitting theoretial curves to the experimental data, essential details of the defect structure such as dumbbell distance, displacements of the nearest neighbours, volume dilatation around interstitial and vacancy and the defect concentration could be determined.

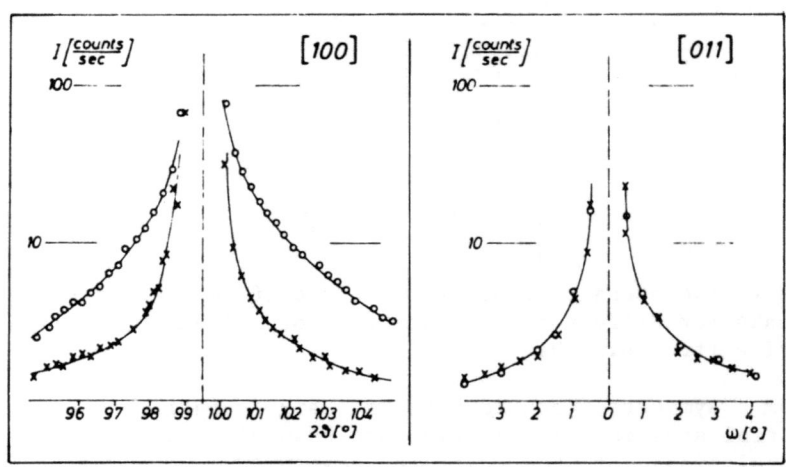

Figure 7. Diffuse scattering intensity measured near the 400 reflection of Al in q-directions along [100] and [011]. (x) are the data points before and (o) the data points after electron irradiation (from P. Ehrhart and W. Schilling [10])

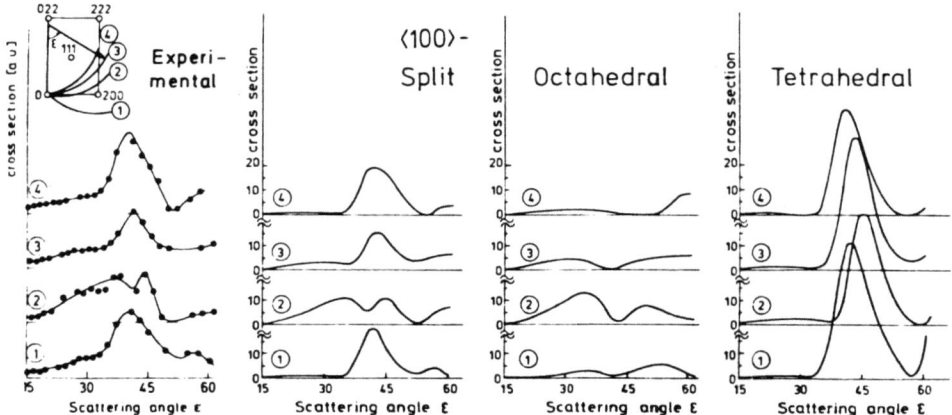

Figure 8. Comparison of experimental and calculated cross-sections of interstitials in electron-irradiated Al for 3 defect configurations. Calculations and measurements are performed for 4 different scans in reciprocal space (from H.G. Haubold [9])

8.3 Study of mosaic structure of large crystals by γ-rays

In the preceding section we have discussed point defects which cause displacement fields decreasing with r^{-2} where r is the distance from the defect centre. Due to this short range displacement field it is relatively easy to produce samples with statistically distributed defects which can be studied by diffraction methods. Additionally it is of great advantage that the very weak diffuse scattering due to point defects can be separated from Bragg scattering and thus can be treated safely within the first Born approximation. On the other hand, new experimental techniques had to be developed because the diffuse scattering due to point defects is extremely weak. For dislocations, which create a displacement field decreasing only with r^{-1}, the situation is much more complicated. Because of this long range displacement field, dislocations tend to interact even for relatively small dislocation densities. However, if one wants to interpret the diffraction pattern from imperfect single crystals on the basis of the kinematical theory the dislocation density has to be rather high. An appreciable amount of work was done on plastically deformed single crystals, e.g. copper, and we refer to an extended abstract by Wilkens [12] which summarizes the state of the art in 1974. Progress in this field is expected from new theoretical approaches [13] as well as from new experimental developments in connection with the dedicated sources of synchrotron radition in the x-ray range which will become available in the near future.

In this section we want to report on diffraction experiments using 0.03Å γ-radiation from strong radioactive sources. These techniques were used to study structural phase transitions and to measure absolute structure factors in imperfect single crystals [14]. In the following, however, we want to present γ-ray diffractometry as a tool for the study of the mosaic structure of large single crystals, with the hope that this technique may become useful in the field of crystal growth by characterizing the quality of the as-grown single crystals while avoiding lengthy preparation.

8.3.1 The γ-ray diffractometer

Instead of an x-ray tube it is possible to use 412 keV γ-radiation from radioactive gold activated in a nuclear reactor. The half-life of the source is 2.7 days and the activity after irradiation in a thermal neutron flux ~10^{14} cm^{-2}s^{-1} is of the order of 100 Ci. The wavelength is 0.03Å and because it is a γ-line the energy spread is $\Delta E/E = \Delta\lambda/\lambda \simeq 10^{-6}$ at room temperature. The best angular resolution in the scattering plane which has been obtained until now is 10sec. of arc [15]. For an incident γ-ray beam of 0.2 x 10 mm^2 cross section we obtained about 5000 photons/sec on the ILL-γ-diffractometer with a new γ-ray source. The way in which the sources are transferred from the reactor to the diffractometer depends very much on facilities such as hot cells installed at the reactor station. The procedure used at the Hahn-Meitner-Institut in Berlin (no hot cell is used) was described recently in some detail [16].

The γ-ray diffractometer is a single crystal diffractometer which consists of the source block, a beam defining slit, a goniometer and two detectors for measuring the transmitted and the Bragg diffracted beam as indicated in Fig. 9. Because of the small Bragg angles of ~1° the crystals are studied in Laue geometry by performing ω-step-scans. Fig. 10 shows the scattering geometry as well as a schematic representation of the quantities measured in γ-ray diffractometry. The measured reflectivity distribution is given by

$$r_m(\omega) = \frac{P_H(\omega) - P_B}{P_T^*}$$

$$P_T^* = P_o\, e^{-\mu_o T_o/\cos\theta_B}$$

with $P_H(\omega)$ and P_B being the Bragg scattered and the background intensity respectively. P_T^* is the transmitted intensity in angular ranges where no Bragg scattering occurs. The integral $\int_m(\omega)d\omega = R_m$ represents as usual the integrated reflecting power which is to be compared with theoretical values for zero absorption.

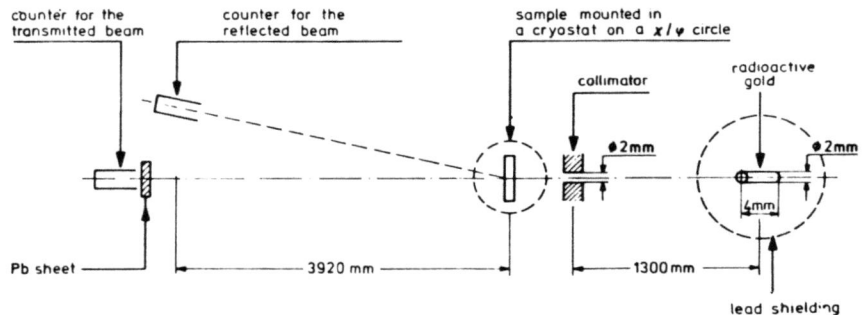

Figure 9. Schematic drawing of the set-up of a γ-ray diffractometer

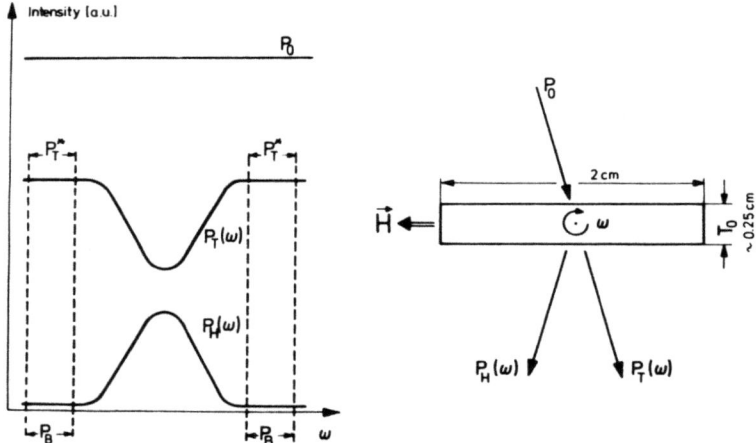

Figure 10. Scattering geometry and schematic representation of the quantities measured in γ-ray diffractometry.

The absorption of 412 keV γ-radiation in matter is rather weak and essentially due to Compton scattering. The mean free path is about 1 cm in copper and 4 cm in aluminium. Therefore large crystals of the size typical for neutron scattering experiments can be investigated by γ-ray diffractometry even if the samples are contained in cryostats, furnaces or high pressure devices without causing additional experimental difficulties. Because of the small Bragg angles essentially only lattice tilts contribute to the broadening of the measured rocking curve, and thus γ-ray diffractometry is complementary to backscattering experiments which are most sensitive to strain causing changes in the lattice parameter [17]. Although the Ewald sphere is very large, possible effects of multiple Bragg scattering on the shape and the integral of the measured rocking curve can be avoided since the atomic scattering factor $f(\underline{K})$ decreases with increasing $\underline{K}$ and because of the generally weak interaction of 0.03Å γ-radiation with crystals [18]. Nevertheless, one is strongly recommended to measure rocking curves for about 3 ψ -settings (i.e. with the sample rotated around the scattering vector).

γ-ray diffractometers are now operational at several research laboratories, most of them using radioactive gold as γ-ray source. Following a proposal by Maier-Leibnitz, the first diffractometer was set up at the Institut Laue-Langevin in Grenoble [15]. Similar machines are available at the research centres in Risø [19] and Jülich [20] and at the University of Munich [21]. In Berlin we have constructed a γ-ray diffractometer especially suitable for absolute structure factor measurements in imperfect single crystals [16]. Two further diffractometers are under construction at the Rutherford Laboratory in the U.K. and in Vienna (Austria). Very recently a new γ-ray diffractometer became operational at the Research Reactor Facillity of the University of Missouri USA [22], which works with 103 keV (λ = 0.125 Å) radiation from ^{153}Sm. Special interest in this γ-ray source arises from the fact that the ^{153}Sm to ^{153}Eu transition at 103 keV is a Mossbauer line.

8.3.2 Interpretation of γ-ray rocking curves

In the following we will call a single crystal imperfect if the integrated reflecting power R_m measured with radiation of wavelength λ at a given reciprocal lattice point $\underline{G}$ is smaller than the theoretical value R_{kin} calculated on the basis of the kinematical theory, and larger than the value R_{dyn} determined from the dynamical theory for a defect free crystal of thickness $t_o \gg t_{ext}$ (= extinction length) i.e.

$$R_{kin} > R_m > R_{dyn}$$

If $R_m = R_{dyn}$ the crystal will be called perfect, if $R_m = R_{kin}$ the crystal will be called ideally imperfect. The difference between R_{kin} and R_m is called extinction and is discussed in terms of an extinction coefficient

$$y = R_m/R_{kin} \leq 1$$

which depends on the defect structure of the sample as well as on the wavelength of the diffracted radiation, the structure factor F_G, the Bragg angle θ_B and on the sample thickness t_o. For simplicity only symmetrical Laue geometry is considered and the radiation is assumed to be unpolarized. For $t_o \ll t_{ext}$ one obtains $R_{dyn} = R_{kin}$.

STUDY OF STATISTICALLY DISTRIBUTED DEFECTS

Perfect single crystals

Starting from the following definition of the extinction length

$$t_{ext} \simeq V/r_o F\lambda$$

we obtain a value $t_{ext} = 66 \mu m$ (corresponding to a Pendellosung distance of $\Lambda_o \simeq 200 \mu m$) for the diffraction of 0.03 Å γ-radiation at the 111 reflection of a perfect Cu crystal. In a crystal of thickness $T_o = 1$ cm we obtain a ratio of $A = T_o/t_{ext} \sim 150$. Due to the Pendellosung effect the integrated reflecting power

$$R_y = \frac{\pi}{2} \int_0^{2A} J_o(x) dx$$

shows fluctuations as a function of $A \propto T_o \lambda$ which are of the order of ± 3% for $A \sim 150$. ($J_o(x)$ is the zero order Bessel function). In general it is assumed that these fluctuations are smeared out due to the wavelength spread in the incident beam and variations in the sample thickness over the beam cross section. If we consider the crystal as being thick for dynamical diffraction theory the FWHH of the corresponding diffraction pattern is

$$2 \cdot W_o = \frac{2 \cdot r_o \cdot F \cdot \lambda^2}{\pi \cdot V \cdot \sin 2\theta_B} = 0.42"$$

The best angular resolution obtained till now with γ-diffractometers is about 10 sec of arc and therefore the shape of γ-ray rocking curves measured with perfect crystals (FWHH of the intrinsic diffraction pattern $\leq 0.5"$) simply reflects the angular divergence of the incident γ-ray beam. If the FWHH of the γ-ray diffraction pattern of a crystal claimed to be perfect is larger than the angular resolution of the diffractometer the crystal must be heavily distorted. Fig. 11 shows such a rocking curve measured at the 111 reflection in a Si platelet. A somewhat better understanding of the origin of this distortion in the crystal is obtained from a series of γ-ray curves measured at reflection $2\bar{2}0$ in neighbouring volume elements also shown in Fig. 11. The FWHH of the individual rocking curves is equal to the instrumental resolution but the displacement of the diffraction patterns on the ω-scale as a function of sample displacement indicates an irregular, macroscopic curvature of the (111) lattice planes.

The integrated reflecting power measured by means of γ-ray diffractometry in thick, perfect single crystals can be compared with the asymptotic theoretical value, provided Pendellosung effects can be ruled out. Fig. 12 shows a series of rocking curves measured in reflection and transmission in various volume elements of an as-grown Cu crystal. The closer one approaches the end of the rod where the growth started, the higher is the integrated reflecting power, which indicates an increasing amount of distortions in the crystal, even though the FWHH of the γ-ray rocking curves is constant.

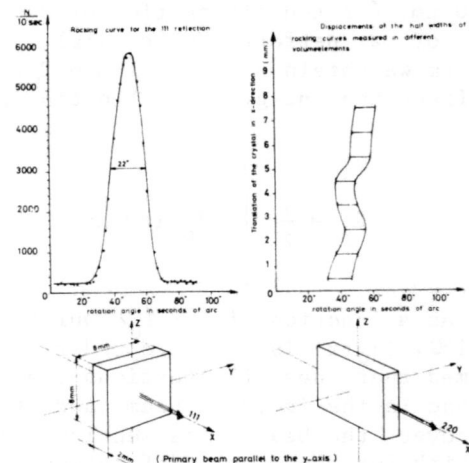

Figure 11. A γ-ray rocking curve measured at the 111 reflection of a platelet-like Si single crystal as well as a series of $2\bar{2}0$ curves obtained after a 90° rotation of the sample (angular resolution 10").

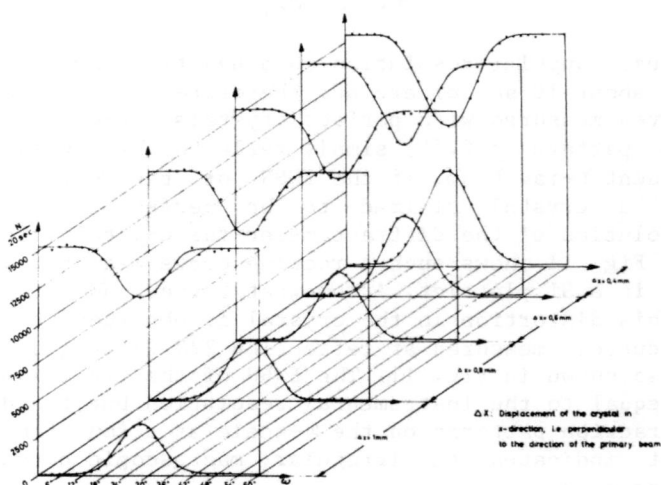

Figure 12. A series of γ-ray rocking curves measured in various volume elements of an as-grown Cu single crystal. Angular resolution of the diffractometer 25".

STUDY OF STATISTICALLY DISTRIBUTED DEFECTS

Transition from perfect to imperfect single crystals

Freund [23] studied Cu single crystals of various dislocation densities with wavelengths ranging from 1.66 Å to 0.03 Å. In Fig. 13 the values of the measured integrated reflecting power are plotted as a function of the quantity λ_{eff} which is essentially proportional to the product $\lambda \ F_G$.

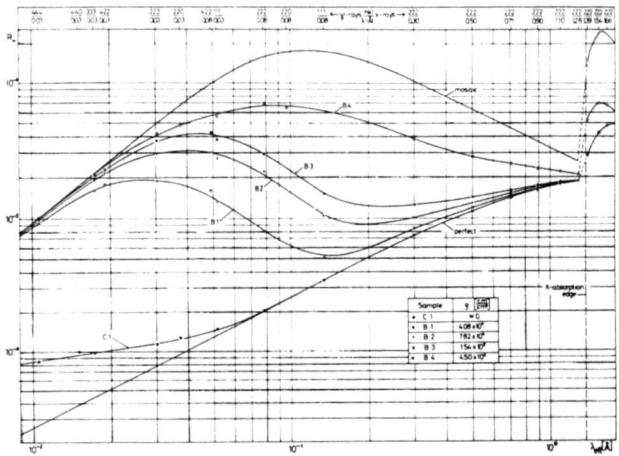

Figure 13. Wavelength dependence of the integrated reflecting power measured in Cu single crystals showing different dislocation densities. The outer curves are calculated from dynamical and kinematical diffraction theory, respectively (from A. Freund [23])

Also in Fig. 13 the theoretical values calculated from the dynamical and the kinematical diffraction theory are plotted. Crystals B_1 to B_4 are Bridgman crystals with dislocation densities of 4.08×10^4, 7.82×10^4, 1.54×10^5 and 4.50×10^5 cm^{-2}, respectively. Crystal C_1 is a dislocation free Czochralski crystal. On the basis of a theory by Authier and Balibar [24], Freund could explain qualitatively the transition from dynamical to kinematical diffraction theory. Once the mean distance between dislocations becomes smaller than the Pendellosung-distance $\Lambda_o = \pi \cdot t_{ext}$, the measured integrated reflecting power deviates from the dynamical value. It is interesting to note that by varying the wavelength of the radiation incident on to the sample, a crystal which behaves like a perfect crystal, and thus can be studied by topography, becomes an ideally imperfect crystal for very short wavelengths. In this Advanced Study Institute Prof. Kato has presented the principles of a new theory which can describe the full transition from dynamical to kinematical diffraction behaviour (that is from coherent to fully incoherent scattering) as a function of two correlation parameters τ and E, the first one describing short range and the second one describing long range correlation. The mathematical problem in describing the transition from amplitude to intensity coupling have now been solved, and a major effort should be made to assign a physical meaning to the two parameters τ and E. It would be most exciting if one could calculate a transition from dynamical to kinematical diffraction behaviour of the type shown in Fig. 13 from a set of physical parameters which were determined in independent measurements. As an example of the very successful interplay between theory and experiment we would like to refer to the experimental work by Larson and Young [25] which proved a statistical, dynamical theory developed by Dederichs [26] about the effect of point defects and clusters of point defects on the Borrmann effect.

Mosaic crystals

γ-ray rocking curves of FWHH much larger than angular resolution of the diffractometer are interpreted in terms of Darwin's mosaic model [27, 29]. In this picture the imperfect crystal of thickness T_0 is assumed to be an aggregate of a great number of independently scattering perfect crystal blocks of thickness $t_0 \ll T_0$. The absorption in a perfect block is assumed to be negligible. The deviation of the lattice-plane orientation ω of one block from the mean lattice plane orientation for the whole crystal is described by means of the so-called mosaic distribution function W(ω), which is a probability function:

$$\int_{-\infty}^{\infty} W(\omega) \, d\omega = 1$$

W(ω) was assumed to be a Gaussian distribution function and its FWHH is called the mosaic spread $\Delta\omega_M$. Darwin had to assume that the half width $\Delta\omega_{dyn}$ of the diffraction pattern for the perfect mosaic blocks is much smaller than the mosaic spread $\Delta\omega_M$ so that one can expect that there is no coherence between neighbouring mosaic blocks.

On the basis of this model Darwin formulated the well-known intensity transfer equations (Laue geometry):

$$\frac{dP_o}{dT} = -\mu_o P_o - \sigma P_o + \sigma P_G$$

$$\frac{dP_G}{dT} = -\mu_o P_o - \sigma P_G + \sigma P_o$$

$P_o(T)$ and $P_G(T)$ represent the power of the incident and diffracted beams respectively at depth T, μ_o is the total linear absorption coefficient. For the coupling constant σ, which has to be a function of the rocking angle ω, one obtains

STUDY OF STATISTICALLY DISTRIBUTED DEFECTS

$$\sigma(\omega) = W(\omega) \cdot \frac{R_{dyn}^*}{\bar{t}_o}$$

$\bar{t}_o$ is the mean block thickness, R_{dyn} is the integrated power calculated with dynamical theory for a perfect plane parallel plate of infinite lateral extension and thickness $\bar{t}_o$. For 0.03 Å γ-radiation the extinction length is in general much larger than $\bar{t}_o$, so that

$$R_{dyn}^* = R_{kin} = Q\bar{t}_o/\cos\theta_B$$

$$Q = r_o^2 \left|\frac{F_g}{V}\right|^2 \frac{\lambda^3}{\sin 2\theta_B} \cdot \frac{1 + \cos^2 2\theta_B}{2}$$

Now the coupling constant $\sigma(\omega)$ becomes independent of the size of the perfect blocks

$$\sigma(\omega) = W(\omega) \cdot \frac{Q}{\cos\theta_B}$$

and the solution of Darwin's intensity transport equations leads to the following expression for the theoretical reflectivity distribution

$$r_{th}(\omega) = \frac{P_G(\omega)}{P_o e^{-\mu_o T_o/\cos\theta_B}} = \frac{1}{2}\left[1 - e^{-2\cdot\sigma(\omega)\cdot T_o}\right]$$

By expanding the exponential to first order one obtains

$$r_{th}(\omega) = \frac{QT_o}{\cos\theta_B} \cdot W(\omega)\left[1 - W(\omega)\cdot\frac{QT_o}{\cos\theta_B}\right]$$

In the so called kinematical limit

$$W(\omega) \cdot \frac{QT_o}{\cos\theta_B} \ll 1$$

and the reflectivity distribution $r_{th}(\omega)$ is proportional to the mosaic distribution function $W(\omega)$

$$W(\omega) = \frac{r_m(\omega)}{R_m}$$

If extinction is present the reflectivity distribution will be flattened. In the peak of the rocking curve, $W(\omega)$ has its maximum and the reduction in reflectivity due to the term $(1 - W(\omega)QT_o/\cos\theta_B)$ will also be strongest.

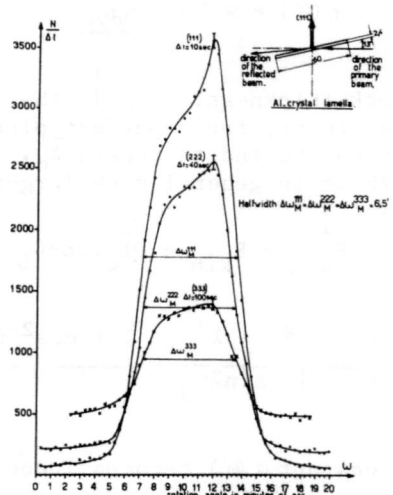

Figure 14. Extinction free γ-ray rocking curves measured at the 111, 222 and 333 reflection of an Al single crystal.

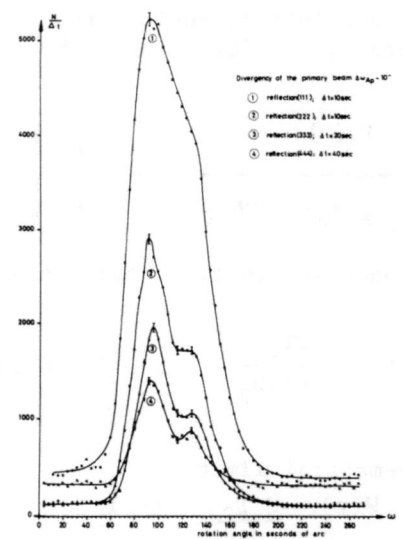

Figure 15. γ-ray rocking curve flattened by extinction for the 111, ..., 444 reflections of Cu single crystal.

In γ-ray diffractometry $Q\alpha\lambda^2$ is about 3 orders of magnitude smaller than in x-ray diffraction, so that extinction will be small even for crystals of thicknesses between 1 and 10 mm. The kinematical limit can be further approached by studying higher orders of reflection because $Q\alpha F_G^2$. Because the Bragg angles are of the order of 1° in γ-ray diffractometry, going to higher order does not alter significantly the irradiated sample volume. Therefore once the kinematical limit is reached, going to a higher order reflection should not change the shape of the measured rocking curve. Fig. 14 shows three rocking curves measured at the 111, 222 and 333 reflection of an Al single crystal. The measured mosaic spread is identical for all three curves and therefore $r_m(\omega)$ should be proportional to the mosaic distribution function. In contrast to this, Fig. 15 shows four rocking curves measured on a Cu crystal where there appears to be a significant difference in the shape of the diffraction patterns for reflections 444 and 111. It is the shape of the 444 rocking curve which is close to the mosaic distribution function $W(\omega)$ whereas the 111 curve is smoothed due to extinction.

Because Darwin's diffraction theory for imperfect single crystals is based on a number of rather crude assumptions concerning the defect structure of the sample and the interaction of the electro-magnetic radiation with crystals, it is tempting to find out to what extent Darwin's theory is reliable for the interpretation of γ-ray rocking curves. For this purpose we measured a series of rocking curves in a 8 mm thick copper crystal for the reflections 111, ..., 444. The 333 and the 444 rocking curves were not affected significantly by extinction. Therefore we could first calculate the mosaic distribution function $W(\omega)$ from the 333 reflection and then the theoretical reflectivity distribution $r_{th}(\omega)$ for two wavelenths 0.03 Å and 0.078 Å [30]. In Fig. 16 the measured reflectivity distributions $r_m(\omega)$ are compared with $r_{th}(\omega)$. For 0.03Å γ-radiation, the agreement between theory and experiment is surprisingly good taking into account that the peak reflectivity of the 111 reflection is about 11 times larger than the value for the 333 reflection. For 0.078 Å-γ-radiation the saturation value of 50% reflectivity for Laue geometry is reached in both the experimental and the theoretical diffraction pattern. The width of the plateau in the measured curve is about 80" whereas for the theoretical curves the plateau is 100" wide. Within Darwin's diffraction theory this discrepancy can be attributed to the neglect of primary extinction in our calculations.

It is difficult to compare x-ray rocking curves with those obtained by γ-ray diffractometry on the same sample, because x-rays mainly see the surface and γ-rays the bulk of the sample. On the other hand, it is instructive to compare results from neutron diffraction with those from γ-ray diffractometry. Fig. 17 shows a typical rocking curve measured with a two axis neutron spectrometer at the 110 reflection of a Be single crystal. The cross section of the incident neutron beam was 1 cm^2 and the wavelength 0.845 Å; the mosaic spread of the copper monochromator was 2.5' and the lattice spacings of sample (d_{110}^{Be} = 1.14 Å) and monochromator did not match. By means of γ-ray diffractometry we investigated in the same Be crystal a volume element of 0.5 x 10 x 8 mm^2 out of the 10 x 10 x 8 mm^2 which was studied by neutrons. The γ-ray rocking curve shows strong fluctuations and is plotted in Fig. 18.

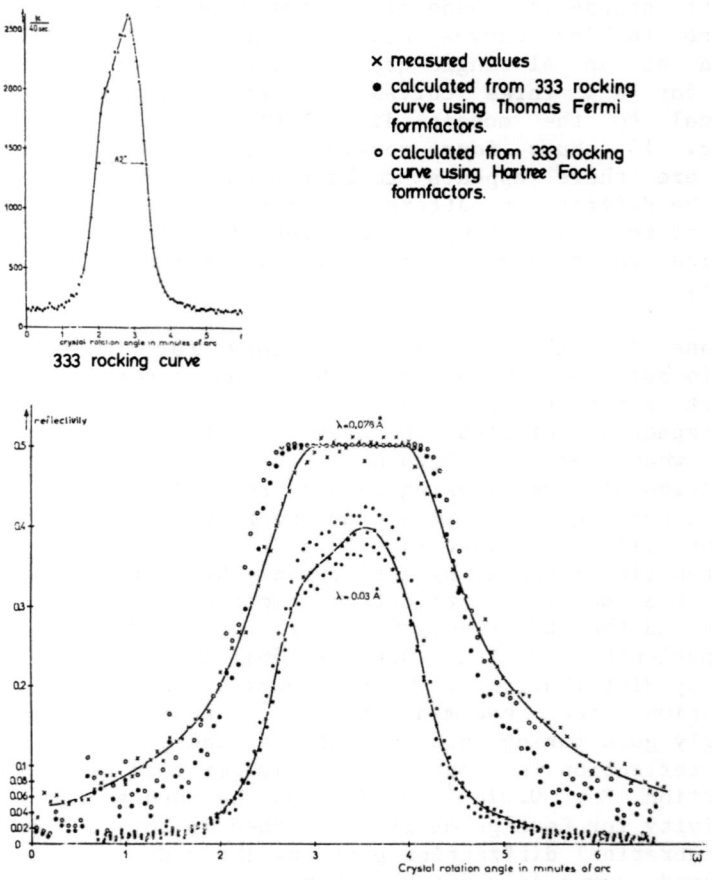

Figure 16. Calculated and measured diffraction patterns of the 111 reflection of a Cu single crystal; wavelength of the γ-ray radiation 0.03 Å and 0.078 Å, respectively.

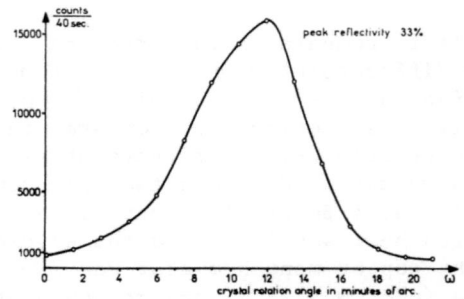

Figure 17. Rocking curve of a Be crystal measured with a two-axis neutron diffractometer. Wavelength 0.845 Å, cross section of the incident beam 1cm^2 reflection 110.

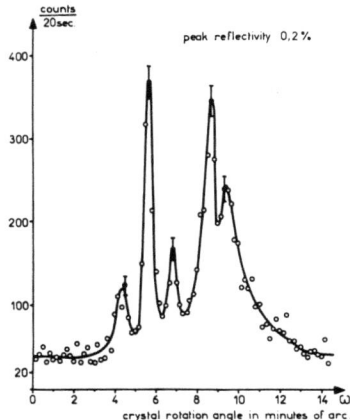

Figure 18. Rocking curve of a Be crystal measured with the γ-ray diffractometer. Cross section of the primary beam 0.5×10 mm^2, angular divergence 25", reflection 110.

There are essentially three reasons for this large difference between the neutron and the γ-ray result:

- The neutrons average over a larger volume than do the γ-rays.

- The neutron diffraction pattern is strongly affected by extinction which smooths possible sharp features in the mosaic distribution.

- The angular resolution in the neutron experiment is much worse but this could be improved by using another monochromator on the neutron diffractometer.

From the γ-ray rocking curve alone one can learn that the investigated volume element contains 5 regions with a mosaic spread not larger than the angular resolution of the diffractometer of 30". The mean lattice plane orientations in these regions are separated from each other by at least 1.5'. Because the diffraction of 0.03 Å γ-radiation in this Be crystal is not affected by extinction, the measured integrated reflecting power is directly proportional to the thickness of the crystal. From the areas of the subpeaks one can thus estimate the amount of material forming these five regions in the investigated volume element of the sample. However, no conclusions can be made concerning the shape of these regions from one measurement.

8.4 Applications of γ-ray diffractometry

Most of the applications of γ-ray diffractometry result from the fact that the instrument is an extremely simple, high resolution diffractometer which enables one to study large single crystals in all kind of environments such as furnaces, cryostats, bending and high pressure devices. Furthermore extinction is in many cases very weak and because of the small Bragg angles mainly lattice tilts contribute to the width of the rocking curve. On the other hand the resolution of the diffractometer in reciprocal space is rather low and therefore one should measure rocking curves for about 3ψ settings. No change in the diffracted beam intensity on rotation about the scattering vector indicates no multiple Bragg diffraction.

8.4.1 Testing of as-grown Al single crystals

In order to allow for systematic studies of large as-grown crystals the sample can be moved automatically in a direction perpendicular to the incident beam and a large number of rocking curves are measured. At the Institut für Festkörperforschung of the KFA Jülich, Al single crystals are grown in the Bridgman technique and at the beginning of the work the rocking curves showed a very irregular shape and the FWHH varied between 20 and 60 sec of arc [20]. After changing from graphite to boron nitride crucibles and mounting the whole growth apparatus on a vibration free base, good single crystals could be produced Fig. 19 shows a series of γ-ray rocking curves and, except for the volume elements close to the ends of the crystal, the FWHH is of the same order as the instrumental resolution of 12 sec. of arc.

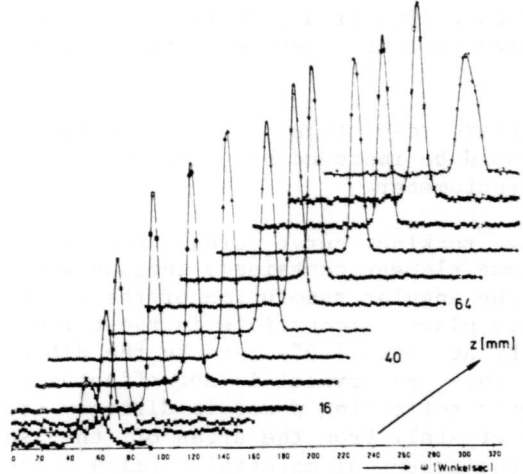

Figure 19. Series of γ-ray rocking curves measured in neighbouring volume elements of an as-grown Al single crystal (from G.G. Mair et al. [20])

8.4.2 Testing of neutron monochromator crystals

Most imperfect single crystals to be used as neutron monochromators show a much lower reflectivity than one would expect according to Darwin's theory for the mosaic spread deduced from the experimental neutron diffraction pattern. In order to study this behaviour we have performed model calculations based on a series of 56 γ-ray rocking curves measured in different volume elements of a large copper single crystal. In Fig. 20 two local mosaic distribution functions ($W_{3,7}(\omega)$ and $W_{4,2}(\omega)$) are shown together with the average over the 56 volume elements, $W_1(\omega)$. After some assumptions concerning primary extinction, reasonable agreement between theory and γ-ray measurements on one hand and the neutron diffraction pattern on the other hand could be obtained [30]. It was shown that the inhomogeneous mosaic structure of the monochromator crystal is the main reason for the relatively low neutron reflectivity. Some of the new developments in the production of neutron monochromators, summarized recently by Freund and Forsyth [31], were stimulated by the results from γ-ray diffractometry.

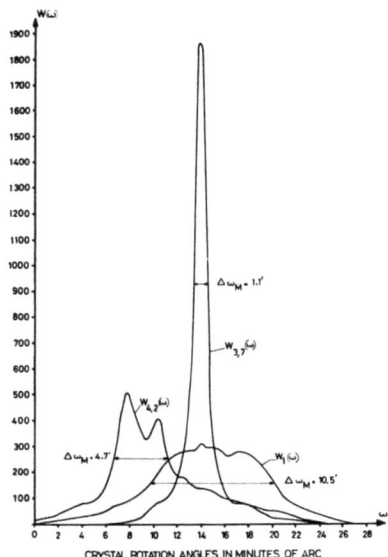

Figure 20. Mosaic distribution functions representative of the mosaic structure of a large Cu single crystal $W_{3,7}(\omega)$ and $W_{4,2}(\omega)$ are the distributions which show the highest or the lowest peak value of all 56 locally measured distributions. $W_1(\omega)$ represents the average distribution for the whole crystal.

8.4.3 Testing samples for neutron scattering experiments

The next example will deal with the investigation of the defect structure of a sample for a neutron diffraction experiment using γ-ray diffractometry. In order to determine the Debye – Waller factor from high resolution, short wavelength diffraction data a perfect single copper crystal was plastically deformed to obtain a well-defined mosaic distribution [32]. The dimensions of the sample were chosen in a range suitable for both γ-ray and neutron diffraction measurements. The samples were rather strongly deformed so that firstly the mosaic spread parameter could be expected to dominate the influence of the particle size parameter, and secondly so that the extinction effect is not bigger than about 30% for the strongest reflection (whereby the difference between the different theories of secondary extinction can be assumed to be rather small).

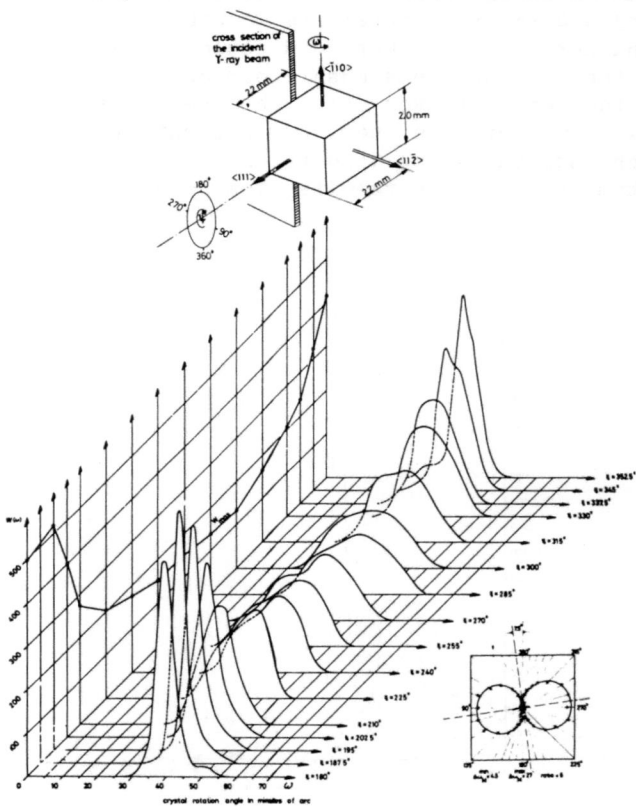

Figure 21. Mosaic distributions $W(\omega)$ deduced from γ-ray rocking curves measured in a 10 mm^3 Cu single crystal at the 111 reflection for different azimuthal angles ψ. The peanut shaped curve represents a plot in polar coordinates of the FWHH = $\Delta\omega_M$ of $W(\omega)$ as a function of ψ.

The sample of about 10 mm^3 was cut so that the (111), ($\bar{1}$10) and (11$\bar{2}$) planes formed the cubic shape of the crystal. γ-ray rocking curves were measured for these three reflections for different settings, i.e. the sample was rotated by 15° or 30° about the scattering vector between two rocking curve measurements. Results are shown in Figs. 21-23. From our experience with imperfect single crystals, the mosaic structure of this sample must be regarded as exceptionally homogeneous. By plotting the FWHH of the mosaic distributions W(ω) as a function of ψ one obtains the characteristic peanut shaped functions also presented in Figs. 21-23. As shown by Thornley and Nelmes in 1976 [33], the observable mosaic distribution is a fourth-order surface when the intrinsic mosaic distribution is ellipsoidal. Therefore the "peanuts" plotted in Figs. 21-23 are not too surprising, although the anisotropy in the FWHH of the measured mosaic distributions is remarkably strong. Next the neutron data were refined with a modified version of the Coppens-Hamilton formalism [34] and later using the Thornley and Nelmes theory. The physical quantities deduced from both refinements were in very good agreement, but we should mention that this is probably the first data set where the classical Coppens-Hamilton approach to correct for anisotropic extinction failed. Therefore we would like to recommend the use of the Thornley-Nelmes formalism which is based on clear physical arguments and not on a somewhat artificial extra parameter.

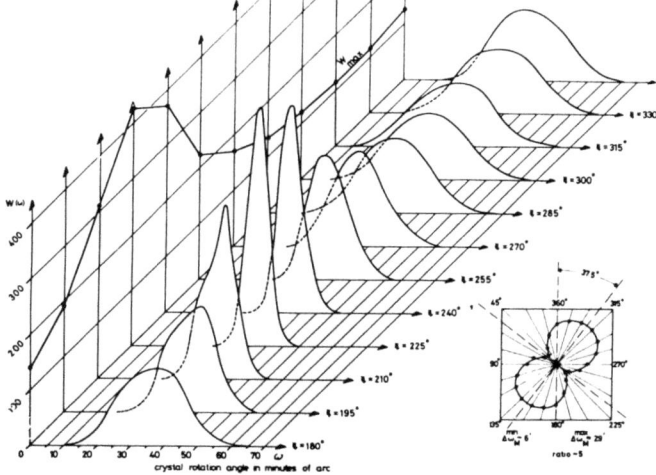

Figure 22. Mosaic distribution W (ω) measured at the 220 reflection for different azimuthal angles ψ. The peanut shaped curve represents a plot in polar coordinates of the mosaic spread $\Delta\omega_M$ as a function of ψ.

The mosaic spread parameter from the refinement of the neutron data was plotted together with the γ-ray results in Figs. 24-25 and again we find peanut shapes but now much smaller anisotropy. After deconvoluting the neutron diffraction profiles measured at 111 for a series of ψ settings from the instrumental resolution, a plot of their FWHH as a function of ψ is close to the γ-ray results and far from the mosaic spread parameter obtained in the refinement. From the shape of some of the γ-ray rocking curves shown in Figs. 21-23 one can suspect that the lattice planes may be bent. This would explain qualitatively the difference between the results from γ-ray diffractometry and the refinement of the neutron intensities, because in the refinement only the local mosaicity will determine the mosaic spread parameter, whereas the γ-ray rocking curves average over the whole crystal. Indeed, by studying a series of neighbouring volume elements on the γ-ray diffractometer this suspicion could be confirmed, since as indicated in Fig. 26 certain lattice planes were bent macroscopically.

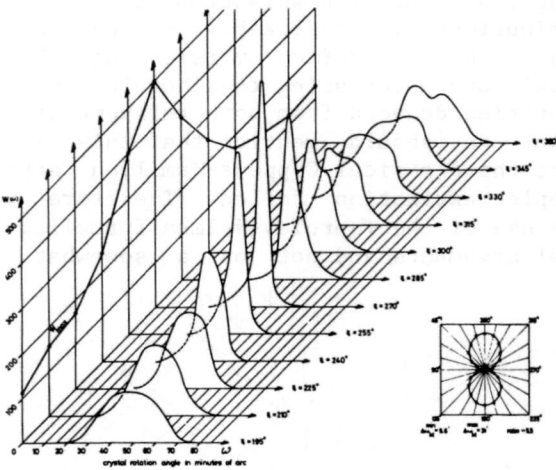

Figure 23. Mosaic distribution W (ω) measured at the $22\bar{4}$ reflection for different azimuthal angles ψ The peanut shaped curve represents a plot in polar coordinates of the mosaic spread $\Delta\omega_M$ as a function of ψ

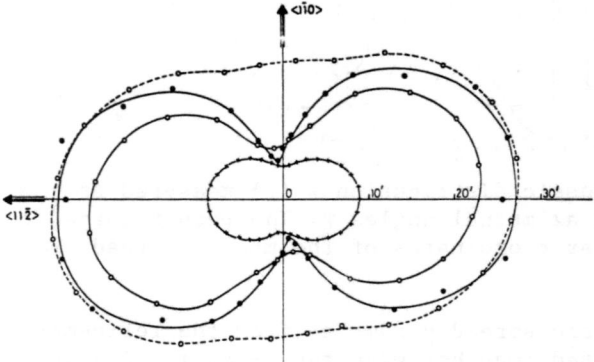

Figure 24. Mosaic spread for the 10 mm^3 crystal cube as a function of rotation around the scattering vector 111: x - x - x: cut through the mosaic distribution deduced from the crystal structure refinement. ●---●---●: result obtained by γ-ray diffractometry. o:::o:::o: FWHH of the measured neutron profiles o---o---o: FWHH of the neutron profiles after a Gaussian deconvolution of the measured profiles with the resolution function of the neutron diffractometer, with an agreement with direct measurement was estimated to have a FWHH of about 16′. The average standard deviation for the values derived from refinement results is 1′

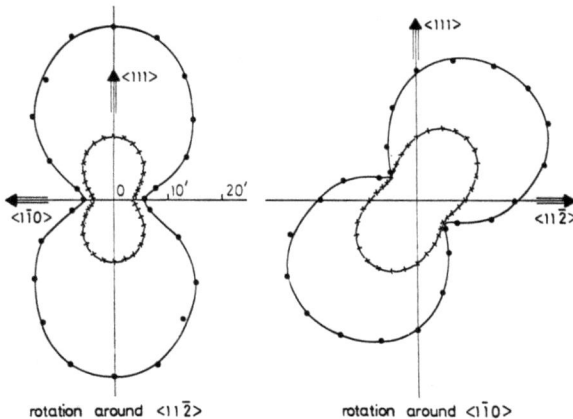

Figure 25. Mosaic spread for the 10 mm^3 Cu crystal as a function of rotation around the scattering vectors $<11\bar{2}>$ and $<1\bar{1}0>$, respectively; x - x - x : cut through the mosaic distribution deduced from the crystal structure refinement, ●---●---● : results obtained by γ-ray diffractometry.

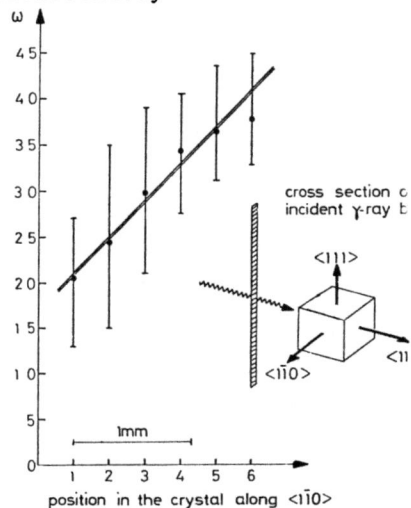

Figure 26. Location of rocking curves indicated by the FWHH as a function of the measurement position in the crystal for a scan of the crystal along $1\bar{1}0$. The dots indicate the centre of gravity of the profiles.

8.4.4 Study of structural phase transitions

The ferroelectric system KH_2PO_4 was studied extensively including a very interesting investigation of the tricritical point by γ-ray and neutron diffractometry as a function of temperature [35-38]. RbH_2PO_4 was studied on the γ-ray diffractometer allowing simultaneous optical observations of the domain structure [39]. The

Jahn-Teller phase transitions in $TmAsO_4$ [40] and $TbVO_4$ [41] were studied on the diffractometers in Risø and Grenoble. The mosaic structure of RbI was studied as a function of pressure [42-43]. In Jülich the precipitation morphology in the $\alpha+\alpha'$ two-phase region of the Niobium-Hydrogen system was studied by γ-ray diffractometry [44]. The "50 - K - Transition" in $PdH_{0.73}$ was investigated by means of γ-ray and neutron diffraction [45].

REFERENCES

1. W. Schmatz (1978) Disordered Structures in Neutron Diffraction (ed. H. Dachs), Springer-Verlag, Berlin-Heidelberg-New York.
2. A.R. Lang (1973) these proceedings.
3. V. Gerold and G. Kostorz (1978) J. Appl. Cryst. 11, 376
4. W. Schmatz (1973) X-ray and neutron scattering studies on disordered crystals in vol. 2 of A Treatise on Materials Science and Technology (ed. H. Herman), Academic Press, New York
5. H. Trinkhaus (1972) Phys. Stat. Sol. (b) 51, 307
6. P.H. Dederichs (1973) J. Phys. F: Metal Phys. 3, 471
7. H. Peisl (1976) Journal de Physique, Colloque C7, Supplement au no 12, 37, 47
8. H. Peisl (1976) Defects and their structure in nonmetallic solids, Plenum, New York, p.381
9. H.G. Haubold (1976) Rev. de Phys. Applique, 11, 73
10. P. Ehrhart and W. Schilling (1974) Phys. Rev. B8, 2604
11. H.G. Haubold, Rep. Kernforschungsanlage Julich, JUL -1090-FF H.G. Haubold (1975) J. Appl. Cryst. 8, 175
12. M. Wilken (1975) J. Appl. Cryst. 8, 191
13. N. Kato (1976) Acta Cryst. A32, 453, 458
14. J.R. Schneider (1976) J. Appl. Cryst. 9, 394
15. J.R. Schneider (1974) J. Appl. Cryst. 7, 541, 547
16. J.R. Schneider, P. Pattison and H.A. Graf (1979) Nucl. Instr. Meth., in press
17. A. Freund and J.R. Schneider (1972) J. Cryst. Growth 13/14, 247
18. J.R. Schneider (1975) J. Appl. Cryst. 8, 530
19. K. Møllenbach (1975) private communication
20. G. Mair, H.J. Fenzl, H. Bleichert and L. Gain (1979) private communication
21. W. Adlhart, F. Frey and J. Schneider (1978) J. Phys. E: Sci. Instrum. 11, 433
22. W.B. Yelon, R.W. Alkire and G. Schupp (1979) Nucl. Instr. Meth., in press W.B. Yelon (1979) private communication
23. A. Freund (1973) Dissertation, Technische Universität Munchen
24. A. Authier and F. Balibar (1970) Acta Cryst. A26, 647
25. B.C. Larson and F.W. Young Jr. (1971) Phys. Rev. B4 1709

26. P.H. Dederichs (1972) Sol. State Phys., 27, 135
27. C.G. Darwin (1914) Phil. Mag 27, 315, 657; C.G. Darwin (1922) Phil. Mag. 43, 800
28. W.H. Zachariasen (1945) Theory of X-ray Diffraction in Crystals, Wiley, New York
29. J.R. Schneider (1977) Acta Cryst. A33, 235
30. J.R. Schneider (1975) J. Appl. Cryst. 8, 195
31. A. Freund and J.B. Forsyth (1979) Neutron Scattering in Materials Science (ed. G. Kostorz) in the series A Treatise on Materials Science and Technology (ed. H. Herman). Chapter X. Academic Press, N.Y.
32. M.S. Lehmann and J.R. Schneider (1977) Acta Cryst. A33, 789
33. F.R. Thornley and R.J. Nelmes (1976) Acta Crst. A30, 748
34. P. Coppens and W.C. Hamilton (1970) Acta Cryst. A26, 71
35. P. Bastie, J. Bornarel, J. Lajzerowicz, M. Vallade and J.R. Schneider (1975) Phys. Rev. B12, 5112
36. P. Bastie, J. Bornarel, J. Lajzerowicz and J.R. Schneider (1976) Ferroelectrics 14, 587
37. P. Bastie, M. Vallade, C. Vettier and C.M.E. Zeyen (1978) Phys. Rev. Lett. 40, 337
38. P.M. Bastie and J. Bornarel (1979) J. Phys. C: Solid State Phys. 12, 1785
39. P. Bastie, J. Lajzerowicz and J.R. Schneider (1978) J. Phys. C 11, 1203
40. K. Møllenbach, K.J. Kjems and S.H. Smith (1977) in Electron-Phonon Interactions and Phase Transitons (ed. T. Riste) NATO ASI Geilo, Plenum, New York p 323.
41. S.R.P. Smith and B.K. Tanner (1978) J. Phys. C11, L 717
42. O. Blaschko, G. Ernst and J.R. Schneider (1977) J. Phys. C 10, 23
43. O. Blaschko, G. Ernst and J.R. Schneider (1977) J. de Physique, 38, 407
44. H.J. Fenzl, M.A. Pick and H. Wenzl (1977) Scripta Metallurgica 11, 271
45. O. Blaschko, R. Klemencic, P. Weinzierl and O.J. Eder (1978) Solid State Comm. 27, 1149

CHAPTER 9

ELEMENTARY DYNAMICAL THEORY

MICHAEL HART

9.1 Introduction

The dynamical theory of X-ray diffraction in perfect crystals was essentially complete long before definitive experiments could be done. There are many summaries available in the literature [1-12] though no single review covers all of the concepts which we need adequately to interpret the phenomena which are observed under the usual spherical wave conditions.

Penning and Polder [13] in 1961 first developed the germ of a ray theory of X-ray wavefield propagation in weakly deformed crystals which led later to a detailed understanding of the problem of calculating the images of imperfections which are obtained in X-ray diffraction topographs. In this article we shall review the development of the dynamical theory of X-ray diffraction in perfect crystals and in homogeneously strained crystals paying special attention to the physical basis of the theory and its verification.

9.2 The Laue condition

In Bragg reflection we are, by definition, interested in the elastic scattering of photons by a periodic medium: the crystal. The incident wave and the diffracted wave have their wavevectors restricted by the conservation of energy and momentum. Thus

$$\hbar\omega_o = \hbar\omega_h$$

and outside of the crystal this implies that $|K_o^i| = |K_h^i| = k$. Inside the crystal the Laue condition must be satisfied

$$\underline{K}_o + \underline{h} = \underline{K}_h \qquad (1)$$

We know that the refractive index for X-rays is very close to unity so that the additional condition

$$|K_o| \simeq |K_h| \simeq |K_o^i| = |K_h^i| = \lambda^{-1}$$

also applies. At this stage we can already understand many of the important features of Bragg reflection in perfect crystals. Inside the crystal there are two waves whose difference in wavevector is exactly equal to the reciprocal lattice vector corresponding to the Bragg-reflecting plane (equation (1)). Therefore, inside the crystal there is set up a standing wavefield whose spacing and orientation are exactly those of the diffracting planes. We easily anticipate that two waves so constrained must be phase coherent, as indeed they are. All that remains then is to determine the relative amplitudes and phases of the two waves. This we do in three stages; first we find a set of waves which can coexist in dynamical equilibrium in an infinite perfect crystal, then we see which of these solutions are excited by the incident wave, and, finally, which of these solutions can survive the absorption of the crystal, apertures or other obstacles which occur in the particular experimental arrangement under discussion. The relative phases and amplitudes of the participating waves are determined because the waves must also satisfy Maxwell's equations.

The familiar case of refraction at the boundary of a dielectric (non-diffracting) medium is presented as an introduction to the dynamical theory in the Appendix.

9.3 Formal definition of the crystal optical parameters

9.3.1 The polarizability of a crystal

The fundamental problem is to find solutions of Maxwell's equations in a medium which has a periodic complex dielectric constant. First, we need to obtain an expression for the dielectric constant of the crystal in a convenient form. The density of scattering matter can be expanded as a Fourier sum over the reciprocal lattice as

$$\rho'(\underline{r}) = \frac{1}{V} \sum_h F_h \exp(-2\pi i \underline{h} \cdot \underline{r}). \qquad (2)$$

In the limiting case of geometrical optics, corresponding to $\lambda \to 0$, the density of scattering matter is the same as the electron density $\rho(\underline{r})$.

By the Fourier integral theorem we can also write the inverse relationship

$$F_h = \int_V \rho'(\underline{r}) \exp(2\pi i \underline{h} \cdot \underline{r}) dV. \qquad (3)$$

Classical dispersion theory gives, for the refractive index, a

Sellmeier type relationship [14]

$$n^2 - 1 = \sum_a \frac{4\pi \rho(r) e^2}{m(\omega_a^2 - \omega^2)}$$

but, because we are concerned only with the two K electrons or eight L_2 electrons in most cases which are of interest, the deviation of $(n^2 - 1)$ from proportionality to the square of the X-ray wavelength is only small. Consequently, it is convenient to write instead

$$n^2 - 1 = \frac{-4\pi e^2 \rho'}{m\omega^2} = \frac{-4\pi e^2}{m\omega^2 V} \sum_a (Z_a + f'_a + if''_a) \quad (4)$$

in which the variation of refractive index (including absorption) with X-ray frequency is accounted for in an equivalent though different way. Both f' and f'' are functions of frequency. Then we can calculate the dielectric constant, neglecting terms of order 10^{-9} to obtain the result

$$n = 1 - \frac{2\pi e^2}{m\omega^2 V} \sum_a (Z_a + f'_a + if''_a)$$

The constitutive relation is

$$D = \epsilon E$$

and the polarization P is defined by

$$D = E + 4\pi P.$$

Since the susceptibility χ is defined by

$$\chi \underline{D} = 4\pi \underline{P} = \frac{\epsilon - 1}{\epsilon} \underline{D}$$

it follows that

$$\chi = \frac{\epsilon - 1}{\epsilon} = \frac{n^2 - 1}{n^2} = \frac{-4\pi e^2}{m\omega^2} \rho' \quad (5)$$

In the same way that the density of scattering matter was expanded as a Fourier sum over the reciprocal lattice so too can we write the susceptibility χ. As in equation (2) we have

$$\chi = \sum_h \chi_h \exp(-2\pi i \underline{h} \cdot \underline{r}) \quad (6)$$

and, using equation (5) with $\epsilon \simeq 1$

ELEMENTARY DYNAMICAL THEORY

$$\chi = \varepsilon - 1 = \frac{-e^2\lambda^2}{\pi mc^2 V} \sum_h F_h \exp(-2\pi i\underline{h}.\underline{r}) \qquad (7)$$

so that, comparing these last two equations,

$$\chi_h = \frac{-e^2\lambda^2}{\pi mc^2 V} F_h. \qquad (8)$$

9.3.2 Properties of the structure amplitude

Let us look in a little more detail at the properties of these Fourier coefficients in the expansion of the susceptibility χ. From equation (8) we see that χ_h is proportional to F_h, the structure amplitude for the Bragg reflection. The structure amplitude can be written as a sum of the atomic scattering amplitudes (with appropriate phase factors) over the unit cell of the structure. Thus,

$$F_h = \sum_a f_a \exp(2\pi i\underline{h}.\underline{r}_a).$$

We assume here that the charge density in the atom is not changed when the atom is incorporated into the crystal. Under the same assumption the detailed physics of the scattering process can also be included by treating the atomic scattering factors as complex quantities

$$F_h = \sum_a (f_a + f'_a + if''_a) \exp(2\pi i\underline{h}.\underline{r}_a). \qquad (9)$$

The structure factor is, in general, a complex number and depends on the choice of the origin of our position vector $\underline{r}_a$. If we shift the origin to $\underline{s}$ so that $\underline{R}_a = \underline{r}_a + \underline{s}$ then we have

$$F_h[\underline{s}] = \exp(2\pi i\underline{h}.\underline{s}) F_h[0]$$

where $F_h(0)$ is the structure factor referred to the origin 0. Because the structure amplitude is so important in determining the phases of diffracted waves, let us briefly digress to establish some of its properties in special cases.

$$F_{\bar{h}} = \sum_a (f_a + f'_a + f''_A) \exp(-2\pi i\underline{h}.\underline{r}_a) \qquad (10)$$

so that

$$F_{\bar{h}}[\underline{s}] = \exp(-2\pi i\underline{h}.\underline{s}) F_{\bar{h}}[0].$$

If the crystal does not absorb X-rays then $f'' \equiv 0$. By inspection of

219

equations (9) and (10) we see that F_h and $F_{\bar{h}}$ are then complex conjugates irrespective of our choice of origin. Thus,

$$F_{\bar{h}} = F_h^*.$$

If the crystal structure is centrosymmetric and the origin from which the structure amplitude is measured is chosen in the symmetry centre we see, again by inspection that F_h and $F_{\bar{h}}$ are equal (because all terms in the summations are even) even if there is some anomalous dispersion so that the atomic scattering factors themselves are complex. In a centrosymmetric crystal with zero absorption F_h and $F_{\bar{h}}$ are also complex conjugate quantities. In that case it follows that they are equal and real. Whether or not the crystal has a centre of symmetry, if the absorption is zero so that F_h and $F_{\bar{h}}$ are complex conjugates, it follows that the product $(F_h F_{\bar{h}})$ is a real quantity.

Let us take two different crystal structures as specific examples; the diamond structure represented by C, Si and Ge and the rock salt structure of LiF and NaCl. Choosing the origin at an inversion centre of the crystal structure midway between two carbon atoms, the coordinates of atoms in the unit cell of the diamond structure are [15]

$(\tfrac{1}{8}, \tfrac{1}{8}, \tfrac{1}{8})$ $(\tfrac{1}{8}, \tfrac{5}{8}, \tfrac{5}{8})$ $(\tfrac{5}{8}, \tfrac{1}{8}, \tfrac{5}{8})$ $(\tfrac{5}{8}, \tfrac{5}{8}, \tfrac{1}{8})$

$(\tfrac{3}{8}, \tfrac{3}{8}, \tfrac{3}{8})$ $(\tfrac{3}{8}, \tfrac{7}{8}, \tfrac{7}{8})$ $(\tfrac{7}{8}, \tfrac{3}{8}, \tfrac{7}{8})$ $(\tfrac{7}{8}, \tfrac{7}{8}, \tfrac{3}{8})$

In the rock salt structure let us choose the inversion centre which is on a sodium atom. The atomic coordinates in the unit cell are

Na at $(0, 0, 0)$ $(\tfrac{1}{2}, \tfrac{1}{2}, 0)$ $(0, \tfrac{1}{2}, \tfrac{1}{2})$ $(\tfrac{1}{2}, 0, \tfrac{1}{2})$

Cl at $(\tfrac{1}{2}, \tfrac{1}{2}, \tfrac{1}{2})$ $(\tfrac{1}{2}, 0, 0)$ $(0, \tfrac{1}{2}, 0)$ $(0, 0, \tfrac{1}{2})$

Taking low order Bragg reflections as examples we find, by direct substitution in equations (9) and (10), the following values for the structure factors

$$F_{220} = F_{\bar{2}\bar{2}0} = -8(f_C + f'_C + if''_C) \tag{11a}$$

$$F_{440} = F_{\bar{4}\bar{4}0} = +8(f_C + f'_C + if''_C) \tag{11b}$$

$$F_{200} = F_{\bar{2}00} = +4\{(f_{Na} + f_{Na'} + if''_{Na}) + (f_{Cl} + f'_{Cl} + if''_{Cl})\} \tag{11c}$$

$$F_{400} = F_{\bar{4}00} = +4\{(f_{Na} + f'_{Na} + if''_{Na}) + (f_{Cl} + f'_{Cl} + if''_{Cl})\} \tag{11d}$$

Notice that in the case of zero absorption (so that $f'' \equiv 0$) these structure amplitudes would all be real. With other choices of origin, the structure factors may of course become complex quantities even when the imaginary parts of the atomic scattering factors are zero. The implicit assumption that $\text{Re}(F_h)$ is positive is usually made. This can always be arranged without loss of generality by a suitable choice of origin.

9.4 Solution of Maxwell's equations

There are several simplifying approximations to be made. First we can assume that the electrical conductivity is zero and that the magnetic permeability μ is unity. In Gaussian CGS units we have

$$\nabla \times \underline{E} = -\underline{\dot{B}}/c$$

$$\nabla \times \underline{H} = \underline{\dot{D}}/c$$

In the solution of the Bragg-reflection problem we need damped plane wave solutions of Maxwell's equations (or sums of plane waves) for which the allowed wavevectors are related to one another by the Laue condition. When two waves are very much stronger than any others the relationship between their amplitudes turns out to be

$$0 = \{k^2(1 + \chi_o) - \underline{K}_o \cdot \underline{K}_o\}D_o + k^2 C\chi_{\bar{h}} D_h \tag{12}$$

$$0 = k^2 C\chi_h D_o + \{k^2(1 + \chi_o) - \underline{K}_h \cdot \underline{K}_h\}D_h.$$

(A detailed solution of Maxwell's equations under these conditions can be found in appendix A of Batterman and Cole's paper and in all of the general references cited earlier). The two equations are independent. For non-trivial solutions we obtain the secular equation

$$\begin{vmatrix} k^2(1 + \chi_o) - \underline{K}_o \cdot \underline{K}_o & k^2 C\chi_{\bar{h}} \\ k^2 C\chi_h & k^2(1 + \chi_o) - \underline{K}_h \cdot \underline{K}_h \end{vmatrix} = 0 \tag{13}$$

The Laue condition, equation (1), places a further constraint on the allowed wavevectors. Since $\underline{K}_o + \underline{h} = \underline{K}_h$ and $\underline{h}$ is by definition real, it follows that

$$\text{Im}(K_o) = \text{Im}(K_h). \tag{14}$$

The secular equation can be conveniently simplified by making the substitutions

$$2k\alpha_o = \underline{K}_o \cdot \underline{K}_o - k^2(1 + \chi_o) \tag{15a}$$

$$2k\alpha_h = \underline{K}_h \cdot \underline{K}_h - k^2(1 + \chi_o) \tag{15b}$$

so that equation (13) becomes

$$\alpha_o \alpha_h = \tfrac{1}{4} k^2 C^2 \chi_h \chi_{\bar{h}} \tag{16}$$

and it is also convenient to define the amplitude ratio $\xi = D_h/D_o$. Using equations (13), (14) and (15) we find in addition that

$$\xi = D_h/D_o = 2\alpha_o/C\chi_{\bar{h}}k = C\chi_h k/2\alpha_h \tag{17}$$

and that

$$\xi^2 = \alpha_o \chi_h / \alpha_h \chi_{\bar{h}} \tag{18}$$

Equation (17) tells us that the phase of the amplitude ratio ξ is determined by the phase of the structure amplitude. If we have a wave incident on a crystal near the Bragg angle we know that the phase of the continuation of the primary wave will not change if the crystal is translated. However, the relative phases of D_h and D_o change by $2\pi \underline{h} \cdot \underline{s}$ where $\underline{s}$ is now the body translation vector for the crystal. The phase shift is just the change in phase of the structure amplitude in the frame of reference of the primary wave.

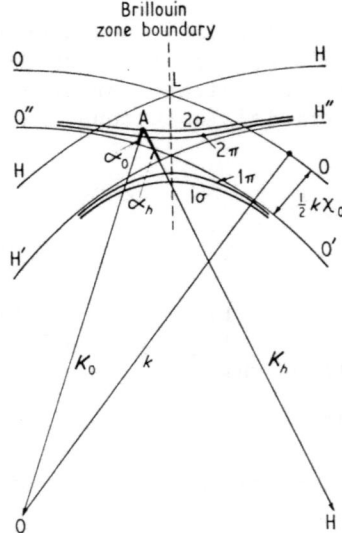

Figure 1. Dispersion surface showing the loci of tie points like A. These define wavevectors whose real parts satisfy both Maxwell's equations and the Laue equation when there are only two strong waves.

ELEMENTARY DYNAMICAL THEORY

Let us temporarily make the assumption that there is no absorption. Then χ_o is real and $\chi_h = \chi_{\bar{h}}$. From equation (16) we see then that the right-hand side is real and that there are two types of solution in which α_o and α_h are either both real or are complex conjugate quantities. Before we consider these two possibilities in detail let us first look at a useful geometric interpretation of these solutions, drawn in fig 1. OO and HH are the traces of segments of spheres of radius k drawn with centres respectively on the origin O and the point H of the reciprocal lattice. When no Bragg-reflected beam is excited, the corresponding situation inside the crystal is that the wavevectors start on spheres O'O" and H'H" which have radii equal to $k(1 + 1/2\chi_o)$, the wavevectors corrected for the mean refractive index of the crystal end at the reciprocal lattice points O and H. Because $|K_o| \approx |K_h| \approx k$, the Laue equation (1) tells us that Bragg reflected waves occur only when the tail of the wavevector for the primary wave lies near the Laue point L which is on the Brillouin zone boundary. The loci of the tails of the wavevectors allowed during Bragg reflection deviate from these spheres in the region near the Laue point and the dispersion equation (16) is the equation of these loci. This convention, in which wavevectors are drawn from the solution surfaces to the reciprocal lattice points, is peculiar to X-ray diffraction but is universally accepted. The opposite convention is commonly used in texts concerned with the theory of solids.

Consider the solution corresponding to the point A on the dispersion surface; α_o is just the perpendicular distance of A from the sphere O'O" and α_h is the corresponding distance measured from the sphere H'H". Thus, as long as the spheres can be approximated by planes, the equation of the dispersion surface approximates the equation of a hyperboloid of revolution with OH as axis. At the diameter points $\alpha_o = \alpha_h$ so that the semidiameter of the hyperbola is given by

$$d_s = \tfrac{1}{2}kC(\chi_h\chi_{\bar{h}})^{\tfrac{1}{2}}\sec\theta \qquad (19)$$

The diameter is approximately $\chi_h k$ and that is $10^{-5}k$. Thus, the region near the Laue point in fig. 1 is greatly magnified with respect to the lengths of the wavevectors. Consider for a moment the case when there is only one wave present in the crystal so that there is no Bragg reflection. If $D_h \to 0$ it follows that $\xi \to 0$ by definition and so too does $\alpha_o \to 0$. The dispersion surface becomes coincident with the sphere O'O", the wave-surface for a single wave. The wavevector is then $k(1 + \tfrac{1}{2}\chi_o)$ and is equal to the refractive index of the crystal. It follows then, using equation (1) that

$$\chi_o = \frac{-e^2\lambda^2}{\pi mc^2 V} \sum_a (Z_a + f'_a + if''_a).$$

Let us return now to the two possibilities which arise in the case of zero absorption when ($X_h X_{\bar{h}}$) is real.

9.4.1 Case 1: α_o and α_h both real

From equations (15a) and (15b) we see that both $\underline{K}_o$ and $\underline{K}_h$ are real. The appropriate solutions are represented by the four branches of the dispersion surface drawn in Fig. 1. While equation (16) describes a hyperbola, there are in fact two hyperbolas with common asymptotes; one for σ-polarized radiation with C = 1 and the other for π-polarized radiation with C = $|\cos 2\theta_B|$. The factor C, which appears first in equation (12), arises directly from the polarization dependence of the scattering amplitude of the Thompson electron. At each point on the dispersion surface the value of α_o is known so that the value of ξ is determined - the solution of Maxwell's equations gives us a complete specification (within the two-wave approximation) of the allowed waves in the crystal. Equations (16) and (17) completely determine the amplitudes and wavevectors of the allowed waves. On the upper two branches of the dispersion surface α_o is positive so that ξ is negative if $X_{\bar{h}}$ is real and negative; the two waves are in antiphase. We will consider the precise meaning of this statement later. On the two lower branches α_o is negative so that ξ is the same for all points on any one branch of the dispersion surface.

9.4.2 Case 2: α_o and α_h are complex conjugates

In this case we find from equations (15a) and (15b) that

$$\underline{K}_o \cdot \underline{K}_o = (\underline{K}_h \cdot \underline{K}_h)^*$$

or

$$|\underline{K}_o| = |\underline{K}_h^*|$$

so that

$$|\text{Re}(\underline{K}_o)| = |\text{Re}(\underline{K}_h)|.$$

From this result it follows that both wavevectors end on the Brillouin zone boundary, shown in Fig. 1 as a broken line. Even when X_h is real, ξ is complex so that the relative phase of the two waves in this case will be a function of α_o. Using equation (18) with $X_{\bar{h}} = X_h$ we find the interesting result that $\xi^2 = 1$ for all possible values of α_o.

9.4.3 Summary of results for zero absorption

The results obtained so far can be very conveniently summarized in two diagrams. Fig. 2 shows an enlarged region of the dispersion surface near the Brillouin zone boundary and also the values of ξ drawn on an Argand diagram.

ELEMENTARY DYNAMICAL THEORY

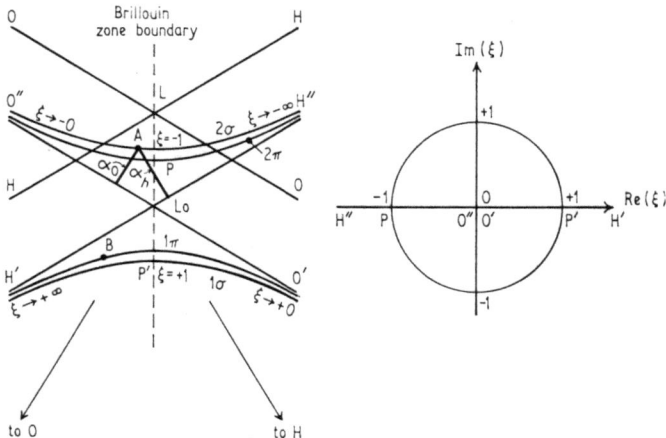

Figure 2. Solution surface diagram showing the real parts of wavevectors and the amplitudes of waves occurring during Bragg reflection.

Consider first a point such as A on the dispersion surface branch labelled 2σ. When A is near O'', α_o is small so that ξ is small and negative (equation (17)). As A is moved to the right on the same branch, $\xi = -1$ at the Brillouin zone boundary and ξ then becomes large and negative as A moves towards H''. The same sequence occurs for points on the 2π branch of the dispersion surface.

On the Argand diagram the corresponding amplitude ratio starts just to the negative side of the origin on the real axis and follows the real axis, crossing the unit circle at P when the solution point on the dispersion surface crosses the Brillouin zone boundary, and thence tending to $\xi = -\infty$.

For points on the lower branches of the dispersion surface, α_o is large and negative when B is near H' so that is large and positive. As B is moved to the right ξ decreases, becoming $+1$ at P' when the solution lies on the Brillouin zone boundary. When B lies near O', α_o is small and negative so that ξ is small and positive. The corresponding path on the Argand diagram is from H' to P' to O'.

Finally, in the case when the allowed wavevectors are not real we find that the real parts of the wavevectors end on the Brillouin zone boundary and then $|\xi| = 1$. The complex amplitudes are represented by points on the unit circle in the Argand diagram.

When the crystal absorbs X-rays the allowed solutions are more complicated. The real parts of the wavevectors of course remain unchanged but the wave amplitudes are altered. In those cases the amplitude diagram is very useful if one is concerned with intensities. Since we are concerned mainly with the phases we will still be able to profit from the dispersion surface showing only the real parts of the allowed wavevectors.

9.5 The Poynting vector and ray optics

The flow of energy in the crystal is described by the Poynting vector for the wavefield in the crystal. The instantaneous Poynting vector $\underline{S} = c\underline{E} \times \underline{H}/4\pi$ is a complicated function and, in practice, we are only concerned with its time and space average. The time average of the Poynting vector is [16]

$$\underline{\hat{S}} = \frac{c \, \text{Re} \, (\underline{E} \times \underline{H}^*)}{8\pi} \tag{20}$$

We recall that the Laue equation imposes the condition $\text{Im}(K_o) = \text{Im}(K_h)$ so that the imaginary parts of the Poynting vector have a common factor. The detailed spatial averaging has been done by von Laue [17] with the result that

$$\underline{\hat{\hat{S}}} = \frac{c}{8\pi} (|E_o|^2 \underline{\hat{s}}_o + |E_h|^2 \underline{\hat{s}}_h) \exp(4\pi \, \text{Im}(\underline{K}_o) \cdot \underline{r}) \tag{21}$$

or

$$\underline{\hat{\hat{S}}} = \frac{c}{8\pi} |E_o|^2 (\underline{\hat{s}}_o + |\xi|^2 \underline{\hat{s}}_h) \exp(4\pi \, \text{Im}(\underline{K}_o) \cdot \underline{r}). \tag{22}$$

In the process of averaging over a large volume of crystal we have of course lost the microscopic details of the standing wave patterns. On the other hand we have gained an important simplification in that we can use the energy flow properties of the wavefields to understand the macroscopic ray optics of the diffraction problem. Except for the evanescent modes which are excited in the Bragg case near the centre of the reflecting range, Penning [18] has shown that geometrical optics can be used to understand the diffraction process inside almost perfect crystals. If one takes the ray direction to be parallel to $\underline{\hat{S}}$ defined in equation (21), then well-behaved rays exist and the concomitant simplifications of geometrical optics can be adopted. On a ray-optical basis we can then decide which components are important in a particular problem, calculate phase shifts along the ray paths and then, where necessary, concentrate on important field components from the microscopic viewpoint.

The usefulness of the dispersion surface constructions now becomes very clear for it can be shown that the Poynting vector $\underline{\hat{S}}$ in equation (22) has the same direction as the normal to the dispersion surface at the point where ξ is defined. Proof that the suitably averaged Poynting vector $\underline{\hat{S}}$ is normal to the dispersion surface was first given by Ewald [19] and Kato [20].

Let us take Fig. 2 as an example. When A is near O" the direction of energy flow in the crystal is normal to the sphere O"O'; in the direction of $\hat{s}_o$. At the centre of the range of Bragg reflection when A is on the Brillouin zone boundary, the energy flow is along the Bragg planes in the direction $(\hat{s}_o + \hat{s}_h)$ because $\xi=1$.

When A is close to H" the energy flow is in the direction of $\hat{s}_h$. As we shall see later, the position of A can be swept across the dispersion surface if the local angle of incidence or the local Bragg angle is changed by a few seconds of arc.

It is sometimes convenient to express equation (22) in another way; if Ω is the angle between $\hat{\underline{S}}$ and the Bragg planes then

$$\tan \Omega = \frac{|\xi|^2 - 1}{|\xi|^2 + 1} \tan \theta.$$

The direction of $\hat{\underline{S}}$ is the ray path in light optics. To establish a ray approximation we must demonstrate that rays can be identified and retain their identity in the crystal.

9.5.1 The ray approximation

Rays can be defined in homogeneous media but after a path length $L = \Lambda^2/\lambda$, where Λ is the width of the ray, the ray is broadened by diffraction. The beam diverges by diffraction through an angle λ/Λ. In practice L is 2 cm for a light beam 0.1 mm wide and thus the ray approach has a very useful range of application in light optics. In the case of X-ray beams in a crystal the ray divergence is greater in the ratio of the curvature of the dispersion surface to the curvature K of the free wave sphere (Figs. 1, 2). That ratio is just $\chi_o (\simeq 10^{-5})$ at the Brillouin zone boundary. Rays 0.1 mm wide therefore propagate for 1 mm or so without losing their form in a Bragg reflecting perfect crystal.

In general the beam must be defined by one wavevector (which is a constant in a homogeneous medium) and wavevector matching is determined by Snell's Law at boundaries, as we shall discuss in the next section. Finally, ray theories permit the calculation of intensities by defining the ray intensity as $I = \int_A \hat{\underline{S}} \cdot d\underline{A}$ where $d\underline{A}$ is the cross-sectional area of the beam.

9.5.2 Phenomenology of absorption along the raypath

We already have the necessary equations that enable us to calculate the influence of absorption on the X-ray wavefields. First, from the Laue equation, we have

$$\text{Im}(\underline{K}_{oj}) = \text{Im}(\underline{K}_{hj}). \tag{14}$$

In addition, when matching wavevectors at the crystal surface we can take $\underline{K}_o{}^i$ as a real quantity without loss of generality and, separating the imaginary part of equation (21), find that

$$\mathrm{Im}(\underline{K}_{oj}) = -k\, \mathrm{Im}(g_j)\hat{\underline{n}} \qquad (23)$$

It should not be surprising to find that the imaginary parts of the wavevectors are always normal to the crystal surface. It means that the planes of constant attenuation are parallel to the crystal surface. The intensity attenuation coefficient μ in the direction of $\hat{\underline{n}}$ is related to the imaginary part of the wavevector in the usual way

$$\mu(\hat{\underline{n}}) = -4\pi\, \mathrm{Im}(\underline{K}).$$

For our development of ray optics we need to know attenuation coefficients, not in the direction of the surface normal, but along the direction of energy flow. The transformation is straightforward (but tedious) using the fundamental equations of the dynamical theory (13) and the above relationships between the imaginary parts of the wavevectors. A convenient and very direct derivation is given in a paper by Bonse [21] with the result

$$\mu(\underline{j}) = \frac{\mu}{B_1}(1 + |\xi|^2 + 2A\, \mathrm{Re}(\xi))$$

with

$$A = \frac{C}{\chi_o{}''}\, \mathrm{Re}(\chi_{ih}) + \frac{\mathrm{Im}(\xi)}{\mathrm{Re}(\xi)}\, \mathrm{Im}(\chi_{ih}) \qquad (24)$$

and

$$B_1{}^2 = 1 + |\xi|^4 + 2|\xi|^2 \cos 2\theta$$

This intractable result is greatly simplified if we restrict the discussion to centro-symmetric crystals and choose our origin at the symmetry centre. Then χ_{ih} is real and

$$A = C\chi_{ih}/\chi_{io}$$

Finally we have that

$$\mu(\hat{\underline{j}}) = \mu(1 + |\xi|^2 + \frac{2C\chi_{ih}}{\chi_{io}}\mathrm{Re}(\xi))(1 + |\xi|^4 + 2|\xi|^2 \cos 2\theta)^{-\frac{1}{2}} \qquad (25)$$

We can see that this result is physically satisfactory by looking at two extreme cases. If ξ is near zero or is very large then there is only one wave in the crystal (equation (17)) and the attenuation coefficient should be simply μ. That this is the case can be seen by inspection of the last equation. Of considerably more interest is the absorption of wavefields near the centre of the Bragg reflection range. As examples let us take $\xi = \pm 1$ so that the wavevectors begin on the Brillouin zone boundary. Then

$$\mu(\hat{\underline{j}}) = \frac{\mu}{\cos\theta}\left|1 \pm C\frac{\chi_{ih}}{\chi_{io}}\right|.$$

The term in parentheses essentially describes the variation with $\lambda^{-1}\sin\theta$ of the absorptive part of the atomic scattering factor. In most X-ray diffraction experiments we are concerned with K- and sometimes L- electron absorption. Consequently the absorption is not particularly dependent on the scattering angle because the electrons are well localized. In fact, Okkerse [22] has found for germanium that $\chi_{ih}/\chi_{io} \simeq \exp(-M)$, the usual Debye-Waller factor. Again, Batterman and Cole [7] have calculated some attenuation coefficients for the 220 reflection of germanium. For a 1 mm thick slab of material and CuKα_1 radiation they found $\mu t = 38$ for the normal photoelectric absorption, but for

Branch 2 $\xi = -1, \sigma$ polarization, $\mu(\hat{\underline{j}})t = 1.9$

$\xi = -1, \pi$ polarization, $\mu(\hat{\underline{j}})t = 12.5$

Branch 1 $\xi = +1, \pi$ polarization, $\mu(\hat{\underline{j}})t = 63.5$

$\xi = +1, \sigma$ polarization, $\mu(\hat{\underline{j}})t = 74$.

Even though the attenuation through normal photoelectric absorption due to the germanium is so large, the crystal is extremely transparent to σ-polarised radiation with $\xi = -1$. This anomalous transparency was first discovered experimentally by Borrmann [23,24] in calcite crystals. The discrimination against π-polarised radiation with $\xi = -1$ has led to the suggestion and demonstration that these thick crystals can be used as X-ray polarizers [25].

9.6 Boundary conditions linking external and crystal waves

We now turn to a consideration of the way in which the waves inside the crystal are excited by a plane wave incident on the crystal from outside. For simplicity we will only consider a linearly polarised incident wave, at the same time noting that this does not result in a loss of generality. By the appropriate coherent or incoherent superposition of the solutions for the two principal orthogonal states of polarization, the wavefield in the general case can be deduced from our restricted solution. As incident wave we take $\mathcal{D}_o^i = D_o^i \exp(2\pi i(\nu\tau - \underline{K}_o^i \cdot \underline{r}))$ but because we are interested only in steady-state solutions and are restricted to elastic scattering by the Laue equation (1), the time-dependent part of the wave amplitude is not of interest.

We avoid philosophical problems by asserting that the crystal surface is a plane, smooth and abrupt even on the scale of X-ray wavelengths. Logically more satisfactory approaches, such as Ewald's boundary wave method, lead to essentially the same conclusions as far as phase matching at the crystal boundaries is concerned. This is not too surprising since real surfaces are abrupt and smooth on a scale which is small compared with the material thickness t_λ which results in a phase shift of 2π. The connection between waves outside and waves inside the crystal can be made in the usual way by matching phases and amplitudes at the crystal boundary. Matching of the field vectors at the crystal surface is almost trivial because the refractive index is so close to unity, while matching of the phase velocity of waves in the crystal surface leads to the usual formulation of Snell's Law.

9.6.1 Amplitude matching

The boundary conditions for field vectors are that the tangential components of both E and H are continuous across the surface. Because the refractive index for X-rays is so close to unity, the reflection amplitude at the crystal surface is quite negligible provided that we are not concerned with glancing angles of incidence comparable with the critical angle for external total reflection.

This fact results in a major simplification in that, to a very good approximation, all of the field vectors are continuous across the crystal surface.

9.6.2 Wavevector matching

The waves inside the crystal must have the same phase velocity parallel to the crystal surface as those outside of the crystal. In other words, the components of the wavevectors in the crystal surface must be the same for both sets of waves. Using $\underline{K}_{oj}$ as the wavevector of any crystal 0 wave we have that

$$\underline{K}_o^i - \underline{K}_{oj} = kg_j \cdot \hat{\underline{n}}$$

where $\hat{\underline{n}}$ is a unit vector normal to the crystal surface, and the subscript j in the last equation takes values 2σ, 2π, 1σ and 1π, corresponding to the four branches of the dispersion surface shown in Fig. 3. For each of the allowed wavevectors $\underline{K}_{oj}$ there will of course be a different value for the accommodation g_j. There are two quite different possibilities shown in projection in Fig. 3.

To distinguish the two cases it is convenient to define two important angle parameters γ_o and γ_h, the direction cosines of the wavevectors with respect to the inward facing surface normal $\hat{\underline{n}}$.

$$\gamma_o = \cos\psi_o = k^{-1} \underline{K}_o \cdot \hat{\underline{n}}$$

$$\gamma_h = \cos\psi_h = k^{-1} \underline{K}_h \cdot \hat{\underline{n}}$$

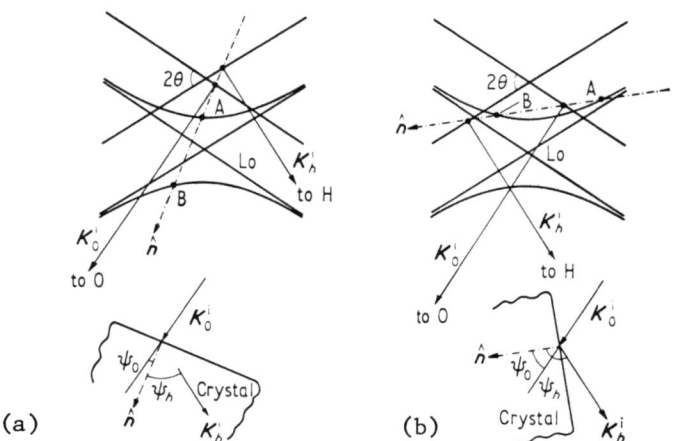

Figure 3. Boundary construction for an incident wave in (a) the Laue case and (b) the Bragg case.

Laue case: $\gamma_o \gamma_h > 0$. The allowed wavevectors must end both on the dispersion surface and on the surface normal drawn through the tip of the incident wavevector $\underline{K}_o^i$. In the Laue case, the geometry is such that there are always two solutions (corresponding to tie points A and B) one belonging to each branch of the dispersion surface. Additional complexity is possible in the Laue case because the waves exit from the crystal through the lower surface which may or may not be parallel to the entrance surface. Repeating the boundary construction at the lower surface may involve an additional parameter $\hat{n}'$, the normal to the lower surface.

Bragg case: $\gamma_o \gamma_h < 0$. In this case the normal $\hat{n}$ drawn through the tip of the incident wavevector $\underline{K}_o^i$ makes either two intersections with the same branch or one intersection with the Brillouin zone boundary. Again, if waves can reach the back surface of the crystal and then return to the front surface, complications can arise if the front and back surfaces are not parallel.

The centre of the range of reflection occurs when the normal to the crystal surface drawn through the tip of the incident wavevector passes through the Lorentz point L_o (Fig. 3). It is interesting and important to notice that the angle of incidence at the centre of the reflection range is different in the Laue case and in the Bragg case. For the symmetric cases, the difference between the Laue angle θ_L and the Bragg angle θ_B is

$$\theta_B - \theta_L = 2\delta \operatorname{cosec} 2\theta. \qquad (26)$$

9.6.3 Spherical waves

We have explicitly assumed that the incident X-ray beam is a monochromatic plane wave. Although that can be approximated, with difficulty, it is much more usual for the incident wave to approximate a spherical wave. Even in the case of synchrotron radiation sources, where one commonly hears the source referred to as a "plane wave source" and where the beam divergence may indeed be quite negligible, one must be careful to distinguish between plane waves and monochromatic plane waves. It is quite clear that almost all of the effects which are observed in X-ray diffraction topography require an understanding of spherical wave dynamical theory either by virtue of an integration or convolution over angle of incidence or because the results are smeared over wavelength (or both) because λ and θ are necessarily connected through Bragg's Law in any crystal system.

We can demonstrate the need for spherical wave theories quite simply. According to Fig. 2, the Bragg diffraction phenomena are observed only when the wavevectors $\underline{K}_o$ and $\underline{K}_h$ terminate within a distance of order d_s of the Brillouin zone boundary. In terms of angles, that is within a range of angles $\Delta\psi_o = 2\, d_s/k\sin\theta$. From equation (19) we find $\Delta\psi_o \simeq 10^{-5}$ in typical cases. On the other hand, the angular diameter of the first Fresnel zone at a distance ℓ from a point source is $\Delta\psi = 2\sqrt{\lambda/\ell}$. At a representative distance of 10 cm from the source the Fresnel zone diameters are larger than the angular range of Bragg reflection $\Delta\psi_o$ as the table shows.

radiation:	WKα_1	AgKα_1	MoKα_1	CuKα_1	CrKα_1
$\Delta\psi$/sec arc	5.96	9.75	11.0	16.2	19.7

Even at synchrotron radiation sources, where ℓ = 10 m might be more appropriate, the Fresnel zone diameter would be only ten times smaller. In most cases of practical interest $\Delta\psi_o$ is never very much smaller than $\Delta\psi$ so that in practice the whole range of Bragg reflection is coherently excited unless special X-ray optical arrangements are made.

9.7 Some demonstrations of the results of the dynamical theory

There have been many experimental demonstrations of the validity of dynamical theory, first for the electron case, where only minute crystals are required, and during the last twenty years for the X-ray case. A complete review is not necessary and I have therefore attempted to select those results which are most relevant to our need to interpret X-ray diffraction topographs.

9.7.1 The Borrmann effect

It is experimentally observed that X-rays may be transmitted at the Bragg angle through crystals whose normal absorption is so high that no detectable transmitted beam would be expected. The effect was first observed in calcite by Borrmann [23,24] though virtually all of the later quantitative work was done with germanium [18,22,26-28] because, at the end of the 1950's, germanium was routinely available as highly perfect crystals and the K-edge at 1.11 Å wavelength makes for straightforward experimentation with copper $K\alpha$ radiation at 1.54 Å.

The normal absorption for 1 mm of germanium with copper $K\alpha$ radiation is $\exp(-38)$ or 3.14×10^{-14}. Thus, no detectable transmitted X-ray beam is expected. As calculated in section 9.5.2 the plane-wave attenuation coefficients are also very large except in the case of σ-polarized radiation near the centre of the dispersion surface, on the upper branch, where $\xi = -1$. These, therefore, are the rays which are responsible for the anomalous transmission. From equation (22) we find $\underline{\hat{S}} \propto \hat{s}_o + \hat{s}_h$ (ie $\perp \underline{h}$) so that the raypath is parallel to the Bragg planes as Borrmann [24] had demonstrated. In 1961 Cole, Chambers and Wood [25] demonstrated that the anomalously transmitted radiation is linearly polarized as expected.

An important physical insight is obtained by calculating the microscopic wavefields which occur in the crystal. Only the waves with amplitude ratios near $\xi = -1$ penetrate the crystal. They are

$$\mathcal{D}_{o2} = D_{h2} \exp 2\pi i (\nu\tau - \underline{K}_{o2} \cdot \underline{r})$$

$$\mathcal{D}_{h2} = D_{h2} \exp 2\pi i (\nu\tau - \underline{K}_{h2} \cdot \underline{r}.$$

The fundamental equations of the dynamical theory provide us with a relatonship between the wave amplitudes

$$\xi_2 = D_{h2}/D_{o2} \tag{17}$$

while their wavevectors are connected by the Laue equation so that

$$\underline{h} = \underline{K}_{h2} - \underline{K}_{o2}. \tag{1}$$

The total field in the crystal is given by

$$|\mathcal{D}_2|^2 = |\mathcal{D}_{h2} + \mathcal{D}_{o2}|^2 \propto 1 + \xi_2^2 + 2\xi_2 \cos(2\pi \underline{h} \cdot \underline{r}). \tag{27}$$

Remembering that we need only consider wavefields with $\xi_2 \simeq -1$ if χ_h is negative, let us now see what this result means in the actual crystal structures and Bragg reflections considered in detail in the discussion of structure factors. If ξ_2 is negative, the electric field is a minimum when $\underline{h} \cdot \underline{r}$ is an integer or zero. The field intensity is modulated with a spatial periodicity $|h|^{-1} = d$ normal to the Bragg planes, as we anticipated earlier.

In the diamond structure, taking the origin of the unit cell at the inversion centre, we have from equation (11a) that $F_{220} = -8f_c$

where f_c is the atomic scattering factor for carbon. Thus the electric field has a maximum value at $\underline{r} = 0$ because F_h is negative, χ_h is positive (equation (8)), α_0 is positive and ξ_2 is positive (equation (17)). In Fig. 4(a) we have drawn a projection of the diamond structure onto the cube plane and have sketched the wavefield $|\mathcal{D}_2|^2$ noting that there is a field intensity <u>maximum</u> at the origin, which is a centre of inversion symmetry in the lattice.

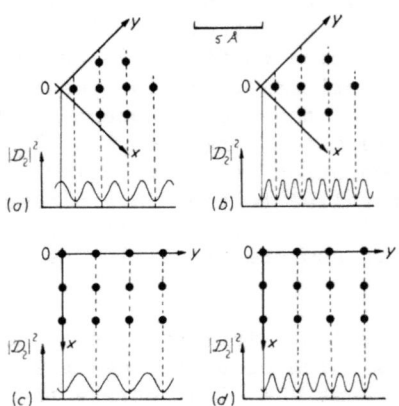

Figure 4. Standing wavefields formed during Bragg reflections: (a) 220 reflection and (b) 440 reflection from the diamond structure; (c) 200 reflection and (d) 400 Bragg reflection from the rock salt structure. Approximately to scale for Si and NaCl.

In the 440 Bragg reflection we have, from equation (11b), that $F_{440} = +8f_c$ so that in that case there is a <u>minimum</u> electric field intensity at the symmetry centre. Again, the physical situation is sketched in Fig. 4(b).

In the same way we can also sketch the wavefields in the rock salt structure. For both the 200 and 400 Bragg reflections F_h is positive (see equations (11c) and (11d)) so that the wavefield has intensity <u>minima</u> at the centre of inversion of the structure, as illustrated in Figs. 4(c) and 4(d). The results presented in Fig. 4 apply to <u>any</u> material crystallizing in the two structures considered and are interesting also from another viewpoint. In every case it turns out that the wavefield has intensity minima in the planes containing atoms. This, physically, is why the branch 2 wavefield suffers less than the normal absorption; the absorbing electrons in the atoms are fairly well localized near the atomic cores and in that region there is very little electric field. In fact the argument can be made semiquantitative as Batterman and Cole [7] have shown. The result is also reasonable from another point of view (Ewald [29]); one measure of the strength of the interaction between the incident wave and the crystal is the effective refractive index, or the distance from the Laue point L (Fig. 2) to the dispersion surface. This distance is a minimum for branch 2σ so that we should expect these wavefields to be weakly absorbed provided that the scattering centres coincide with or lie in the same planes as the absorption centres. By making judicious use of absorption edges it is possible to find crystal structures in which the planes of absorbing atoms lie between the principal planes of scattering so that the 2σ and 2π branch wavefields suffer more absorption than the 1σ and 1π branch wavefields [30,31].

We can also gain some insight on the influence of heat motion on the anomalous transmission effect. Thermal vibration of the atoms will result in them sampling larger electric fields than those at the atomic sites. Consequently we might expect that the absorption would increase with increasing temperature. Qualitatively then Okkerse's result cited earlier is consistent with the standing wavefield picture.

In all cases ξ_1 and ξ_2 have opposite signs so that the standing wave patterns for branch 1 waves are exactly complementary to those sketched in Fig. 4. On any one branch of the dispersion surface the sign of ξ is constant and the <u>position</u> of the standing wavefield in the crystal structure is independent of the magnitude of ξ. As $|\xi|$ varies, the modulation depth of the intensity field changes from zero when $|\xi|$ is very large or very small (only one beam present) to 100% for the σ-state of polarization with $|\xi| = 1$.

9.7.2 Ray tracing in perfect crystals

We have already found that the ray approximation may be justified provided that a sufficiently small bundle of wavevectors can be identified (sections 5.1, 5.2). In practice that implies the isolation of narrow-angle wave groups from the spherical wave source. In laboratory experiments collimation with slits or pinholes to 0.1 seconds of arc of less is impracticable so that some angular demagnification is necessary. Two different approaches have been made [32,22].

In the first arrangement (Fig. 5a) the angular magnification between the rays outside the crystal and the anomalously transmitted rays is equal to the curvature of the dispersion surface divided by the curvature of the free wave sphere (Fig. 1); say 10^5, so that in practice the transmitted beam slit need only be 150 µm wide to select a ray bundle whose effective divergence was 0.05 seconds of arc. Authier [32] used the scheme shown to verify the ray propagation characteristics in perfect crystals. The angular demagnification which can be obtained in oblique Bragg reflections (Fig. 5b) was used by Bonse [33] to study ray propagation in the Bragg case. Here the angular magnification is γ_h/γ_o and values up to 10^2 are reasonably easy to achieve. Thus, with a 10 µm slit, the effective ray divergence can be as small as 1 second of arc.

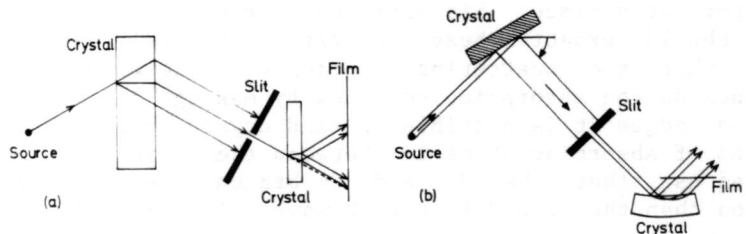

Figure 5. Schemes used to isolate narrow bundles of rays (a) using the angular magnification of the Borrmann fan [32] and (b) using oblique reflection [33]

Ray tracing experiments have shown that beams can be isolated and propagate, maintaining their form, in perfect crystals and that their path is parallel to the spatially averaged Poynting vector $\hat{\underline{S}}$ (equation (22)). More detailed measurements also showed [32] that the phenomenological treatment of absorption (section 9.5.2) is quite adequate to determine ray intensities.

9.7.3 Crystal reflection profiles

In the non-dispersive (or parallel) double crystal setting it is easily shown [1, 34, 35] that the "rocking curve", obtained as the variation of doubly diffracted intensity $I(\alpha)$ when the second crystal is rotated, is just the convolution of the reflectivity functions $R_1(\theta)$ and $R_2(\theta)$ for the two crystals. Thus

$$I(\alpha) = \int_{-\infty}^{\infty} R_1(\theta) \, R_2(\theta-\alpha) \, d\alpha \, . \tag{28}$$

Throughout the early days of X-ray spectroscopy these rocking curves were used to test the "perfection" of spectroscopic crystals. The two Bragg reflections were commonly from the opposite surfaces of a cut or cleaved crystal sample so that $R_1 \equiv R_2$. Since the self convolution is in principle symmetric such a procedure gives no *a priori* information about the shape of the reflection profile in the Bragg case but does conveniently characterise the crystal in terms of the rocking curve width and the integrated reflectivity, often quoted in terms of the "per cent reflection". Assuming the results of the dynamical theory, optical parameters can be obtained from equal crystal rocking curves by profile fitting. In practice the profile width is dominated by χ_{rh} and the reflectivity by χ_{io} and the parameter χ_{ih}/χ_{io} which determines the magnitude of the Borrmann effect. In the Bragg case, such analyses have been made, for example, by Bonse [36].

It is beyond the scope of this short introduction to dynamical theory to derive expressions for the crystal reflectivity R as a function of angle of incidence. In principle the derivations are straightforward, but they are extremely complicated when oblique reflections are considered and in the Laue case. Detailed derivations are given in several of the general references [2,6,9]. Archetypal rocking curves are shown in Fig. 6 in the approximation or zero X-ray absorption.

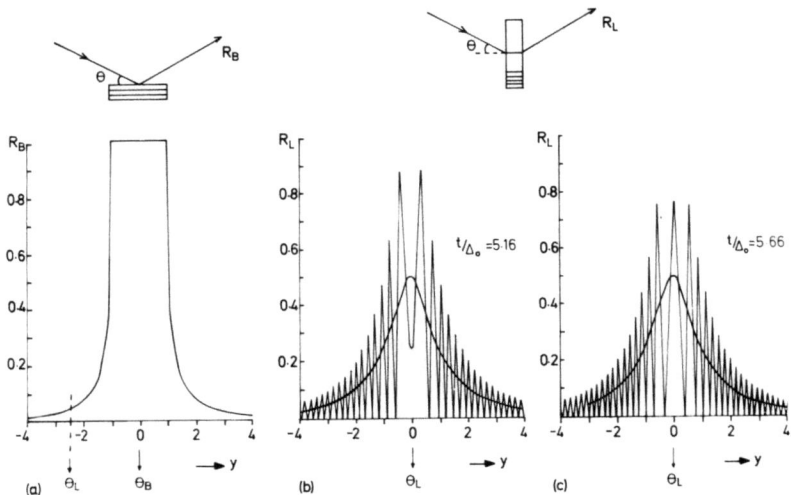

Figure 6. Zero absorption plane wave reflection curves in (a) the Bragg-case and, (b) and (c), in the Laue case for two different crystal thicknesses.

In the symmetric Bragg case the centre of the range of reflection is shifted from the Laue angle θ_L (2d sin $\theta_L = \lambda$) by $2\delta \cos 2\theta$ (equation (26)). Generally, the shift is given by

$$\theta_B - \theta_L = \frac{|\chi_{ro}|}{2 \sin 2\theta} \left(1 + \frac{|\gamma_h|}{\gamma_o}\right) \tag{29}$$

The width of the Bragg peak in the symmetric Bragg case is given by $\Delta\psi_o = 2|C|\chi_{rh}/\sin 2\theta$ and in general by

$$\Delta\psi_o = \frac{2|C| |\chi_{rh}|}{\sin 2\theta} \sqrt{\frac{|\gamma_h|}{\gamma_o}} . \tag{30}$$

It is customary and convenient to work in terms of a parameter y which varies from -1 to +1 in the range of total reflection. On the y-scale the Bragg reflection curve is given by

$$R_B(y) = \left[|y| - \sqrt{y^2 - 1}\right] \tag{31}$$

In the symmetric Laue case the situation is much more complicated because coherent wavefields from both branches of the dispersion surface contribute to the reflectivity. In the symmetric Laue case the centre of the Bragg reflection occurs at θ_L and, as one would expect, the peak is symmetric. On the y-scale the Laue case reflection curve is given by

$$R_L(y) = \frac{\sin^2(\pi t\sqrt{1+y^2}/\Delta_o)}{(1+y^2)} . \qquad (32)$$

Note that the crystal thickness is now an important parameter and that the oscillations in the numerator are scaled with Δ_o, the reciprocal of the diameter of the dispersion surface ($2d_s = \Delta_o^{-1}$). As Figs. 6b,c show [37], the Laue case reflection profiles are quite complicated and, bearing in mind that the Pendellosung distance Δ_o normally lies in the range 1 μm and 100 μm, quite small variations in thickness result in observations of the averaged reflection curve which is indicated by dashed curves in Fig. 6b. It is

$$\bar{R}_L(y) = (2(1+y^2))^{-1} \qquad (33)$$

The characteristic Pendellosung distance Δ_o is different for the two polarization states and is given by equation (19)

$$\Delta_o = \lambda \cos\theta/C(\chi_h\chi_{\bar{h}})^{\frac{1}{2}} . \qquad (34)$$

Typical values for silicon are [38,39] for $AgK\alpha_1$ radiation

hkl	111	220	311	400	331	422	333	440
Δ_o^σ/μm	52.70	46.79	71.81	55.38	81.09	62.68	93.32	70.90

Equal crystal rocking curves permit direct observation of the plane wave Pendellosung fringes as Bonse et al. [37] have shown, in the Laue case.

Most of the work which has been directed to measurements of Bragg reflection profiles has been done under conditions where one profile (R_2) is much narrower than the other. Nevertheless, non-dispersive settings of the double crystal arrangement are essential. Then $R_2(\theta-\alpha) = \delta(\theta-\alpha)$ to an appropriate degree of approximation (equation (28)) and $I(\alpha) = R_1(\alpha)$ so that the intrinsic single crystal reflection curve is directly measured. The techniques used to create narrow reflection widths are exactly the same as those used to isolate pseudo-plane-waves in the ray tracing experiments (section 9.7.2 and Fig. 5).

In the earlier work [40-44] oblique Bragg reflections were used to narrow the angular spread of the exploring beam in double or triple crystal arrangements and almost all of the measurements were made in the Bragg case. Later workers used the angular magnification of the wavefield fan in transmission (Fig. 5a) to create suitably narrow reflection profiles with which to explore crystal reflection profiles [32,45]. More recently, multiple asymmetric Bragg reflections have been used to create extremely narrow angular widths for detailed profile analysis [46]. A fairly complete bibliography of profile analysis by double and multiple Bragg reflections is given by Kohra [47] whose group has pioneered the application of profile analysis as a means of measuring structure amplitudes (see, for example, [48-50]).

9.7.4 Pendellosung fringes

In practice the Pendellosung fringes which occur as a result of interference between waves from different branches of the dispersion surface are observed under spherical wave conditions rather than in the specialized rocking curves described in section 9.7.3.

Fundamental to a description of defect image contrast in X-ray diffraction topographs is an understanding of energy flow and interference in section patterns. In Fig. 7a a monochromatic beam, perhaps 10 μm wide, is incident at the Bragg angle. Nevertheless, the beam divergence will be sufficiently large to excite the complete range of Bragg reflecton and energy flows in the crystal throughout the fan bounded by $\underline{K}_o$ and $\underline{K}_h$. The width of the base of the triangle of energy flow is $2t \sin \theta$ which may easily amount to several hundred μm in typical cases. Thus, the details of the X-ray energy flow pattern can be recorded on high resolution films quite easily.

It is clear from Fig. 7 that rays which overlap on the entrance surface of the crystal can become spatially separated by large distances at the exit surface. Since only those rays which are superimposed can give rise to interference effects, interference is observed between rays with the same energy flow directions ($\hat{S} = \underline{\nu}$ in Figs. 7a,b) which are related to the conjugate points D and $\bar{D}$ on the dispersion surface. Thus, the interference length $(D\bar{D} \cdot \underline{\nu})^{-1}$ depends upon the direction of energy flow in the crystal and the loci of the fringes are hyperbolas as shown. On the centre line of the section pattern the relevant rays relate to the diameter points of the dispersion surface so that the Pendellosung distance there is <u>exactly</u> Δ_o as in the plane wave case at the centre of the range of Bragg reflection (equation (32) and Fig. 6b). In any experiment one observes a projection of the intersection of the crystal wavefields interference pattern with the exit surface of the crystal. If the crystal is wedgeshaped, as in Fig. 7c, then characteristic hyperbolic fringes are observed, whereas in a parallel sided slab of crystal the Pendellosung fringes appear as parallel stripes.

Following the first exploration of Pendellosung fringes in X-ray diffraction [51] the spherical wave theory was developed in detail by Kato [8,11,52,55]. At the same time many of the features of Pendellosung fringe patterns were demonstrated in detail so that it became feasible to use the Pendellosung fringes to measure structure amplitudes with unprecedented precision.

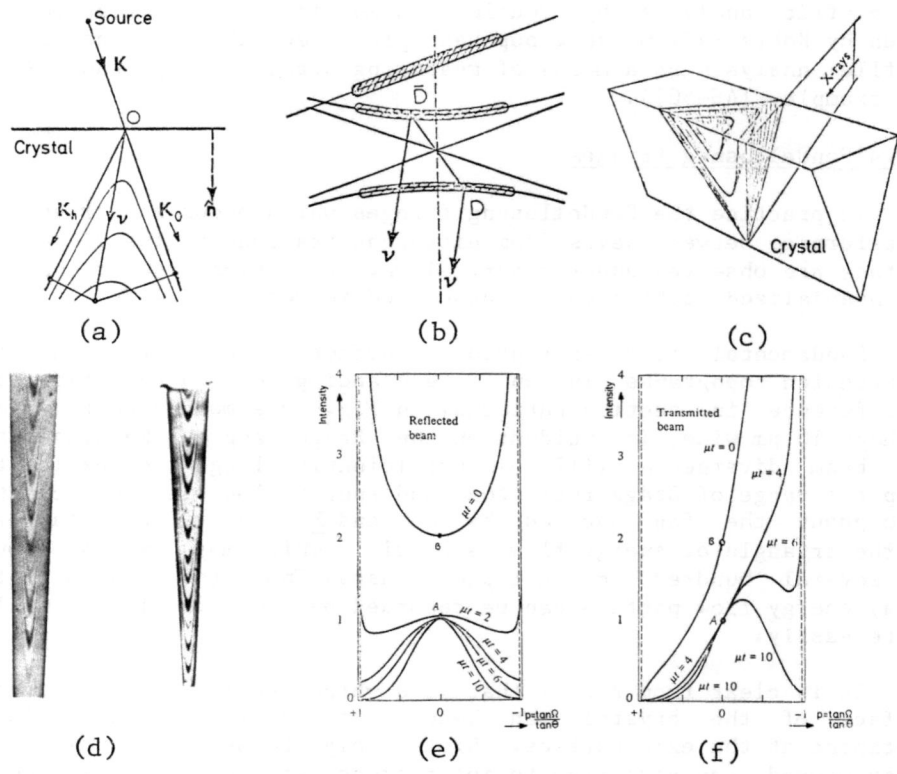

Figure 7. (a) Geometry of section patterns in real space and (b) in reciprocal space showing the region of coherent illumination for spherical waves. (c) Three dimensional geometry and (d) section patterns for wedge-shaped samples. (e) (f) show the spatial intensity distributions averaged over the Pendellosung fringe patterns.

Hattori and Kato [56] carefully measured the fringe shape to confirm that they are hyperbolic with the edges of the section pattern as asymptotes. This amounts to a demonstration that the dispersion surface is hyperbolic. As in light optics, the change from plane wave to spherical wave conditions results in terms such as the trigonometric function in equation (32) being replaced by the corresponding Bessel function. In the asymptotic form, this change manifests itself as a phase shift of $\pi/4$ in the fringe pattern and this too has been verified [58,58]. With unpolarized sources, or with linearly polarized sources when the electric vector is neither in nor normal to the plane of dispersion, two sets of fringes are obtained. In the first case the fringe patterns are incoherently superimposed and their properties have been thoroughly explored [59-61]. Coherent superposition of the Pendellosung fringe patterns can be used to generate elliptically polarized radiation [62-64]. A large number of isolated measurements of structure amplitudes have been made using Pendellosung fringes [51,65-68], some with an estimated precision of 0.1%. Only in the case of silicon [38,39] has a sufficiently large number of structure factors been measured for a detailed analysis of the results to be attempted. In that case R-factors near 0.1% are obtained when experimental results are compared with the available structure factor models [39,69].

9.7.5 Spatial distribution of intensity in section patterns

Particularly when the section pattern is too small to observe at high resolution, or when the fine features of interference fringes are not interesting, it is still important to be aware of the gross features of the energy flow distribution in the section pattern. That is obtained by averaging over the oscillatory terms in the spherical wave theory (analogous to the corresponding plane wave rocking curve result in equation (23)) and gives the results shown in Figs. 7e,f [52]. In the case of very low absorption ($\mu t < 2$) the reflected beam section pattern is characterized by an excess of intensity in its edges, which are clearly visible in Fig. 7d too. With very high absorption ($\mu t \gg 10$) the intensity is concentrated in the centre of the section pattern where the anomalously transmitted energy flow occurs (section 9.7.1). Since the anomalous transmission depends on a fundamental crystal property rather than upon any special conditions at the entrance surface it should not be surprising that the transmitted and reflected beam section patterns become identical when the absorption is high enough. Other results can be deduced in special cases, for example, on the centre line of the section pattern where $p \equiv y \equiv 0$. When the absorption is very high only one wavefield beam (with $\xi = 1$) contributes to the intensity so that the normalised intensity at A is 1 in both the reflected and transmitted beam patterns. When the absorption is zero both branches of the dispersion surface contribute equally (with $\xi = 1$ and $\xi = +1$) wavefields so that the normalised intensity at B is 2 in both patterns.

Although these averaged intensity distributions have never been explicitly confirmed in quantitative detail they have been qualitatively confirmed in a wide variety of experimental results and are routinely recognised in section pattern topographs.

9.8 Wavefield propagation in distorted crystals

The chapter given by Professor Kato (chapter 10) is concerned with the problem of relating diffracted intensities and *images* to the microscopic structure of crystals. That is a far cry from the understanding of dynamical theory of diffraction in perfect crystals which we have just briefly summarized. We still have no indication of how to proceed to investigate images since, in an imperfect crystal, the key feature of translational symmetry does not exist. Equation (1) is irrelevant, the structure amplitude (equation (3)) no longer has discrete values with simple phases and Maxwell's equations cannot be solved! New assumptions are necessary.

Provided that the crystal is sufficiently homogeneous that, for example, the Bragg angle is constant to within $\Delta\psi_0$ (the range of Bragg reflection) over the characteristic distance Δ_0 (the Pendellosung distance, equal to $1/2d_s$) one can proceed to develop a phenomenological theory in an obvious way by reference to the existing optical theory of ray propagation in inhomogeneous media. This is what Penning and Polder [13,18] have done. Here we will briefly outline Penning and Polder's theory and then we will summarize the available results for homogenously deformed crystals in sufficient detail to form an introduction to the theory of defect images in X-ray diffraction topographs. In practice the ray or Eikonal approximation is adequate to determine the images of defects to within about 10 μm of a stress discontinuity such as the core of a dislocation or the edge of mismatched film on the surface of a crystal.

9.8.1 Ray propagation and the refractive index gradient ∇n

We assume throughout that the crystal strains will always be so small that the optical parameters can be regarded as constants. Then the <u>local dispersion surface</u> corresponding to the <u>local reciprocal lattice vector</u> h' is a unique surface whose orientation and position change to follow the strainfield. Our task here is to find the equations governing the locus of the dispersion surface in the deformed crystal so that the propagation of rays can be evaluated. We assume as before (in the perfect crystal case) that the ray intensity is determined by

$$dI/I = -\mu(\hat{j}) \cdot d\underline{\ell} \qquad (35)$$

where $d\underline{\ell}$ is along the ray path and $\mu(\hat{j})$ (equation (24)) will now be a function of position in the crystal.

Consider the propagation of a ray from a point P in a perfect crystal region with reciprocal lattice $\underline{h}'$ to a point Q in another perfect crystal region with reciprocal lattice $\underline{h}$ (fig. 8). Since, far off the Bragg position, the wave vector $\underline{K}_0$ will remain unaffected by the strain we can assume that the two dispersion surfaces have a common O-wave assymptote. If the boundary between the two regions is abrupt, then two wavefields exist at Q; one closely resembles the original ray and the new ray is related to the other branch of the dispersion surface. If $\underline{h} - \underline{h}'$ is small then the intensity of the new wave is small and when writing equations we shall ignore it altogether. This is a **key assumption** which breaks down when the strain gradients are large.

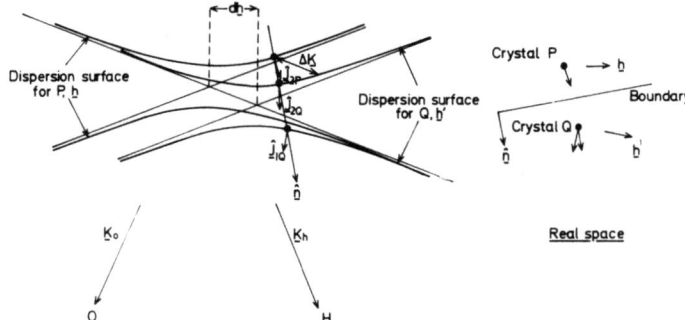

Figure 8. Propagation parameter for a ray travelling from P to Q in an inhomogeneous crystal

Let us first look at those strainfields which have no effect — those for which $d\xi \equiv 0$ and the wave vector changes by $\Delta \underline{K}$

we have $\quad \Delta \underline{K} \cdot \underline{K}_0 = 0$

and $\quad \Delta \underline{K} \cdot \underline{K}_h = \underline{K}_h \cdot d\underline{h}$.

For example, if $\Delta \underline{K} = 0$ then $d\xi = 0$ and $\underline{K}_h \cdot d\underline{h} = 0$. This defines a plane within which $\underline{K}_0$ is constant i.e. a plane within which the refractive index is effectively constant. By analogy with the optical case we suppose that changes in wavevector $d\underline{K}_0$ will be normal to this plane, that is in the direction of $\nabla(\underline{K}_h \cdot \underline{h})$.

9.8.2 The characteristic equation relating $d\xi$ to the deformation

We assume that
$$d\underline{K}_0 = \alpha \nabla (\underline{K}_h \cdot \underline{h}) \quad . \tag{36}$$

The constant α is determined by the condition that at both P and Q the wavefield must lie on the local dispersion surface. For simplicity we will assume that $\chi_h = \chi_{\bar{h}}$. The fundamental equations (13) are

$$K_0^2 = k^2 + k^2 \chi_0 + k^2 C \chi_{\bar{h}} \, \xi \tag{37}$$

$$K_h^2 = k^2 + k^2 \chi_0 + k^2 C \chi_h \, \xi^{-1} \tag{38}$$

From equation (37) we have, at P

$$2 \underline{K}_0 \cdot d\underline{K}_0 = k^2 C \chi_h \, d\xi$$

and from equation (38) we have, at Q with $\underline{K}_h + d\underline{h} = \underline{\hat{K}}_o + \underline{h}'$

$$2(\underline{K}_o + \underline{h}') \cdot (d\underline{K}_o + d\underline{h}) = -k^2 C \chi_h \, d\xi/\xi^2$$

or

$$\xi^2 (\underline{K}_o + \underline{h}') \cdot (d\underline{K}_o + d\underline{h}) = -\underline{K}_o \cdot d\underline{K}_o .$$

This result can be substituted in equation (36) to find α and after a good deal of manipulation [13,18] one finds the required result

$$\frac{d\xi}{d\ell} = -\frac{2}{k^2 C \chi_h} \cdot \frac{\xi^2}{kB_1} (\underline{K}_o \cdot \nabla)(\underline{K}_h \cdot \nabla)(\underline{v} \cdot \underline{h}') \tag{39}$$

where $\underline{v}$ is the atomic displacement vector which is defined by $\underline{h}' = \underline{h} + \nabla(\underline{h}' \cdot \underline{v})$ and B_1 is defined in equation (24), by $B_1^2 = 1 + 2|\xi|^2 \cos 2\theta + |\xi|^4$. If we define axes with z along the trace of the Bragg planes and x normal to the Bragg planes and take $\underline{h} \cdot \underline{v} = u \cdot 2k \sin \theta$ then

$$\frac{d\xi}{d\ell} = -\frac{4\xi^2 \sin\theta}{C \chi_h B_1} \left[\cos^2\theta \frac{\partial^2 u}{\partial z^2} - \sin^2\theta \frac{\partial^2 u}{\partial x^2} \right] \tag{40}$$

In practice it is not very convenient to work in terms of the moving dispersion surface in Fig. 8 but to consider instead movements of the tiepoints on the universal fixed dispersion surface corresponding to the perfect crystal. Let us look first at some limiting cases:

(a) Suppose $u \propto x$; then $d\xi = 0$. This represents a uniform expansion of the crystal in the x direction so that the crystal remains perfectly periodic and we would therefore expect rays to propagate unchanged.

(b) Suppose $\xi = 0$ or $\xi = \pm \infty$ so that we are far from the Bragg position. Then, in both cases $d\xi \to 0$ as expected, since we require that a single wave with wavevector $\underline{K}_o$ is uninfluenced by crystal strainfields.

(c) Suppose that the atomic displacements $\underline{v}$ are all contained within the Bragg planes. Then $\underline{v} \cdot \underline{h}' \equiv 0$ and $d\xi/d\ell = 0$ so that no effect is expected. On this basis one can often detect unique axes or planes in the strainfields, and for example, thereby determine the Burgers vectors of dislocations.

(d) Suppose $u = xz/\ell$ where ℓ is large. Then equation (40) shows that $d\xi = 0$. This deformation changes the shape of the Bragg planes, leaving them flat but with a varying spacing so that they are fanned. This situation was explored in detail by Cole and Brock [70]. In some cases they were able to measure Bragg angle differences of up to 100 seconds of arc between the entrance and exit surfaces of an anomalously transmitting bent germanium crystal.

Having determined the change in amplitude ratio $d\xi$ the properties of the ray, for example, intensity and phase shift are determined easily. Exactly the same results were obtained by Kato [71-73] in an elegant series of papers concentrating on phase changes and Pendellosung phenomena rather than upon ray tracing. His starting point was concerned with Fermat's principle for Bloch waves which is closely related to Penning and Polder's [13] assumption about ray trajectories.

9.8.3 Ray paths in a homogeneously deformed crystal

The energy flow direction $\hat{S}$, parallel to $\hat{j}$, is given by $\tan \Omega = \tan \theta \, (\xi^2-1)/(\xi^2+1)$. Using a convenient parametric form, we have therefore

$$\frac{dx}{d\ell} = (\xi^2-1) \frac{\sin\theta}{B_1} \qquad (41)$$

$$\frac{dz}{d\ell} = (\xi^2+1) \frac{\cos\theta}{B_1}$$

and, following Penning and Polder, we define a deformation constant β as

$$\beta = \frac{2 \tan \theta}{C \chi_h} \left[\cos^2\theta \, \frac{\partial^2 u}{\partial z^2} - \sin^2\theta \, \frac{\partial^2 u}{\partial x^2} \right] \qquad (42)$$

so that equation (40) becomes

$$\frac{d\xi}{d\ell} = - \frac{2 \xi^2 \cos\theta}{B_1} \beta \qquad (43)$$

and therefore

$$\frac{d\xi}{dz} = - \frac{2 \xi^2 \beta}{\xi^2+1} , \qquad \frac{d\xi}{dx} = - \frac{2 \xi^2 \beta}{\xi^2 -1} \cdot \cot\theta$$

These two equations can be integrated within the limits of the crystal surfaces; from $x = 0$ to x, $z = 0$ to z and $\xi = \xi_i$ to $\xi = \xi_e$ with the result [13]

$$\left(\frac{2\beta x}{\tan\theta} + \xi_i + \frac{1}{\xi_i}\right)^2 - \left(2\beta z + \xi_i - \frac{1}{\xi_i}\right)^2 = 4 . \qquad (44)$$

Thus, the path is a hyperbola; once again we have an "image" of the hyperbolic dispersion surface. The asymptotes of the family of ray path hyperbolas are the edges of the section pattern.

The paths are shown schematically in a number of simple cases [21] in Fig. 9 for one state of polarization. It is interesting to note that different polarization states take different paths through the crystal as Fig. 10 shows.

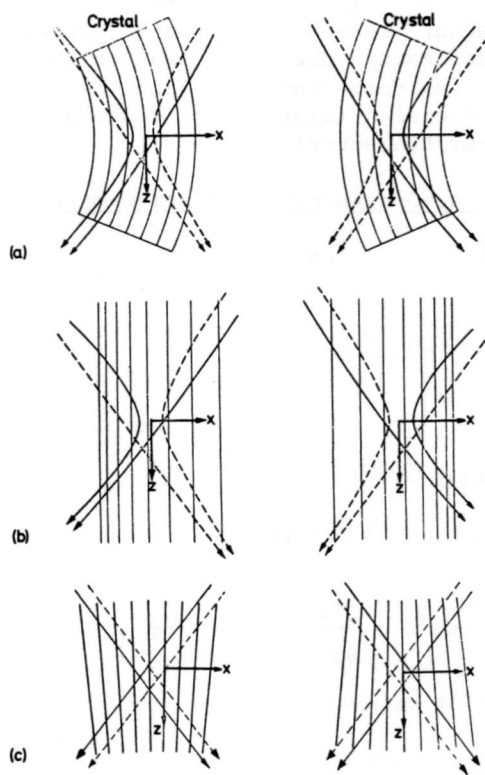

Figure 9. Raypaths in distorted crystals ⟶ upperbranch ($\xi<0$) solutions ⇢ lower branch solutions ($\xi>0$). Note that reversing the sign of the deformation interchanges the roles of the two branches of the dispersion surface. (a) pure bend (b) a linear gradient of lattice parameter and (c) a "fanned" crystal for which $d\xi = 0$.

It is instructive to calculate the distance through which the tiepoints move under the influence of a uniform deformation. It is convenient to rewrite the equation of the dispersion surface (equation (15)) in terms of parameters X and Y as

$$\alpha_o = Y \cos\theta - X \sin\theta$$

$$\alpha_h = Y \cos\theta + X \sin\theta$$

ELEMENTARY DYNAMICAL THEORY

then

$$X = \frac{1}{2\sin\theta}(\alpha_h - \alpha_o) = \frac{\tfrac{1}{2}k\,C\,\chi_h}{2\sin\theta}\left(\frac{1}{\xi} - \xi\right)$$

and the distance (parallel to $\underline{h}$) moved by the tiepoint in the deformed crystal is

$$\Delta X = \frac{\tfrac{1}{2}k\,C\,\chi_h}{\sin\theta}\left[\left(\frac{1}{\xi_i} - \xi_i\right) - \left(\frac{1}{\xi_e} - \xi_e\right)\right]$$

$$= \frac{\tfrac{1}{4}k\,C\,\chi_h}{\sin\theta} \cdot 2\beta t$$

or, in terms of the diameter of the dispersion surface

$$\frac{\Delta X}{d_s} = \tfrac{1}{2}\beta t\,\cot\theta = p\,\cot\theta\,. \tag{45}$$

Thus, the distance moved by the tiepoint is independent of ξ. It is also the same for all branches of the dispersion surface since ΔX is independent of the state of polarization of the ray. Typical raypaths are shown in Fig. 10.

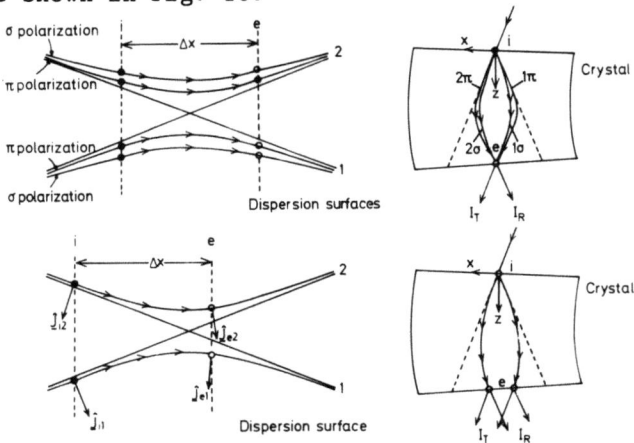

Figure 10. Raypaths in a homogeneously deformed crystal showing the different paths followed by the two polarization states (upper diagrams) and the paths of two plane wave components where wavevectors remain related by the plane wave boundary conditions even in the distorted crystal.

9.8.4 Intensity of rays in homogeneously deformed crystals

The intensity change as the ray propagates is determined from equation (35). Thus

247

$$dI/I = -\mu(\hat{j}) \cdot d\underline{\ell}$$

We have $\mu(\underline{j})$ as a function of ξ in equation (24) and $d\xi/d\ell$ in equations (40) or (43) so that the intensity equation can be directly integrated to give

$$I_e/I_i = \exp - \frac{\mu t}{\cos\theta}(1 + \frac{|\chi_{ih}/\chi_{io}|C}{\beta t} \ln \frac{\xi_i}{\xi_e}) \qquad (46)$$

This result of course applies to the ray path <u>inside the crystal</u>. If we wish to calculate the intensities in the external beams then the amplitude splitting factors arising from the boundary conditions must be included. The results for the diffracted beam I_R and the transmitted beam I_T are [13] shown in Fig. 11a and

$$I_T/I_0 = \frac{1}{1+\xi_i^2} \cdot \frac{1}{1+\xi_e^2} \frac{I_e}{I_i}$$

$$I_R/I_0 = \xi_e^2 \, I_T/I_0 \, . \qquad (47)$$

9.9 Experimental results from homogeneously deformed crystals

Theoretical interest was first aroused by a series of experiments which were designed to measure the influence of strain on the Borrmann effect [70, 74-78]. Later more detailed experiments concerned with ray tracing and Pendellosung fringes confirmed the theoretical predictions in quantitative detail.

9.9.1 The Borrmann effect in homogeneously deformed crystals

In practice the integrated reflection has been measured as a function of deformation in a crystal which was either mechanically bent [70, 74, 75, 77] or elastically deformed by a temperature gradient normal to the Bragg planes [77, 78]. Mechanical bending is not a very reliable method for introducing deformations in the Laue-case because the major part of the strain field is in principle harmless and the important part arises only as a result of deviations from symmetrical reflection. Thus small imperfections such as rough surfaces, and the influence of anisotropic elasticity, have large parasitic influences on the active deformation. On the contrary, a temperature gradient normal to the Bragg planes is ideal experimentally and the most convincing experimental results were obtained with that method [77]. The deformation parameter is given by

$$\beta = \frac{2 \tan \theta}{C \chi_h} \alpha \frac{dT}{dx}$$

where α is the thermal expansion coefficient and dT/dx is the temperature gradient. The calculated integrated intensities are shown in Fig. 11b. After a series expansion (with $\mu t >> 1$) these theoretical curves are obtained by integrating equation (47) over the angle of incidence to obtain the integrated reflection power.

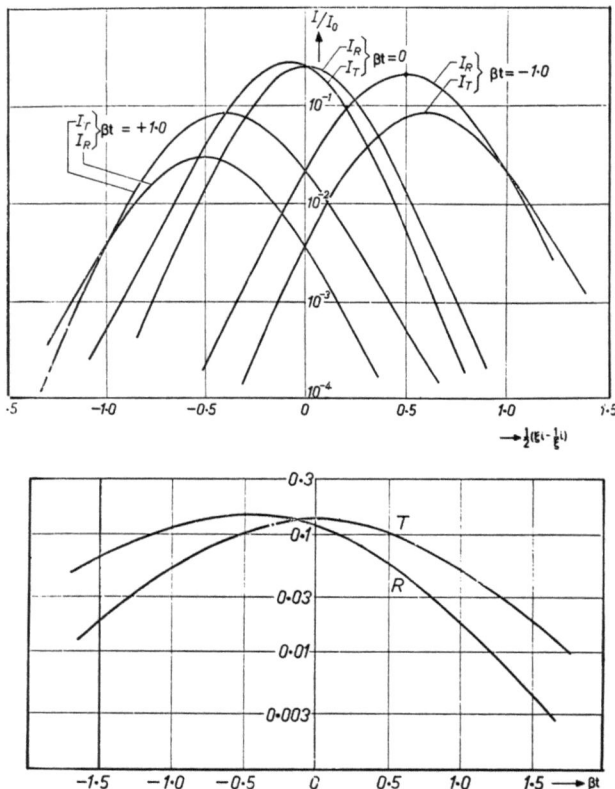

Figure 11. (a) Plot of the intensities of the transmitted (I_T) and reflected (I_R) beams in a thick crystal as a function of angle of incidence for several values of the deformation parameter. βt (b) The integrated intensities T, R (equation 48)) as a function of the deformation.

Writing T and R for the integrated reflections and using a subscript 0 for the undeformed crystal we have [18]

$$\ln T/T_0 = -\frac{1}{6} p^2 (\frac{3}{2} + a)$$

$$\ln R/R_0 = 2 \ln(\sqrt{1+p^2} - p) - (2\sqrt{1+p^2} - p)a^{-1} - \frac{1}{6} p^2 (\frac{3}{2} + a) + 2a^{-1}$$

where $p = \frac{1}{2}\beta t$ and $a = (\mu t/\cos\theta)(\chi_{ih}/\chi_{io})$. With a view to the interpretation of the contrast in the diffraction images of defects it is useful to note the result for small deformations, i.e. for a >> 1 but p << 1. The contrast in the integrated reflection is

$$\frac{R - R_O}{R_O} = -p \text{ and } \frac{T - T_O}{T_O} = -\frac{1}{6} p^2 a \tag{48}$$

as can be seen from the integrated reflection curves in Fig. 11b. The image contrast in the diffracted beam is reversed when the sign of the deformation is reversed and the image contrast is stronger in that beam than in the transmitted beam if we assume that $p \simeq 0.2$ and $a \simeq 20$.

9.9.2 Ray tracing experiments

Ray tracing experiments, which verify a number of the results obtained earlier, have been performed in both the Laue-case [79,80] and in the Bragg-case [81,82]. Each of these experiments used the methods described in section 9.7.2 and in Fig. 5 for ray tracing. Temperature gradient deformations were studied in the Laue-case and four-point elastic bending was used in the Bragg-case experiments.

Agreement with theory, both in the Laue-case and in the flanks of the Bragg case rocking curve, (when the same ray paths are generated) indicates very clearly the value of concentrating upon the details of wavefield propagation rather than upon the detailed diffraction geometry, which is relatively unimportant.

Malgrange [79] measured ray profiles for both branches of the dispersion surface and demonstrated the modified Borrmann absorption (equation (47)) which results in the deformed crystal. By recording both beams which are split off at the exit surface of the deformed crystal Hart and Milne [80] were, in addition, able to show that all tiepoints move the same distance in the same direction on the dispersion surface, in agreement with equation (45).

9.9.3 Pendellosung fringes in deformed crystals

Although the bending of Pendellosung fringes in the elastic strainfields of dislocations had been observed by 1960 [83], no experiments with controlled deformations were performed until 1965 [84]. Quantitative measurements, showing agreement to within a few per cent with the dynamical theory, were made in silicon crystals deformed by a uniform temperature gradient normal to the Bragg planes [85] and the contraction of the fringe spacing which occurs in both homogeneous and inhomogeneous deformations was observed in section patterns from beryllium oxide [84] and in integrated reflection topographs from silicon [86].

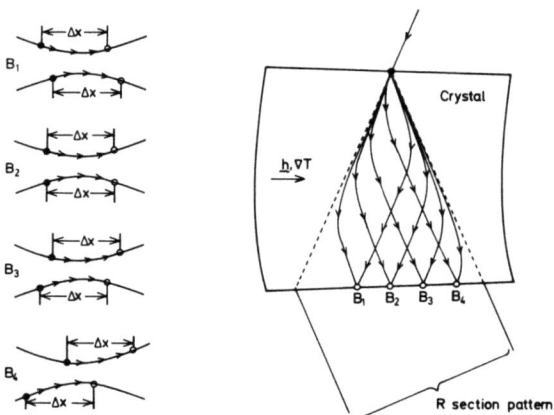

Figure 12. Raypaths of the interfering beams in section patterns obtained with a homogeneously deformed crystal as shown in Figure 13.

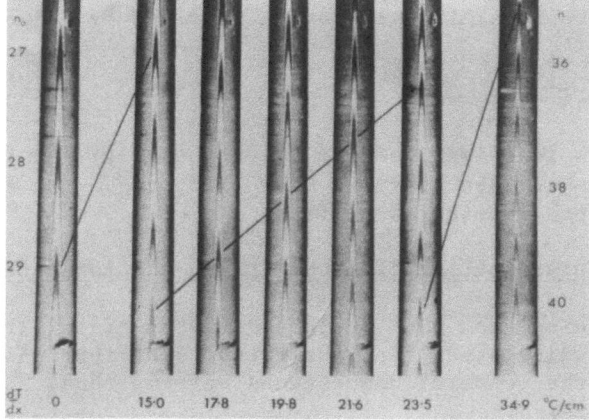

Figure 13. Section patterns obtained in silicon deformed by a constant temperature gradient. 220 reflection AgKα₁

The phase integrals, measured along the ray paths, are given in Kato's papers setting out the Eikonal theory [71-73] and in the experimental paper by Hart [85]. In essence one has to calculate the phase difference $2\pi n$ along two paths which intersect at a point B in the exit surface of the crystal, having originated at the vertex A of the section pattern in the entrance surface (Fig. 12). Thus

$$2\pi n = 2\pi \left[\int_A^B \underline{K}_{O2} \cdot \hat{\underline{j}}_2 \, d\ell - \int_A^B \underline{K}_{O1} \cdot \hat{\underline{j}}_1 \, d\ell \right] \tag{49}$$

where the subscripts refer to the two branches of the dispersion surface. Since the total change ΔX in the wave vector component parallel to $\underline{h}$ is independent of the path in the crystal, it follows that the interfering beam paths will always be skew symmetric with respect to the section pattern edges as Fig. 12 indicates. Some of the experimental results are shown in Fig. 13. It is immediately obvious from the shape of the ray paths [85] that, independently of the sign of the deformation, the effect of a deformation is to increase the interference order n and to decrease the local fringe spacing. At the centre of the section pattern we find [73,85]

$$n = \frac{1}{2} n_o \left[\sqrt{1 + p^2} + p^{-1} \text{ arc sinh } p \right] \quad (50)$$

where n_o is the Pendellosung fringe order in the perfect unstrained crystal. Since $p = \frac{1}{2}\beta t$ is proportional to the crystal thickness it follows that the fringes are no longer (quasi-) uniformly spaced if the crystal is deformed by a constant temperature gradient, or indeed by any constant deformation β. Differentiating equation (50) we find that the local fringe spacing is given by

$$\frac{dt}{dn} = \frac{\Delta_o}{\sqrt{1 + p^2}} \le \Delta_o \quad . \quad (51)$$

The contraction in fringe spacing is very easily seen in Fig. 13. At the highest temperature gradient $\beta t = 3.5$ and, as equation (51) shows, the fringe spacing is halved.

9.9.4 Spatial distribution of intensity in section patterns

Corresponding to the perfect crystal case (Fig. 7e,f), Kato [73] has also calculated the intensity distributions in section patterns in the case of homogeneous deformations. The results are shown in Fig. 14.

A reasonable qualitative understanding of the main features of the section patterns can be obtained by superimposing our results for the modified Borrmann effect (equations 41,47) on the ray paths on Fig. 13. For zero deformation, $\beta t = 0$, the section patterns are the same as in Fig. 7e,f for $\mu t = 0$ though the intensities are now normalised to unity at the centre of the pattern since we sketch contributions of the two branches of the dispersion surface separately. Fig. 7e,f showed the sum of the two intensities. At first it is surprising that the forward diffracted wave has the same intensity distribution from each branch of the dispersion surface, but this is essentially connected with the necessary skew symmetry of the ray paths in Fig. 13 and the consequent equality of the surface splitting factors in each case. That one intensity distribution generally increases and the other decreases with deformation in the diffracted beam section pattern follows

ELEMENTARY DYNAMICAL THEORY

qualitatively from equation (46). Remembering that the Pendellosung fringes result from interference waves belonging to the upper and lower branches of the dispersion surface, the section pattern profiles show that the fringe contrast in the diffracted beam will become smaller as the deformation becomes larger, as observed in practice [85]. Finally, the representative ray paths in Fig. 12 show that in the deformed crystal the "pile-up" of the ray paths near the edges of the section pattern (the so-called margin enhancement effect) is reduced. Numerical examples of all these phenomena are given in the Table adjacent [73].

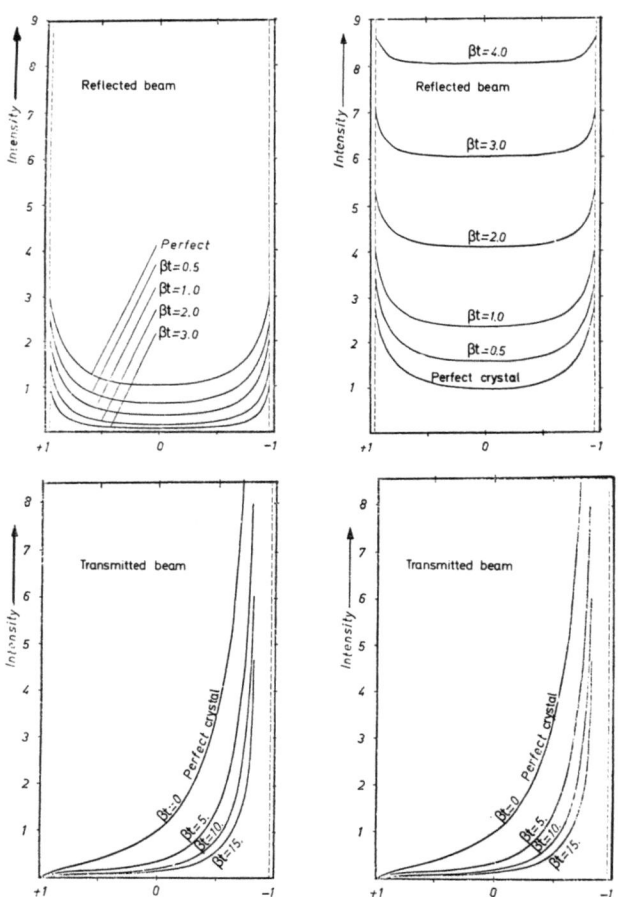

Figure 14. Spatial intensity distribution in section patterns for a homogeneously deformed crystal. The uppermost pair of patterns relate to the reflected beam and the lower pair are for the transmitted beam. For $\beta t>0$, the left-hand diagrams relate to the upper branch of the dispersion surface ($\xi<0$) and the right-hand diagrams are for the lower branch of the dispersion surface ($\xi>0$). These roles are reversed when $\beta t<0$.

βt	Margin effect	Integrated intensity	Fringe spacing	Contrast
0	110	1.571	1.00	100
1	83	1.844	0.898	67
2	37	2.086	0.707	17
3	17	2.308	0.555	8
4	8	2.510	0.447	6
5	4	2.696	0.371	4
10	~0	3.493	0.196	~0
20	~0	4.708	0.099	~0

At large values of the strain parameter, say $\beta t > 5$ the diffracted beam pattern shows no visible fringes and has approximately uniform intensity. The forward beam pattern is concentrated in the region so close to the continuation of the primary beam that it cannot be observed. Under those conditions one can still measure the integrated intensity as Janacek et al ([87] have done and thus the breakdown of the ray-optical theories can be observed.

9.10 Images of crystal defects in real crystals

A wide range of defects can be analysed on the basis of the elementary theory which we have already outlined. For example, the images of stress-free planar defects such as cracks (far from the tip), gaps, stacking faults and non-diffracting regions such as twin lamellae can all be interpreted in terms of the perfect crystal spherical wave theory in which the phase of the structure amplitude varies by $\underline{h}\cdot\underline{u}$ where $\underline{u}$ is the translation vector between the two parts of the crystal (in one part of the crystal equation (9) defines the structure amplitude while in the other part ($\underline{r}_a$ is replaced by $\underline{r}_a + \underline{u}$). Discontinuities in structure simply give rise to new waves whose amplitudes and phases are determined by Snell's Law (section 9.6.1 and 9.6.2). Provided that the strain field is sufficiently weak for the ray approximation to be valid, then the procedures outlined for the case of a homogeneous solution no longer permit analytic solutions but they are straightforwardly used in computer calculations, though it should be noted that the calculation of a section pattern requires a significant amount of computer time, while the computation of the integrated reflection pattern, obtained in a traverse topograph, takes a substantial length of time. Indeed, one finds that the time required to take a Lang topograph of 1 mm^2 of crystal containing ten dislocations is far less than the time required to compute the answer at the same resolution! In many cases it is possible to find regions in the defect strainfield where the deformation parameter β is substantially constant over the volume of the section pattern. In such case (the far fields of localised defects in thin crystals when the Bragg angle is small) a semiquantitative estimate of the image contrast can often be obtained from the results obtained for homogeneous deformations.

9.10.1 Qualitative interpretation of defect images

There are a large number of approximate results and "rules of thumb" in general use. In this section some useful results are summarised and an indication of their theoretical origins is given. Let us first consider planar defects and then line defects.

<u>Stacking faults</u> separate two perfect regions of a crystal and are characterised by a fault vector $\underline{f}$. The structure amplitudes in the two regions of crystal (equations 9,10) are $F_h[0]$ and $F_h[\underline{f}]$ for the $\underline{h}$-order Bragg reflection. We have

$$F_h[\underline{f}] = F_h[0] \exp(2\pi i \underline{h} \cdot \underline{f})$$

When $\underline{h} \cdot \underline{f}$ = integer or zero the two parts of the crystal are indistinguishable and therefore the stacking fault is invisible. In other Bragg reflections the stacking fault acts as a crystal boundary so that new waves are generated when a wavefield crosses the fault boundary. If the stacking fault divides the crystal into two wedge-shaped regions we easily anticipate the appearance of Pendellosung-like fringes in the image. The detailed spherical wave theory is given in reference [88].

<u>Bicrystals</u> with reciprocal vectors $\underline{h}$, $\underline{h}'$ can result if there is a discontinuous change in impurity concentration in a crystal or if one crystal is laid upon another. The structure amplitudes are related by

$$F_h[\underline{h}'] = F_h[\underline{h}] \exp(2\pi i (\underline{h}-\underline{h}') \cdot \underline{r})$$

These are indistinguishable when $\underline{h}-\underline{h}' \cdot \underline{r}$ is an integer or zero and at these places in the boundary the boundary is invisible. Between those places fringe patterns are observed which may with difficulty be distinguished from the images of an array of parallel dislocations (which may also separate one part of a bicrystal from the other). In <u>twinned crystals</u> one may have common interplanar spacings in both parts so that the boundary may become invisible in Bragg reflections from those planes. However, it is more common for $|F_h|$ to be different in the two twin-related reflections and then the boundary remains visible. When only one component of the twin is in the Bragg reflection position then the shape of that component <u>only</u> is seen in the topograph.

<u>Twin Lamellae</u> and non-diffracting zones such as gaps introduce new raypaths and phase shifts in the section pattern. If the lamellae are too thin to be resolved, twin lamellae produce

images which are very similar to the images produced by stacking faults. A detailed theory is given in reference [89].

It has long been realised that atomic displacements in the Bragg planes have no influence on the Bragg reflection. This result is obtained in the dynamical theory in 8.2(c) where we found that, if $\underline{v} \cdot \underline{h}' = 0$, $d\xi = 0$. Let us write $\underline{v} = [u,v,w]$.

The atomic displacements around a <u>screw dislocation</u> are given in terms of the Burgers vector $\underline{b}_s$, parallel to $\underline{x}$ and to the line vector $\underline{L}$ as

$$u = 0$$
$$v = 0$$
$$w = b_s \cdot \theta/2\pi$$

If $\underline{h} \cdot \underline{b}_s = 0$ it follows that $d\xi = 0$ and the screw dislocation vanishes. Thus, if a series of different topographs is taken, the Burgers vector can be determined by $\underline{b}_s // \underline{h}_1 \times \underline{h}_2$ where $\underline{h}_1$ and $\underline{h}_2$ are the reciprocal vectors for two Bragg reflections in which the dislocation vanishes.

<u>Edge dislocations</u> are characterised by the Burgers vector $\underline{b}_e$, line vector $\underline{L}$ and atomic displacements

$$u = \frac{b_e}{2\pi} \left[\tan^{-1} \frac{y}{x} + \frac{1}{2(1-\nu)} \frac{xy}{x^2+y^2} \right]$$

$$v = \frac{-b_e}{8\pi(1-\nu)} \left[(1-2\nu)\ln(x^2+y^2) + \frac{x^2-y^2}{x^2+y^2} \right]$$

$$w = 0$$

Only when $\underline{h} // \hat{\underline{z}}$ does the edge dislocation vanish precisely so that it is impossible in principle to determine the Burgers vector of an edge dislocation from its vanishing conditions. However, it is commonly found that the atomic displacements parallel to the Burgers vector are predominant in determining the image contrast of edge dislocations. Under appropriate conditions one may then find that edge dislocations give only weak images when $\underline{h} \cdot \underline{b}_e = 0$. Then the Burgers vector can be determined if several Bragg reflections are used to obtain weak images. As in the previous example $\underline{b}_e // \underline{h}_1 \times \underline{h}_2$ where $\underline{h}_1$ and $\underline{h}_2$ are the reciprocal lattice vectors of the reflections which produce "weak" images.

<u>General dislocations</u> may give weak images when $\underline{h} \cdot \underline{b} = 0$ and thereby their Burgers vectors may be determined. In general detailed quantitative calculations are required.

9.10.2 Semi-quantitative interpretation of defect images

Fairly homogeneous deformations are often found in the far elastic fields of defects. For example, between 10 µm and 100 µm from a localised defect in a thin crystal. Under such conditions we may use the results of the dynamical theory for homogeneously deformed crystals with average values of the strain gradient parameter [59,90]. We recall that

$$\beta = \frac{2\tan\theta}{C\,\chi_h}\left[\cos^2\theta\,\frac{\partial^2 u}{\partial z^2} - \sin^2\theta\,\frac{\partial^2 u}{\partial x^2}\right] \qquad (42)$$

The Bragg angle in transmission topography is often small and if $\partial^2 u/\partial x^2$ is not large compared with $\partial^2 u/\partial z^2$ then

$$\beta \simeq \frac{\sin 2\theta}{C\,\chi_h}\,\frac{\partial^2 u}{\partial z^2} \simeq \frac{\sin 2\theta}{C\,\chi_h}\,\rho$$

where ρ is the average curvature of the Bragg planes. We also note that if β is small then

$$(R-R_o)/R_o = -\tfrac{1}{2}\beta t = -p \qquad (48)$$

and the Pendellosung fringe spacing is

$$\Delta = \frac{dt}{dn} = \frac{\Delta_o}{\sqrt{1+p^2}} \qquad (51)$$

In many situations, at distances greater than (say) 10 µm from a localised defect such as the edge of an oxide film or a dislocation, we find $0 < p \lesssim 0.5$. Thus, we observe up to 50% area contrast if only one branch of the dispersion surface is important and a few extra Pendellosung fringes would appear if both branches of the dispersion surface contribute to the integrated intensity and the crystal thickness lies in the range $10\Delta_o < t < 100\Delta_o$.

These effects are commonly seen in section topographs of the far strainfields of single dislocations [83] near growth horizons and imperfect planar defects [84] and near localised strain discontinuities such as oxide film edges [91-94]. Independently of the sign of the strain gradient the local Pendellosung fringe order increases ($n=n_o(1+(1/6)p^2)$), equation (50) so that new fringes appear. From the increased number of fringes, or better by an exact calculation based on a model strainfield, the value of p can be determined as Kato and Patel [8,91-93] have done.

An excess or deficiency of intensity can be used to determine not only the magnitude but also the sign of the deformation as equation (48) shows. The sign of β is determined by the sign of $\underline{v}\cdot\underline{h}'$ in equation (39). Thus the sign of the contrast can be inverted by changing from the $\underline{h}$ to the $\underline{\bar{h}}$ Bragg reflection or by inverting the sign of the deformation. If $\underline{h}//+x$ in Fig. 9 then the rule is that excess intensity occurs in topographs of the deformations shown in Fig. 9(a) and Fig. 9(b) on the right hand side of the diagram (for the upper branch ($\xi<0$) branch of the dispersion surface). In unusual situations the sign may be reversed as Blech and Meieran [30,31] have demonstrated. Again, with $\underline{h}//+x$ the strainfields on the left hand side of Fig. 9(a) and (b) are imaged with diminished intensity in the integrated reflection. The sign of the contrast is opposite for the lower branch of the dispersion surface. In weakly absorbing crystals the net contrast is very low because the two branches of the dispersion surface make opposite contributions to the integrated intensity. This can be inferred from equation (48) or from the spatial intensity patterns shown in Fig. (14) where for $\beta t < 0.5$, the gain of intensity in one pattern is approximately offset by a similar loss of intensity in the other section pattern. For $\mu t \simeq 2$ large effects can be observed since the upper branch wavefields dominate the intensity pattern. At very large values of μt the images of strained regions always show an intensity deficiency as equation (46) shows.

While these qualitative descriptions can be improved in quantitative detail in simple cases, a full description of defect image contrast requires lengthy computations based on a theoretical model of the defect. Since the path is determined by the deformation parameter it is not possible to determine directly the deformation from the intensity distribution in either section topographs or in the integrated reflection. With very hard radiation, such as can be conveniently obtained with some synchrotron radiation sources, this situation may change if the Bragg angle is so small that the column approximation is justified.

REFERENCES

1. R.W. James (1948) The optical principles of the diffraction of X-rays. Bell, London.
2. W.H. Zachariasen (1945) Theory of X-ray diffraction in crystals. Wiley, New York.
3. J.C. Slater (1958) Rev. Mod. Phys. 30, 147
4. G. Borrmann (1959) Rontgenwellenfelder in Beitrage zur Physik und Chemie des 20 Jahrhunderts, Vieweg und Sohn, Brunswick
5. M. von Laue (1960) Rontgenstrahl Interferenzen, Akademische Verlag, Frankfurt

6. R.W. James, (1963) Solid State Physics 15, 53
7. B.W. Batterman and H. Cole (1964) Rev. Mod. Phys. 36, 681
8. L.V. Azaroff, K. Kaplow, N. Kato, R.J. Weiss, A.J.C. Wilson and R.A. Young (1974) X-ray diffraction, McGraw Hill, New York
9. Z.G. Pinsker (1978) Dynamical scattering of X-rays in crystals. Springer, Berlin
10. A. Authier (1970) Advances in structure research by diffraction methods 3, 1
11. N. Kato (1963) Crystallography and crystal perfection p 153 Academic Press, New York
12. N. Kato (1968) Acta Geologica et Geographica (Bratislava) 14, 43
13. P. Penning and D. Polder (1961) Philips Res. Repts. 16, 419
14. M. Born and E. Wolf (1959) Principles of Optics Pergamon, London
15. R.W.G. Wyckoff (1948) Crystal structures vol. 1. Wiley, New York
16. J.C. Slater and N.H. Frank (1947) Electromagnetism, McGraw Hill, New York
17. M. von Laue (1952) Acta Cryst. 5, 619
18. P. Penning (1966) Proefschrift. Technische Hogeschool te Delft
19. P.P. Ewald (1958) Acta Cryst. 11, 888
20. N. Kato (1958) Acta Cryst. 11, 885
21. U. Bonse (1964) Z. Phys. 177, 385
22. B. Okkerse (1962) Philips Res. Repts. 17, 464
23. G. Borrmann (1941) Z. Phys. 42, 157
24. G. Borrmann (1950) Z. Phys. 127, 297
25. H. Cole, F.W. Chambers and C. Wood (1961) J. Appl. Phys. 32, 1942
26. L.P. Hunter (1959) J. Appl. Phys. 30, 874
27. G. Borrmann and G. Hildebrandt (1959) Z. Phys. 156, 189
28. B.W. Batterman (1962) Phys. Rev. 126, 1461
29. P.P. Ewald (1963) Fifty years of X-ray diffraction p248 Oesthoek, Utrecht
30. E.S. Meieran and I.A. Blech (1968) Phys. Stat. Sol. 29, 653
31. I.A. Blech and E.S. Meieran (1969) Phys. Rev. 179, 731
32. A. Authier (1961) Bull. Soc. Franc. Miner. Crist. LXXXIV, 51
33. U. Bonse (1964) Z. Phys. 177, 529
34. A.H. Compton and S.K. Allison (1934) X-rays in theory and experiment, Van Nostrand, New York
35. L.V. Azaroff (1974) X-ray spectroscopy, McGraw Hill, New York
36. U. Bonse (1961) Z. Phys. 161, 310
37. U. Bonse, W. Graeff, R. Teworte and H. Rauch (1977) Phys. Stat. Sol. a43, 487
38. P.J.E. Aldred and M. Hart (1973) Proc. Roy. Soc. A332, 223
39. P.J.E. Aldred and M. Hart (1973) Proc. Roy. Soc. A332, 239
40. M. Renninger (1953) Naturwiss. 40, 50
41. M. Renninger (1955) Acta Cryst. 8, 597
42. M. Renninger (1961) Z. Naturforsch. 16a, 1110
43. R. Bubakova (1962) Czech. J. Phys. B12, 776
44. K. Kohra (1962) J. Phys. Soc. Japan 17, 589
45. S. Nakano (1965) Dr. Thesis University of Tokyo

46. K. Kohra and S. Kikuta (1968) Acta. Cryst. A24, 200
47. K. Kohra (1975) International Summer School on X-ray dynamical theory and topography. Limoges
48. S. Kikuta (1971) Phys. Stat. Sol. 45, 333
49. N. Nakayama, S. Kikuta and K. Kohra (1971) Phys. Lett. 37A, 29
50. T. Matsushita and K. Kohra (1974) Phys. Stat. Sol. 24, 531
51. N. Kato and A.R. Lang (1959) Acta. Cryst. 12, 787
52. N. Kato (1960) Acta. Cryst. 13, 349
53. N. Kato (1961) Acta. Cryst. 14, 627
54. N. Kato (1968) J. Appl. Phys. 39, 2225
55. N. Kato (1968) J.Appl.Phys. 39, 2231
56. H. Hattori and N. Kato (1966) J. Phys. Soc. Japan 21, 1772
57. H. Homma, Y. Ando and N. Kato (1966) J. Phys. Soc. Japan, 21, 1160
58. M. Hart and A.D. Milne (1968) Phys. Stat. Sol. 26, 185
59. M. Hart (1963) Ph.D. Thesis, University of Bristol
60. H. Hattori, H. Kuriyama and N. Kato (1965) J. Phys. Soc. Japan 20, 1047
61. M. Hart and A. R. Lang (1965) Acta. Cryst. 19, 73
62. P. Skalicky and C. Malgrange (1972) Acta. Cryst. A28, 501
63. M. Hart (1978) Phil. Mag. B38, 41
64. M. Sauvage, J.F. Petroff and P. Skalicky (1977) Phys. Stat. Sol. a43, 473
65. M. Hart and A.D. Milne (1969) Acta. Cryst. A25, 134
66. H. Hattori, H. Kuriyama, T. Katagawa and N. Kato (1965) J. Phys. Soc. Japan 20, 988
67. N. Kato and S. Tanemura (1967) Phys. Rev. Lett. 19, 22
68. S. Tanemura and N. Kato (1972) Acta Cryst. A28, 69
69. P.F. Price, E.N. Maslen and S.L. Mair (1978) Acta Cryst. A34, 183
70. H. Cole and G.E. Brock (1959) Phys. Rev. 116, 868
71. N. Kato (1963) J. Phys. Soc. Japan 18#n, 1785
72. N. Kato (1964) J. Phys. Soc. Japan 19, 67
73. N. Kato (1964) J. Phys. Soc. Japan 19, 971.
74. L.P. Hunter (1958) Proc. Kon. ned. Akad. Wetensch. Amst. B61, 214
75. L.P. Hunter (1959) IBM J. Res. 3, 106
76. G. Hildebrandt (1959) Z. Krist 112, 312
77. B. Okkerse and P. Penning (1963) Phil. Res. Repts. 18, 82
78. G. Borrmann and G. Hildebrandt (1959) Z. Phys. 156, 189
79. C. Malgrange (1968) Acta Cryst. A24, 126
80. M. Hart and A.D. Milne (1971) Acta Cryst. A27, 430
81. U. Bonse (1964) Z. Phys. 177, 529
82. U. Bonse and W. Graeff (1973) Z. Phys. 28a, 558
83. N. Kato (1963) Crystallography and crystal perfection. Academic Press, London
84. M. Hart (1965) Appl. Phys. Lett. 7, 96
85. M. Hart (1966) Z.Phys. 189, 269
86. Y. Ando and N. Kato (1966) Acta Cryst. 21, 284

87. Z. Janacek, J. Kubena and V. Holy (1978) Phys. Stat. Sol. a50, 285
88. A. Authier (1968) Phys. Stat. Sol. 27, 77
89. A. Authier, A.D. Milne and M. Sauvage (1968) Phys. Stat. Sol. 26, 469
90. C. Malgrange (1975) International Summer School on X-ray dynamical theory and topography. Limoges.
91. J.R. Patel and N. Kato (1968) Appl. Phys. Lett 13, 40
92. N. Kato and J.R. Patel (1973) J. Appl. Phys. 44, 965
93. J.R. Patel and N. Kato (1973) J. Appl. Phys. 44, 971
94. N. Kato and Y. Ando (1966) J. Phys. Soc. Japan 21, 964

APPENDIX: ONE-WAVE RESULTS REVISITED

Let us consider solutions of Maxwell's wave equation in the "one-wave" case. We will do that using the notation and methods of the dynamical theory as developed in Sections 3 to 6. The solutions are applicable (when no Bragg reflections are excited) if the response of the medium can be represented by an isotropic scalar refractive index $n = 1 + \chi_0/2$. The results are also applicable to amorphous materials and when the X-ray wavelength lies outside the limiting sphere.

9.11 A solution of Maxwell's equations

The transverse wave equation permits time harmonic plane-wave solutions of the form

$$\mathcal{E} = \underline{E} \exp 2\pi i (\nu\tau - \underline{K}\cdot\underline{r}) , \mathcal{H} = \underline{H} \exp 2\pi i (\nu\tau - \underline{K}\cdot\underline{r})$$

subject to the constraints $\underline{E}\cdot\underline{K}=\underline{H}\cdot\underline{K}=\underline{E}\cdot\underline{H} = 0$. In practice we are only interested in $\mathcal{E}$ and the real part of the wavevector has magnitude $k = 1/\lambda$ in vacuum and $n_r k$ in the medium. n_r is the real part of the refractive index. In Fig. A1 the set of allowed wavevectors drawn to the point O terminate on a sphere, shown in projection, of radius $n_r k$. In the one-wave case the dispersion surface is a sphere. A typical plane wave in the medium, with wave vector $\underline{K}_o = \vec{AO}$ is shown together with a vacuum plane wave with wavevector $\underline{K}^i = \vec{VO}$.

Maxwell's wave equation places no restriction on the wave amplitudes $\underline{E}_o$ and $\underline{E}^i$ and the amplitudes are independent of the orientation of the corresponding wavevectors. In the two-wave case, and indeed in many-wave cases, the wave amplitudes vary as the orientation of the wavevector changes with respect to the crystal axes (Section 9.4). In the m-wave case (m-1) wave amplitude ratios are fixed by the dispersion relation inside the medium.

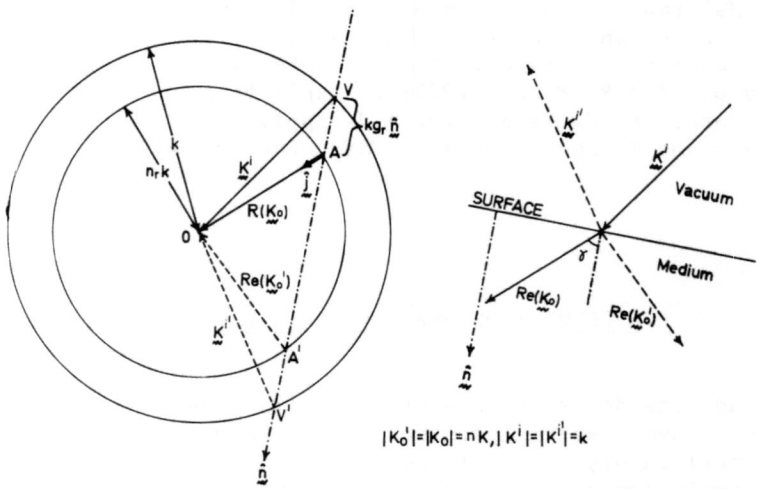

Figure A1. Raypaths and wavevectors in the one-beam case when no Bragg reflections are excited showing the free wave spheres in reciprocal space (left) and ray directions in real space (right).

9.12 The Poynting vector and ray optics

The time-averaged Poynting vector is defined in equation (20). For a single plane-wave in an isotropic medium it is given by

$$\hat{\underline{S}} = \frac{c}{8\pi k} |\underline{E}|^2 \mathrm{Re}(\underline{K}_o) \exp(4\pi i \, \mathrm{Im}(\underline{K}_o) \cdot \underline{r}) \tag{A1}$$

Thus the direction of energy flow $\underline{i}$ is parallel to $\mathrm{Re}(\underline{K}_o)$ and hence it is normal to the wave sphere (or dispersion surface) at A (Fig. A1).

9.13 Boundary conditions linking external and internal waves

We consider only the artificial case of a smooth and abrupt boundary between the medium and the vacuum. The component of phase velocity in the surface of the medium must be the same for both internal and external waves. Therefore

$$\underline{K}^i - \underline{K}_o = k\, g\, \hat{\underline{n}}$$

and remembering that $\underline{K}^i$ is real we can separate the real and imaginary parts of the equation as

$$\underline{K}^i - \mathrm{Re}(\underline{K}_o) = k\, g_r\, \hat{\underline{n}} \quad \text{and} \quad \mathrm{Im}(\underline{K}_o) = k\, g_i\, \hat{\underline{n}} \qquad (A2)$$

In Fig. A1 we note that there are two pairs of waves which satisfy both Maxwell's equations and the boundary conditions. Their wavevectors are VO and V'O in vacuum and AO and A'O in the medium. From equations (A1) and (A2) we note that the planes of constant attenuation are parallel to the surface of the medium.

Since, for X-rays, the refractive index is very close to unity we may ignore reflection at the surface and the boundary condition reduces to the requirement that $\underline{E}$ shall be continuous. Thus $|\underline{E}^i| = |\underline{E}_o|$ and $|\underline{E}^{i'}| = |\underline{E}_o'| = 0$ so that the two waves shown dashed in Fig. A1 are absent. In the corresponding optical case when $(n - 1)$ is not small, three of the four possible waves have finite amplitudes which are given by the Fresnel formulae.

9.14 Absorption of rays

We have seen that whereas the energy flow or ray direction is parallel to $\mathrm{Re}(\underline{K}_o)$, the planes of constant attenuation are normal to the surface of the medium since $\mathrm{Im}(\underline{K}_o) = k\, g_i \hat{\underline{n}}$ (Equation (A2)). In any ray theory we need to evaluate the absorption coefficient μ along the ray; that is in the direction of $\underline{j}$. Since $\mu(\hat{\underline{n}}) = 4\pi\, \mathrm{Im}(\underline{K}_o)$ we find immediately that $\mu(\underline{j}) = \mu(\hat{\underline{n}})/\cos\gamma$ where γ is the angle between the ray and the surface normal (Fig. (A1)). The corresponding result for the two-wave case is given in equation (25).

CHAPTER 10

PERFECT AND IMPERFECT CRYSTALS

N. KATO

10.1 Introduction

In this chapter, the author intends to present the basic concepts to understand the diffraction phenomena observed in crystals of various degrees of perfection. As in any other fields, the diffraction theory or the understanding of diffraction phenomena has experienced a spiral growth or a swing of pendulum.

Immediately after the discovery of X-ray diffraction, Darwin [1], Ewald [2] and Laue [3], the eminent physicists of those days, presented the <u>perfect</u> theory of dynamical diffraction for perfect crystals. Soon after, however, people realised that no perfect crystals exist in nature. Most of real crystals were thought to be "mosaic crystals", an aggregation of small crystallites. The model itself is not exact but well represents the essence of real crystals. For this reason, the interests of diffractionists swung towards the kinematical theory which is much simpler than the dynamical theory in style and very adequate in dealing with small crystallites of less than a few microns in size. The theory for otherwise perfect crystals gave a firm foundation to the structural analysis of single crystals. The theory for imperfect crystals also afforded much information on the crystal imperfections. The knowledge of thermal vibrations, or lattice dynamics, and the modulated structures of alloys, polytypes and others are examples.

It is also worth mentioning that the theory takes an optical formalism which is called <u>Fraunhofer diffraction</u>. Usually, the diffraction theory of this style concerns scattered waves from a small crystallite at a distance far from the scatterers.

When the direct observation of lattice defects became possible by means of electron microscopy and X-ray diffraction topography in the end of 1950's, the interests in the dynamical theory swung back with realistic applications. At this stage, however, we had to deal with imperfect crystals. Contrary to the kinematical theory, dynamical theory takes an optical formalism, which is convenient to describe <u>wave propagation</u> inside and outside the crystal. The situation, therefore, is very adequate for understanding the direct image of defects. The image of defects can be observed as a disturbance from the standard wave propagation expected in perfect crystals. Also, usually, large crystals of mm size have to be dealt with, so that the kinematical theory loses its validity. Thus, the use of dynamical theory is not due to pendantry but necessity.

In the past two decades, the theoretical understanding of this field has been considerably advanced. Accordingly, some of the classical concepts were required to be revised. One of them is the introduction of a spherical wave theory. Namely, the incident wave has to be regarded as a spherical wave. The main topics of this article are the achievements of this period.

Although X-ray diffraction topography increased and is now going to increase knowledge of crystal imperfections the poorness of spatial resolution is a kind of Achilles' tendon. For this reason, so far the topographic method cannot be used for small defects of less than microns in size. Under this circumstance, it is very desirable to use topography in combination with goniometry; namely diffuse scattering and small-angle scattering in a broad sense. At this stage, again, our interests may swing back to the kinematical theory, but not in the same manner of the classical period.

When we are concerned with small invisible defects or highly distorted crystals, the observed intensity is postulated to be a statistical average of the intensities over an ensemble of possible arrangements of lattice defects. For this reason, it is desirable to take into account the statistical nature of defects also in the dynamical theory. In fact, the kinematical theories of diffuse scattering and small-angle scattering include such statistical operations. In the latter part of this article, some recent developments of these topics will be reviewed.

10.2 Ray optical considerations

Obviously, crystal diffraction is a phenomenon typical of wave optics. Nevertheless, as in usual optics, most of the phenomena can be interpreted by ray-optical considerations with some supplementary concepts on the phase and amplitude associated with a ray. Since the rays are easily visualized, this approach or way of understanding is useful even when the exact wave-optical solutions are available. In this section, some typical cases of diffraction appearing in topography will be discussed in terms of ray optical considerations. More emphasis is put on the interpretation of the wavefield rather than the justification of ray-optics. Mathematical details involved will be seen in the references and the text book of Azaroff [4]. (Throughout this manuscript g and g are to be taken as identical: Eds.)

10.2.1 Ray optics in vacuum; kinematical theory

In the kinematical theory of diffraction (scattering), the medium is assumed to be vacuum except taking an elementary scatterer. The ray optics in vacuum is consisted of three principles.

1) The phase increment along a ray path of length ϕ is

$$\phi = K\ell \quad (1)$$

where $K = 2\pi/\lambda$ is (angular) wave number. This principle is justified by the fact that the plane wave expression

$$D(\underline{r}) = D \exp i\,(\underline{K}\cdot\underline{r}) \quad (2)$$

is a characteristic solution of the wave equation in vacuum.

2) The amplitude of the wave associated with a ray is determined by energy conservation

$$\mathrm{div}\,\underline{S} = 0 \quad (3)$$

where $\underline{S}$ is the energy flow vector and has the expression

$$\underline{S} = c\,K|A|^2 \underline{\nu}/4\pi \quad (4)$$

where A is the amplitude and $\underline{\nu}$ is the unit vector of the ray direction. This is the case of non-absorbing crystals. The effect of absorption will be dealt with separately. (see Eq. (25)) Again the justification is given by the fact that the expression of $\underline{S}$ for the plane wave (2) is

$$\underline{S} = c\,\underline{K}|D|^2/4\pi \quad (5)$$

In this case, the directions $\underline{\nu}$ and $\underline{K}$ are identical. The basic idea of ray optics is to regard any wave as a plane wave in a local sense.

3) Generalised Huygen's principle. The total sum of the possible wavelets arriving at an observation point P gives the wave field at P. The physical implication of this principle is deep. If one admits the wavelets as a correct solution of the wave equation, the principle is the principle of superposition in linear differential equations.

Now, we shall consider a specific problem of crystal diffraction; namely Thomson scattering due to an isotropic (spherical) wave. Taking optical paths from the source S to an observation point P via scattering points Q (Fig. 1), by using the principles mentioned above, we can write down the wave field

$$D(P) = \int_V \left[\frac{e^{iKR_1}}{4\pi R_1} \right] Z(Q) \left[\frac{e^{iKR_2}}{4\pi R_2} \right] dv_Q \tag{6}$$

where $R_1 = \overline{SQ}$, $R_2 = \overline{QP}$, Z is the scattering length per unit volume and V implies the domain of the crystal. In Thomson scattering

$$Z(Q) = 4\pi e^2 \, \rho(\underline{r}) C/mc^2 \tag{7}$$

where ρ is the density of electrons, C is the polarisation factor and the others are notations in standard usage.

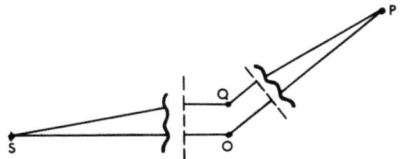

Figure 1. Thomson scattering in the form of Fraunhofer diffraction

The amplitude decay factors $(1/R_1)$ and $(1/R_2)$ are deduced from Eq. (3) because its integral form is $R^2|A|^2 =$ constant in isotropic space. The factor $1/4\pi$ is a matter of convenience.

When R_1 and R_2 are sufficiently large compared with the size of scatterer, one can use the Fraunhofer approximation.

$$\frac{e^{iK|R-r|}}{4\pi|R-r|} = \frac{e^{iKR}}{4\pi R} \exp -i(\underline{K}^* \cdot \underline{r}) \tag{8}$$

where $\underline{K}^*$ is the wave vector having the direction of $\underline{R}$. Hereafter, however, we often omit the asterisk. The spherical wave is essentially a plane wave if the area concerned is far from the source. Then, Eq. (6) can be written as

$$D_s(P) = \frac{e^2}{mc^2} C \frac{e^{iKR_s}}{R_s} D_e \int_V \rho(\underline{r}_Q) \exp i\{(\underline{K}_e - \underline{K}_s) \cdot \underline{r}_Q\} dv_Q \tag{9}$$

where the suffices e and s indicate the incident and the scattered waves respectively, and D_e is the amplitude of the incident wave at the origin O.

When the crystal is perfect, $\rho(\underline{r})$ is periodic and has the form

$$\rho(\underline{r}) = \sum_g \rho_g \exp 2\pi i \, (\underline{g} \cdot \underline{r}) \tag{10}$$

If we pick up the diffracted wave owing to the g-component, Eq. (9)

can be written as

$$D_g(P) = \frac{e^2}{mc^2} \frac{1}{V} (F_g C) \frac{e^{iKR_s}}{R_s} D_e \int_V \exp{-i(\underline{q}\cdot\underline{r}_Q)} \, dv_Q \qquad (11)$$

where V is the volume of unit cell, F_g is structure factor and $\underline{q}$ is the deviation vector from the exact Bragg condition, which is defined by

$$\underline{K}_s = \underline{K}_e + 2\pi\underline{g} + \underline{q} \qquad (12)$$

The integral involved in Eq. (11) implies the broadening of the Bragg-reflected wave owing to the finiteness of the crystal.

When the crystal is deformed by the displacement $\underline{u}(\underline{r}_Q)$ from its perfect state, one can write the electron density as follows with a good approximation

$$\rho(\underline{r}_Q) = \rho_{perfect}(\underline{r}_Q - \underline{u}) \qquad (13)$$

Therefore, the crystal can be treated as a perfect crystal provided that the structure factor F_g is multiplied by the lattice phase factor defined by

$$G = \exp{-2\pi i(\underline{g}\cdot\underline{u})} \qquad (14)$$

where $\underline{g}$ is the reciprocal lattice vector in a perfect state. G is $\underline{r}_Q$-dependent so that the integral in Eq. (11) must be replaced by

$$\int_V \rightarrow G(\underline{q}) = \int_V G(\underline{r}_Q) \exp{-i(\underline{q}\cdot\underline{r}_Q)} \, dv_Q \qquad (15)$$

In general, $G(\underline{r}_Q)$ reduces the coherent size of the crystal so that more broadening of the Bragg reflection is expected. In the book of Azaroff [4], this point is discussed for a few typical distortions in terms of ray considerations.

10.2.2 Plane wave theory and spherical wave theory

Any X-ray source is a spherical wave because the coherent source size is of atomic scale. Nevertheless, since the specimen is very far from the source in usual experiments, it seems to be a reasonable approximation to regard the incident wave as a plane wave. In fact, Eq. (11) is derived with this approximation. When we observed the details of the diffraction phenomena in Si perfect crystals by means of topography [5], it turned out that this traditional view was to be revised [6].

In the kinematical case, the Bragg reflection occurs over an angular range $\Delta\theta \simeq \lambda/\tau$ in the order of magnitude, τ being the coherent crystal size (cf. Eq. (11)). $\lambda/\tau = 10^{-3}$ for $\tau = 10^3 \text{Å}$. It is, however, not difficult to make a parallel wave with less divergence by using a slit system.

In dynamical cases, we are interested in perfect crystals of mm in size. Then, $\Delta\theta$ is an order of 10^{-7} if we estimate it kinematically. From a dynamical viewpoint, this estimation is too small but the actual range is still 10^{-5} at largest. The diffraction condition is drastically changed when the incident direction is changed by a fraction of this angle. Parallel waves of this accuracy are not attainable by a simple slit system. For this reason, it is necessary to develop the spherical wave theory as an alternative to the plane- wave theory. They are two extreme theories and complementary to each other. In fact, with a good approximation, the wave field of the former is the Fourier transform of the wave field of the latter and vice versa. In this article, this relation is called "FT relation".

10.2.3 Ray optics in the crystalline medium

In this section, ray optics in crystalline media will be described. The principles are very analagous to those in vacuum described in section 2.1.

The phase increments

They are given by

$$\phi_o = (\underline{k}_o \cdot \underline{v})\ell \quad , \quad \phi_g = (\underline{k}_g \cdot \underline{v})\ell \tag{16a,b}$$

where $\underline{k}_o$ and $\underline{k}_g$ are the wave vectors of the Bloch wave described below. The ray direction $\underline{v}$ is different from $\underline{k}_o$ and $\underline{k}_g$, in general cases.

The expressions (16) can be justified by the fact that the Bloch wave having the form

$$d(\underline{r}) = d_o \exp i\,(\underline{k}_o \cdot \underline{r}) + d_g \exp i\,(\underline{k}_g \cdot \underline{r}) \tag{17}$$

$$\underline{k}_g = \underline{k}_o + 2\pi \underline{g} \tag{18}$$

is a characteristic solution of the wave equation in perfect crystals ($\underline{r} = \ell\underline{v}$). The geometrical properties of the wave vectors $\underline{k}_o$ and $\underline{k}_g$ are well illustrated by the construction in reciprocal space (Fig. 2). (The readers are assumed to be familiar with this construction to some extent. The details will be seen in Azaroff's book, for example.) They can be written as

$$\underline{k}_o = \underline{\bar{K}}_o + \Delta\underline{K} + \Delta\underline{k} \quad , \quad \underline{k}_g = \underline{\bar{K}}_g + \Delta\underline{K} + \Delta\underline{k} \qquad (19a,b)$$

$$\underline{\bar{K}}_g = \underline{\bar{K}}_o + 2\pi\underline{g} \qquad (19c)$$

The hyperbolic surface on which the tail D of Δk lies is called "dispersion surface" and D is called "dispersion (tie) point".

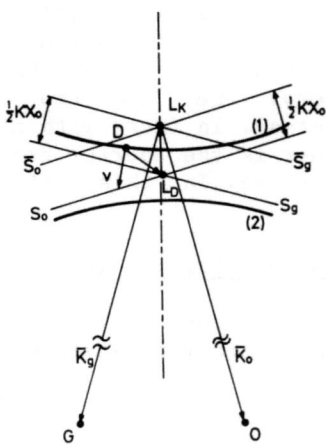

Figure 2. The construction of wave vectors by using the dispersion surface. Eqs. (19a, b and c) denote the same relations.

The expression of $(\Delta\underline{k}\cdot\underline{v})$ which is important in calculating ϕ_o and ϕ_g will be explained in the next section. Only, let it be mentioned that it changes critically depending on the deviation of the wave vector from the exact Bragg condition. The deviation is specified by the position of D on the dispersion surface. When D is on the Brillouin zone boundary, the Bragg condition is exact.

The amplitude d_o and d_g

The same principle (Eq. 3) can be applied also to the crystalline medium. However, the expression of $\underline{S}$ associated with the Bloch wave (17) must be

$$\underline{S} = \frac{c}{4\pi} \{\underline{\bar{K}}_o |d_o|^2 + \underline{\bar{K}}_g |d_g|^2\} \qquad (20)$$

The ratio d_g/d_o is fixed in Eq. (17), so that Eq. (3) determines d_o and d_g completely (To be exact, $\underline{k}_o$ and $\underline{k}_g$ must be used instead of $\underline{\bar{K}}_o$ and $\underline{\bar{K}}_g$, respectively).

Eq. (20) can be visualised by Fig. 3. Depending on the ratio

d_g/d_o, the direction of $\underline{S}$ denoted by $\underline{\nu}$ spans over all directions between $\underline{\overline{K}}_o$ and $\underline{\overline{K}}_g$. In connection with this, an important thing is that the ray direction $\underline{\nu}$ is normal to the dispersion surface at D [7]. This is a general property in ray optics of any type.

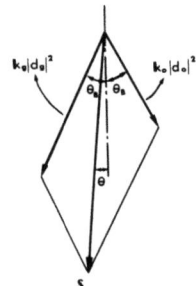

Figure 3. Energy flow vector $\underline{S}$.

So far, the deviation from the Bragg condition has been specified by the geometrical position of the dispersion point D. For analytical treatments, the parameter in real space

$$p = \tan \theta / \tan \theta_B \qquad (21)$$

is very useful where θ is the angle between $\underline{\nu}$ and the reflection plane. From Eq.(20), it is easily seen that

$$|d_g/d_o|^2 = (1-p)/(1+p) \qquad (22)$$

Also, from the geometrical form of the dispersion surface (hyperbolic) and the fact that $\underline{\nu}$ is perpendicular to the dispersion surface, one can derive

$$(\Delta \underline{K} \cdot \underline{\nu}) = \frac{1}{2} K \chi_o (\cos \theta / \cos \theta_B) \qquad (23a)$$

$$(\Delta \underline{k} \cdot \underline{\nu}) = \pm \frac{1}{2} KC \, \text{Re} \, (\chi_g \chi_{-g})^{\frac{1}{2}} (1-p^2)^{\frac{1}{2}} (\cos \theta / \cos \theta_B) \qquad (23b)$$

where χ_o and χ_g are the Fourier components of the polarisability χ of the medium, which is

$$\chi(\underline{r}) = -\frac{1}{\pi} (\frac{e^2}{mc^2}) \lambda^2 \rho(\underline{r}) \qquad (24)$$

(Thomson scattering is assumed).

In Eq. (23b), the double sign appears. The upper and lower correspond to the branches (1) and (2) of the dispersion surface (Fig. 2.) respectively. Because of this branching, in general, two types of Bloch waves appear in the crystal. They are denoted by (1) and (2), henceforth.

The generalised Huygens principle

In contrast to the kinematical scattering, several discrete waves arrive at an observation point in dynamical wave propagation. For this reason, interference is sigificant instead of diffraction. Nevertheless, the principle is applied in the same way to calculate the wave field.

Borrmann's anomalous transmission [8]

So far, we have not mentioned absorption. Phenomenologically, the effects can be described by assuming the imaginary component in the charge distribution $\rho(r)$ due to anomalous dispersion. Then, it is easily seen in Eq. (23a) and the remark below Eq. (27) that the normal aborption coefficient is

$$\mu_o = K \chi_o^i \tag{25}$$

Similarly, one can expect from Eq. (23b) the additional absorption coefficient

$$\mu_B = \pm KCJ_m \left[(\chi_g \chi_{-g})^{\frac{1}{2}} \right] (1-p^2)^{\frac{1}{2}} \tag{26}$$

The characteristic point is that, owing to this term, the wave (1) is less absorbed and the wave (2) suffers higher absorption. (In usual cases $J_m (\chi_g \chi_{\bar{g}})^{\frac{1}{2}}$ is negative). As a result, X-rays penetrate through the crystal when the Bragg condition is nearly satisfied (p=0). This is called the Borrmann anomalous transmission. The readers can find the references on these topics in Festschrift to Professor Borrmann's 65th birthday (Zeit. f. Naturforschung (1973) 28a; particularly, the article of Hildebrandt, Stephenson and Wagenfeld).

10.2.4 Perfect crystals

Firstly, we shall consider the Laue case which is simplest and useful. In Fig. 4, starting from a point source S, a wavelet arrives at an entrance point E'. The wavelet creates two Bloch waves. They propagate along two different directions, unless the Bragg condition is satisfied exactly (p=0). If the position E' is suitably chosen, the Bloch wave of type (1) will pass through an observation point P on the exit surface. Outside the crystal, the wave splits into O and G plane waves. Similarly, when the wavelet arriving at a suitable position E", the Bloch wave of type (2) passes through the same position P. Then, appreciable interference occurs between the waves of type (1) and (2) both in O and G waves.

The connection between the directions of SE' and E'P are determined through the relation between $\underline{K}_e$ (the incident wave vector) and $\underline{k}_o$ (the crystal wave vector of O wave). The relation is called "law of refraction"; namely the tangential components of two wave vectors on the entrance surface are identical to each other. Geometrically, the dispersion point E of $\underline{K}_e$ and D of $\underline{k}_o$ align on a line having the direction normal to the crystal boundary. In the Laue cases, two dispersion points $D^{(1)}$ and $D^{(2)}$ are excited. SE' is parallel to $\underline{K}_e$ whereas E'P has the direction normal to the dispersion surface at D.

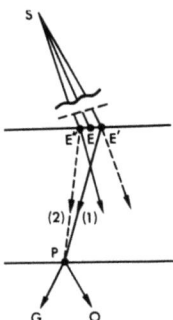

Figure 4. The ray picture of the dynamical wave fields (Laue case). S - source; P - observation point; E = entrance point of the ray satisfying the exact Bragg condition. E' and E" = entrance points of the rays arriving at P.

This is the exact picture of rays. In practice, however, E' and E" are approximately assumed to be E. In fact, the distance is less than or comparable with our spatially resolvable distance; a few microns under the ordinary experimental conditions. The phase difference between (1) and (2) waves, then, is calculated by the use of Eqs. (23b).

$$\Phi = \phi^{(1)} - \phi^{(2)} = 2(\frac{e^2}{mc^2})(\frac{\lambda}{v})C\text{Re}(F_g F_{-g})^{\frac{1}{2}}\sqrt{1-p^2}(\ell_o+\ell_g) \quad (27)$$

where $\ell(\cos\theta/\cos\theta_B)$ is replaced by $(\ell_o+\ell_g)$, ℓ_o and ℓ_g being the kinematical optical paths illustrated in Fig. 11.

At this stage, one needs a remark. In ray optics, sometimes a bundle of rays cross each other on a surface or a line or at a point. They are called caustics. If this situation happens, a constant phase $\pi/2$ has to be added to the rays after the caustics. This behaviour can not be derived from the simple ray optics but is well justified from wave-optical considerations [10] (see Azaroff's book, appendix of Chap. 3).

In our present problem, actually the rays of type (1) make caustics as shown in Fig. 5. Qualitatively, it is easily figured out by constructing rays, by using the fact that $\underline{v}$ is perpendicular to the dispersion surface and "refraction law" connects the vacuum and crystal waves. The crystal surface plays the role of a convex lens for the rays (1) and a concave lens for the rays (2). Thus, one can expect fringe patterns inside the triangular (Borrmann) fans :

Fringe shape : hyperbolic ($x_o x_g$ = const) (28a)

Fringe spacing : along $\underline{v}$ direction $\Lambda^{-1} = \frac{1}{\pi} (\frac{e^2}{mc^2}) (\frac{\lambda}{v}) \text{CRe}(F_g F_{-g})^{\frac{1}{2}} (\frac{\cos\theta}{\cos\theta_B}) \sqrt{1-p^2}$ (28b)

The additional phase $\pi/2$ introduces a shift of the fringe positions by a quarter of Λ. These types of fringe were observed in X-ray topography and called Pendellosung fringes [5].

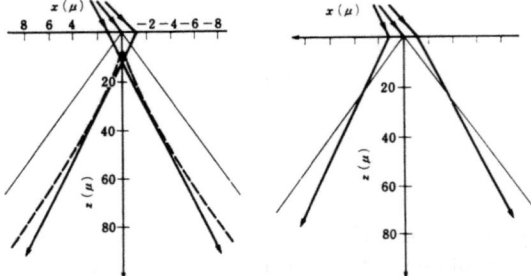

Figure 5. The exact ray pictures in the cases of rays of type (1) and (2); the left and right, respectively.

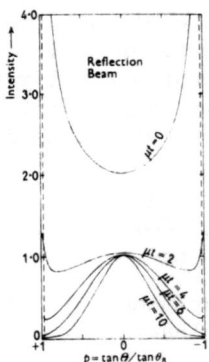

Figure 6. The intensity distribution of the Bragg reflected beam on the exit surface; the parameter p being defined by Eq. (21).

The intensity distribution at the exit surface is obtained by calculating the energy flow associated with each ray specified by p or the angle θ. Since some lengthy calculations are involved, here only the overall view omitting the interference term is shown in Fig. 6. Historically, this intensity distribution was obtained by a ray theory [11]. After observing the P-fringes, the purely wave-optical solution has been obtained by using FT relation [6,12]. Then employing a mathematical technique called the stationary phase method, the ray-optical properties mentioned above are derived. The results are practically correct except in a region of the first few fringes.

For the Bragg case, Uragami [13] obtained the wave field as the solution of Takagi-Taupin equations which will be mentioned below. Saka et al [14] also derived the same result by FT relation from the plane wave solution. Experimentally, Uragami [15] has studied and claimed fringe structure in the topograph of Berg-Barrett (reflection) type. Recently, a more clear-cut experiment was performed by Lang and Mai [16]. Fig. 7 is their result.

Figure 7. Bragg-case Pendellosung fringes from diamond (001) surface. Reflexion 113, CuKα_1, incidence plane vertical on figure. Going upwards, markers point to 1st. minimum, 2nd., 3rd., 4th. and 5th. maxima. Field width 165 µm.

10.2.5 Plate-like defects

The theory of the perfect crystal is immediately applied to the interpretation of the images of a single plate-like defect. As shown in Fig. 8, we shall consider the crystal divided by a defect plane.

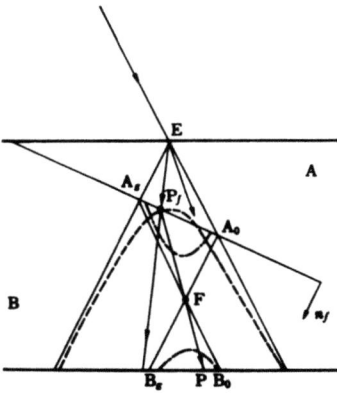

Figure 8. The ray pictures in the crystal having a stacking fault.

From the view of diffraction theory, it is convenient to classify the plate-like defects into three categories:

a) Stacking fault: $g_B = g_A$, $\chi_{g,B} = \chi_{g,A} \exp-2\pi i(g \cdot u)$ (29a,b). Here, u is the displacement in stacking.

b) Twin : $g_B = g_A$, $|\chi_{g,B}| \neq |\chi_{g,A}|$ (30a,b). The Dauphine twin of quartz is a typical one.

c) Misorientation boundary : $g_B \neq g_A$, $\chi_{g,B} = \chi_{g,A}$ (31a,b)

Actual defects are not as simple as classified here. There is an indication that the Dauphine twin of natural quartz is associated with a misorientation. Moreover, it is very plausible that a long range deformation is associated with it. Nevertheless, it is very useful to consider the above typical cases. In this article, however, mainly the stacking fault is discussed [17,18]. The details of other cases are referred to Azaroff's book.

As explained in section 2.3 the incident plane wave excites two Bloch rays (1) and (2) at the entrance surface. Each ray is consisted of O and G waves. They are denoted by (1,0), (2,0), (1,G) and (2,G). Generally, at the defect plane, each wave excites four waves in the same way as the entrance surface. Thus, in total, we have to consider 16 waves. They can be denoted by a set of notations

$\{^{1,1}_{O,G}\}$, $\{^{1,1}_{G,G}\}$, $\{^{1,2}_{O,G}\}$, $\{^{1,2}_{G,G}\}$

} For G wave in the crystal B,

$\{^{2,1}_{O,G}\}$ $\{^{2,1}_{G,G}\}$ $\{^{2,2}_{O,G}\}$ $\{^{2,2}_{G,G}\}$

where the first column indicates the mode of the wave in the crystal A and the last column does the mode in the crystal B. The similar notations in which the last G is replaced by O can be used for O waves in the crystal B. The one-to-one combination of O and G waves makes an independent Bloch wave, namely the ray. Thus, we must take 8 rays.

This is the situation in the misorientation boundaries (c). In the cases of (a) and (b), some of them degenerate and the number of the independent waves and rays are reduced. In the case of stacking fault, the number reduces to four because whether the mode is O or G waves in the crystal A is irrelevant to the mode of the waves in the crystal B. This situation is realised by "refraction law" on the defect plane. Thus, the waves in the crystal are specified by

P-type : (1.1), (2,2) and Q-type : (1,2) and (2,1)

The dispersion points of the waves of P type are identical to those of the waves (1) and (2) in the crystal A, respectively. Therefore, the rays penetrate through the defect plane without deflection. On the other hand, the dispersion points of the waves of Q-type are different from either one of (1) and (2) waves. Thus, the rays deflect at the defect plane. In this particular case, they focus as shown in Fig. 8. Moreover, since the dispersion points of (1,2) and (2,1) waves are conjugate to each other, they propagate in the same direction in real space and the new type of Pendellosung fringes appear along the rays. Calculation shows that the fringes are also hyperbolic as illustrated in Fig. 8. The theoretically expected topograph is what is illustrated in Fig. 9.

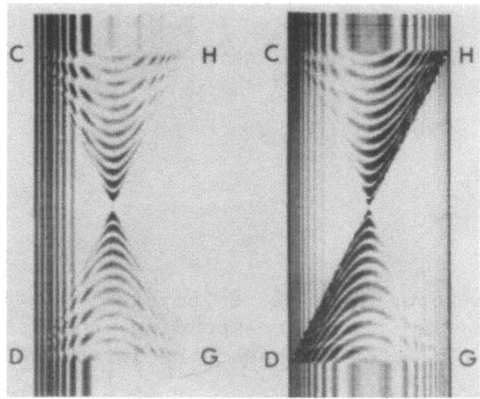

Figure 9. The computer simulation of the section topographs of O (the left) and G (the right) reflections for the crystal including a stacking fault.

Experimentally, the typical topograph expected by the theory was not obtained before Wonsiewicz and Patel took it in the case of silicon [19]. Real topographs indicate the importance of the interference of the waves of P and Q type, as well as the effects of Borrmann absorption [20]. The theory neglecting absorption gives the same intensity distribution for $\pm(\underline{g}\cdot\underline{u})$ on the exit surface. In real cases, they are different. Using this information, one can determine the sign of $\underline{u}$; consequently one can distinguish whether the fault is intrinsic or extrinsic.

In the cases of twins and misorientations the topographs are rather complicated, although the principles of calculating the intensity distribution on the exit surface of the crystal B are similar. Then, one needs a computer in practice. Katagawa has made a universal program which enables us to simulate the image of a single plate-like defect of any kind. Fig. 10 is one example in the case of a Dauphine twin. The parameter of misorientation is adjusted, there, to fit the experimental topographs. The agreement is reasonably satisfactory. The misorientation is an order of a few tenths of a second and changes from area to area even in the same twin boundary.

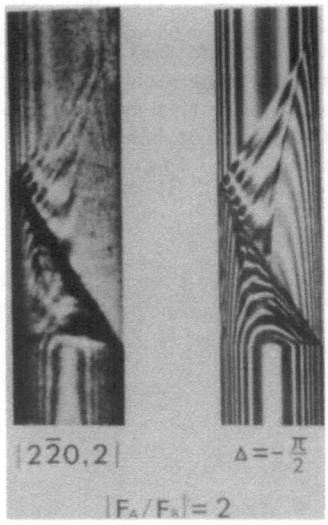

Figure 10. Section topographs of a crystal including a Dauphine twin. The right is the corresponding computer simulation. A misorientation of $\Delta\theta/\Delta\theta_B = 1/8$ has to be introduced.

10.2.6 Long range distortion

Another type of lattice imperfection is a continuous distortion in an extended area. Inhomogeneous distribution of impurities may cause such a distortion. The surrounding area of any dislocation is another example.

The ray-optical approach is useful also in this case provided that the associated wave can be regarded locally as a plane wave. As exemplified in the section 2.5 this condition is not necessarily satisfied, always. There, we have seen that a ray creates new rays when it hits a plane defect. For this reason, the lattice distortion must be reasonably moderate. Sometimes, the theory described in this section is called "Eikonal theory". In fact, it is very analogous to the theory having the same name in visible ray optics.

First, we have to know the ray trajectory. Now, we cannot expect it being a straight line as in perfect crystals. Before going into this problem, however, we shall discuss the phases assuming that the trajectory was known. The expressions (16), then must be modified in the form

$$\phi_o = \int_Q^P (\underline{k}_o \cdot \underline{v}) d\ell \qquad \phi_g = \int_Q^P (\underline{k}_g \cdot \underline{v}) d\ell \qquad (32a,b)$$

where Q and P are the start and end points of the ray trajectory. Similarly, Eqs. (17) and (18) are also to be revised in the forms

$$d(\underline{r}) = d_o(\underline{r}) \exp i\phi_o(\underline{r}) + d_g(\underline{r}) \exp i\phi_g(\underline{r}) \qquad (33)$$

$$\underline{k}_g(\underline{r}) = \underline{k}_o(\underline{r}) + 2\pi \underline{g}(\underline{r}) \qquad (34)$$

where

$$\underline{k}_o = \text{grad } \phi_o(\underline{r}) \qquad \underline{k}_g = \text{grad } \phi_g(\underline{r}) \qquad (35a,b)$$

and

$$\underline{g}(\underline{r}) = \text{grad}(G(\underline{r})) \qquad (36)$$

which is the local reciprocal lattice vector, $G(\underline{r})$ being defined by Eq. (14). Here, we shall not mention the justification of these modifications, but you will easily see that they are reasonable ones, if you admit that the wave behaves locally as a plane wave. The wave of the form (33) is called a "modified Bloch wave".

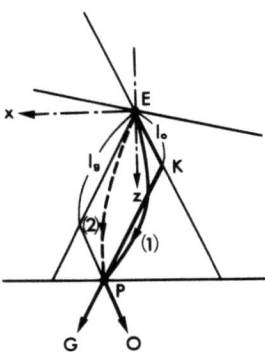

Figure 11. The ray picture in the crystal having a long-range distortion. EKP is the optical path expected in the kinematical theory.

Since the vector $\underline{g}$ changes from position to position, Eq. (19) based on Fig. 2 also must be modified. Traditionally, the origin O of the reciprocal lattice is fixed in the construction of the dispersion surface. Under the present situation, it is more convenient to fix the Lorentz point L_K. Then, the same dispersion surface can be used regardless of the change of $\underline{g}$ vector. As the result, however, the reciprocal lattice points O and G are moved on the surfaces perpendicular to $\bar{\underline{K}}_o$ and $\bar{\underline{K}}_g$ respectively. Thus, from Fig. 11, Eq. (19) must be modified as follows.

$$\underline{k}_o = \bar{\underline{K}}_o + \Delta \underline{K} + \Delta \underline{k} + \Delta \underline{g}^o \tag{37a}$$

$$\underline{k}_g = \bar{\underline{K}}_g + \Delta \underline{K} + \Delta \underline{k} + \Delta \underline{g}^g \tag{37b}$$

The expressions (23) for $(\Delta \underline{K} \cdot \underline{v})$ and $(\Delta \underline{k} \cdot \underline{v})$ need not be revised as mentioned above, although p and θ are implicitly the functions of position. The expressions of Δg^o and Δg^g are given in terms of $\underline{g}(\underline{r})$. After some mathematical manipulations finally one can write

$$\phi_o = [\bar{\underline{K}}_o \cdot (\underline{r}_P - \underline{r}_Q) + \tfrac{1}{2} K \chi_o (\ell_o + \ell_g)] + T + N_o \tag{38a}$$

$$\phi_g = [\bar{\underline{K}}_g \cdot (\underline{r}_P - \underline{r}_Q) + \tfrac{1}{2} K \chi_o (\ell_o + \ell_g)] + T + N_g \tag{38b}$$

where

$$T \equiv \int_Q^P (\Delta \underline{k} \cdot \underline{v}) d\ell = \pm [\tfrac{1}{2} K C \, \mathrm{Re}(\chi_g \chi_{-g})^{\tfrac{1}{2}}] (\cos\theta_B)^{-1} \int_Q^P \sqrt{1 - (\tfrac{dx}{dz})^2} \, dz \tag{39}$$

$$N_o \equiv 2\pi \int_Q^P (\Delta \underline{g}^o \cdot \underline{v}) d\ell = \pi (\underline{g} \cdot (\underline{u}_P - \underline{u}_Q)) + N \tag{40a}$$

$$N_g \equiv 2\pi \int_Q^P (\Delta \underline{g}^g \cdot \underline{v}) d\ell = -\pi (\underline{g} \cdot (\underline{u}_P - \underline{u}_Q)) + N \tag{40b}$$

and

$$N = \pi \int_Q^P \{ \cot\theta_B \tfrac{\partial}{\partial z}(\underline{g} \cdot \underline{u}) \tfrac{dx}{dz} + \tan\theta_B \tfrac{\partial}{\partial x}(\underline{g} \cdot \underline{u}) \} \, dz \tag{41}$$

The coordinates (x,z) are defined in Fig. 12.

Now we shall return to the problem to find the ray trajectory. Again, we shall not mention the justification of our result. Instead, from the analogy of the ordinary optics or Lagrange formalism of mechanics, we admit Fermat's principle or Lagrange's variational principle. They state that, if the phase (or action) integral is given as Eq. (32), the trajectory is given by the variational principle

$$\delta \phi_o = 0 \quad \text{and} \quad \delta \phi_g = 0 \tag{42a,b}$$

or the equivalent differential equations

$$\frac{d}{dz}\left(\frac{\partial \phi}{\partial (dx/dz)}\right) - \frac{\partial \phi}{\partial x} = 0 \tag{43}$$

Inserting the expressions (39) and (41) into this (other terms are irrelevant to the variation), one obtains the ray equation as

$$\pm m_o c \frac{d}{dz}\left\{\frac{p}{\sqrt{1-p^2}}\right\} = f \tag{44}$$

where

$$m_o = \frac{1}{2} KC \, \text{Re} \, (\chi_g \chi_{-g})^{1/2} / \sin \theta_B \tag{45}$$

$$c = \tan \theta_B$$

$$f = \frac{2\pi}{\sin 2\theta_B} \left\{ \frac{\partial}{\partial z_o} \frac{\partial}{\partial z_g} (\underline{g} \cdot \underline{u}) \right\} \tag{46}$$

$$p = \tan \theta / \tan \theta_B = \frac{1}{c} \frac{dx}{dz} \tag{47}$$

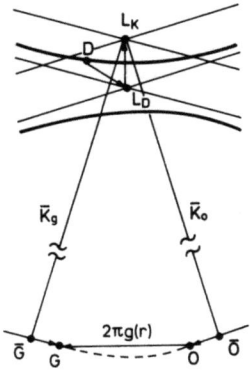

Figure 12. The wave-vector construction for distorted crystals. The dispersion point D is moved on the dispersion surface according to the variation of the reflection vector $\underline{g}(r)$; see Fig. 2.

It is interesting that the equation of the Bloch ray has exactly the same form as that of the Einstein particle in mechanics. The waves of type (1) and (2) behave as if they are positively and negatively charged particles under an electric field, respectively. However, the discussion of this topic is not the purpose of this article. Nevertheless, it is practically useful to apply the mechanical analogy to understanding our diffraction phenomena. For example, (i) the Bloch wave (particle) is limited within the Borrmann fan (light cone); (ii) the smaller structure factor χ_g (mass m_0) the more sensitive are the diffraction phenomena (bending of the particle) to the strain field (electric field).

The ray pictures in distorted crystals are given in Fig. 12. Usually, the rays of type (1) and (2) arrive at an observation point P. They are bent in opposite directions. They make interference fringes unless the lattice is heavily distorted. The phase difference between the two rays is increased because of the increase of the path length. Then, usually, the fringe spacing is decreased compared with that of perfect crystals. In general, also, the fringe shape is distorted and the higher order fringes appear locally in the case of inhomogeneous strain gradient (f in Eq.(44)). The phenomena might be called "photoelasticity" in X-ray region.

For the reason of theoretical importance, the case of a constant strain gradient is extensively worked out ([21]). The contraction of the fringes are confirmed experimentally by Hart [22] and Ando and Kato [23]. In particular, the former showed good agreement between theory and experiment.

In this article, a little is mentioned on the intensity associated with a ray. The splitting of the incident energy into (1) and (2) rays on the entrance surface, the splitting into O and G beams on the exit surface and the divergence of the ray tube along the trajectory (Eq.(3)) are the key factors to be considered. In general, the ray having a trajectory closer to the kinematical case (ray (1) in Fig.12) has a strong intensity of G-beam. The other is usually weak. For this reason the intensity variation of this origin is called "Extinction effect".

In connection with this situation, the effect of the Borrmann absorption is important. The attenuation of the intensity associated with a ray is given by $\exp{-B^{(i)}}$ [i=1 or 2], where

$$B^{(i)} = \pm KCJ_m[(\chi_g\chi_{-g})^{\frac{1}{2}}] \int_Q^P \sqrt{1-p^2} \, dz \qquad (48)$$

(see Eq. (26) and Eq. (39)). Therefore, it is easily anticipated that the Borrmann anomalous transmission occurs along the raypath where the modified Bloch wave satisfies nearly the Bragg condition ($p \simeq 0$). If the crystal is heavily distorted, the effects disappear because, then, the integral in Eq. (48) is practically zero.

Combining "Extinction effect" and "Borrmann anomalous effect", one can expect the variety of diffraction intensity. Compared with the intensity in perfect crystals, often it is stronger but sometimes weaker, depending on the sign and the magnitude of

($\underline{g}\cdot\underline{u}$). These phenomena are called "Black and White" phenomena. They are useful to detect the existence of a long-range distortion inside the crystal. An extensive study of this subject will be seen in the paper of Ando and Kato [24]. Some interesting examples will be seen, for example, in the case of the strain induced by the growth sector boundaries in NaCl [25] and $NaClO_3$ [26] and quantitative evaluation of the strain induced by oxide film on Si [27 - 29].

Finally a little will be mentioned on the historical development of this subject. The concept of the bending of Bloch rays was first introduced by Hildebrandt and Borrmann based on the experiment on the anomalous transmission in distorted crystals [30]. Penning and Polder took up the subject and developed a kind of geometrical optics of the Bloch rays [31]. Kato [32] and Kambe [33] gave the wave-optical foundation to their theory and enabled us to deal with the wave-optical nature of the Bloch waves, as well as the anomalous transmission in the crystal of any thickness.

All these theories are concerned with the Laue cases. Bonse, on the other hand, developed the theory for the Bragg cases [34]. He and his colleagues analysed extensively the images of dislocations near the crystal surfaces with this theory [35, 36]

10.3 Wave-optical considerations

Obviously, it is desirable to obtain the exact wave field, if it is possible. At present, however, it seems formidable to find the exact solution of the Maxwell equations in the exact sense even for perfect crystals. A more practical approach is to find the exact solution for an approximate wave equation. From this viewpoint the equations of Takagi-Taupin type are most useful. We shall derive them and discuss the related topics.

10.3.1 Wave equations of Takagi-Taupin type

We shall start with Eq. (6). It can be written in the following form, if the scattered wave owing to the $\underline{g}$-component of $\rho(\underline{r}_Q)$ is picked up as in the case of Eq. (11),

$$D_g(\underline{r}) = z_g \int \frac{\exp iK|\underline{r}-\underline{r}_Q|}{4\pi|\underline{r}-\underline{r}_Q|} G(\underline{r}_Q) \exp 2\pi i(\overline{\underline{g}}\cdot\underline{r}_Q) D_o(\underline{r}_Q) dv_Q \quad (49)$$

where D_o and G stand for the incident wave and the lattice phase factor (14), respectively. Here, we are interested in the wave field at a position $\underline{r}$ which is not necessarily very far from $\underline{r}_Q$. Thus, the Fraunhofer approximation (8) can not be used.

Eq. (49) has the form of convolution of a spherical wave and a few quantities at its origin. Therefore, using the convolution theorem of Fourier transform, we shall have

$$D_g(\underline{k}) = Z_g (\underline{k}^2 - K^2)^{-1} \{G(\underline{q}) * D_o(\underline{k} - 2\pi\underline{g})\} \quad (50)$$

where * indicates the convolution and $D_o(\underline{k})$, $D_g\underline{k}$ and $G(\underline{q})$ are the Fourier transform of $D_o\underline{r}$, $D_g(\underline{r})$ and $G(\underline{r})$, respectively.

$$\left[\begin{array}{l} \int \dfrac{e^{iK|\underline{r}|}}{4\pi|\underline{r}|} \exp - i (\underline{k} \cdot \underline{r}) \, dv = (2\pi)^{-3} (\underline{k}^2 - K^2)^{-1} , \int F(\underline{r}) G(\underline{r}) e^{-i(\underline{k} \cdot \underline{r})} dv \\ = (2\pi)^3 \int F(\underline{k}-\underline{q}) G(\underline{q}) \, d\underline{q} = (2\pi)^3 F(\underline{k}) * G(\underline{q}) \end{array}\right]$$

It is worth noticing that $G(\underline{q})$ is the same as the expression (15) appearing in the kinematical theory. If $\underline{k}$ is rewritten as

$$\underline{k} = \overline{\underline{K}}_o + 2\pi\underline{g} + \underline{q} = \overline{\underline{K}}_g + \underline{q} \quad (51)$$

and use is made the notations $D_o(\overline{\underline{K}}_o + \underline{q}) = D_o(\underline{q})$ and $D_g(\overline{\underline{K}}_g + \underline{q}) = D_g(\underline{q})$, we shall have

$$[(\overline{\underline{K}}_g + \underline{q})^2 - K^2] \cdot D_g(\underline{q}) = Z_g \cdot G(\underline{q}) * D_o(\underline{q}) \quad (52)$$

When the lattice distortion is not very heavy, the appreciable regions of $D_o(\underline{q})$, $D_g\underline{q})$ are confined within a domain of small $\underline{q}$. Thus the approximation

$$(\overline{\underline{K}}_g + \underline{q})^2 - K^2 \approx 2(\overline{\underline{K}}_g \cdot \underline{q}) \quad (53)$$

is allowed. Then, the Fourier inverse transformation of Eq. (52) gives

$$\dfrac{\partial}{\partial s_g} D_g(\underline{r}) = i \kappa_g \exp -2\pi i (\overline{\underline{g}} \cdot \underline{u}) D_o(\underline{r}) \quad (54a)$$

$$\left[\int (\overline{\underline{K}}_g \cdot \underline{q}) D_g(\underline{q}) \exp i (\underline{q} \cdot \underline{r}) \, d\underline{q} = (i)^{-1} (\overline{\underline{K}}_g \cdot \underline{\nabla}) D_g(\underline{r}) \right]$$

where $\dfrac{\partial}{\partial s_g}$ is the differentiation along $\underline{K}_g$ direction and

$$\kappa_g = \dfrac{e^2}{mc^2} (\lambda/v) F_g C \quad (55)$$

Similarly, we shall obtain

$$\dfrac{\partial}{\partial s_o} D_o(\underline{r}) = i \kappa_{-g} \exp 2\pi i (\overline{\underline{g}} \cdot \underline{u}) D_g(\underline{r}) \quad (54b)$$

regarding D_g and D_o as the incident and reflected wave, respectively.

In this formalism, it is clear that Eqs. (54) imply the kinematical reflection of G-wave from O-waves and vice-versa. The combination describes the dynamical relation of multiple reflections. Underlying approximation is the same as that used in the dynamical theory; namely the sphere of $K = |\underline{K}|$ is approximated by a tangential plane in the relevant regions around L_K, $\bar{O}$ and $\bar{G}$ in Fig. 12.

The functions $D_0(\underline{r})$ and $D_g(\underline{r})$ in Eqs. (54) represent amplitude modulation in the crystal. The true waves are $D_0(\underline{r})\exp\{i(\underline{K}_0\cdot\underline{r})\}$ and $D_g(\underline{r})\exp\{i(\underline{K}_g\cdot\underline{r})\}$. To be exact, one has to take also O-component of $\rho(\underline{r}_Q)$ in addition to the G-component in Eq. (49). After the similar manipulation described above, it gives the term $i\kappa_0 D_g$ on the right of Eq. (54a) where κ_0 is $(e^2/mc^2)(\lambda/v)F_0$. This term is irrelevant to the lattice distortion and polarisation. Similarly, the term $i\kappa_0 D_0$ must be added on the right of Eq. (54b). These additional terms, however, can be eliminated if one takes into account the effect of the refractive index at the beginning by replacing $\bar{\underline{K}}_0$ and $\bar{\underline{K}}_g$ by $\underline{k}_0$ and $\underline{k}_g$, respectively.

A set of Eqs. (54) is somewhat different from the original Takagi-Taupin equation [37, 38]. Nevertheless, one can show that they are equivalent. The derivation presented here is essentially the same as described in Kato's paper [39]

10.3.2 Perfect crystals

When the crystal is perfect, the lattice phase is constant. In this case, the integration of Eqs. (54) is straightforward. The incident wave is assumed to have the form

$$D_0(0, s_g) = A \delta(s_g) \tag{55a}$$

$$= A \sin 2\theta_B \delta(x_0) \tag{55b}$$

where (s_0, s_g) are the oblique coordinates having the origin at the entrance point E. and the coordinates x_0 is perpendicular to s_0-axis. Recalling the familiar expression

$$\delta(x_0) = \frac{1}{2\pi} \int \exp i (K_x \cdot x_0) \, dK_x \tag{56}$$

one can see that $D_0(0, s_g)$ is an approximate expression of a spherical wave [6]. Another boundary condition is, obviously,

$$D_g(s_o, 0) = 0 \tag{57}$$

The Laplace transform of Eqs. (54a and b) wih respect to the variable s_g gives

$$\frac{\partial}{\partial s_o} D_o(s_o, p) = i\kappa_{-g} D_g(s_o, p) \tag{58a}$$

$$p\, D_g(s_o, p) = i\kappa_g D_o(s_o, p) \tag{58b}$$

Eliminating $D_g(s_o, p)$, one obtains

$$\frac{\partial}{\partial s_o} D_o(s_o, p) = -(\kappa_g \kappa_{-g}/p) D_o(s_o, p) \tag{59}$$

The solution of this under the boundary conditions (55a) is given by

$$D_o(s_o, p) = A \exp -(\kappa_g \kappa_{-g}/p) s_o \tag{60a}$$

$$\left[D_o(0, p) = \int D_o(0, s_g) e^{-p s_g} ds_g = A \right]$$

Inserting this into Eq. (58b) one obtains

$$D_g(s_o, p) = A(i\kappa_g/p) \exp -(\kappa_g \kappa_{-g} s_o/p) \tag{60b}$$

The Laplace inversion transform gives immediately $D_o(s_o, s_g)$ and $D_g(s_o, s_g)$. For example,

$$D_g(s_o, s_g) = \frac{(i\kappa_g)}{2\pi i} A \int_{\gamma-i\infty}^{\gamma+i\infty} \frac{1}{p} \exp\left[p s_g - (\kappa_g \kappa_{-g} s_o)/p\right] dp \tag{61a}$$

$$= i\kappa_g A \sum_{n=0}^{\infty} \frac{1}{n! n!} (-\kappa_g \kappa_{-g} s_o s_g)^n \tag{61b}$$

$$= i\kappa_g A J_o(2\kappa_g \kappa_{-g})^{\frac{1}{2}} \sqrt{s_o s_g}) \quad \text{for } s_o s_g > 0 \tag{61c}$$

$$= 0 \quad \text{for } s_o s_g < 0 \tag{61d}$$

where J_o is the Bessel function of the zeroth order. Here elementary knowledge is assumed of the Laplace transform, the contour integrals in complex planes. γ is called "convergence length". For $p > \gamma$, the integrand must be regular. In our problem, therefore, $\gamma = 0$. This expression was obtained first by Kato [6] by Fourier transform of the plane wave solution, and by Takagi [37] as the solution of Takagi-Taupin's equation.

10.3.3 The case of constant strain gradient

The above method can be extended easily to the case of constant strain gradient. Then, the lattice phase has the form

$$2\pi (\bar{g} \cdot u) = \alpha s_o^2 + f s_o s_g + \beta s_g^2 \qquad (62)$$

where α, β and f are constants. Since the function $g(s_o)$ or $h(s_g)$ in the lattice phase gives simply the phase change, one can assume only the second term, without loss of generality. In this case, Eqs. (58) are modified as follows

$$\frac{\partial}{\partial s_o} D_o(s_o, p) = i\kappa_{-g} D_g(s_o, p - if s_o) \qquad (63a)$$

$$p D_g(s_o, p) = i\kappa_g D_o(s_o, p + if s_o) \qquad (63b)$$

Therefore, the combination of these gives the differential equation corresponding to Eq. (59) as follows

$$\frac{\partial}{\partial s_o} D_o(s_o, p) = -\left[(\kappa_g \kappa_{-g})/(p - if s_o)\right] D_o(s_o, p) \qquad (64)$$

The solution with the boundary condition (55a) is

$$D_o(s_o, p) = A \left[\frac{p - if s_o}{p}\right]^\sigma \qquad (65)$$

where

$$\sigma = \kappa_g \kappa_{-g}/if \qquad (66)$$

Using the relation (63b) we shall have

$$D_g(s_o, p) = A \frac{i\kappa_g}{p} \left(\frac{p}{p + if s_o}\right)^\sigma \qquad (67)$$

Thus, we shall have the wave field

$$D_g(s_o, s_g) = \frac{A}{2\pi i} \int_{\gamma-i\infty}^{\gamma+i\infty} \frac{i\kappa_g}{p} \left(\frac{p}{p+ifs_o}\right)^\sigma e^{ps_g} dp \qquad (68a)$$

$$= i\kappa_g A F(\sigma, 1, -ifs_o s_g) \qquad (68b)$$

where $F(\sigma, 1, -ifs_o s_g)$ is a confluent hypergeometric function.

$$F(a,c,z) = \frac{\Gamma(c)}{\Gamma(a)} \sum_{n=0}^{\infty} \frac{\Gamma(n+a)}{\Gamma(n+c)} \frac{z^n}{n!} \qquad \text{where } \Gamma(x) \text{ is the gamma function.}$$

The result was obtained by Katagawa and Kato [4], Chukhovskii [41] and Petrashen [42] independently.

The importance of obtaining the exact solution is that the solution enables us to estimate the validity of any approximate method. Katagawa and Kato discussed the validity of the Eikonal theory in detail. In the case of lattice bendings, the Eikonal theory is valid provided that

$$|S| = \frac{4\pi^2 d}{\Lambda_o^2} R > 1 \qquad (69)$$

where d is the net-plane placing, R is the radius of curvature of the net plane, and Λ_o is the fringe spacing defined by Eq. (28) ($\theta = 0$, $p = 0$). For reasonable figures (d = 1Å, Λ_o = 50 μm) R amounts to 2 m. Thus in a bulk crystal of 2 mm in thickness the theory can be applied to the strain of 10^{-3} (nearly the strain of break down) in the order of magnitude. Thus, it turns out that the Eikonal theory is applicable almost in any practical cases. Nevertheless, the theory is doubtful for a localised high strain such as the core of dislocations.

Recently, Chukhovskii and Petrashen [43] revised the asymptotic formulae of $D_g(s_o, s_g)$ by the use of Olver's method. By this way, one can introduce reasonable attenuation factor to the wave field, which enables us to avoid the divergence in the integrated intensity which was a defect of the simple Eikonal theory. The similar problem was also treated by Chukhovskii, Gabrielyan and Petrashen [44] in the Bragg case.

10.3.4 Computer simulation

In any diffraction theory, whether it is exact or not, the analytical solutions are available only in limited cases. The problem of practical importance are usually complex, so that one needs numerical solutions. Many programs of computer simulation are presented for this purpose. At present, the following three types of the program are available.

1) The simulation directly from Takagi-Taupin equations for any kind of distortions. (Epelboin, Sauvage and Authier).

2) The calculation of the images of a plate-like defect of any kind (Katagawa).

3) The calculations of the contrasts due to a long range distortion based on the Eikonal theory (Ando, Patel and Kato).

Excellent review articles on this subject can be seen in Acta Cryst. A35 (1974) presented by Patel and Epelboin. The valuable references also are collected in this article.

10.4 Statistical theory of dynamical diffraction

When the crystal is nearly perfect but include small invisible defects (thermal vibrations, clusters of impurities). What diffraction effects are expected in topography? In another opposite case, when many lattice defects of visible size are included in the crystal they can not be individually distinguished. Then, what distribution of intensity is expected in the section topographs? For dealing with these problems, one needs a statistical theory of dynamical diffraction which enables us to characterise the statistical nature of crystalline media.

In the kinematical theory, the observed intensity is the Fourier transform of the correlation function of the scattering amplitude at two different positions [45], which is essentially a generalised Patterson function. Historically, Laue [3, 46] has opened the door of the theory of this type. Van Hove's theory is a modern version of his theory.

In the dynamical theory, Darwin's work [47] on secondary extinction can be regarded as an embryo of the statistical dynamical theory. He treated the intensity balance of O and G beams on an intuitive assumption of incoherence of the waves. On the other hand, the traditional dynamical theory, which is sometimes called the theory of primary extinction, assumes perfect coherence of O and G waves. To build a bridge between the two schemes is desirable for the understanding of the physics of extinction phenomena.

With these motivations, the author developed a statistical dynamical theory in a series of papers [48]. We shall describe the basic ideas of this approach and summarize some of the results in the following sections.

10.4.1 Derivation of energy transfer equations

The current theories on secondary extinction are formulated on the energy transfer equations of the forms [49]

$$\frac{\partial I_o}{\partial s_o} = -(\mu_o + \sigma) I_o + \sigma I_g \qquad (70a)$$

$$\frac{\partial I_g}{\partial s_g} = -(\mu_o + \sigma) I_g + \sigma I_o \qquad (70b)$$

where μ_o is the normal absorption coefficient and the coupling constant σ is the reflection strength in intensity per unit distance. You will easily find a similarity with the wave equations of Takagi-Taupin type (Eqs. (54)).

Any optical theory should be justified by wave optics. For this reason, we shall start with (54). The physical implications of these equations are easily visualised by drawing zig-zag paths in the Borrmann triangle (Fig. 13). Here, a sufficiently narrow wave is taken as the incident wave. It is worthwhile to remind that the assumed incident wave is a polydirectional wave. (see Eq. (56)). At each kink point of the path the factor $i\kappa_{\pm g} \exp{\pm 2\pi i (\underline{g} \cdot \underline{u})}$ has to be multiplied.

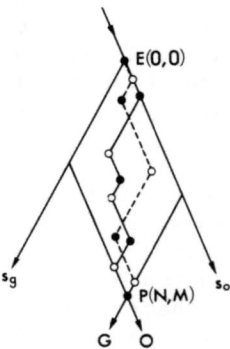

Figure 13. The zig-zag paths of the wavelet (full line) and the complex-conjugate wavelet (dotted line).

Recalling the generalised Huygens' principle, one can write down the wave field at P in the form

$$D_g(P) = \sum_R A_R \exp iP_R \tag{71}$$

where A_R and P_R are the amplitude and the phase of the wavelet specified by R, the symbolic notation of an optical path connecting E and P. (Henceforth, only G waves and G beams are discussed). The observed intensity must be

$$\langle I_g(P) \rangle = \sum_R \sum_{R'} A_R A_{R'}^* \langle \exp i(P_R - P_{R'}) \rangle \tag{72}$$

where R' indicates a path of the complex conjugate wavelet. The expression of A_R is given by

$$A_R = (i\kappa_g a)^{k+1} (i\kappa_{-g} a)^k (A/a) \tag{73}$$

where k is the number of the kinks at which the path changes from G to 0 direction, and "a" is a unit increment of distance, which is introduced for numbering the possible kink points. After calculating $\langle I_g \rangle$, it is reduced to zero. The total number of possible kink points along the distance (s_o, s_g) are $(N=s_o/a, M=s_g/a)$. The phase P_R ($P_{R'}$) depends on a set of the lattice phases at the kink points. The statistical operation, therefore, is applied only to the phase factor as indicated in Eq. (72). Here we shall consider only the lattice defects of displacement type. If the defects are constitutional type as alloys, the average has to be applied also to the amplitudes.

In giving a feeling of confidence on the procedure of calculation, we shall consider a perfect crystal, in which P_R and $P_{R'}$ are constant. Then, without any loss of generality, they can be assumed to be zero. The problem is reduced to calculating Eq. (71) with the expression (73); namely

$$D_g(P) = (i\kappa_g A) \sum_k \alpha_k (-\kappa^2 a^2)^k \tag{74}$$

where α_k is the number of the possible routes having kink points of (G→0) type. It is easily shown that

$$\alpha_k = \frac{N!}{k!(N-k)!} \frac{M!}{k!(M-k)!} \simeq \frac{(s_o/a)^k (s_g/a)^k}{k!k!} \quad (a \to 0) \tag{75}$$

Then, we shall have the same expressions as Eq. (61).

Another interesting case is highly distorted crystals, in which the phases P_R and $P_{R'}$ are completely random. Then, $\langle \ \rangle$ on the right of Eq. (72) takes unity only when R=R'. Otherwise they are null. Inserting the expression (73) into Eq. (72), we shall have

$$I = \sum_k \alpha_k |A_k|^2 \tag{76a}$$

$$\langle I_g \rangle = |\kappa_g|^2 |A|^2 + O(a^2) \tag{76b}$$

The first term corresponds to k=0 and k'=0; namely the kinematical intensity.

When the crystal is less distorted, the ensemble average < > of the phase factor will take an appreciable value provided that the following conditions are satisfied.

1a) A pair of the neighbouring kinks along R is close enough.

1b) similarly, a pair of the neighbouring kinks along R' is close enough.

2. Remaining isolated kinks of R and R' make one-to-one correspondence and the routes are close enough with each other. Then, a pair of routes R and R' form an appreciable beam route.

The reason for this is that the phases of the paired kink points are cancelled out and the averaged phase factors are appreciable. Otherwise, the phase factor $\exp i(P_R - P_{R'})$ is zero after taking the average.

For quantitative arguments, we need the correlation functions of the lattice phase factors. For example, the second order correlation function must be

$$f(z) = \langle \exp iG(s_o, s_g) \cdot \exp -iG(s_o+z, s_g) \rangle \tag{77a}$$

$$= \langle \exp iG(s_o, s_g) \cdot \exp -iG(s_o, s_g+z) \rangle \tag{77b}$$

Also, we shall define the correlation length,

$$\tau_2 = \int_0^\infty |f(z)|^2 dz \tag{78}$$

The length τ_2 is a measure of separation within which two kinks can correlate. Beyond the length τ_2, the phases of the two kinks are essentially independent.

In order to obtain concrete results, we shall take into account only the second order correlation or assume that the higher order correlation functions are factorised into a product of the second order correlation functions. Also, the condition

$$\Lambda_o \gg \tau_2 \qquad (79)$$

is assumed, where Λ_o is the fringe spacing (extinction distance) defined by Eq. (28b) (p=0). Under these conditions, one can show that

$$\langle I_o(s_o,s_g)\rangle = |\kappa_g \kappa_{-g}||A|^2 \sqrt{\frac{s_o}{s_g}} I_1(2|\kappa_g \kappa_{-g}|\tau_2 \sqrt{s_o s_g}) \exp{-\mu_e(s_o+s_g)} \qquad (80a)$$

$$\langle I_g(s_o,s_g)\rangle = |\kappa_g|^2 |A|^2 I_0(2|\kappa_g \kappa_{-g}|\tau_2 \sqrt{s_o s_g}) \exp{-\mu_e(s_o+s_g)} \qquad (80b)$$

where I_0 and I_1 are the modified Bessel functions and

$$\mu_e = \mu_o + 2\mathrm{Re}(\kappa_g \kappa_{-g})\tau_2 \qquad (81)$$

These expressions, in fact, satisfy the energy transfer equations

$$\frac{\partial \langle I_o \rangle}{\partial s_o} = -(\mu_o + 2\mathrm{Re}(\kappa_g \kappa_{-g})\tau_2)\langle I_o\rangle + 2|\kappa_{-g}|^2 \tau_2 \langle I_g\rangle \qquad (82a)$$

$$\frac{\partial \langle I_g \rangle}{\partial s_g} = -(\mu_o + 2\mathrm{Re}(\kappa_g \kappa_{-g})\tau_2)\langle I_g\rangle + 2|\kappa_g|^2 \tau_2 \langle I_o\rangle \qquad (82b)$$

The expressions are very similar to Eq. (70), but it is significant that the coupling constants are derived from the wave equations without any specific model of the lattice distortion, such as "mosaic crystals". Also, it is significant that the intensities discussed here are excited by a poly-directional wave.

The integrated intensity R_g is given simply by integrating the intensity given by Eq. (80b) over the exit surface. In the simplest case of a parallel-sided crystal [48], it is given by

$$R_g/R_g^K = \frac{\sinh(N^2+L^2)^{1/2}T}{(N^2+L^2)^{1/2}T} \exp{-MT} \qquad (83)$$

where R_g^K is the integrated intensity in the kinematical theory, T is the thickness of the crystal. The notations M, N, L are defined by

$$M = \frac{1}{2}\left(\frac{1}{\gamma_o} + \frac{1}{\gamma_g}\right)\mu_e \qquad N = \frac{1}{2}\left(\frac{1}{\gamma_o} - \frac{1}{\gamma_g}\right)\mu_e \qquad (84a,b)$$

$$L^2 = (2\tau_2|\kappa_g \kappa_{-g}|)^2/\gamma_o \gamma_g \qquad (84c)$$

where γ_o and γ_g are the direction cosines of the directions $\underline{s}_o$ and $\underline{s}_g$ with respect to the normal of the crystal.

The topography people will be interested more in the intensity distribution given by Eq. (80), whereas the crystal analysis people will be interested in Eq. (83). In fact, R_g/R_g^K is the "Extinction Coefficient".

The present theory is not very satisfactory in a sense that the applicable range is still limited by the condition (79). One of the reasons of this limitation lies in the assumption of taking only the second order correlation. One can relax the condition (79) to some extent as

$$\Lambda_o \simeq \tau_2 \tag{85}$$

by taking higher order correlations. Then the correlation length τ_2 must be reduced by multiplying a suitable factor [48]. Another reason of the limitation is the implicit assumption of the integrability of Eq. (78). The more general approach will be given by taking the correlation function in the form

$$f(z) = E^2 + (1-E^2)g(z) \tag{86}$$

where E is $\langle \exp iG \rangle$, and the function $g(z)$ is the correlation function of $(\exp iG - E)$ at two different positions separated by z. Now, it is enough to assume the integrability of $g(z)$. The correlation length is denoted by τ.

Then, one can show that the intensity consists of three terms

$$I_g = E^2 I_g^{(c)} + E^2(1-E^2) I_g^{(m)} + (1-E^2) I_g^{(i)} \tag{87}$$

The similar equation is also obtained for the O-beam. $I_g^{(i)}$ in the third term is essentially the intensity distribution (80b) which is the perfectly incoherent component. [Rigorously speaking, τ_2 in Eq. (80) and also in Eq. (81) must be replaced by τ_e, where τ_e is given by $\tau_e = E^2(\Lambda_o/2) + (1-E^2)\tau$]. The first term is the perfectly coherent component, which is given by

$$I_g^{(c)} = |\kappa_g|^2 |A|^2 |J_o(2(\kappa_g \kappa_{-g})^{\frac{1}{2}} E\sqrt{s_o s_g})|^2 \exp -\mu_e(s_o+s_g) \tag{88}$$

The extra absorption coefficient $2\text{Re}(\kappa_g \kappa_{-g})\tau$ in μ_e is to be introduced since the coherent component transforms to the incoherent component during the wave propagation. By that amount, the second term which is partially coherent is created. The expression $I_g^{(m)}$ is not simple as $I_g^{(c)}$ and $I_g^{(i)}$. Essentially, however, it is represented by the convolution integral of coherent and incoherent intensity distributions. Fig. 14 illustrates diagrammatically what happens in the crystal.

The diffraction phenomena in the topics which are raised by the first question at the beginning of this section will be described by taking $E \simeq 1$ in Eq. (87). The visibility of Pendellosung fringes will be reduced. On the other hand the intensity in the second question will be answered by taking $E \simeq 0$.

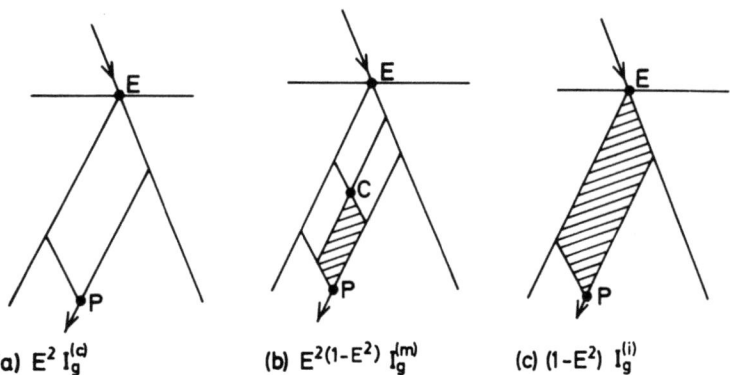

(a) $E^2 I_g^{(c)}$ (b) $E^2(1-E^2) I_g^{(m)}$ (c) $(1-E^2) I_g^{(i)}$

Figure 14. Schematic diagram of the coherent and the incoherent intensities (see Eq. (87)). The hatched area indicates the incoherent area whereas the white area indicates the coherent area. E = entrance point; P = observation point; C = position where the conversion from the coherent wave to incoherent beam takes place.

REFERENCES

1. C.G. Darwin (1914) Phil. Mag. <u>27</u>, 315
2. P.P. Ewald (1917) Ann. der Physik, <u>54</u>, 519
3. M.V. Laue (1918) Ann.der Physik, <u>56</u>, 497
4. L.V. Azaroff (1974) X-ray Diffraction, McGraw Hill, New York
5. N. Kato and A.R. Lang (1959) Acta.Cryst. <u>12</u>, 787
6. N. Kato (1961) Acta. Cryst. <u>14</u>, 526 and 627
7. N. Kato (1958) Acta. Cryst. <u>11</u>, 885
8. G. Borrmann (1941), Z. Physik <u>42</u>, 157
9. G. Borrmann (1950) Z. Physik <u>127</u>, 297

10. P. Debye (1909) Ann.de.Phys. **30**, 755
11. N. Kato (1960) Acta. Cryst. **13**, 349
12. N. Kato (1968) J. Appl. Phys. **39**, 2225 and 2231
13. T. Uragami (1969) J. Phys. Soc. Japan **27**, 147; ibid (1970) **31**, 1141
14. T. Saka, T. Katagawa and N. Kato (1972) Acta. Cryst. **A28**, 102 and 113; ibid **A29**, 192
15. T. Uragami (1971) J. Phys. Soc. Japan **28**, 1508
16. A.R. Lang and C.H. Mai (1979) Proc. Roy. Soc. in press.
17. N. Kato, K. Usami and T. Katagawa (1967) Adv. X-ray Analysis, Plenum, New York
18. A. Authier (1968) Phys. Stat. Sol. **27**, 77
19. B.C. Wonsiewicz and J.R. Patel (1975) J. Appl. Cryst. **8**, 67
20. J.R. Patel and A. Authier (1975) J. Appl. Phys. **36**, 3162
21. N. Kato (1964) J. Phys. Soc. Japan **19**, 971
22. M. Hart (1966) Z. Phys. **189**, 269
23. Y. Ando and N. Kato (1966) J. Phys. Soc. Japan, **21**, 964
24. Y. Ando and N. Kato (1970), J. Appl. Cryst. **3**, 74
25. S. Ikeno, H. Maruyama and N. Kato (1968) J. Cryst. Growth, **3,4**, 683
26. I. Kito and N. Kato (1974) J. Cryst. Growth, **24**, 25
27. N. Kato and J.R. Patel (1968) Appl. Phys. Letters, **13**, 42; (1973) J. Appl. Phys. **44**, 965
28. J.R. Patel and N. Kato (1968) Appl. Phys. Lett. **13**, 40; (1973) J. Appl. Phys. **44**, 971
29. Y. Ando, J.R. Patel and N. Kato (1973) J. Appl. Phys. **44**, 4405
30. G. Hildebrandt (1959) Z. Krist. **112**, 312
31. P. Penning and D. Polder (1961) Philips Res. Rep. **16**, 419
32. N. Kato (1963) J. Phys. Soc. Japan **18**, 1785; (1964) J. Phys. Soc. Japan, **19**, 67
33. K. Kambe (1965) Z. Naturforsch., **20a**, 770; (1968) Z. Naturforsch. **23a**, 25
34. U. Bonse (1964) Z. Phys. **177**, 385
35. U. Bonse, E. Kappler and A. Shill (1964) Z. Phys. **178**, 221
36. U. Bonse and W. Graeff (1973) Z. Naturforsch. **A28**, 558
37. S. Takagi (1962) Acta Cryst. **15**, 1311; (1969) J. Phys. Soc. Japan, **26**, 1239
38. D. Taupin (1964) Bull. Soc. Fr. Mineral Cristallogr. **87**, 469
39. N. Kato (1973) Z. Naturforsch. **28a**, 604
40. T. Katagawa and N. Kato (1974) Acta Cryst. **A30**, 830
41. F.N. Chukhovskii (1974) Soviet Phys. Cryst. **19**, 301
42. P.V. Petrashen (1973) Fizika Tverdovo Tela, **15**, 3131
43. F.N. Chukhovskii and P.V. Petrashen (1977) Acta Cryst. **A33**, 311
44. F.N. Chukhovskii, K.T. Gabrielyan and P.V. Petrashen (1978) Acta Cryst. **A34**, 610
45. Van Hove (1954) Phys. Rev. **95**, 249
46. M.v. Laue (1925) Ann. d. Physik, **78**, 167
47. C.G. Darwin (1922) Phil. Mag. **43**, 800

48. N. Kato (1976) Acta Cryst. $A32$, 453 and 458; (1979) ibid. $A35$, 9; (1980) ibid. $A36$, 171
49. W.C. Hamilton (1957) Acta Cryst. 10, 629
50. W.H. Zachariasen (1967) Phys. Rev. Lett. 18, 195
51. W.H. Zachariasen (1967) Acta Cryst. $A23$, 558

CHAPTER 11

X-RAY SOURCES

U. BONSE

11.1 Definition of a 'powerful source'

The development of more powerful sources plays a major role in the overall improvement of X-ray techniques in general and of topography in particular. In recent years some progress has been made to increase the brightness and also the total photon flux of conventional sealed-off X-ray tubes. Rotating anode tubes have been developed to greater variety, to more reliability and to higher brightness and flux. Synchrotron radiation has entered the field of X-ray diffraction and, with the increasing availability of storage rings dedicated to the production of synchrotron X-rays will presumably replace conventional X-ray sources in many experiments in the near future. Compared with a conventional source a synchrotron source is rather costly and, in general, requires the individual experimenter to leave his home laboratory, transfer his equipment to the synchrotron radiation centre and start using X-rays as member of a 'users community' at the storage ring. Considering the amount of financial and organisational disadvantages experienced by the newcomer to the big facility it is very important to find out clearly whether the synchrotron source is better for the particular problem to be solved and how much it is better. With this respect the statement, the synchrotron source is 'more intense' is certainly not adequate, since, with this generality, it may not even be true in the particular case. Hence it is highly desirable to define the powerfulness of a source precisely.

Certainly a 'better' (more powerful) source is one which delivers a larger number of photons per time interval to the detecting system of a particular experiment under otherwise equal parameters of geometry and resolution, where resolution refers to spatial, angular, spectral and/or time resolution depending whatever the method is specifically aiming at. From this definition of powerfulness it is obvious that it cannot be given for the source alone but that the acceptance and transmittance properties of the measuring apparatus have to be included.

X-RAY SOURCES

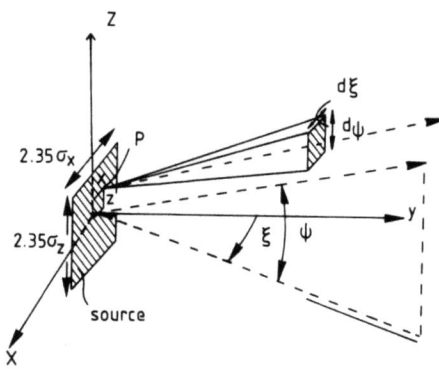

Figure 1. Geometry used for defining source characteristics. X-ray source extending in X,Z plane. The direction of emission is given by angles ψ and ξ. σ_x, σ_z source dimensions in X- and Z- directions, respectively.

For a quantitative discussion we define the following source characteristics (Fig. 1):

brightness n (x, z, ξ, ψ, E, t) [phot. s^{-1} mm^{-2} $mrad^{-2}$ per 0.1% $\Delta E/E$], gives the number of photons emitted at time t from the source point (x,z) along the direction (ξ,ψ) with energy E per time interval, unit source area, unit solid angle and 0.1% bandwidth.

intensity N (ξ,ψ, E, t) $\equiv \int_{source}$ ndxdz (phot. s^{-1} $mrad^{-2}$ per 0.1% $\Delta E/E$)

partially integrated intensity $N'(\xi, E, t) \equiv \int_{-\pi/2}^{\pi/2} N d\psi$ (phot. s^{-1} $mrad^{-1}$ per 0.1% $\Delta E/E$). With a synchrotron radiation source the integration is normal to the plane of the electron orbit.

spectral flux $\Phi_S(E,t) \equiv \int_\Omega N d\xi d\psi$ (phot. s^{-1} per 0.1% $\Delta E/E$)
Ω is solid angle of beam.

total flux $\Phi_T(t) \equiv 10^3 \int_{>0}^{\infty} E^{-1} \Phi_S dE$ (phot.s^{-1})
σ_x σ_y (mm) is standard deviation of horizontal, vertical source or beam size, respectively. Source area A $\equiv \sigma_x \sigma_y$ $(2.35)^2$ (mm^2) (Full width at half maximum is 2.35σ). σ'_x σ'_z (mrad) is standard deviation of horizontal, vertical beam divergence respectively. $\Omega = \sigma'_x \cdot \sigma'_y (2.35)^2$

horizontal emittance $\varepsilon_x \equiv \sigma_x \sigma'_x$ (mm mrad)

vertical emittance $\varepsilon_z \equiv \sigma_z \sigma'_z$ (mm mrad)

The parameters defined above can be marked with subscripts σ and π if σ- and π-polarization states have to be distinguished. It should be noted that n, N, N', Φ_S and Φ_T give number of photons per sec and not directly radiated power.

For a source to be 'powerful' or to be better than another source, different techniques usually require different source parameters to be maximized. We shall illustrate the situation by giving a few examples:

(a) Fluorescent spectroscopy: Φ_T should be large over a certain energy range. The values of n, N, A and to some extent also of Ω are not important.

(b) Focusing X-ray optics, techniques employing narrow slit systems and topographic techniques, where the geometrical resolution depends on the source size: n should be large.

(c) Perfect crystal techniques: N should be large and σ_z' equal to or smaller than the perfect crystal diffraction range. n need not necessarily be large since mostly relatively wide beams can be used.

(d) Time resolved studies: a sharp pulse structure of n, N, Φ_T and Φ_S is advantageous.

We shall see below that with respect to the above requirements synchroton sources and conventional X-ray tubes can score with techniques totally different.

In general, X-rays can be produced in at least three principally different ways: by electron impact on (usually) solid targets (EI), by fluorescent excitation (FE) and as synchroton radiation (SR) from electrons or positrons orbiting in synchrotrons and storage rings or following curved paths in special magnet systems like wigglers and undulators set up at storage rings. Due to their generally low intensity, FE sources play no role in topography, and we shall not discuss them further.

11.2 X-rays generated by electron impact on solid targets (EI)

The most common method to produce X-rays is to employ an X-ray tube. Electrons emitted from a hot tungsten wire (cathode) are accelerated towards the target (anode) by applying a suitable DC voltage from 10 kV to 100 kV between cathode and anode. In rare cases in which extremely hard X-rays are wanted a betatron operated in the MeV range [1,2] can be employed for electron acceleration. For X-ray diffraction topography X-ray energies above 100 keV are practically not needed. As is well known, the spectrum obtained with EI consists of two parts: the continuous bremsspectrum and the spectrum determined by the atomic level system of the element (s) composing the target.

X-RAY SOURCES

11.2.1 Bremsspectrum

For the thick targets which have to be used in high power tubes, it is practically impossible to calculate precisely the characteristics of the continuum radiation like polarization, spectral and space distribution. In order to estimate the intensity for a comparison with SR sources we use the approximate expression which was derived by Kulenkampff and co-workers [3, 4, 5] from experimental data

$$I_\nu = Cz\,(\nu_0 - \nu) \quad \text{(erg s)} \qquad (1)$$

where I_ν is the energy radiated by one electron at frequency ν per unit frequency into 4π solid angle. z is the atomic number of the target material and $\nu_0 = E_e/h$ the upper frequency limit as determined by the electron energy $E_e = eU$. U is the tube voltage, e the electron charge and h Planck's constant. $C = (5 \pm 1.5) \times 10^{-50}$ (erg s^2) is an empirically found constant.

The actual continuum spectrum differs from that given by equation (1) mainly due to absorption in the target. The effect is considerable at the high energy side of the absorption edge of the target material and varies also with the take off angle δ of the X-ray beam. Nevertheless the total intensity is fairly well given by integration of I_ν over ν. We set $E = h\nu$ the photon energy, normalize to an energy band $\Delta E/E = \Delta\nu/\nu = \Delta\lambda/\lambda = 10^{-3}$ and obtain per electron and 4π solid angle

$$I_{0.1\%} \simeq 10^{-3}\,C(z/h^2)\,E_e E\,(1 - E/E_e) \quad \text{(erg)} \qquad (2)$$

From (2) we calculate the spectral flux Φ_{SB} generated by a current i (mA) (i.e. $i \times 6.42 \times 10^{15}$ electrons s^{-1}) with $C = 5.75 \times 10^{-50}$ erg s^2

$$\Phi_{SB}(E) = 1.3 \times 10^7\,zUi\,(1 - E/Ue) \quad \text{(phot.s}^{-1}\text{ per 0.1\% }\Delta E/E) \qquad (3)$$

where U is in kV and E in keV. It should be noted, that in the low energy limit (long wave-length) the number Φ_{SB} of bremsstrahlung photons per 10^{-3} band-width becomes independent of E.

We shall use (3) to make a rough estimate of the number N_B of bremsstrahlung photons diffracted by a perfect crystal in a typical topography experiment. Let the source (Fig. 1) and the slit have a size of $\sigma_x \cdot \sigma_z = 5$ mm $\times$ 0.5 mm each and be 10^3 mm apart. Then $\sigma'_z = 1$ mrad and with an assumed Bragg angle $\theta_B = 45°$ we have $\sigma'_z = \Delta\lambda/\lambda = \Delta E/E = 10^{-3}$. Thus the energy band is of the order assumed in deriving equation (3). However, for a fixed λ and hence fixed E the accepted divergence is determined by the perfect crystal diffraction range $\Delta\theta_p = 0.1$ mrad which is considerably less than σ'_z. Furthermore $\sigma'_x = 10$ mrad, hence the solid angle accepted by the crystal at fixed E is $\Delta\Omega = \sigma'_x \Delta\theta_p = 1$ mrad^2 i.e. N_B is in effect the source intensity N defined at the beginning. It follows

$$N_B = Uiz(1 - E/Ue) \quad [\text{phot.s}^{-1}\,\text{mrad}^{-2}\text{ per }0.1\%\ \Delta E/E]$$

Values of N_B for $U = 40$ kV, $i = 40$ mA are given in Table 1 for typical target materials at their corresponding $K\alpha$-energies for comparison with estimates of characteristic line intensities (section 2.2). It may be noted that for photographic exposure these intensities are extremely low. As will be shown, with characteristic lines and with SR much higher intensities can be obtained.

The bremsstrahlung is found to be slightly polarized parallel (i.e. π) to the direction of the electron beam [6, 7]. Very close to the short wavelength limit the polarization P is largest, where as usual we define

$$P = (N_\pi - N_\sigma) / (N_\pi + N_\sigma) \tag{5}$$

Values of P between 10% [6] and 46% [7] have been observed. Observations on Pt, Ag, Cu and Fe targets failed to reveal any significant dependence of polarization on atomic number. With thick targets, the variation of emitted intensity with angle is not very pronounced. Nevertheless there is a maximum of emission at right angles to the incident electron beam (if the detection is not impossible because of absorption). With increasing voltage the maximum shifts somewhat towards the direction of the electron beam [8].

11.2.2 Characteristic line spectrum

The question of how many characteristic X-rays will be generated when one electron hits the target was studied by Green [9] who measured line intensities for a number of target elements at different tube voltages U and compared them with the theory by Green and Cosslett [10]. Satisfactory agreement of experiment and theory was observed for the K-series but not for the L-series. According to Green's experimental results, the total number P_j of quanta generated per electron in a line j (j = $K\alpha$, $K\beta$, $L\alpha$, etc.) is

$$P_j = K_{jz}(E_e - E_{ejz})^{1.63} \tag{6}$$

E_e is the electron energy and E_{ejz} is the minimum excitation energy in keV for the element with atomic number z and the series to which the line j belongs. Values of E_{ejz} as well as of E_{ejz} the energies of characteristic lines, are tabulated in [11]. K_{jz} are experimental quantum efficiencies of characteristic X-ray production given in [9]. For a particular line j, K_{jz} depends on z but not on the tube voltage U. From the data given in ref. [9] we have derived K_{jz} values for some common target materials (Table 2). Observed is not directly P_j but rather $P_j f(\delta)/4\pi$, where $P_j/4\pi$ is the number of quanta generated by one electron per steradian. $f(\delta)$ is a factor taking into account the absorption within the target. $f(\delta)$ depends on U, on the take off angle δ between X-ray beam and surface and on the incidence angle of the electron beam. Green [12] measured $f(\delta)$ at normal incidence for the $K\alpha$ lines of C, Al, Ti, Fe, Cu, Ge, Mo and Ag and for the $L\alpha_1$ lines of Nd, Ta and Au. Values taken from [12] for $\delta = 6°$, the usual take off angle with X-ray tubes, and for different tube voltage are given in Table 2.

For the comparison with the continuous spectrum (equation (4)) we derive the observable intensity N_L of characteristic lines per $mrad^2$ generated by a current i (m A).

$$N_L \simeq 5 \times 10^8 \, K_{jz} \, f(\delta) \, i \, (E_e - E_{ejz})^{1.63} \quad (7)$$
[photons s^{-1} $mrad^{-2}$ per 0.1% $\Delta E/E$]

where E_e and E_{ejz} are in keV. Values of N_L calculated for U = 40 kV, i = 40 mA, $\delta = 6°$ are given in Table 1 for five different lines. When comparing N_L with corresponding intensities N_B of the continuum is has to be remembered that the bandwidth $\Delta E/E \simeq 3$ to 5×10^{-4} for characteristic lines (natural line width of Mo $K\alpha_1$ (Mo $K\alpha_1$) is 3.79×10^{-4} (3.93×10^{-4}), respectively [13] whereas N_B is normalized to $\Delta E/E = 10^{-3}$. This means that N_B is in effect even about 1/3 to 1/2 lower when normalized to the same bandwidth. Furthermore, the same solid angle $\Delta\Omega = 1 \, mrad^2$ can be realized with a diffracting perfect crystal by making $\sigma'_x = 10$ mrad as before. Again for a fixed λ, $\sigma_z' = \Delta\theta_p = 10^{-1}$ mrad is determined by the perfect crystal diffraction range $\Delta\theta_p$, hence $\Delta\Omega = 1 \, mrad$. As seen by comparison of N_L and N_B of Table 1, the line intensities are roughly 1000 times larger than the corresponding bremsstrahlung intensities.

Regarding the relative intensities of different lines of the same series the following empiric rules hold [13]

(a) for most elements $N_{K\alpha_1} = 2 N_{K\alpha_2}$

(b) $N_{K\alpha_1} : N_{K\alpha_2} : N_{K\beta_2} = 100 : 50 : 35$ for W and $100 : 50 : 25$ for Cu

11.2.3 Efficiency of X-ray production by EI

The total intensity I_T per incident electron contained in the continous spectrum is obtained by integrating equation (1)

$$I_T = \int_0^{\nu_o} I_\nu d\nu = \tfrac{1}{2} \, C \, z \, e^2 \, U^2/h^2 \quad (8)$$

Dividing by eU, the energy applied by the impinging electron we obtain the efficiency η_B for bremsstrahlung generation

$$\eta_B = CzeU/2h^2 \simeq 10^{-6} \, zU \quad (9)$$

where U is in kV. Equation (9) is in agreement with measurements and calculations by Rump [14], Kramers [15] and Kulenkampff and Schmidt [5]. With $z = 42$ (Mo) and $U = 40$ kV we have $\eta_B \simeq 1.7 \times 10^{-3}$ which is very low.

The energy of the characteristic radiation per incident electron is calculated by multiplying equation (6) with E_{jz}, the quantum energy of the emitted radiation

$$I_L = K_{jz} E_{jz} (E_{ej} - E_{ejz})^{1.63} \tag{10}$$

Dividing again by $eU = E_e$, the energy applied by the incident electron, we obtain the efficiency η_L for the generation of characteristic radiation.

$$\eta_L = K_{jz} E_{jz} (eU - E_{ejz})^{1.63} / eU \tag{11}$$

For a typical line like Mo Kα with $E_{jz} = 17.5$ keV, $E_{ejz} = 20$ KeV, $K_{jz} = 6.4 \times 10^{-6}$ (Table 2) generated with a tube voltage of $U = 40$ kV we calculate $\eta_L = 3.7 \times 10^{-4}$ which is even lower than η_B. It follows, that practically the total energy of the electron beam is converted into heat. Thus the problem of anode cooling sets in effect the limit to the total flux Φ_T of an X-ray tube.

<u>11.2.4 X-ray tubes</u>

There are two principal objectives in designing an X-ray tube: first to conduct away as much power as possible in order to obtain a high total flux Φ_T and, secondly, to keep the size of the focus small in order to increase the brightness n, which, as is known from general principles cannot be increased by optical instruments. Since the maximum specific load, i.e. the dissipated power per focal area, increases with decreasing focus size, Φ_T and n cannot be optimized simultaneously. Consequently it depends on the major intended use of the tube whether the design favours Φ_T (spectroscopy tube) or n (fine focus diffraction tube). A certain flexibility is achieved when the focal size can be changed by focussing the electron beam on the anode more or less. However, focussing electron guns require to apply to the Wehnelt cylinder (Fig. 2) an adjustable potential close to the cathode potential which usually is at high potential against ground. Hence the provision of a focussing potential complicates the design of the high voltage generator and is thus available with more expensive units only.

The specific load can be increased considerably by placing the focus always on a fresh, cooled face, i.e. by moving the anode away under a stationary focussed electron beam. The technical solutions to this idea are X-ray generators with rotating anode tubes. A sealed off tube with stationary anode is illustrated by Fig. 2. Fig.

3 shows the principal construction features of a pumped tube with a water cooled rotating anode.

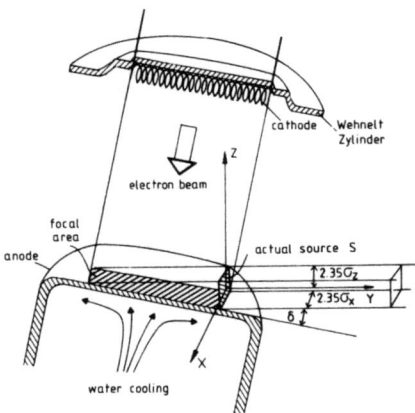

Figure 2. Conventional sealed off X-ray tube with stationary anode and water cooling. The Wehnelt cylinder may be used for focusing the electron beam on the anode. The X-ray beam is taken off at an angle δ against the anode surface. The size of the actual source S may be varied by altering δ and also by taking the beam off along the X-direction resulting in a line source.

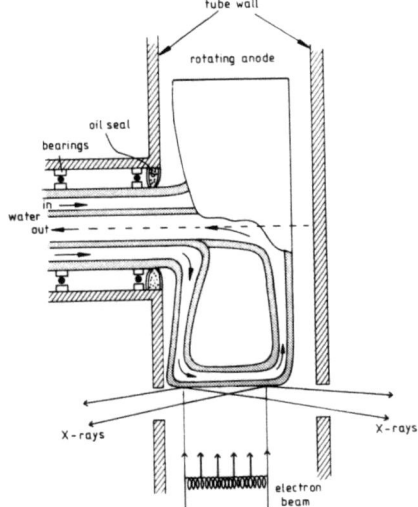

Figure 3. Permanently pumped rotating anode tube with water cooled anode.

It is obvious that the operation of sealed off tubes is very much simpler than that of rotating anode tubes where the coolant has to be supplied through a rotating shaft and all the means of making and controlling an excellent vacuum are needed. Nevertheless, because of the considerable gain in Φ_T and/or n, there are a number of powerful rotating anode generators on the market. An excellent description of rotating anode tubes which includes also aspects of design has recently been given by Yoshimatsu and Koza [16].

We give a survey of presently available tubes by listing in Table 3 and 4 their main parameters like focus size, maximum load and maximum specific load, brightness n_B and n_L and intensities N_B, N_L for bremsstrahlung and characteristic radiation respectively. n_B, n_L, N_B and N_L are given for E = 17.5 keV (Mo Kα-radiation) and take off angle $\delta = 6°$ with the tube operated at U = U_{max} and maximum total load. The data are specifically useful for the comparison with SR sources (Table 5).

11.3 Synchrotron X-ray sources

11.3.1 General

As is well known, charged particles orbiting in circular accelerators emit synchrotron radiation (SR). The total emitted intensity is proportional [17] to γ^4 where

$$\gamma = E_e/mc^2 \tag{12}$$

E_e is the particle energy and m the particle mass. It follows from equation (12) that, for a given energy E_e, SR from light particles like electrons and positrons is much more intense than that from heavier particles like protons etc. SR has a continous spectrum with a maximum which shifts to shorter wavelengths proportional to E_e^{-3}. Over the last decade, E_e in electron (positron) accelerators increased more and more. As a consequence SR also in the X-ray region below 10A wavelength became available.

For diffraction topography SR was first used in 1973 by Tuomi, Naukkarinen, Laurila and Rabe [18,19], who made topographs of a silicon crystal with SR from DESY. Hart [20] obtained topographs of silicon and lithium fluoride crystal with SR from NINA. Early uses of SR X-rays included X-ray interferometry [21], EXAFS measurements [22], small angle scattering [23] and photoelectron spectroscopy (XPS) [24]. Since then it became more and more apparent that, because of the unique properties of SR namely, continous spectrum, high Φ_S, N and n, extreme collimation, sharp time structure and defined polarization states, SR is not only very well suited for diffraction topography and other known X-ray techniques but may open up entirely new experimental possibilities in the X-ray range. A survey of X-ray investigations performed with SR may be found in the last section of the article by Gudat and Kunz [25]. In table 5 we give a list of twelve storage rings which, as far as their machine parameters are concerned, are in principle capable of generating SR in the X-ray range. Except the European Synchrotron Radiation Facility (ESRF), for which there is only a study report so far [26], all other rings are either in operation, under construction or in the planning stage.

That up to now so little SR topography has been employed is mainly due to the limited availability of synchrotron X-rays in the past. The situation no doubt will change when more storage rings exist for SR dedicated operation in the X-ray region.

11.3.2 Calculation of the radiation properties of a SR source

The radiation properties of a SR source can be calculated from exact analytical expressions, which, as we have seen, is impossible with EI X-ray sources. Thus a further advantage of SR is that it can be used for calibration purposes over a wide range of wavelengths [27, 28, 29]. We briefly summarize the theoretical expressions describing SR Further details may be found in reviews in the literature [30, 31, 32].

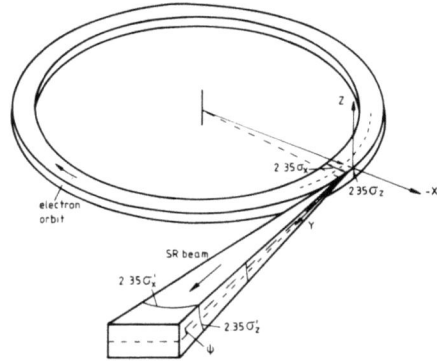

Figure 4. Geometry of a storage ring source. For a given tangent point at the orbit the radiation is contained within a cone of opening σ_z which narrows with E_e and E. The effective source size $\sigma_z \times \sigma_x$ is determined jointly by the cross-sectional size of the electron beam and the divergence σ_x' in the orbital plane accepted by the measuring apparatus.

The instantanous power $I(\lambda, \psi)$ radiated by a monoenergetic electron along a circular orbit per unit wavelength and radian of angle ψ against the orbital plane is [33] (Fig. 4)

$$I(\lambda, \psi) = \frac{27}{32\pi^3} \frac{e^2 c}{R^3} \left[\frac{\lambda_c}{\lambda}\right]^4 \gamma^8 \left[1 + (\gamma\psi)^2\right]^2 \cdot \left\{ K_{2/3}^2(\zeta) + \frac{(\gamma\psi)^2}{1+(\gamma\psi)^2} K_{1/3}^2(\zeta) \right\} \quad (13)$$

R is the radius of the orbit. λ_c is the so called 'critical wavelength' given by

$$\lambda_c = (4\pi R/3)\gamma^{-3} \tag{14}$$

or

$$\lambda_c (\text{Å}) = 5.59 R (\text{m}) E_e^{-3} (\text{GeV}) = 12.4/E_c (\text{keV}) \tag{15}$$

$K_{2/3}$ and $K_{1/3}$ are modified Bessel functions of the second kind wih argument ζ defined by

$$\zeta \equiv \frac{\lambda_c}{2\lambda} (1 + (\gamma\psi)^2)^{3/2} \tag{16}$$

SR is elliptically polarized and the terms

$$K_\pi \equiv K_{2/3}^2 (\zeta), \quad K_\sigma \equiv \frac{(\gamma\psi)^2}{1 + (\gamma\psi)^2} K_{1/3} (\zeta) \tag{17}$$

correspond to the components which are polarized parallel π and perpendicular σ to the electron orbit. The polarization is

$$P = (I_\pi - I_\sigma)/(I_\pi + I_\sigma) = (K_\pi - K_\sigma)/(K_\pi + K_\sigma) \tag{18}$$

For all λ SR is completely polarized in the plane of the orbit. The dependence of the two components on ψ for an arbitrary pair λ, E_e is qualitatively shown in Fig. 5.

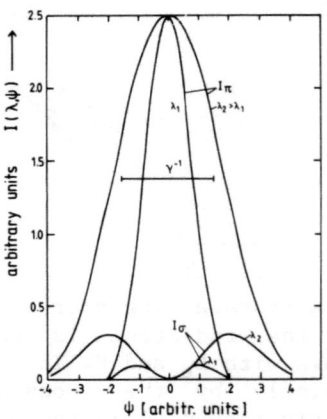

Figure 5. Dependence of polarization of SR on the angle ψ against the electron orbit. Note that shorter wavelengths are more collimated near $\psi = 0$.

As is seen, I_π is largest for $\psi = 0$, whereas I_σ has its maximum value at a certain angle ψ_m above and below the orbit plane. The total power $I(\psi)$ radiated at elevation angle ψ is obtained by integrating equation (13) over λ [33]

$$I(\psi) = \frac{7}{16} \frac{e^2}{R^2} c \gamma^5 \left[1 + (\gamma\psi)^2\right]^{-5/2} \left[1 + \frac{5}{7} \frac{(\gamma\psi)^2}{1 + (\gamma\psi)^2}\right] \tag{19}$$

where again the first term in the big bracket corresponds to parallel and the last term to perpendicular polarization. From (19) one calculates that the intensity of the parallel component falls to one-half within a cone of $1.13\,\gamma^{-1}$ opening angle e.g. with $E_e = 5$ GeV $\simeq 0.1$ mrad. Similarly one finds the perpendicular component to have maximum intensity at $\psi_{\sigma max} \simeq \pm\, 0.63\,\gamma^{-1}$. Short wavelengths are even more collimated than the above values which apply to the λ-integrated distribution [30]

The spectral distribution $I(\lambda)$ of SR is obtained by integrating equation (13) over ψ [34]

$$I(\lambda) = \frac{3^{5/2} e^2 c}{16 \pi^2 R^3} \gamma^7 \left(\frac{\lambda_c}{\lambda}\right)^3 \int_{\lambda_c/\lambda}^{\infty} K_{5/3}(\eta)\, d\eta \qquad (20)$$

From (20) we obtain $N'(\lambda)$ for a current i [mA] and an X-ray beam of 1 mrad divergence in the y, x-plane (Fig. 4)

(21)
$$N'(\lambda) = 7.87 \times 10^8 \lambda^2\, i\, \frac{E_e^7}{R^2} \left(\frac{\lambda_c}{\lambda}\right)^3 \int_{\lambda_c/\lambda}^{\infty} K_{5/3}(\eta)\, d\eta \;\; [s^{-1} mrad^{-1} 0.1\% \tfrac{\Delta E}{E}]$$

E_e is in GeV, R in m and λ in Å. We have calculated the distribution $N'(\lambda)$ of equation (21) for the 12 storage rings mentioned above. The results are shown in Fig. 6 together wih a list of operational data for which the calculations have been made. In Fig. 6, the energies of typical X-ray lines, namely, FeKα (6.40 keV, 1.936 Å), CuKα (8.05 keV, 1.54 Å), MoKα (17.5 keV, 0.71 Å) and AgKα (22.2 keV, 0.56 Å) have been indicated, and the corresponding SR intensities N' are numerically given in Table 5. As mentioned in the beginning, N' is obtained by integration of N with respect to ψ (Fig. 1 and Fig. 4) over the range of the order of σ'_z of vertical divergence. σ'_z as function of the wavelength can be estimated [35] according to

$$\sigma'_z \simeq \frac{0.511}{E_e} \left(\frac{\lambda}{\lambda_c}\right)^{1/3} \qquad \text{for } \lambda \gg \lambda_c \qquad (22)$$

$$\sigma'_z \simeq \frac{0.295}{E_e} \left(\frac{\lambda}{\lambda_c}\right)^{1/2} \qquad \text{for } \lambda \ll \lambda_c \qquad (23)$$

with E_e in GeV. σ'_z is defined as the angular range outside which the intensity has fallen below e^{-1} of the peak at $\psi = 0$. Values of σ'_z at E = 10 keV (λ=1.24 Å) calculated for the 12 storage rings are also listed in Table 5. In all cases $\sigma'_z < 0.1$ mrad which implies that the intensities N' are available within the very small solid angle $\Delta\Omega = \sigma'_z\, \sigma'_x < 0.1$ mrad x 1 mrad = 0.1 mrad^2. This means that in the orbit plane of SR sources N is more than ten times larger than the N' intensities given in Table 5. However, above and below this plane N falls off rapidly. Both facts have to be remembered, when we compare N' of the SR sources with N_B, N_L of the X-ray tubes as given in Table 3 and Table 4.

11.3.3 Time structure of SR from storage rings

In a storage ring only discrete stable positions occur around the orbit with a separation determined by the microwave frequency. A pulse structure with intervals t_p of the order of ns results if all these pockets (some 100 to 500) are filled with electron bunches. The pulse length t_b is 30 to 200 ps. If only one bunch is in the ring the repetition time t_p is about one to five microseconds. For time decay studies or live topography it may be desirable to have t_b as short and t_p as long as possible, e.g. t_b = 30 ps and t_p = 2 s. Live topography requires the intensity within one or a few bunches to be large enough to give a single picture. On the other hand, when the variations of the structure to be investigated can be controlled at will, e.g. by varying a magnetic field or similar, then a stroboscopic imaging technique becomes possible by phase locking the variations to the bunches. At present time resolved studies seem to be of interest in Biology [36,37].

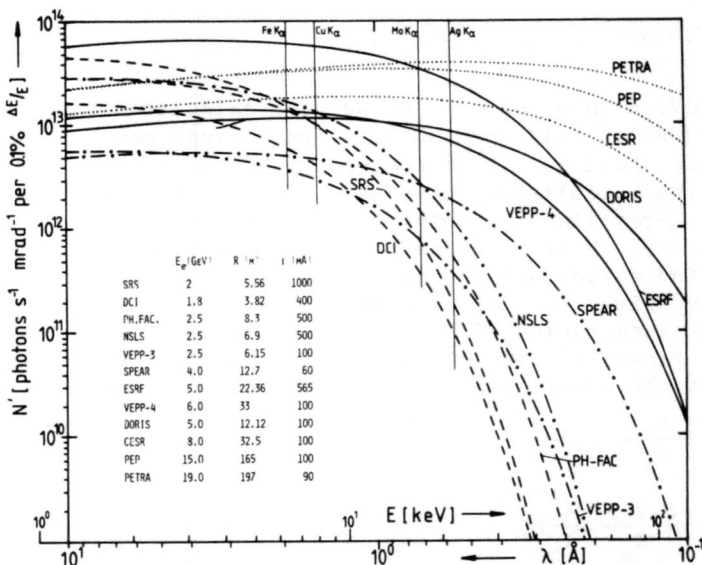

Figure 6. Calculated intensities N' for the twelve storage rings listed in Table 5. E_e is electron energy, R magnetic radius of orbit and i the electron current.

It should be noted that with single or few bunch operation the beam current i is considerably less than the maximum value given in the data list of Fig. 6.

11.3.4 SR from wigglers and undulators

With a given storage ring, E_{CB} can be increased if an odd number, e.g. 3 to 11, of alternatively magnetized dipole magnets, a so-called 'wiggler', with a larger field B_W than the field B_B of the ordinary bending magnets is introduced into an otherwise straight section of the electron orbit (Fig. 7). SR from the wiggler has a higher E_{CW} according to [38].

$$E_{CW} = E_{CB} \ (B_W/B_B) \qquad (24)$$

The ratio B_W/B_B can be as large as 4 with superconducting magnets [39]. Wigglers are planned or in operation with 5 of the storage rings of Table 5. Besides shifting E_C the wiggler increases also the intensity by a factor which is roughly given by the number of radiating turns, i.e. 2 in Fig. 7. At the same time the beam is widened which for certain applications may be disadvantageous.

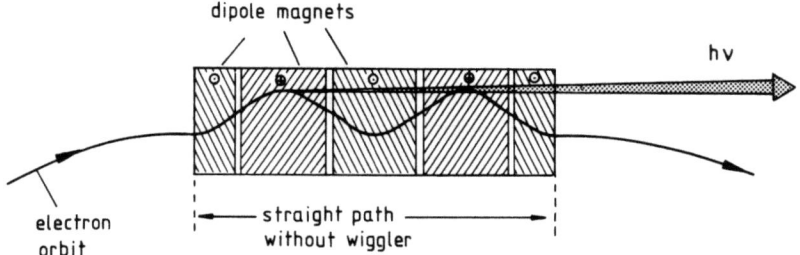

Figure 7. Wiggler with five dipole magnets.

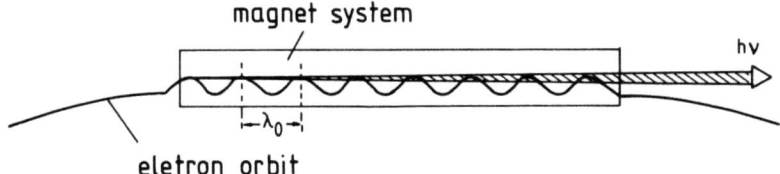

Figure 8. Principle of an undulator with a large number (typically 100) of dipoles for coherent emission of SR for any individual electron.

Undulators are wiggler-like magnet systems with a large number of m of dipole units, where m is typically of the order of 100 (Fig. 8). The principal new effect occuring in the undulator is the coherent emission of SR by the individual electron from all turns simultaneously. The condition for coherence is that the time difference for an electron and the light at the distance λ_0 of two wiggles is equal to the period of the light wave. The emitted spectrum is no longer continuous but consists of a set of discrete wavelengths. The wavelength λ_i (ith harmonic) which can be obtained at an angle θ from the trajectory is [26]

$$\lambda_i = \lambda_0 \ (1 + \gamma^2\theta^2 + \alpha^2\gamma^2/2)/2i \ \gamma^2 \qquad (25)$$

The fundamental is with E_e in GeV and λ_o in cm

$$\lambda_1 = 13\lambda_o/E_e^2 (\text{Å}) \qquad (26)$$

α is the maximum deflection angle of the electron beam. The undulator provides a set of lines which are very strong because of all the intensity of the previous continuum is concentrated in the lines. For the ESRF various undulators have been studied [26]. One example gives with 90 dipoles and

$$K \equiv \alpha\gamma \equiv eB_o\lambda_o/(2\pi mc) = 0.9337\, B_o\lambda_o, \quad B_o(\text{Tesla})\; \lambda_o(\text{cm}) \quad (27)$$

where B_o is the magnetic field and m the electron mass, for $K = 2$, $\lambda_o = 5.6$ cm a band ranging from 8.7 Å to about 0.46 Å (19th harmonic). The total emitted power is about 6 kW and 3×10^{16} photons are emitted per sec at 1A at $\Delta\lambda/\lambda \simeq 0.05$ in the 9th harmonic. The whole set of lines can be shifted by changing the magnetic field B_o and hence the vaue of K. The brightness n is expected to be improved by three to four orders of magnitude compared with nomal bending magnets.

A different type of undulator is based on the use of an helical magnetic field instead of an array of dipole fields [43,44,45]. The helical undulator has properties similar to those of the plane undulator.

11.4 Summary and conclusions

The main categories by which we have discussed the properties of the various kinds of X-ray sources were intensity N (or N'), brightness n and their dependence on direction and/or wavelength. Within the limitations of the approximations used we may summarize the results very roughly by comparing the data obtained for MoKα wavelength of all sources directly. For more detailed comparison the reader is referred to the data given in Tables 1, 3, 4 and 5 or encouraged to calculate himself intensities from the expressions given as equations (4), (7) and (21).

sealed off X-ray tube	$N_B < 10^5$ $N_L < 6 \times 10^7$	$n_B < 2 \times 10^5$ $n_B < 8.5 \times 10^7$
rotating anode tube	$N_B < 2.7 \times 10^6$ $N_L < 1.2 \times 10^9$	$n_B < 2.7 \times 10^6$ $n_L < 1.2 \times 10^9$
storage ring (DORIS)	$N' < 8 \times 10^{12}$	$n' < 4 \times 10^{12}$ ($\sigma_x \sigma_y = 2$ mm^2)

units: $N(\text{ph.s}^{-1}\, \text{mrad}^{-2}/0.1\%\Delta E/E)$, $n(\text{ph.s}^{-1}\, \text{mrad}^{-2}\, \text{mm}^{-2}/0.1\%\Delta E/E)$

X-RAY SOURCES

With all X-ray tubes N_L is peaked at characteristic wavelengths, but not with respect to direction. N_B is neither peaked in direction nor in wavelength. With storage rings N' is strongly collimated but not peaked with respect to wavelength.

From the data given we conclude that at MoKα wavelength a SR source like DORIS gives about 6×10^3 photons more than the characteristic line generated with a rotating anode tube, provided the acceptance of the apparatus to be supplied with photons is no more than about 1 mrad^2. With increasing σ'_z the photon number obtained with X-ray tubes increases correspondingly whereas with the SR source it remains constant for a fixed energy band. On the other hand, with a characteristic line the energy band is limited to about 3 to 5×10^{-4}. If the experiment can afford a wider band then with a SR source the intensity will increase whereas it will remain constant with a line source.

Another extremely important feature of the SR source is that, when combined with a suitable narrow band monochromator [46,47], it is in effect a tunable line source. In a similar way with bremstrahlung also a tunable line source can be realized, however, as has been shown with the data of Tables 3, 4 and 5, such a source has extremely low intensity when compared with the SR source.

Table 1

Intensity N_L of characteristic lines and intensity N_B of bremstrahlung at the line energy E_{jz} for some typical target materials. E_{ejz} is the minimum excitation energy of the line. Tube voltage $U = 40$ kV, tube current $i = 40$ mA. Intensities in photons s^{-1} mrad^{-2} per 0.1% $\Delta E/E$; for N_L $\Delta E/E$ is the natural width, which is roughly $(0.5 \text{ to } 0.3) \times 10^{-3}$.

Target material	z	Line	E_{ejz} (keV)	E_{jz} (keV)	$N_L \times 10^{-6}$ ($\delta = 6°$)	$N_B \times 10^{-3}$
Fe	26	Kα	7.10	6.40	99	36
Cu	29	Kα	8.86	8.05	77	39
Mo	42	Kα	20.0	17.5	13	39
Ag	47	Kα	25.5	22.2	4.5	35
Au	79	Lα$_1$	11.9	9.73	13	100

Table 2

Experimental quantum efficiencies K_{jz} and absorption factor $f(\delta)$ for characteristic line production after Green [12] for typical lines and tube voltage U. z is the atomic number.

Line	z	$K_{jz} \times 10^6$ (keV$^{-1.63}$)	$f(\delta)$ ($\delta=6°$) for U (kV) = 10	20	30	40
CrKα	24	34				
FeKα	26	30	0.98			0.56
CoKα	27	27				
NiKα	28	25				
CuKα	29	22		0.86	0.75	0.65
MoKα	42	6.4				0.78
AgKα	47	4.0			0.84	0.73
WKα	74	0.31				
AuKα	79	0.19				
AuLα$_1$	79	7.0		0.70	0.54	0.42

Table 3

Characteristics of sealed off X-ray tubes with Mo target. N_L, N_B are intensities for $K\alpha$ radiation and bremsstrahlung at the $K\alpha$ wavelength calculated with equations (7) and (4), respectively.

Type		01	F 60 – 04	10	20	FA 100
Focus l	[mm]	8	8	10	12	
Focus w	[mm]	0.15	0.4	1	2	
Max. load P_m	[kW]	0.8	2.0	2.4	2.7	3.0
U_{max}	[kV]	60	60	60	60	100
i at P_m	[mA]	13	33	40	45	30
source size ($\delta=6°$)	[mm^2]	0.12	0.32	1	2.4	
max. specific load p_m	[kW mm^{-2}]	0.67	0.63	0.24	0.11	
$N_L \times 10^{-6}$ [ph. s^{-1} mrad^{-2} per 0.1% $\Delta E/E$]		10	33	32	35	61
$N_B \times 10^{-3}$ "		23	59	71	80	104
$n_L \times 10^{-6}$ [ph. s^{-1} mrad^{-2} mm^{-2} per 0.1% $\Delta E/E$]		85	79	32	15	
$n_B \times 10^{-3}$ "		192	184	71	33	

Table 4

Characteristics of rotating anode tubes with Mo target. N_L, N_B are intensities for $K\alpha$ radiation and bremsstrahlung at the $K\alpha$ wavelength calculated with equations (7) and (4), respectively

Type			RU 200	200	500	1000	1500	GX-6
Focus l	[mm]		10	3	10	10	10	3
Focus w	[mm]		0.5	0.3	0.5	1	1	0.3
Max. load P_m	[kW]		12	5.4	30	60	90	4.0
U_{max}	[kV]		60	60	60	60	60	50
i at P_m	[mA]		200	90	500	1000	1500	80
source size ($\delta=6°$)	[mm^2]		0.5	0.09	0.5	1	1	0.09
max. specific load p_m	[kW mm^{-2}]		2.4	6	6	6	9	4.4
$N_L \times 10^{-6}$ [ph. s^{-1} mrad^{-2} per 0.1% $\Delta E/E$]			150	70	390	780	1170	40
$N_B \times 10^{-3}$	"		357	160	892	1784	109	
$n_L \times 10^{-6}$ [ph. s^{-1} mrad^{-2} mm^{-2} per 0.1% $\Delta E/E$]			300	780	780	780	1170	440
$n_B \times 10^{-3}$	"		714	1778	1784	2676	1211	

X-RAY SOURCES

Table 5

Storage rings suited for X-ray generation.
Comparison of intensities N'

Name	Location	Remarks*	E_c [keV]	λ_c [Å]	σ'_z at E=10keV [µrad]	N' $[10^{11} ph.s^{-1} mrad^{-1} per 0.1\% \Delta E/E]$ at E [keV] =				
						6.42 (FeKα)	8.06 (CuKα)	17.52 (MoKα)	22.21 (AgKα)	
SRS	Daresbury, UK.	dedicated, under construction, also wiggler	3.2	3.88	83	130	95	6.6	1.9	
DCI	Orsay, France	SR Lab	3.39	3.67	96	58	39	3.3	0.98	
Photon Factory	Tsukuba, Japan	dedicated, under construction	4.18	2.97	76	134	97	14	4.4	
NSLS	Upton, MY, USA	" wiggler	5.0	2.48	83	163	127	26	12	
VEP P3	Novosibirsk, USSR	SR Lab. wiggler	5.6	2.22	88	36	29	7.2	3.5	
SPEAR	Stanford, Cal. USA	2 SR Labs wiggler	11.2	1.1	76	50	46	25	18	
ESRF	Europe	study, also undulator	12.4	1.0	63	599	562	335	248	
VEPP4	Novosibirsk,USSR	under constr. SR Labs	14.5	0.86	57	131	125	83	65	
DORIS	Hamburg, F.R.G.	2 SR Labs	23	0.54	82	113	112	93	81	
CESR	Ithaca, NY., USA	under constr. SR Labs	35	0.36	68	176	180	171	160	
PEP	Stanford, Cal.USA	under construction	45.4	0.27	53	320	329	333	321	
PETRA	Hamburg, F.R.G.	no SR Lab	77.2	0.16	57	331	346	383	385	

* The operating parameters E_e, R[m] and i [mA] are given in Fig. 6

REFERENCES

1. R. Wideroe (1951) Brown Boveri Mitt. 38, 260
2. K. Gund (1953) Stahl u. Eisen 73, 710
3. H. Kulenkampff (1922) Ann. Phys. 69, 548
4. H. Kulenkampff (1926) Handbuch der Physik, edited by Geiger and Scheel, 23, pp. 433
5. H. Kulenkampff and L. Schmidt (1943) Ann. Phys. (5) 43, 494
6. P. Kirkpatrick (1927) Phys. Rev. 22, 226
7. E. Wagner and P. Ott (1928) Ann. Phys. 85, 425
8. W.W. Loebe (1914) Ann. Phys. 44, 1033
9. M. Green (1963) in X-ray Optics and Microanalysis, (ed. H.H. Pattee, V.E. Cosslett and A. Engström) Academic Press, New York and London, p.185
10. M. Green and V. E. Cosslett (1961) Proc. Phys. Soc. (London) 73, 924
11. J. A. Bearden (1974) in International Tables for X-ray Crystallography Vol. IV (ed. J.A. Ibers and W.C. Hamilton), The Kynoch Press, Birmingham, p.20
12. M. Green (1963) in X-ray Optics and Microanalysis, (ed. H.H. Pattee, V.E. Cosslett and A. Engstrom) Academic Press, New York and London, p.361
13. A.E. Sandstrom (1957) in Handbuch der Physik 30, edited by S. Flugge, Springer Verlag Berlin, Gottingen, Heidelberg, pp.78
14. W. Rump (1927) Z. Physik 43, 254
15. H.A. Kramers (1923) Phil. Mag. 46, 836
16. M. Yoshimatsu and S. Kozaki (1977) in X-ray Optics, Topics in Applied Physics Vol. 22 (ed. H.-J. Queisser) Springer Verlag Berlin, Heidelberg, New York, p.7
17. J. Schwinger (1949) Phys. Rev. 75, 798
18. T. Tuomi, K. Naukkarinen, E. Laurila and P. Rabe (1973) Acta polytech. Scand. Ph100, 1
19. T. Tuomi, K. Naukkarinen and P. Rabe (1974), Phys. Stat. Sol. (a) 25, 93
20. M. Hart (1975) J. Appl. Crystallogr. 8, 436
21. U. Bonse and G. Materlik (1975) in Anomalous Scattering (ed. by S. Rameseshan and S.C. Abrahams) Mungsgaard Kopenhagen, p.107
22. P. Eisenberger and B.M. Kincaid (1975) Chem. Phys. Lett. 36, 134
23. J. Barrington Leigh and C. Rosenbaum (1974) J. Appl. Crystallogr. 7, 117
24. L. Lindau, P. Pianetta, K.Y. Yu and W.W. Spicer (1976) Phys. Rev. BB13, 492
25. W. Gudat and C. Kunz (1979) in Synchrotron Radiation, Techniques and Applications (ed. by C. Kunz), Topics in Current Physics, Vol. 10, Springer Verlag Berlin, Heidelberg, New York, p.55
26. Y. Farge (ed.) (1979) Feasibility Study for a European

Synchrotron Radiation Facility, European Science Foundation, Strasburg
27. J.R. Stevenson, H. Ellis and R. Bartlett (1973) Applied Optics 12, 2884
28. D. Einfeld and D. Stuck (1976) Opt. Comm. 19, 197
29. E. Pitz (1969) Appl. Opt. 8, 255
30. R. Haensel and C. Kunz (1967) Z. Angew. Phys. 23, 276
31. R. Godwin (1969) in Springer Tracts in Modern Physics Vol. 51 (ed. by G. Hohler) Springer Verlag Berlin, Heidelberg, New York
32. C. Kunz (ed.) (1979) Synchrotron Radiation, Techniques and Applications, Topics in Current Physics Vol. 10, Springer Verlag Berlin, Heidelberg, New York
33. J. Schwinger (1949) Phys. Rev. 75, 1912
34. D.H. Tomboulian and P.L. Hartmann (1956) Phys. Rev. 102, 1423
35. J.D. Jackson (1975) Classical Electrodynamics, John Wiley, New York, London, Sydney, Toronto, p.676
36. H.B. Stuhrmann (1978) Quarterly Review of Biophysics II, I, 71
37. J. Barrington Leigh and C. Rosenbaum (1976) Ann. Rev. Biophysics and Bioengineering 5, 239
38. H. Winnick and R. Hehn (1978) Nucl. Instr. Meth. 152, 9
39. D.E. Baynham, P.T.M. Clee and D.J. Thompson (1978) Nucl. Inst. Meth. 152, 31
40. H. Motz (1951) J. Appl. Phys. 22, 527
41. D.F. Alferov, Yu A. Bashmakov and O.G. Bessenov (1976) in Synchrotron Radiation, Lebedev Phys. Inst. Series 80 (ed. N.G. Basov) New York Consultants Bureau, p.97
42. A. Hofmann (1978) Nucl. Instr. Meth. 152, 17
43. D.F. Alferov, Yu A. Bashmakov and E.G. Bessonov (1974) Sov. Phys. Tech. Phys. 18, 1336
44. L.R. Elias et al. (1976) Phys. Rev. Lett. 36, 717
45. B.M. Kincaid (1977) J. Appl. Phys. 48, 2684
46. J. A. Beaumont and M. Hart (1974) J. Phys. (E) 7, 823
47. U. Bonse, G. Materlik and W. Schröder (1976) J. Appl. Cryst. 9, 223

The following review references were added in proof:

48. J.V. Gilfrich, P.G. Burkhalter, R.R. Whitlock, E.S. Warden and L.S. Birks (1971) Anal. Chemistry 43, 934
49. D.B. Brown and J.V. Gilfrich (1971) J. Appl. Phys. 42, 4044
50. D.B. Brown, J.V. Gilfrich and M.C. Peckerar (1975) J. Appl. Phys. 46, 4537
51. C.M. Dozier, D.B. Brown and J.W. Criss (1978) IEEE Transact. on Nucl. Science ns-25, 1634

CHAPTER 12

X-RAY DETECTORS

A.R. LANG

12.1 Introduction

12.1.1 X-ray photon counters

Finding the diffracted beam and adjusting the crystal are the standard preliminaries before taking a topograph. A convenient X-ray detector and associated circuitry are a great help, if not essential. We will commence by reviewing X-ray photon detectors, and will outline the type of detector and counting circuit combination that satisfies the topographer's needs in the simplest way. In our discussion of photographic recording in Section 12.2 we will deal most fully with the Ilford L4 nuclear emulsion which has served X-ray topography well for over two decades, but we shall also consider recording means having resolving powers both lower and higher than this nuclear emulsion. In addition, we will offer practical advice on the photomicrography of topographs. Next, in Section 12.3 we will discuss electronic recording methods which make 'instant topography' possible. In all attempts at quantitative discussion, we shall not hesitate to use approximations and rounding-off of numbers in order to simplify the presentation.

Primary X-ray photon detectors function by converting the X-ray photon into ionization in a gaseous, liquid or solid medium, or into a number of lower-energy photons (in the visible or near ultraviolet, for example). The average energy needed to form a primary ion pair in rare gases such as argon, krypton and xenon is about 26 eV; whereas that needed to form an electron hole pair in a semiconductor is about three times the bandgap energy, and is thus abut 3 eV in silicon and 2 eV in germanium. Consequently, a MoKα photon with all its energy, 17.5 keV, absorbed can produce 670 electric charge pairs in a gas-filled counter and about 10 times as many in a Ge semiconductor detector. This difference is important with regard to energy resolution, as explained below. Phosphors which convert X-rays into visible radiation, and which are much used for this purpose, have energy conversion efficiencies of about 10% in the case of NaI(Tl), used in scintillation counters, and about

X-RAY DETECTORS

25% in the efficient (but slowly decaying) activated ZnS phosphors which are used in fluorescent screens (and sometimes as the primary detector in electronic X-ray image intensification systems). Consequently on average, a MoKα photon will be converted into 560 'blue' photons ($\lambda \sim 400$ nm) by NaI(Tl) and into 1800 'green' ($\lambda \sim 500$ nm) by ZnS.

Gas-filled counters are divided into ionization chambers (in which no deliberate ionization amplification by multiple collisions of ions occurs), proportional counters (in which controlled amplification by multiple collisions does take place), and Geiger counters (in which very great but indeterminate internal amplification occurs). Only recently have high-gain, low-noise pulse amplifiers developed to the stage where counting of photons of as low energy as MoKα and CuKα in an ionization chamber is a feasibility. The wavelength discrimination is limited by the variance, $N^{\frac{1}{2}}$, in the number, N, of primary ion pairs (as discussed further below), and the coefficient of variation, $N^{\frac{1}{2}}/N$, of pulse height in the case of MoKα radiation would be about 4%, ideally. In view of the advantages of having high amplification in the detector (as in the proportional counter) or indeed very high amplification (in the case of the Geiger counter or scintillation counter), on the one hand, and of better energy discrimination (as in the semiconductor detector), gas ionisation counters are not used in X-ray work like they are in neutron diffraction. It is not easy to make a compact gas-filled counter which will have uniformly good absorption efficiency for all the X-ray wavelengths used in X-ray topography. Here the scintillation counter incorporating standard commercially available hermetically sealed NaI(Tl) scintillator crystals with thin Al or Be windows facing the X-rays does best. If the dead-time of a counter and its associated circuitry is τ then, in first approximation, the fraction of counts lost therefrom is $n\tau$ where n is the mean incident counting rate. With laboratory sources and not too imperfect crystals, one does not often have to deal with beams as strong as 10^5 counts/sec. Hence, for topographic needs, a net dead-time of 1 μsec is low enough. A low detector background counting rate is a great help when searching for weak reflections, but for standard topographic experiments there is not much point in striving for a detector background rate much lower than that corresponding to the minimum background scatter from the specimen and from the air path of the X-ray beam. A good scintillation counter will have a background rate not more than about 1 count/sec when used with a simple discriminator to cut off small pulses. The level can be reduced to about 0.1 count/sec by employment of a single channel pulse height selector. Given sufficiently stable circuits, it is practicable to improve signal-to-background ratio at modest cost in counting efficiency by narrowing the window between upper and lower discriminator levels and thereby rejecting some of the wings of the pulse-height distribution. For standard topographic setting-up experiments, use of a semiconductor detector with full cryogenic paraphernalia is not sensible, but with a simple thermoelectric cooler it would be practicable. Risking over-simplification, a rating of salient performance features of the counters considered is set out in Table 1. The scintillation counter is our first choice.

Table 1

PROPERTIES OF X-RAY DETECTORS

Type	Geiger	Gas proportional	Scintillation	Semiconductor
Characteristic Absorption efficiency	moderate	moderate	good	good (Ge)
Wavelength discrimination possible	none	moderate	poor	good
Background	high	low	moderate	low
Dead-time approximate	100µs	1µs	1µs	<1µs
Durability	fair	poor	good	fair

12.1.2 Counting statistics and counting ratemeters

Suppose random events are being counted, N being the mean number occurring in a given time interval. The probability $P(n)$, of counting N events in this interval is given by the Poisson distribution $P(n) = (n!)^{-1} N^n \exp(-N)$. For $N > 50$ this distribution is closely approximated by the Gaussian $P(n) = (2\pi N)^{-1} \exp\{-(N - n)^2/2N\}$. Both distributions have a standard deviation of $N^{\frac{1}{2}}$: thus the coefficients of variation of counts of 100, 10^3, 10^4, etc. are 10%, 3%, 1% etc. This illustrates the importance of having an efficient X-ray detector. In a laboratory experiment a typical beam intensity on a Bragg peak will be 10^4 photons/sec. A scintillation counter will count substantially all of them, a Geiger counter perhaps only 10%. Locating the peak is much less tedious with the more efficient detector. If counting losses due to dead-time are substantial, the form of $P(n)$ is significantly changed. This topic, incuding the effect of a time-dependent variation of source intensity of period small compared with the counting interval, is of concern in X-ray diffractometry, and is discussed in [1].

Counting ratemeters contain integrating circuits to smooth out statistical fluctuations in the incoming count rate. The conventional method is an RC 'tank circuit' of which the time constant T is the product of the (effective) resistance R and (effective) capacitance C. The effective 'memory' of this circuit is 2RC. Therefore the coefficient of variation in a reading of a countrate r is $(2rRC)^{-\frac{1}{2}}$. The response at time t when the incoming count rate increases step-wise from 0 to r is $r[1-\exp(-t/T)]$, and so is initially rt/T. Another type of integration is provided by a linear memory circuit that records all incoming counts, and also stores them for time T. Counts held for more than T are subtracted from those incoming.(This circuit, a dream 30 years ago, is now very much a practicality). Rate of response to an incoming count rate increasing step-wise from 0 to r is of course linear, being rt/T up to the maximum, attained when $t = T$; but the coefficient of variation at countrate r is $(rT)^{-\frac{1}{2}}$ and hence is $\sqrt{2}$ that of an RC tank circuit giving the same initial response to a step-function change in count-rate. The linear memory type of circuit has advantages when a profile analysis of scanned peaks is to be performed, as in powder diffractometry; but for the X-ray topographer looking for a peak, the quicker <u>initial</u> response of the RC circuit giving the same effective smoothing as the linear memory circuit is an advantage. The X-ray topographer's ratemeter must provide an adequate counting rate range. A standard operation in crystal alignment requires measuring, quickly and conveniently, the intensities of α_1 peak, α_2 peak, trough between the peaks and the background. Linear rather than logarithmic scales are more suitable, and each decade must be subdivided : a system giving full scale deflections at count rates of 100, 200, 500, 1000, 2000 etc. simplifies the arithmetic. All such range changing can be done digitally, by scaling circuits preceding the analogue integrator. A few time constants, from 1 sec to 5 sec as maximum, suffice. It is an interesting phenomenon, that when random events are divided by various scaling factors, and presented as audible clicks, a raucous noise transforms fairly abruptly into an acceptable note of recognisable pitch as the scaling factor is increased to above about 100 [2]. Consequently, a loudspeaker emitting the output from the ratemeter scaling stages is a very helpful adjunct to the counting equipment.

12.2 Photographic recording methods

12.2.1 Photographic densitometry and sensitivity

As an amplifier and recorder of low-energy signals, the photographic process based on the chemical development of the latent

image stored in silver halide grains remains unsurpassed in power
and versatility. It is a fortunate circumstance that the active
elements, silver and bromine, have a good stopping power for the
X-ray wavelengths used in crystallography and topography. The
accepted measure of blackness of a processed photographic emulsion
is the density D, defined by

$$D = \log_{10}(I_o/I) = -\log_{10}T$$

where I_o is the intensity of light incident, I the intensity
transmitted, and the transmittance, T, is the fraction I/I_o. In
photography with visible wavelengths it is customary to display the
characteristics of a photographic emulsion when processed a
particular way by a plot of D versus logE, where E = exposure, =
intensity of illumination x duration of exposure. This has been the
practice since the work of Hurter and Driffield in 1890, and these
plots are often still called 'H and D curves'. Their advantage is
that they display an important straight segment which corresponds to
the working range of the emulsion, and the slope of this straight
segment is the 'gamma' or Contrast. Curves of D versus E do not show
a straight segment. The linearity in the D versus logE plot arises
because a multiplicity, p, of hits by photons (in the visible energy
range) is required, on average, to render a halide grain
developable. Indeed, from the analysis of D versus logE curves,
estimates of p can be derived; typically p will range from 10 to 100
in a given emulsion [3,4]. By contrast, both theory and experiment
make clear that, at X-ray energies, the ionization by photoelectrons
and delta rays consequent upon absorption of a single X-ray photon
renders not one but indeed many grains developable (the number being
greater the more dense and fine-grained the emulsion, for a given
X-ray energy). From such a different statistical situation follows a
different law: D α E, over the working range of exposures. This is
well known to radiographers and crystallographers; and it was shown
long ago [5] that with X-rays a plot of D versus E was more useful
than a plot of D versus log E. Surprisingly, a recent publication
relating to X-ray topographic photographic technique disregarded
this experience [6]. The departure from linearity at the lower end
of the linear segment of the D versus E curve in the X-ray case
arises from 'chemical fog' (with a contribution from cosmic ray
exposure in the case of old emulsions), and the background density
(i.e. that at E=0) can be unpleasantly high (> 0.3) in the case of
some fast X-ray films. At what higher value of D the straight
segment starts to bend over, i.e. the emulsion starts to saturate,
depends upon many factors, including the photometric procedure
adopted; but values of D between 3 and 4 would be representative in
the case of thick emulsions such as are used in X-ray topography.

12.2.2 Statistical limitations on resolution

It is evident, from examining X-ray topographs at a

magnification of a hundredfold or more, that a serious defect impairing image resolution is a high 'apparent granularity' in the structure of the image. The qualification 'apparent' is introduced because we are referring to topographs taken with emulsions (such as L4 nuclear emulion) in which diameters of developed grains should not be more than about 1/4 μm, and such would generally be well below visual resolution in the system used for viewing the topographs. This apparent granularity must be blamed upon statistical fluctuations in the small number of photons absorbed in image area elements on the size scale we wish to resolve. The photographic emulsion, even in fine-grain form, is too powerful an amplifier. If we made the numbers of photons absorbed per picture element high enough to render their statistical fluctuations unobtrusive, our emulsion would be totally black. This problem has been analyzed before and criteria have been stated for confident detection of image density differences in face of statistical 'photon noise' [7]. Here we will just highlight the situation by a brief discussion from first principles. Let us divide our image into picture elements (pixels) 10μm x 10μm in size. Suppose, for simplicity, that one photon absorbed always produces one developable grain, and that all developed grains have the same diameter, 1μm. Let 100 photons be absorbed in the pixel. These will produce 100 developed grains distributed within the emulsion (assumed several tens of micrometers thick) under the pixel area, and these in turn produce a photographic density, D, (which might be about 0.6). The standard deviation of the number of photons absorbed is 10, and hence the coefficient of variation is 0.1. We may choose to regard as significant a departure of k standard deviations on one or other side of the mean value. Then the densities of adjacent pixels (which have received the same average photon dose) can vary within the range $D(1-0.1k)$ to $D(1+0.1k)$, in this example, without exciting suspicion of representing a real local difference in diffracted intensity. If we took k=2, then 2.3% of pixels would, by chance flucuation, exhibit densities higher than 1.2D. For k=3 or 4 the corresponding probabilities of by chance exceeding densities 1.3D and 1.4D are 1.3×10^{-3} and 3×10^{-5}, respectively, which are sufficiently small to justify choosing one such value of k (say k=4) to determine our threshold for detection of a true signal above statistical fluctuations. Let us develop the argument to estimate how many photons we must absorb, on average, to detect a contrast, C, where C is a relative increment in density $\Delta D/D$. From above, we see that $C = kN^{-\frac{1}{2}}$ where N is the number of absorbed photons in the pixel. But for a given photographic density, $N = \eta P d^2$ where η is the absorption efficiency of the emulsion, P is the incident dose per unit area and d is the edge length of the pixel. So we get

$$C^2 d^2 \eta P = k^2$$

a simplified statement of the de Vries - Rose law [8,9]. Note the constraints it imposes; we shall run into them again in Section 12.3

when dealing with electronic recording of topograph images. We have chosen k; ηP is limited. To gain higher contrast sensitivity we must sacrifice spatial resolution, and vice versa. Summarising the analysis of detectability we performed in [7] (where we took the probability of reaching our contrast level by chance to be as high as 10%), we found that with normally processed L4 nuclear emulsion a pixel diameter of 2μm and C = 0.3 was a border-line case with CuKα, and a pixel diameter of 10μm and C = 0.1 was a border-line case with MoKα.

12.2.3 Fast X-ray films

As a measure of film speed we shall use the quality 'Photographic effectiveness' (PE), proportional to the ratio of photographic density to the X-ray dose per unit area: PE α (D/P). We define PE, for a given radiation and processing, as

$$PE = \eta\, m A$$

in which η is the absorption efficiency, m is the number of developed grains produced per absorbed photon, and A is the average projected area of a developed grain. For some fast films (which are double-coated) η may be as high as 0.75 for CuKα. The number m will depend upon the emulsion thickness, the density of packing of grains, the undeveloped grain size, and the extent of 'infectious development' by which developing grains induce development of adjacent grains in which no ionization has been produced. With higher photon energies (e.g. AgKα) a fair fraction of the energy of an absorbed photon may be lost through escape of fluorescent X-rays and photoelectrons, or by the latter spending energy uselessly in the gelatin. Thus we cannot take m as proportional to photon energy. Note that in all emulsions m is subject to statistical fluctuation; and the range of sizes of A may have a wide spread. Fast films gain their high speed relative to L4 nuclear emulsion by virtue of very much larger values of A, which overrides their generally smaller values of η and m (excluding the multiplication of m by infection that occurs in fast films). Thus fast films contain intrinsically much worse 'photon noise' fluctuation than nuclear emulsions, but this is generally masked by the even more obtrusive true granularity. We may divide emulsions used for topography into those in which true granularity, or apparent granularity due to photon noise, dominates. The faster and coarser-grained end of the range of films commercially available is covered in a recent compilation [10]. For topographic application, a brief summary of the situation is as follows. Kodak 'No-Screen' is too coarse, and has too high background fog for any topographic use. Dropping a factor of about 2 in speed, we come to high speed dental films which have convenient format, but are still too coarse for other than 'test' role to check crystal alignment, etc. Dropping another factor of 2 in speed brings us to Kodak Industrex C, which is sufficiently fine-grained to be useful in diffuse-reflexion topographs, for example. This last-mentioned film is probably comparable with Agfa-Gevaert Structurix D7 film. The Agfa series D7, D5, D4 and D2 offers a sequence of emulsions of decreasing grain-size, and decreasing speed, which has X-ray topography applicability, but awaits quantitative assessment with regard to the factors η, m and A

12.2.4 Properties and processing of nuclear emulsions

For high-quality topographs, the Ilford L4 nuclear emulsion is the recommended recording medium. Attributes that render it eminently suitable are small grain size ($\sim$0.15 μm before development), high X-ray stopping power (emulsion density 3.82 kgm^{-3} with halide weight fraction as high as 83%), sensitivity to low-energy electrons such as the photoelectrons and Auger electrons produced when X-rays are absorbed, and availability in a range of emulsion thicknesses well attuned to the needs of topographers. Thicknesses of L4 emulsion needed to give η=0.5 are roughly 100μm, 50μm and 12 μm for AgKα, MoKα and CuKα, respectively. Hence the available thicknesses 100 μm, 50μm and 25 μm are appropriate for use with these three radiations, respectively. Peculiarities of technique attending the processing of nuclear plates stem from the much greater thickness of emulsion compared with that used in normal photographic work, and are simply those precautionary measures needed to ensure uniform development throughout the emulsion (so as to give equal statistical weight to absorption events at all depths in the emulsion) and to avoid damage and distortion of the emulsion when it is in its wet, swollen state. Processing instructions have been given in [7]. Here we will simply make six recommendations applicable to emulsion thicknesses of 25 μm and upwards. (In the case of 25 μm emulsion used for CuKα and softer radiations, some relaxation towards a more standard processing procedure is allowable when there is pressure to reduce processing time.) The recommendations are as follows:

1. Use the X-ray developer diluted, i.e. 1 part 'D19' to 3 parts of water, and do not use 'Amidol' developer though the latter has been employed with thick emulsions in nuclear research

2. Soak the emulsion in water before development so that it can swell and allow rapid subsequent inward diffusion of developer.

3. Develop at a carefully monitored temperature below room temperature, such as 0° or 5°C. This slows down development rate relative to diffusion rate and allows more uniform and more controlled development.

4. Fix for long enough: not less than the time to clear the emulsion plus 50% longer.

5. Wash long enough i.e. 1 hr. for 25μm, 2 hrs for 50 μm, 3-4 hrs for 100μm thick emulsions.

6. Dry slowly, at room temperature: no forced draught to be used.

A useful characteristic of Ilford L4 emulsion is that after an initial period (very roughly 10 minutes at $5^{\circ}C$) density increases linearly with development time, for a given exposure. Often there is needed a set of topographs of similar mean density for comparison. If one topograph has been unavoidably underexposed, it can be given extra development to bring up the mean density (and hence the contrast) to the desired level. But quite cursory examination will pick out the 'underexposed, overdeveloped' plates in the set. It will show the most obtrusive 'photon noise'.

12.2.5 Handling and preservation of nuclear plates

Before use, nuclear plates should be stored away from ionising radiations and in a cool, dry place: $10^{\circ}C$ and 50% relative humidity are advised. It is desirable to let the plate warm up to room temperature in a ventilated enclosure before confining it in its cassette, so that any condensed moisture can evaporate. On no account must the area of emulsion to be exposed be touched by fingers or other objects. After processing and complete drying the emulsion surface must be protected by a cover glass, and between glass and emulsion should be put an inert medium. The air-hardening embedding medium 'Eukitt' is recommended (supplied by I. Hecht-Mertens, Kiel-Hassee, Stadtrade 27, German Federal Republic).

12.2.6 Ultra high resolution image recorders

For a recording resolution higher than even L4 nuclear emulsion can give, one looks either to special photographic plates or to the so-called 'grainless' recorders. The latter are used as 'X-ray resists' in X-ray lithography: the best known is PMMA (polymethylmethacrylate). For wavelengths in the range 0.5nm to 5nm X-ray lithography using PMMA and related substances is now well established [11]. In passing to shorter wavelengths such as are needed for X-ray topography one must remember that with organic resists, not only is the X-ray stopping power very low, but the longer ranges of photoelectrons compared with those in nuclear emulsions will nullify the attempts to achieve resolutions of 1μm or better. The image diffusion by long photoelectron tracks can be counteracted by making the detector very thin, but this (with PMMA say) would involve prohibitively low values of η. The hope, for topographic usefulness, lies in resists containing heavy elements, such as barium lead acrylate. With very fine grain photographic emulsions one is on surer ground. Kodak High-Resolution Plates Type 649-0, emulsion thickness 10μm, are about 10 times slower than 25μm Ilford L4 plates when used to record Cu $K\alpha$ radiation, and when processed similarly and to a similar density, as the L4 plate. The

Type 649-0 plate has a resolution of 2,000 lines/mm for visible light and will give sub-micrometre resolution in X-ray applications [12].

12.2.7 Photomicrography of X-ray topographs

It is perhaps a statement of the obvious to remind the topographer that to produce a good print requires a good negative and a good negative requires a good original topograph. Nevertheless, the difficulty of making good enlargements of topographs must not be minimised: arguably it demands more skill, and certainly it demands more care and patience than the production of good originals. The photographic drill is an exercise in contrast compression. X-ray topographs, like radiographs, are essentially 'contrasty' by visual photographic standards. This follows from the linear relation between D and E as opposed to the proportionality of D to log E that holds at visible wavelengths (as discussed in Section 12.2.1), quite apart from the circumstance that (at shorter wavelengths and low X-ray absorption) there may be a 100-fold difference in integrated X-ray reflection from a volume of perfect crystal compared with a volume with kinematic diffracting behaviour. As regards equipment, our maxim is 'simplest is best': the fewer optical elements in the system, the fewer the interfaces to collect dust and introduce artefacts. For those whose topographs do not exceed about 25mm x 25mm in size, we could indeed claim that not more than two lenses will meet all needs: one lens to use in a standard '35 mm' enlarger to enlarge fields ranging from the maximum down to about 1 1/2 mm in diameter, and the second lens a flat-field objective (10x magnification, and numerical aperture (NA) of 0.2 or 0.25) for producing higher-magnification negatives of fields less than 1 1/2 mm in diameter. For making lantern slides, or when rapid enlargement of many small fields is needed, 35 mm film can be used; but for serious work the use of sheet film is necessary. The latter can be obtained with orthochromatic fine-grain emulsions with which a red safelight of brightness adequate for effective monitoring of the progress of development is permitted. In choosing his lenses and objectives, the X-ray topographer must consider both the theoretical resolution (in air), R, of his lens and the depth of focus, H, within his object:

$$R = \lambda/2NA$$

$$\text{and } H = \lambda/4\mu \sin^2(\alpha/2)$$

In these formulae, the wavelength λ can be taken as 0.55nm; μ the refractive index of processed nuclear emulsion, can be taken as 1.51 and α the semi-angle of marginal rays in the object, is arc sin (NA/μ). Table 2 shows the connection between 'f numbers' of camera and enlarger lenses and their NA values; and it gives the theoretical minimum resolvable distance in air and the theoretical depth of focus in processed nuclear emulsion. Remember that processed nuclear emulsions shrink down to about 0.4 times their original thickness.

Table 2

THEORETICAL RESOLUTION IN AIR, AND DEPTH OF FOCUS IN PROCESSED NUCLEAR EMULSION, OF LENSES

f number	NA	Resolution μm	Depth of focus, μm
4.5	0.110	2.5	68
3.5	0.141	1.9	41
2.8	0.175	1.6	26
-	0.25	1.1	13
	0.5	0.55	3

12.3 Electronic recording methods

12.3.1 Basic problems

When does the X-ray topographer really benefit from rapid (indeed effectively instantaneous) viewing of topographic images? The answer is: when a large number of images (probably of different specimens, as in process control) must be examined as speedily as possible, or when the topographic feature of interest is in motion. In the latter category fall moveable objects such as dislocations, grain boundaries, twin boundaries, phase boundaries (including crystal-melt and crystal-solution interfaces), domain boundaries, X-ray Moiré fringes. Whether or not an electronic image intensification system will satisfy the topographer's needs depends upon how well it copes with the fundamental physical factors which together circumscribe its performance. These are (1) statistical limitations of resolution (2) instrumental limitations of resolution, (3) instrumental noise and (4) dynamic range. We commence by considering (1), and restate the deVries-Rose law, equation (2), in the form appropriate to 'real-time' viewing:

$$C^2 d^2 \eta I \tau = k^2$$

in which the incident dose per unit area, P, which appeared in equation (2) is replaced by $I\tau$ where I is the incident flux of photons per unit area per unit time and τ is the integration time.

If τ were determined solely by the eye, we could substitute values such as 0.2 sec for the dark-adapted eye or 0.1 sec for normal laboratory light levels (in practice, the use of phosphor viewing screens may justify us in taking a minimum value of τ to be 0.5 sec in discussing photon statistics). Larger values of τ can of course be provided by electronic integration (digitally or in analogue storage tubes); and values of τ between 2 and 10 seconds could still be regarded as providing 'instant' topography (they would not be too long for use in inspecting silicon wafers, say, or in observing recrystallisation; but would be generally rather long for providing a smooth continuous record of dislocation or domain motion). Now a typical value of I when taking a topograph (section or projection) with a 1kW laboratory source would be 10^3 photons mm^{-2} sec^{-1}. This, with $\eta=1$, will give us 10 perceived photons/sec in a pixel of 100 μm edge, and, accordingly, a resolution severely limited by photon noise. Synchrotron sources now or shortly in operation should offer about a hundredfold increase in I, which would give us 10 perceived photons/sec in a pixel of 10 μm edgelength. We take 10 μm as being on the frontier of instrumental resolution; and note that with this instrumental capacity, and assuming $\eta=1$, we will then need integration of some seconds duration to detect features of pixel dimension having $C \lesssim 1$.

Remarking further on instrumental resolution, factor (2), we just make three points. Firstly, early image intensification systems such as that of Lang and Reifsnider [13] which had instrumental resolutions at about the 100 μm level could not have made effective use of finer resolutions because of photon noise as already discussed. Secondly, systems which use phosphor screens to convert X-rays to light, and which view the phosphor with high-aperture optics (such as fibre-optic coupling to the phosphor) will have resolution limits comparable with the phosphor thickness. Reducing phosphor thickness may reduce η unacceptably for harder wavelengths. A possible method for combating this dilemma is the use of columnar-structured phosphors, and the most successful attempt so far is that of Stevels & Kuhl [14]. Thirdly, we do not see system instrumental resolutions of $\lesssim 10$ μm being attained in the immediate future.

Regarding instrumental noise, factor (3), these can cover a wide range, from the very low value of ~ 10 photons mm^{-2} sec^{-1} [13] achieved over ten years ago to the unacceptably high equivalent of 10^5 photons mm^{-2} sec^{-1} in early systems using the PbO target X-ray sensitive vidicon [15]. In practice, as stated in Section 12.1 in another context, there is not much point in trying to make background noise substantially less than the intensity level of diffuse X-ray scatter from the specimen.

On factor (4) we would point out that, up to now, experience with X-ray topographic image intensification systems is so limited that few are aware of the difficulties these system have in faithfully recording the wide contrast variation that topographs exhibit. At present only the channel-plate image intensifier has the ability to deal with the dynamic range involved.

Several reviews of electronic X-ray imaging systems exist [7, 16, 17], to which reference may be made for descriptions of the many types of system that have been tried. The development of instrumentation proceeds steadily, but basic difficulties and design conflicts remain. Further discussion of these problems in relation to design of system elements will be found in [7].

REFERENCES

1. H.P. Klug and L.E. Alexander (1979) X-ray Diffraction Procedures for Polycrystalline and Amorphous Materials, 2nd edition Wiley-Interscience, New York - London - Sydney - Toronto.
2. B.L. Cardozo and R.J. Ritsma (1968) IEEE Transactions in Audio and Electroacoustics AU-16 159
3. H.J. Zweig (1961) J. Opt. Soc. America 51, 310
4. R.C. Valentine (1966) in Advances in Optical and Electron Microscopy, Vol. 1 (ed. R. Barer and V.E. Cosslett) Academic Press, London New York.
5. A. Charlesby (1940) Proc.Phys. Soc. Lond. 52, 657
6. Y. Epelboin, A. Jeanne-Michaud and A. Zarka (1979) J. Appl. Cryst 12, 201
7. A.R. Lang (1978) in Diffraction and Imaging Techniques in Material Science, Volume II: Imaging and Diffraction Techniques; Second revised edition (ed. S. Amelinckx, R. Gevers and J. Van Landuyt) North-Holland Pub. Co. Amsterdam-New-York-Oxford, p 623
8. H. DeVries (1943) Physica 10, 553.
9. A. Rose (1942) Proc. Inst. Radio Engrs. 30, 295.
10. M. Elder and O.S. Mills (1979) A Survey of X-ray Films, Report to the Commision on Crystallographic Apparatus of the International Union of Crystallography.
11. E. Spiller and R. Feder (1977) in X-ray Optics (ed. H.J. Queisser) Springer -Verlag, Berlin -Heidelberg - New York, p. 35.
12. S. Mardix and A.R. Lang (1979) Rev. Sci. Instrum. 50, 510
13. A.R. Lang and K. Reifsnider (1969) Appl. Phys. Lett. 15, 258
14. A.L.N. Stevels and W. Kuhl (1974) Medicamundi 19, 3
15. J.I. Chikawa and I. Fujimoto (1968) Appl. Phys. Lett. 13, 387
16. R.E. Green, Jnr. (1971) in Advances in X-ray Analysis (ed.C.S. Barrett, J.B. Newkirk and C.O. Ruud) Plenum Press, New York, p. 311
17. W. Hartmann (1977) in X-ray Optics (ed. H.J. Queisser) Springer-Verlag, Berlin - Heidelberg - NewYork, p. 190

CHAPTER 13

SAMPLE PREPARATION

D.KEITH BOWEN

13.1 INTRODUCTION

Specimens examined by topographers range from that hardest of materials, diamond, to soft, plastic crystals such as molecular organic crystals or dislocation-free high-purity copper crystals which must be picked up by both ends simultaneously or else they bend under their own weight. Sometimes the crystals are examined whole, and at other times accurate sections of particular orientations must be taken. In all cases the information within the crystal must be transferred to the X-ray beam without the introduction of artefacts before it is worth taking a topograph, let alone interpreting it. The extreme sensitivity of most X-ray topographic methods to surface and internal strains, and the low magnifications usually employed grossly complicates this task; thus, the slovenly methods adequate for preparation and handling of transmission electron microscope specimens will not do for X-rays. The familiarity with which one recognizes 'technical terms' from the literature, such as the apocryphal collection in Table 1, indicates how often even experienced topographers do not always excel in specimen preparation.

Regrettably, one can only learn the techniques of preparation the hard way; but I shall attempt in this chapter to provide a guide to the beginning of the way. The first topic to be considered is that of crystal orientation. This is followed by, to use engineer's jargon, the forming and shaping processes (mechanical and solvent cutting, for example) and then the finishing process which in our case consists of preparing a strain-free surface to the required flatness and if necessary protecting it against deterioration. Finally, the sample must be mounted in a strain-free manner on the X-ray camera, and the mounting must not interfere either with the topographic method or with whatever environmental experiments are desired, such as heating or cooling the specimen, applying a stress to it or subjecting it to corrosive attack. So far as my knowledge permits, I shall include a wide range of materials in this survey of techniques, though the reader should not expect to find herein a recipe for handling his latest impossible crystal. Some remarks will be made about the contrast to be expected from various types of artefacts, but more details and examples on that topic will be found in Professor Lang's tutorial (Appendix 2).

Table 1

SOME 'TECHNICAL TERMS' AND THEIR INTERPRETATION

Jargon	Interpretation
The specimen was accidentally strained during mounting	The specimen was dropped on the floor
Extreme care was taken in all stages of specimen handling	The specimen was not dropped on the floor
Some polishing marks can be seen	I didn't etch it enough
Contrast-insensitive residue	Dirt
A Fiducial Reference Mark	A scratch

13.2 Determining crystal orientation

The requirement is not only to determine the orientation of the crystal to the necessary accuracy, but also to be able to cut it parallel to a given (hkl) plane without losing accuracy. Nor need the plane be a low-index one. The general method is to use a robust goniometer which fits in the same way on both an orienting device and a cutting device. Robust it must be since cutting devices are by nature aggressive, and the goniometer must be capable of immersion in paraffin (for spark erosion) or be immune to corrosive attack by solvents used in chemical saws, or be unaffected by abrasive slurries used in other methods. Ordinary crystallography goniometers will not survive, but special ones are made for example by Metals Research in England for spark erosion and by South Bay Technology in the USA for chemical machining. The latter has the particularly nice feature that it can easily be dismantled for cleaning. The machining required to make one is not beyond a standard laboratory workshop, and use of 18% Cr 10% Ni stainless steel is usually adequate to ensure a reasonable life. The mounting method should ideally be kinematic to obtain the highest accuracy, but this is usually not obtainable in the cutting process and V-block type mountings are adequate and quite cheap.

SAMPLE PREPARATION

In some cases, such as strongly facetted crystals or ones with deep, facetted etch pits, one may use purely optical orientation methods. For an externally facetted crystal a standard optical goniometer is used. However, one point is not obvious. If a large crystal is being used, the facet planes will deviate more and more from their nominal low-index orientation the further they are away from the centre of the crystal. This can easily be seen in large sonar KDP crystals or large hydrothermal quartz crystals. Whether this is serious in a particular application depends of course on the precision required and on the deviations found for a given facet. The latter depends on the anisotropy of surface energy; if this is known then the precise form of the facet can be predicted from the Wulff construction (see for example [1]). If the whole crystal is available then all measurements must be made at the centres of the faces.

When facetted etch-pits are produced on a crystal surface they must have at least the symmetry of that surface. The triangular etch pits on (111) silicon surfaces are well known. If such a surface is slightly off (111) the crystal can easily be oriented by an optical back-reflection method. A collimated beam of light passes through a small hole in a screen set normal to the beam, and falls on the crystal surface. The latter is held in a goniometer whose mounting is referenced to the beam. When the (111) plane is normal to the light beam the reflected pattern on the screen (from the etch-pit facets) shows exact three-fold symmetry. The precision of this method is not as great as that of X-ray methods since the pattern is usually somewhat diffuse, but say 1^0 accuracy can be obtained with great speed.

The majority of crystal orientation work will be performed by X-ray methods, however, and a summary of the techniques appears in Table 2.

Commercial Laue or flat-plate cameras are often the poor relation of the available diffraction camera range, with poor collimation, inadequate referencing of specimen to film and too small a film. They also rarely meet the new radiation safety regulations unless the whole camera and generator is enclosed, which makes specimen adjustment a trifle inconvenient. Fortunately this is again an item that is within the competence of laboratory workshops, and appropriate specifications are given below:

(1) The incident beam divergence must be low, for example by using a source not greater than 400 µm square (the improvement with 100 µm is very noticeable) with a collimator at least 2 - 3 cm long of diameter 100 µm. An interchangeable coarser collimator is useful for preliminary work. Lead glass capillary tubes and hypodermic needles are both satisfactory.

Table 2

X-RAY METHODS FOR CRYSTAL ORIENTATION

Method	Accuracy	Application
Commercial Laue back-reflection camera	60'	Orientation of bulk crystal prior to cutting
Precision Laue back-reflection camera	6'	" " " "
Transmission Laue camera	60'	Orientation of thin slices; useful for white-beam synchrotron radiation specimens
V-block goniometer	<1'	Precision cutting or lapping of low-index planes
4-circle precision goniometer; measure angles to <100>s	<1'	Precision cutting or lapping of arbitrary planes

(2) There must be accurate identification of a specimen/goniometer direction with a direction on the film. Commercial cameras are most deficient in this respect. Methods include the X-ray shadow of a sharp edge on the camera, small external apertures drilled accurately in line with the collimator (putting black dots on the film through visible light leakage in a known place), fluorescent dots inlayed in the cassette in line with the collimator as in the Polaroid system, or indenters to emboss or scribe the film <u>in situ</u>.

(3) There must be accurate location of the beam centre on the film. A hypodermic needle which cuts its own hole is probably the best way of ensuring this.

(4) Adjustment of the specimen-film distance to the standard (usually 30 mm) must be within $\pm$ 0.1 mm.

(5) A large film should be used, for example 20 x 12.5 cm.

An illustration of our solution to these constraints [2] is shown in Fig. 1. The camera is totally shielded, all the adjustments can be made externally and it sits on a simple optical bench for permanent alignment. The collimator is a hypodermic needle which makes its own hole in the film. The film is covered with black paper on one side only; four 1 mm apertures in the cassette backplate are accurately referenced to the hypodermic hole and give small black dots symmetric about the centre of the beam. Joining the dots in pairs gives a cross centred on the beam and accurately aligned with the goniometer horizontal and vertical. An externally-controlled rod can be lowered to put a small fluorescent screen in the beam for alignment (observed through the PVC or glass side pieces – note that Perspex (plexiglass) – does not stop scattered radiation adequately); the rod also carries a needle with a soft fibre end that can be trimmed at 30 mm from the film to set the specimen-film distance. This camera has proved very satisfactory in use and gives a precision of about $0.5°$ with graphical interpretation and about $0.1°$ with direct measurement on the film.

Figure 1. Precision Laue camera on optical bench; goniometer for spark erosion in position.

For higher precisions still, such as are involved in making monochromator crystals (which are in effect topographic specimens on which one hopes not to see contrast) one must mount the cutting goniometer on something that is still more accurately referenced to the X-ray beam. If the plane to be cut is low-index (more precisely, produces a fairly strong Bragg diffraction peak), and a preliminary cut within $~3°$ can be made, then one can use the simple, cheap and effective V-block method [3], which uses inexpensive engineering components. The equipment needed is an X-ray tube with two manual circles on a goniometer, but these need only be accurate enough to find the Bragg reflection from the plane that is to be cut. A jig is needed which is capable of adjusting the orientation within about $3°$ [3, 4] and which is contained within an accurate cylinder. This cylinder is placed on the V-block and the latter and the counter adjusted until the reflection is found. The cylindrical jig is then rotated on the V-block, by hand in the past but with a motor and friction mechanism if safety regulations are in force. A stop or shaft collar is needed to prevent the jig sliding up and down the V-block, and this can be set using a light probe to get the specimen in the same place each time. The effect of the rotation is to cause the diffracted beam to oscillate in intensity if the Bragg plane is not exactly normal to the axis of rotation, and setting the crystal is then a matter of adjusting the specimen orientation until a steady reading is obtained upon rotation. The specimen stays in the jig for further treatment; accuracies much better than 1' can be obtained this way but can only be preserved by lapping treatments rather than cutting processes.

When the plane to be cut does not produce a decent diffracted beam the only solution is to measure the orientation precisely on a four-circle goniometer, preferably by finding the three <100> directions and their angular relationship with a 'fiducial reference mark' on the specimen or with the cutting goniometer. Fortunately this is not a common requirement, since two of the circles must be absolutely calibrated to $0.001°$ and also capable of carrying the robust cutting goniometer; a possible but expensive specification.

If a high throughput of oriented specimens is required one may adopt standard industrial techniques used for example in the production of piezoelectric transducers. The X-ray orientation process is followed by a single cut (or double cut if two or more planes must be produced on the specimens) to produce one (or two) reference faces on what might be a large crystal. The crystal is then detached from the cutting goniometer, which is expensive and needed for the next crystal boule, and attached to cheap cutting or lapping blocks. The reference face is then aligned by standard metrological techniques with the cutting or lapping tools and the subsequent shaping done by dead reckoning. The realignment methods involve use of dial gauges (DTI's) or, for higher precision, autocollimators (which essentially are an optical method for detecting the alignment between the instrument axis and the normal to a surface. Since they can without great difficulty go down to seconds of arc in precision, there is no great danger of losing the orientation accuracy, which is usually much worse than this. An excellent introduction to engineering metrology applied to crystal cutting (and indeed a detailed discussion of many of the shaping, forming and mechanical finishing treatments mentioned in this chapter) is given by Fynn and Powell [4]. Newcomers to this field could do worse than go to an ordinary mechanical workshop and watch a good machinist 'picking up' the centre of a hole or the alignment of a face for subsequent reference.

13.3 Shaping and forming processes

Table 3 summarises the main shaping and forming methods and includes a crude selection guide for various classes of material. The distinction between 'ductile' and 'brittle' refers essentially

SAMPLE PREPARATION

to the ease of dislocation propagation in the crystal compared to the ease of crack propagation; that between 'strong' and 'weak' refers to the ease of movements of either dislocations or cracks, whichever it is.

Table 3

SUMMARY OF SHAPING AND FORMING PROCESSES

Specimen type:	strong, brittle	weak, brittle	strong, ductile	weak ductile	comments
examples:	Si, Ge	organic crysts.	Fe-Si crysts.	Cu, Zn	
Method					
sawing	n	n	(y)	n	jeweller's saw
Abrasive disc	y	n	y	n	
Free abrasive on eg string saw	y	n	y	n	
diamond disc	y	n	(y)	n	pref. annular
machine grinding	y	n	y	n	take fine cut
diamond turning and milling	y	?	y	n	untried as yet
hand grinding	y	(n)	y	n	
sonic/ultrasonic	y	(n)	n	n	
fine abrading/ lapping	y	(n)	y	n	
spark erosion	n	n	y	n	conductors only
electrolytic	n	n	y	y	conductors only
chemical, solvent	y	y	y	y	
ion beam	y	y	y	y	untried as yet

This guide should help to steer newcomers in the right direction. Solving a shaping problem with a particular specimen may still be an arduous process - in particular the selection of a good chemical or electrolytic cutting/polishing agent seems to require the services of a friendly witch. Further technical details and recipes can be found in the following publications:

General forming and precision shaping: Fynn and Powell [4], Bond [3].

Grinding and mechanical polishing of brittle materials especially ceramics: Schneider and Rice [5].

Chemical and electropolishing: principles, Tegart [6], recipes, Brammar and Dewey [7].

Some remarks about the physical processes involved may be useful in considering applications. The distinction between ductile and brittle materials is pertinent as follows. Ductile materials will flow in front of mechanical cutting media (abrasive particles or the edge of a saw) through dislocation motion; ridges will be left at the sides of particle tracks, and dislocations will propagate to a greater or lesser extent inside the crystal. Brittle materials, in contrast, will abrade by fracture; the locally high stresses set up by indentation of fine abrasive particles will cause Herzian cracks, which travel first perpendicularly down into the surface and then spread out roughly parallel to it. Such cracks from adjoining sites join up and a very small piece of surface falls off. In principle therefore, there is far less damage to the interior of a brittle material than to a ductile one during abrasion, and materials such as high-purity copper or plastic organic crystals are the severest test of specimen preparation technique.

It is not always obvious, however, whether a material will be ductile or brittle under abrasive conditions. For example, silicon is thoroughly brittle at room temperature under ordinary conditions, as anyone who has 'accidentally strained the specimen during mounting' is painfully aware The statement is often made that dislocations cannot move in silicon below about 600°C; however, Stickler and Booker [8] showed quite clearly that dislocations were introduced near the surface in initially dislocation-free Si by abrasion at room temperature. This accounts for the width of rocking curves in Si and similar materials after abrasion, together no doubt with an effect due to abrasive wedging open small cracks in the surface. The mechanism of this dislocation generation poses no problem in metallurgical terms since the local stress is very high, the local region is constrained by its surroundings, and there may be a little adiabatic heating caused by the friction of the particle.

SAMPLE PREPARATION

Spark machining or spark erosion equipment is available both on a laboratory scale and for industrial forming processes. It has the merit that the erosion of the work follows the shape of the tool, hence very complex shapes in two dimensions (with simple extension in the third dimension) can be made quite readily; for example, tensile specimens or the part-annulus shapes that can be used to minimise mounting strains discussed below. Publicity material often refers to it as 'strain-free machining' and it is certainly much gentler than most mechanical methods, but a purely spark machined surface does not even give a good Laue picture let alone a topograph. Some chemical treatment is also needed as discussed in the next section. It is only applicable to conductors - essentially to metals - and relies on passing a spark of controlled energy across a gap of controlled size between the tool and the work, with both immersed in a dielectric such as paraffin. The mechanism of erosion is one of local melting or vaporization of material, and a spark-machined surface has characteristic bumps and craters when viewed under a microscope. The energy can be chosen on the machine, and normal practice is to 'rough out' the part on a medium range, then approach the final dimensions on a fine setting, being sure to take enough off to remove the damage caused at the coarse setting. Even so, the method is best reserved for robust crystals such as iron-silicon alloys, rather than the delicate pure metal crystals such as copper, aluminium or cadmium.

The mechanisms of chemical and electromachining will be considered briefly in the next section, as they are most commonly used for finishing rather than forming. They are, however, quite suitable for forming if suitable equipment is used rather than the common beaker, and the most delicate crystals can be shaped by these methods only. Electrolytic machining (for conductors) is probably the ultimate method since with a suitable servo a constant gap can be kept between the tool and work and no abrasion takes place. With chemical machining a light contact with a cloth or brush loaded with solvent helps to preserve the desired surface figure and this can damage some crystals (eg organic molecular crystals). In each case the method is similar to lapping a specimen on a cloth (or roller to obtain a cylinder, etc.) but the solvent takes the place of the abrasive in the case of chemical forming; for electrolytic forming the 'lap' is a conductor supporting a fairly thin film of electrolyte. With careful design of polishing machines and experimental control, optical quality flat surfaces can be prepared by either method [9].

Ion beam machining is a last resort, but it could be invaluable. It only affects the surface layers and does not create internal damage, and with a scanning system could produce any shape. Its disadvantage is that it is extremely slow, and for average-size topographic specimens the forming might take several days - time that is usually unavailable on ion-beam machines since they tend to be fully used by the electron microscopists who probably bought the only machine in the neighbourhood. I know of no cases where it has yet been applied to topography, but there seems no objection in principle.

13.4 Finishing processes: polishing and etching

The essential purpose of the finishing process is to remove the damage caused by the cruder forming methods. As already mentioned, sometimes one has to do the whole preparation by a finishing method. The information required before deciding on a method is the depth of damage caused in a material by the various processes. Of course, this will depend on the method in detail, on the material and its prior treatment, on the skill of the operator and on pure luck, but a rough idea can be given as shown in Table 4.

Table 4

APPROXIMATE DAMAGE DEPTHS (um) AFTER VARIOUS SURFACE TREATMENTS

Method	Brittle material	Ductile material
'heavy' mechanical (cutting, grinding)	cracking	100 - 1000 or more
'light' mechanical (cutting, hand grinding)	10 - 50	50 - 200
fine abrading, lapping	0.5 - 5	20 - 100
spark erosion	usually non-conductors	50 - 500
electrolytic	"	oxidation (?)
chemical or solvent	monolayer (?) oxidation (?)	surface layer (?) etch-pitting
ion beam	~ 0.01	~ 0.01

In the case of silicon the damage depth can be quantified more precisely from the work of Stickler and Booker [8], who measured the damage depths in both abraded (SiC paper) and lapped (free diamond particles) silicon crystals. For a given particle size the free particles of diamond caused about one third of the damage of the bound silicon carbide ones, and the depth of damage also increased about twice as rapidly with particle size with the bound particles. Whilst some of this may depend on the different nature of the particles, this result does give a valuable insight into the abrasion process and confirms the general view that free-particle cutting is a gentler method. With 10 μm diamond (which cuts quite rapidly) the damage depth was only ~2 μm whereas with 600 grit carbide paper (about 30 μm) it was ~12 μm. The consequence is that cutting processes using free particles, such as a string saw (in which a string charged with abrasive slurry is pulled or reciprocated across the specimen) or lapping on a plate using a slurry, are to be preferred to those with embedded abrasives, to ease the finishing process. So far as flat surfaces are concerned this is fortunate since these lapping methods are the most precise available; string saws are less precise than rigid saws, but a rigid blade can replace the string in a properly designed apparatus to produce good results. The techniques are discussed in detail in Fynn and Powell [8]. Similar results should hold for other brittle materials such as most minerals and ceramics, although in ceramics with some ductility, e.g. MgO, the abrasion damage will be nearer 100 μm even with fine abrasive.

A sawn surface can clearly be improved by mechanical lapping with fine particles, if the material has little ductility. The finest diamond particles obtainable are about 0.25 μm, but further improvement can be obtained by using alumina or silica particles, which are available in slurries down to about 0.025 μm. An excellent commercial product, developed for and widely used by the semiconductor industry, is 'Syton', an alkaline silica sol with approximately 0.025 μm particles. For silicon, this actually combines chemical polishing with free-abrasive lapping but it is remarkably effective on other materials including metals. Polypropylene laps are used and the method (followed by a quick dip in chemical polish or etch) probably produces the best surfaces that are routinely available.

Nevertheless, unless one confines his topography to whole natural diamond crystals, a chemical or electrolytic polish or etch is normally the final step. In the electrolytic process, only really applicable to metals, the specimen is made the anode in an electrolytic cell, an inert cathode (Pt, graphite, stainless steel) is used, the right electrolyte is selected after an extensive literature search and trial and error, and the right potential applied. In an 'ideal' cell the current-voltage relationship looks something like Fig. 2; in practical cells the distinction between the ranges is often blurred. However, low potentials generally give etching (selective attack at dislocation outcrops and grain boundaries, surface facetting) and higher ones polishing until gas evolution sets in, which usually causes major pitting. If a good electrolyte is found, the polishing really is polishing, with surface bumps being levelled out and a specular surface produced from a coarsely-lapped starting condition; special precautions (outlined in the previous section) are needed to get the surface flat over large areas as well as smooth. The mechanism is the formation of a viscous layer of anode products at the surface (often aided by choice of viscous components in the electrolyte); projections beyond this layer are much more rapidly attacked hence polishing results.

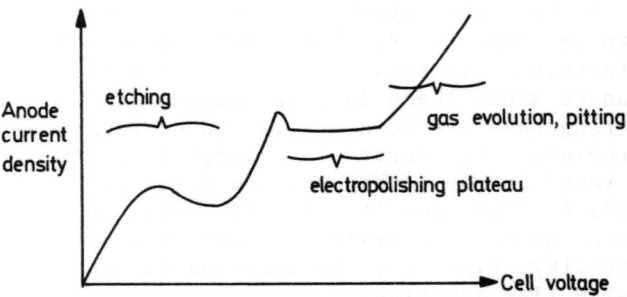

Figure 2. Ideal electrolytic cell potential-current relationship

The chemical polish is usually much less of a true polish, and one must generally start with a flat surface. The best that is usually obtained is even removal of material from bumps and hollows alike, and it is often difficult to avoid etching effects, especially when there is any discontinuity such as a dislocation or fine precipitate outcrop. Hence the method is often indiscriminately referred to as 'etching' or 'polishing'. In any case, one does not have any choice for the final treatment for non-metals, and it is worth the trial of finding a suitable reagent. One hazard is that the technique is often very sensitive to the precise grade or manufacturer of reagents; I was once baffled for some time by a sudden failure to polish niobium with a chemical polish that had worked routinely in the past, until I discovered that the laboratory had, in an effort to improve quality, started buying analytical grade HF rather than the dirty technical grade that worked very well. An illustration of the effect of this final treatment on topographic quality is shown in Fig. 3.

Ion beam etching was mentioned in the previous section, and with its very narrow depth of damage could be an important finishing method which also produces atomically clean surfaces. However, to my knowledge it has not yet been tried for topography, so comments must be speculative. It is probable that it would be very satisfactory as an intermediate process, removing the bulk of mechanical damage and figuring the surface to the desired shape. However, the damage in the top 10 nm could be quite severe, with implanted ions (possibly clustered) and perhaps a thin amorphous or highly damaged layer in the last few atoms. This may or may not affect the contrast, but even if it does not would possibly affect processes of interest in the specimen such as surface corrosion, or dislocation generation.

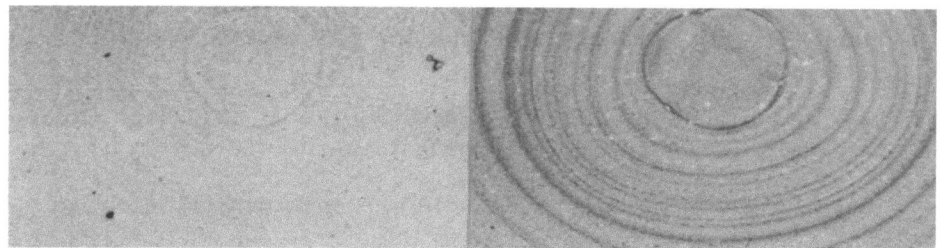

Figure 3. Effect of chemical polishing (etching) of yttrium aluminium garnet crystal; left, mechanically lapped with diamond (~1 µm) abrasive; right, same crystal after 15 minutes in hot orthophosphoric acid, revealing growth banding. Synchrotron radiation reflection topograph (VEPP-3, Novosibirsk) [K.J. Roberts, S.T. Davies and D.K.Bowen, unpublished]

13.5 Specimen mounting on the topographic camera

After the perfect specimen, which truly samples the crystal under study, has been made one has to get it into the X-ray beam without further damage. This may be the hardest step of all, since, almost by definition, adhesives do not have the same elastic and thermal properties as the samples under test. The following methods are available:

(1) Kinematic mounting

(2) Wax mounting, with various grades of wax

(3) Adhesive mounting, with a vast range of adhesives

(4) Mylar sandwich mounting

(5) With any of the above, special specimen shapes can be used to minimise strain

With the standard Lang method, a specimen curvature with radius greater than ~50 m will give a band of contrast on the specimen with no contrast elsewhere; and a specimen strain of $10^{-3} - 10^{-4}$ will be clearly visible as a strong black region. The problems are severe, and most topographers have encountered them.

Kinematic mounting, where the specimen rests on three points, is probably the ultimate in mounting if conditions permit, since the specimen is minimally restrained and free to expand if the temperature changes without introducing inhomogeneous strain. This method has been used very successfully by Kohra's group in Tokyo in their detailed investigations of topography using very highly collimated and monochromated incident beams. It is obviously restricted in angular range and has the danger that a chance breeze on a light specimen could easily misalign it or even deposit it on the floor.

Wax mounting is probably the most popular method. With this and the adhesive methods there are three golden rules:

Use the softest wax that will hold the specimen still enough

Use as little of it as possible

ATTACH THE SPECIMEN AT ONE POINT ONLY

The softest wax is probably the lowest melting, so will again damage the specimen least. It is well worth while making up a simple jig to position the specimen correctly relative to the stub it is to be attached to, rather than use the usual odd nuts, bits of metal and matchboxes that happen to be lying around. A minimal quantity of wax can then be melted onto the specimen with a low-temperature soldering iron (e.g. a standard one run at a much lower voltage than normal).

There are very many varieties of wax, and supplies vary considerably between countries. Low-melting paraffin wax is fairly widely available and is recommended - but in a hot climate, such as my own laboratory, it becomes too soft to hold the specimen up. It is usually possible to find an intermediate wax that will suffice; beeswax is rather hard and so are sealing wax and Tan wax (a high-melting filled wax) though these are excellent for fixing specimens on the cutting goniometer. Plasticene, or modelling clay, is a favourite standby of crystallographers, but it will not do for most topographic specimens. It requires too much force to make a firm joint, and, being microcrystalline, can give undesirable scatter if it gets in the X-ray beam.

Adhesives are often much slower than waxes, but they do avoid thermal damage to low-melting samples. Ordinary Araldite (two-component epoxy resin) is quite good - it is necessary to select adhesives with a low contraction on setting, which argues for polymerising glues rather than those which rely on solvent evaporation. The new 'superglues' (cyanoacrylic and cyanonitrile adhesives) have low shrinkages and could be very good (if the fixture is to be permanent) but are as yet untried.

SAMPLE PREPARATION

Problems with waxes and glues are avoided entirely by making two discs each carrying a 'drumhead' of a light plastic such as Mylar or polyethylene. The discs are bolted together with the specimen in between, with suitable spacer pieces to cater for different specimen thicknesses. This method has been used for delicate organic and inorganic crystals in Klapper's laboratory in Aachen for some years, and is most satisfactory. The mounting is quite rigid enough, the plastic contributes negligible scatter and holds the specimen so gently that mounting strains are undetectable.

Where strain must be minimised and adhesives or waxes must be used, one may endeavour to decouple the mounting point from the important part of the specimen. Fig. 4 shows Miltat's solution for the study of magnetic domains in iron-silicon. The spark-eroded almost-complete annular hole effectively isolates the mounting point from the central region. This principle can often be applied; another example, though not on specimens, is Hart's use of decoupling lengths of silicon for mounting and moving his monochromators and interferometers.

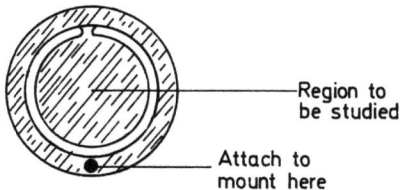

Figure 4. Cutting of a specimen to isolate the mounting strains from the region of interest; after J. Miltat

Finally, there are special problems in the mounting of specimens for dynamic experiments. There are no general solutions, but a few examples may help to generate ideas:

(1) For applying a 400 N force to tensile topographic specimens, Miltat and I [10] used ordinary Araldite heated to about 60°C to increase fluidity and improve curing. Also, the flat surface of the specimen was placed in direct contact with the reference flat of the chuck and the glue applied to the other side of the specimen, with a short piece of metal transmitting the stress affixed to both the chuck and this glued face. The first method, in which the glue was placed between the specimen and reference face, was a disaster and resulted in a specimen bent to <50 m radius hence giving banded images.

(2) For thicker and stiffer tensile samples, Young and Sherrill [11] and George [12] have used hooks on the shaped ends of the specimens. These provide self-alignment and minimal bending.

(3) For high-temperature dynamic studies of melting in silicon, Chikawa [13] again 'decoupled' the specimen and mount, simply by using the centre of the specimen as the part to be melted and attaching the specimen to the goniometer at a single remote point.

(4) For low temperature work a variety of adhesives can be tried: low-T strain-gauge cement, varnishes, glycerine or even ice.

(5) For high temperatures, ceramic cements are available; these have again been developed for strain gauges.

(6) For electrical conductivity, resins impregnated with copper or silver are available (from suppliers of electron-probe microanalysis materials) - but the second electrical contact must be freely suspended or strain will result.

A general reference on adhesives may be useful [15].

After this catalogue of difficulties the would-be topographer might feel despondent about his or her chances of getting an informative diffracted beam as far as the nuclear emulsion. But experience teaches rapidly; and the serendipitous topographer will even use his mistakes. The literature abounds with interesting science discovered by acute observation of a specimen 'accidentally strained during mounting'!

REFERENCES

1. J.W. Christian (1975) The Theory of Transformations in Metals and Alloys (2nd. edn.) Pergamon, Oxford, p153
2. D.K. Bowen and M. Dudley, to be published
3. W.L. Bond (1976) Crystal Technology, John Wiley, New York
4. G.W. Fynn and W.J.A. Powell (1979) The Cutting and Polishing of Electro-Optic Materials, Adam Hilger, Bristol
5. S.J. Schneider and R.W. Rice (eds.) (1972) The Science of Ceramic Machining and Surface Finishing, NBS Special Publication #348
6. W.J.M. Tegart (1959) The Electrolytic and Chemical Polishing of Materials, Pergamon, Oxford
7. I.S. Brammar and M.A.P. Dewey (1966) Specimen Preparation for Electron Metallography, Blackwell Scientific Publications, Oxford
8. R. Stickler and G.R. Booker (1963) Phil. Mag. $\underline{8}$, 859
9. J.W. Mitchell, J.C. Chevrier, B.J. Hockey and J.P. Monaghan Jr. (1967) Can. J. Phys. $\underline{45}$, 453
10. D.K. Bowen and J. Miltat (1976) J. Phys. E: Scientific Instr. $\underline{9}$, 868
11. F.W. Young and F.A. Sherrill (1971) J. Appl. Phys. $\underline{42}$, 230
12. A. George, C. Escaravage, W. Schroter and G. Champier (1973) Cryst. Lattice Defects $\underline{4}$, 29
13. J.-I. Chikawa (1980) this volume, chapter 15
14. C.J. Smithells (1967) Metals Reference Book vol. I, Butterworths, London

CHAPTER 14

LABORATORY TECHNIQUES FOR X-RAY REFLECTION TOPOGRAPHY

R.W. ARMSTRONG

14.1 The nature of reflection topography

When a real crystal is set at the Bragg condition, the surface layer reflects X-rays non-uniformly, to an extent which depends on the deviation from flatness of the crystal surface and on the microstructural features of the sub-surface crystal volume. To make use of this occurrence, a number of experimental techniques have been developed for obtaining topographic images of the X-ray intensity reflected over any crystal surface area and for tracing local variations in the reflected intensity on a point by point basis back to the combined surface features and internal microstructure of the material. The full range of wave length conditions accessible in the laboratory are covered by the techniques involving, say, characteristic K radiation (the Berg [1] - Barrett [2] technique, penetrating polychromatic radiation)the Schultz [3] technique), or, most sensitively, crystal monochromated radiation, say, as utilized by Bonse [4].

The effect on the reflected intensity of X-rays of surface steps and a dislocation subgrain boundary is shown schematically in Fig. 1 after the detailed description of the Berg-Barrett technique given by Newkirk [5]. The figure has been applied by Farabaugh, Parker and Armstrong [6] to the interpretation of results obtained on a vapour grown sapphire crystal. For the straightforward application of this technique to produce a zero layer reflection, the characteristic K radiation is chosen to be incident at a small angle, say, $< 10°$, to the specimen surface while a Bragg angle near to $45°$, just larger than the angle between the reflecting planes and the crystal surface, is selected for reflecting planes with a structure factor giving a reasonable intensity of reflected X-rays now leaving the crystal nearly perpendicular to its surface.

The crystal and X-ray geometry required to obtain a Berg-Barrett topograph of a zinc crystal through the (001) surface with cobalt $K\alpha$ radiation is shown on a stereographic projection basis in Fig. 2 as described by Armstrong and Wu [7]. A Bragg angle of $41.6°$ is computed for diffraction from the $(\bar{1}013)$ inclined at an angle of 35.5° to the (0001). The condition of rotating the crystal through an angle $\alpha_o = 6.1°$ about the [1̄2̄10] so as to obtain the (1̄013) zero layer reflection directed at an angle of 12.9 to [0001] surface normal is shown along the equatorial (1̄2̄10) plane of diffraction. The relative positions of the X-ray source, diffraction crystal and recording film are indicated for this condition in the inset diagram.

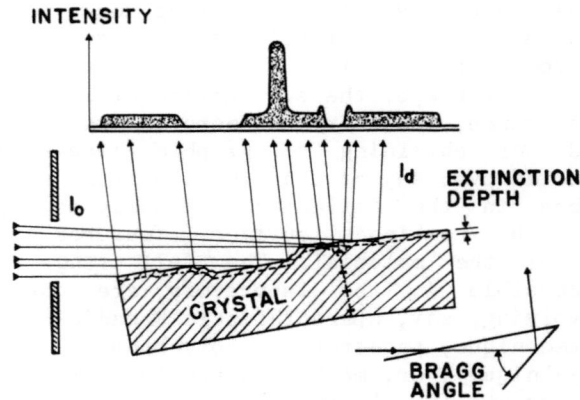

Figure 1. Schematic drawing of surface steps and a subgrain boundary observed within a Berg-Barrett reflection topograph [6].

Also shown in Fig. 2 is the larger rotation angle, α_s, required to obtain a skew plane reflection from the (01̄12) by direct rotation of the crystal about the vertical [1̄2̄10]. In fact, α_s can be made very much smaller than is shown in Fig. 2 by first employing a clockwise rotation of the crystal about [0001] until the angle, $(\pi/2) - \theta_{BS}$, is met for the Bragg condition with a chosen minor second rotation of the crystal about its new vertical axis. Farabaugh [6,8] has described the employment of such skew plane reflections in his study of the surface features observed for a vapour grown sapphire crystal. He has given emphasis to the minimized distortion which is achievable in the X-ray image by making α_s almost arbitrarily small.

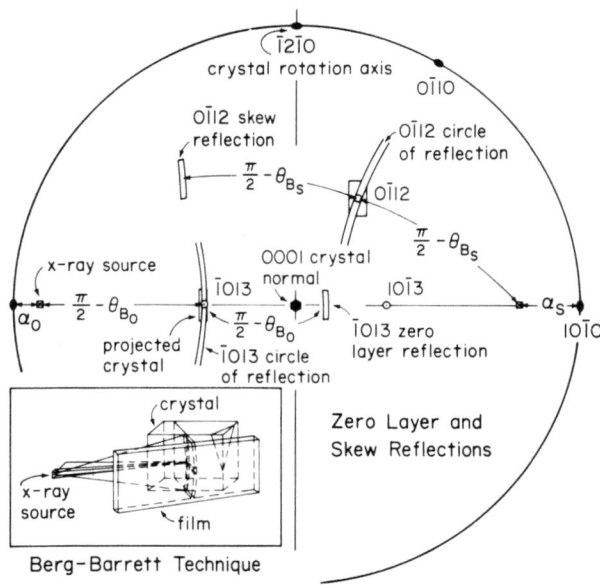

Figure 2. Stereographic projection analysis of zero-layer and skew reflection topographs obtained of a zinc crystal with the Berg-Barrett technique.

14.2 Diffraction contrast in reflection topographs

As described by Newkirk [5], the resolution of details within an X-ray topograph of a directly reflected image is strongly affected both by the (geometrical) vertical divergence of the probing X-ray beam and by local changes in the spacing and orientation of the crystal diffracting planes. Resolution is best controlled by placing the recording film as close to the specimen surface as possible. A ratio on the order of 0.005 for the specimen-to-film distance and X-ray source-to-specimen distance is normally employed with a conventional X-ray tube aligned to give a 1.0 mm spot focus size for the X-ray target.

Individual dislocations and dislocation subgrain boundaries are capable of giving direct images of themselves in reflection topographs by inputing an enhanced reflecting power to crystal volume elements centred on the defect positions. This imaging is termed EXTINCTION contrast after its description by Newkirk [5] and Barrett [2]. Fig. 3 shows a view of this contrast effect for non-planar dislocation spirals emanating from an apparent inclusion laying just below the (001) surface of a zinc crystal, as described by Schultz and Armstrong [9]. Turner, Vreeland, Jr., and Pope [10] have described various experimental aspects of enhancing the conditions for observing dislocations in metal single crystals, including copper and aluminium crystals, by giving special consideration to filtering the diffracted intensity of X-rays and involving the use of a stereographic projection analysis for tracing multiple reflections at a single crystal setting.

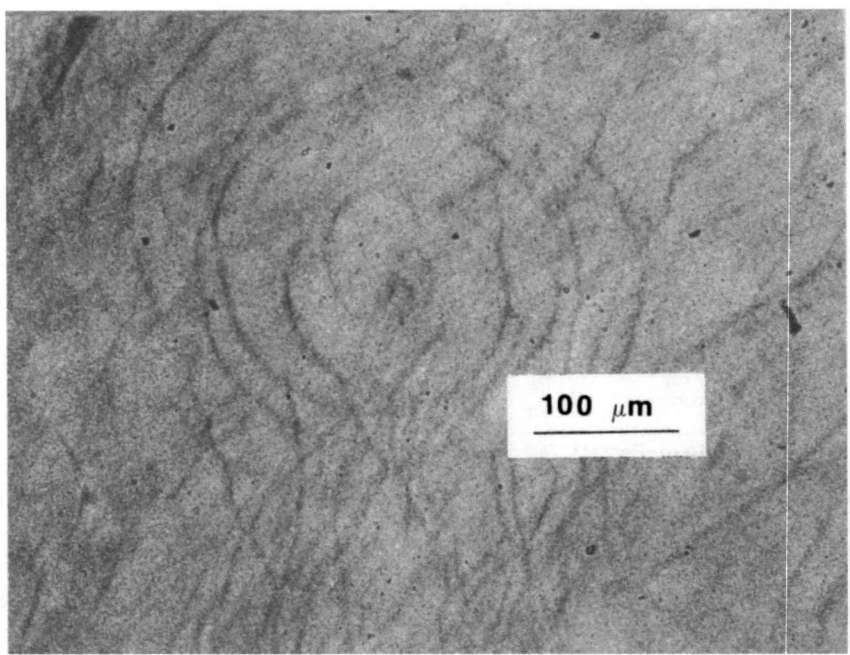

Figure 3. Observation of dislocation loops about an inclusion in a zinc crystal [9].

Individual dislocations, dislocation subgrain boundaries, and various types of domain structures are capable of being observed in X-ray reflection topographs for the additional reason of different beam directions being associated with diffraction from the changed orientation or interplanar spacing of crystal regions surrounding these defects. This consideration of giving emphasis to the effective crystal orientation which is involved in the diffraction process leads to predictable alterations in the position and amount of the diffracted intensity recorded in the X-ray image, as described by Bonse [4]. Fig. 4 shows a comparison of this ORIENTATION - related effect, in 4(a), with the previously mentioned EXTINCTION - related effect, in 4(b), as determined by Bonse [4] for individual dislocations in a relatively perfect crystal of germanium.

Both of the topographs in Fig. 4 were obtained with crystal monochromated Cu $K\alpha_1$ radiation. For Fig. 4(a), the crystal specimen was positioned at the midpoint of intensity on the edge of its rapidly changing dependence on angular position near to the Bragg condition so that the dislocation strain field could produce effective crystal orientations giving reduced as well as enhanced reflected intensities in the topograph. The "integral" reflection of Fig. 4(b), giving EXTINCTION contrast, was produced by rotating the crystal through the total angular range of reflection able to satisfy the Bragg condition.

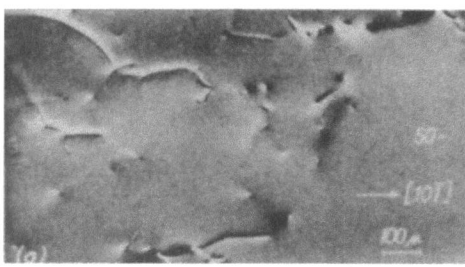 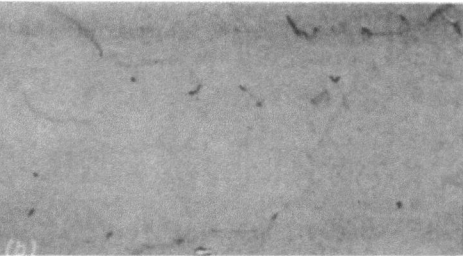

Figure 4. Single image and 'integral' image of dislocations in a germanium crystal, obtained with monochromated radiation [4].

The EXTINCTION and ORIENTATION contrast effects which have been described for these preceding X-ray topographs are, of course, not really independent of each other. Lang [11] and Authier [12] have discussed in some detail various aspects of these two considerations for understanding diffraction contrast in X-ray topographs, in particular, in relation to the dynamical theory of X-ray diffraction, as described by Batterman and Cole [13] and by James [14]. Of critical importance in understanding diffraction contrast is the X-ray extinction distance, ξ, (c.f. Fig. 1) which is measured on a reciprocal scale as the minimum separation along the major axis of the (hyperbola) branches of the dispersion surface.

$$\xi = \frac{mc^2}{e^2} \cdot \frac{\pi V}{|C|\lambda|F|} \cdot \left[|\sin(\theta_B - \chi)| \sin(\theta_B + \chi) \right]^{\frac{1}{2}} \qquad (1)$$

where m is the electron mass, e its charge, c is the velocity of light, V is the volume of the unit cell, C is the polarization factor, λ is the X-ray wave length, F is the structure factor for the cell, θ_B is the Bragg angle, and χ is the negative angle between the crystal surface and the diffracting planes. Lang [11] and colleagues [15,16] have proposed that the direct image width w, of a dislocation in a transmission topograph can be estimated as

$$w \sim (\xi/2\pi) \, \underline{g} \cdot \underline{b} \qquad (2)$$

where $\underline{g}$ is the reciprocal diffracting plane vector and $\underline{b}$ is the dislocation Burgers vector.

Roessler and Armstrong [17] have described on a dynamical theory basis for a perfect crystal a number of the X-ray parameters which seem important to observing crystal defects by reflection topography and these parameters can be related to ξ as follows:

$$\sigma_{e_{max}} = (2\pi/\xi), \tag{3}$$

$$\Delta\theta_A = 2\lambda \, |\sin(\theta_B - \chi)|/\xi\sin 2\theta_B, \tag{4}$$

$$\Delta\theta_R = 2\lambda \, [\sin(\theta_B + \chi)]/\xi\sin 2\theta_B; \tag{5}$$

where $\sigma_{e_{max}}$ is the maximum value of the primary extinction coefficient determining the reflected intensity of X-rays and $\Delta\theta_A$ and $\Delta\theta_R$ are the angular ranges for total acceptance and reflection of X-rays, respectively, at the Bragg condition.

The integrated intensity for asymmetric diffraction from a perfect non-absorbing crystal affected only by primary extinction is given by [14]

$$\rho_P = \frac{8}{3\pi} \cdot \frac{\lambda^2}{\sin 2\theta_B} \cdot \frac{|F|}{V} \cdot \frac{e^2}{mc^2} \left[\frac{|\sin(\theta_B - \chi)|}{\sin(\theta_B + \chi)}\right]^{\frac{1}{2}} \cdot |\bar{c}|. \tag{6}$$

Equation (6) may be compared with the integrated intensity for asymmetric diffraction from a crystal composed of mosaic blocks undergoing secondary extinction as [18]

$$\rho_S = \frac{1}{2\mu} \cdot \frac{\lambda^3}{\sin 2\theta_B} \cdot \frac{F^2}{V^2} \cdot \frac{e^4}{m^2c^4} \cdot \left[1 - \cot\theta_B \tan \chi\right] \cdot \overline{c^2}, \tag{7}$$

where μ is the linear absorption coefficient, in which case, the ratio of intensities is also connected with ξ by

$$\frac{\rho_S}{\rho_P} = \frac{3\pi^2}{16} \cdot \frac{1}{\mu\xi} \cdot \left[1 - \cot\theta_B \tan \chi\right] \sin(\theta_B + \chi) \cdot \frac{\overline{c^2}}{\bar{c}^2}. \tag{8}$$

Table 1 gives a comparison of diffraction conditions for obtaining reflection topographs of Zn, Ge, LiF, and Ni, crystals. The values of ξ may be compared with those given for transmission topographs by Lang [11]. In the reflection case, the penetration of X-rays is exponentially reduced with increasing depth within the crystal according to the factor $\mu/\sin(\theta_B + \chi)$ for absorption alone as compared with $(2\pi/\xi)$ for primary extinction due to the Bragg reflection condition. For the diffraction conditions described in Table 1 for zinc, less than 1% of the incident intensity of Bragg wave length X-rays is lost by absorption before the beam is reflected. James [19] has described this type of comparison for rock salt crystals on the basis of the number of lattice planes traversed. A similar absorption and primary extinction result to that of zinc is obtained for lithium fluoride in Table 1. Schultz and Armstrong [20] and, most recently, Vreeland, Jr., [21] have determined that dislocations are observed within zinc crystals to a depth of approximately 5 microns for the diffraction conditions described in Table 1.

Table 1

DIFFRACTION CONDITIONS FOR REFLECTION TOPOGRAPHY OF
ZINC, GERMANIUM, LITHIUM FLUORIDE AND NICKEL CRYSTALS

Material	Surface	Radiation	Reflection	Reciprocal Absorption Coefficient μ^{-1}, cm.	Extinction Distance ξ, cm.
Zn	(0001)	Co Kα	{10$\bar{1}$3}	1.5×10^{-3}	2.0×10^{-4}
Ge	(111)	Cu Kα_1	(444)	2.5×10^{-3}	11×10^{-4}
LiF	(001)	Cr Kα_1	($\bar{2}$02)	9.9×10^{-3}	14×10^{-4}
Ni	(1$\bar{1}$0)	Cu Kα_1	(2$\bar{2}$0)	2.5×10^{-3}	3.1×10^{-4}

The diffraction conditions in Table 1 were employed to determine the angular ranges for acceptance and reflection of X-rays and, also, to determine the ratio of integrated intensities of X-rays on a secondary and primary extinction basis as shown in Table 2. The proposed benefit of reducing $\Delta\theta_R$ with the asymmetric diffraction conditions normally employed for Berg - Barrett topography [17] is indicated for the zinc and lithium fluoride reflections as compared with the symmetric germanium and nickel ones. The first column of ratio values for the integrated intensities, $\rho_s^\perp/\rho_p^\perp$, was obtained for the polarization factor C = 1 describing the electric vector being perpendicular to the plane of diffraction. The second column in parentheses is for the ratio of equations (7) and (6) as in equation (8).

Table 2

COMPUTED X-RAY PARAMETERS FOR REFLECTION TOPOGRAPHS OF ZINC, GERMANIUM, LITHIUM FLUORIDE AND NICKEL CRYSTALS

Material	Condition	ξ (cm)	$\Delta\Theta_A$ (rad.)	$\Delta\Theta_B$ (rad.)	$\dfrac{\rho_s^\perp}{\rho_p^\perp}$, $\left(\times \dfrac{\overline{c^2}}{c^2}\right)$
Zn	{10$\bar{1}$3}CoKα	2.0×10^{-4}	1.8×10^{-4}	2.0×10^{-5}	2.8, (4.6)
Ge	(444)CuKα_1	11×10^{-4}	4.0×10^{-5}	4.0×10^{-5}	3.8, (3.9)
LiF	($\bar{2}$02)CrKα_1	14×10^{-4}	3.3×10^{-5}	5.0×10^{-6}	3.4, (4.4)
Ni	(2$\bar{2}$0)CuKα_1	3.1×10^{-4}	6.2×10^{-6}	6.2×10^{-5}	8.9, (12.5)

Equation 8 should give an indication of the maximum enhancement of intensity which might be produced by the dislocation strain field according to its extent within the surrounding crystal volume. The difficulty with this comparison is that the dislocation strain field appears to affect the integrated intensity of X-rays within a region scaled to the size of ξ, according to Lang [11], but ξ is matched, also, with the scale of an individual mosaic block among the many which are involved in the model of the secondary extinction calculation. This implies that the intensity enhancement can never be as large as indicated by equation (8). Kuriyama and Miyakawa [22] have given a dynamical theory description of the extinction process within an imperfect crystal containing a general type of inhomogeneous strain and it may be that this description can be applied to computing from first principles the direct effect of a dislocation on the diffracted intensity for the reflection case.

14.3 Dislocation subgrain boundaries

The observation of dislocation subgrain boundaries within crystals has been an important accomplishment of the Berg-Barrett and Schulz reflection techniques beginning with the initial results reported by each of these investigators [1-3]. Such boundaries are easily observed because of their orientation contrast. A significant refinement in the observation of such substructural entities was made by Weissmann [23,24] who employed crystal monochromated radiation to obtain reflection topographs of polycrystalline metals by utilizing the X-ray alignment of a Debye-Scherrer cylindrical camera.

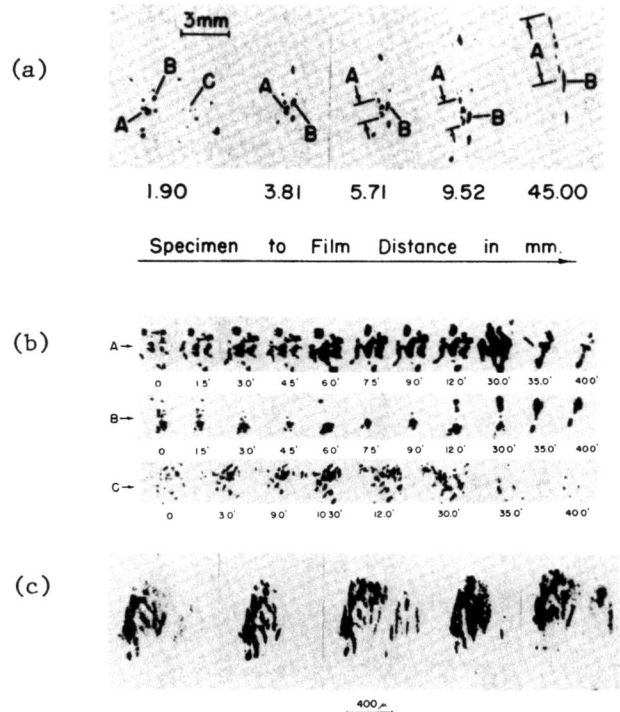

Figure 5. First and second order substructural entities within the grains of polycrystalline nickel: (a) at one specimen setting the outward tracing of grain reflections toward the film position of a Debye-Scherrer cylindrical camera; (b) lattice misalignments obtained at the specimen surface for different angular settings of the specimen; and (c) a sequence of reflections for a particular grain at 30" of arc intervals of specimen rotation [23,24].

Fig. 5 shows a composite example of the observations by Weissmann of first and second order substructural entities within the grains of polycrystalline nickel. It is interesting that the collections of subgrains within grains A, B, and C at the top of the figure show at increasing specimen-to-film distances the pronounced effect even for crystal monochromated radiation of the vertical divergence within the beam spreading the reflected intensity of a relatively equiaxed subgrain shape into that of a circular arc (cf. Fig. 2). The remainder of the figure shows at relatively low topographic magnification the inclusive grain images obtained at the specimen surface for different angular settings of the specimen, first, at intervals of 1.5 minutes or more and, secondly, at 30 second intervals. A detailed study of such images and the measurement of their variation with the specimen setting led to the specification of several orders of subgrain structures within the material according to the subgrain size, the misorientation between adjacent subgrains, and the distortion within subgrain volumes.

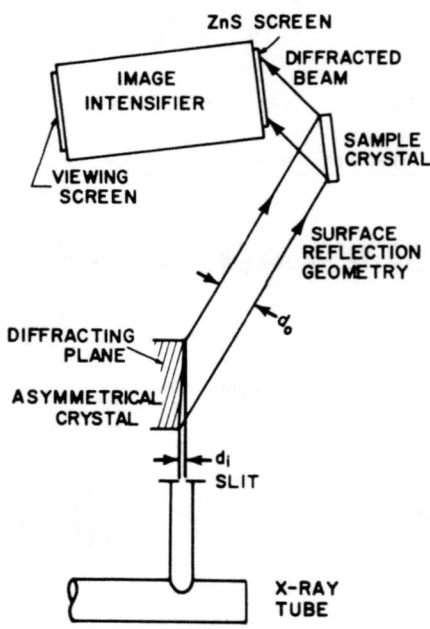

Figure 6. General optical alignment of asymmetric crystal topography (ACT) showing positions of first and second crystals and X-ray image intensifier in surface reflection geometry [25].

Boettinger et al. [25] have recently described some advantages of employing an asymmetric (double) crystal topography (ACT) technique for obtaining topographs of large crystal surfaces, after previous results obtained by Kohra, Haiizume and Yoshimura [26]. A schematic diagram of the general optical alignment for this technique for surface reflection is shown in Fig. 6, involving the positions of the first and second crystals and an image intensifier. An ACT ($\bar{2}$20) surface reflection topograph of a nickel crystal is given in Fig. 7 after crystal growth results reported by Kuriyama, Boettinger and Burdette [27]. Fig. 7 shows a number of prominent subgrain boundaries exhibiting appreciable lengths parallel to particular <111> traces in the ($\bar{2}$20) surface.

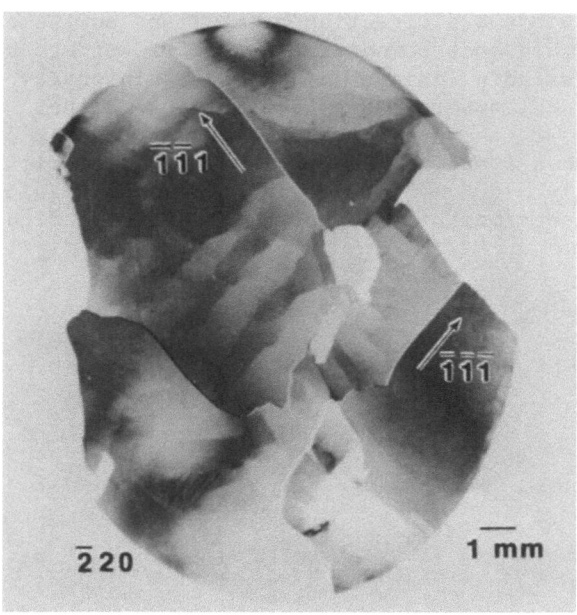

Figure 7. ACT $\bar{2}20$ surface reflection topograph of a nickel crystal showing subgrain boundaries, some of which are parallel to <111> directions [27].

A detailed study by Armstrong, Boettinger and Kuriyama [28] of these first order subgrain boundaries with [$\bar{1}\bar{1}1$] and [$\bar{1}1\bar{1}$] traces in the figure has shown they lay in ($\bar{1}01$) and ($10\bar{1}$) surfaces, respectively, both containing the [010] crystal grown axis. Within the larger subgrain volumes enclosed by these boundaries a second order subgrain structure is observed, also, due to its orientation contrast. These two orders of substructural entities, though observed here in a specially grown nickel single crystal, appear to match nicely the substructural description given by Weissmann [23,24] in his pioneering topography study of polycrystalline nickel material. Nakayama, Weissmann and Imura [29] have previously reported detailed topographic observations obtained for a tungsten crystal which exhibited this sort of first and second order subgrain boundary structure earlier described by Weissmann.

In Figs. 5 and 7 the combined conditions of the dislocation spacings and the diffraction parameters for these topographic images have prevented the observation of the individual dislocations composing the subgrain boundaries. Lang and Miuscov [30] have estimated, for magnesia crystals examined by transmission topography, that individual dislocations could be resolved within boundaries having misorientations up to about 10 seconds of arc. This misorientation is typically comparable to the angular range of total reflection for a diffracting crystal and so the observation of dislocations within such boundaries can occur under conditions in which no appreciable change in diffracted intensity is associated with the individual subgrain reflections on either side of the boundary. Fig. 8 shows a Berg-Barrett image obtained by Burns and Birau [31], with a (vertical and horizontal) geometrical divergence of approximately 100 sec, revealing individual dislocation within tilt and twist sections of a boundary in a lithium fluoride crystal.

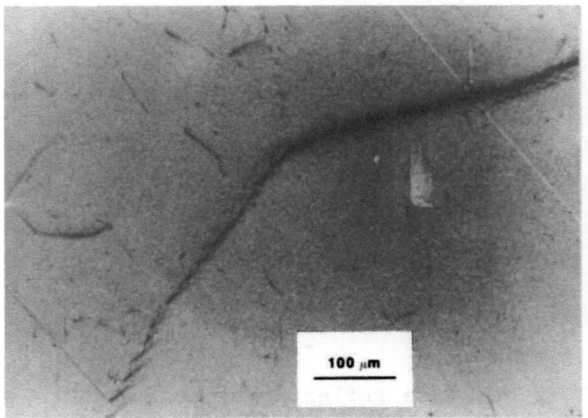

Figure 8. Berg-Barrett image, with a divergence of 5×10^{-4} radians, revealing individual dislocations within tilt and twist sections of a boundary in a lithium fluoride crystal; $\bar{2}02$ reflection obtained with $CrK\alpha_1$ radiation. S.J. Burns (University of Rochester) and O. Birau (University of Timisoara), unpublished results.

An example of a segmented dislocation subgrain boundary, which is observed principally because of its orientation contrast, is shown for the Berg-Barrett topograph of a zinc crystal in Fig. 9, after results obtained by Wu and Armstrong [32]. In this case, the boundary is of the order of 1° rotation about the [0001] crystal surface normal and, because this rotation axis is in the plane of diffraction (c.f. Fig. 2), no appreciable intensity change occurs for the subgrain reflections on either side of the boundary. The adjacent subgrain reflections overlap each other to produce darkening along their mutual boundary interface except for the separation of reflections which occurs for one direction of the boundary trace to produce an absence of reflected intensity along the white boundary segment. Of special significance for this (white) separation of subgrain reflections is the darkening which is observed on either side of the white band. The enhanced reflected intensity responsible for this darkening is attributed to the very localized extinction contrast of the dislocations responsible for the boundary misorientation even though the dislocations themselves are not resolvable. The situation is like that drawn in Fig. 1.

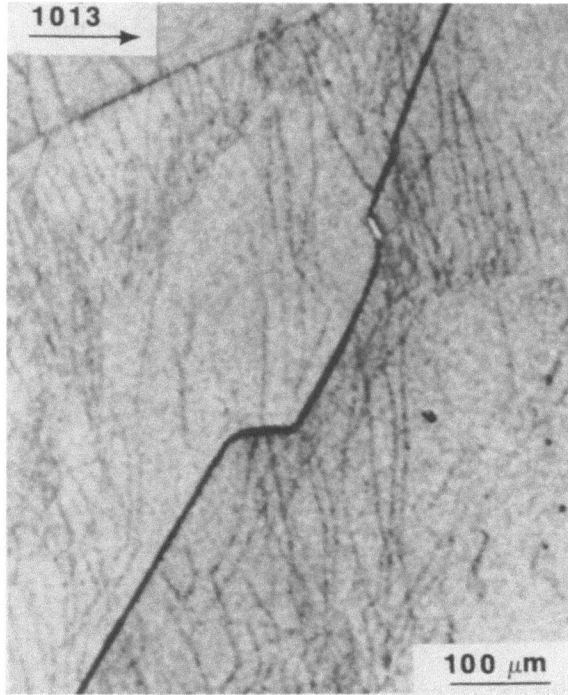

Figure 9. Segmented subgrain boundary in a zinc crystal with an [0001] rotation axis for the misorientation showing orientation and extinction contrast [32].

14.4 Applications to crystal growth processes

The several techniques of X-ray reflection topography have been particularly usefully applied to providing information about the growth processes of crystals and about the defect structures which are produced within the crystals. The reflection techniques lend themselves especially to characterizing the defect structures of reasonably perfect crystals because the techniques require a minimum amount of specimen preparation.

The rotation axes for the misorientation of dislocation subgrain boundaries within single crystals have been determined from reflection topographs to be closely associated with the imposed direction of crystal growth. Fig. 10 shows, for example, an unusual alignment of dislocations forming a number of subgrain boundaries with [0001] rotation axes within a zinc crystal grown by solidification along the difficult [0001] growth axis (perpendicular to the hexagonal crystal face). In this case, the crystal was grown onto an [0001] oriented seed and solidified at a sufficiently slow growth rate, say, 1 cm/hr, that dislocation climb could occur to allow this arrangement of subgrain boundaries to form a lineage structure along the total length of the crystal [32]. The junction of subgrain boundaries at A in Fig. 10 is shown to involve dislocation lines bending into the [0001] from the (0001). Bollmann [33] has described some basic aspects of the formation of subgrain boundary junctions in crystals and these considerations could possibly be fruitfully studied by the X-ray reflection topography techniques.

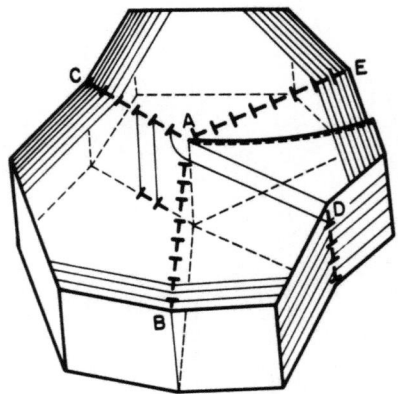

Figure 10. Three-dimensional geometry of dislocation subgrain boundaries within a zinc crystal solidified along [0001] [32].

Fehmer and Uelhoff [34] have observed by double crystal (monochromatic) X-ray reflection topography of copper crystals that subgrains appeared to form preferentially in crystals grown along <110>. In their study, dislocation-free crystals were produced by growth along <100>. In addition, in certain crystals, prismatic dislocation loops were observed to form around copper oxide particles. In several recent studies involving double crystal reflection topography, Bye [35] has observed subgrain boundaries within cadmium mercury telluride crystals; and, Bye and Cosier [36] have observed (solute-associated) subgrain boundaries in quartz crystal resonators.

Fig. 11, from Achter, Vold and Digges, Jr., [37] shows the subgrain structure observed within a (100) surface reflection topograph they obtained by applying the Schulz technique, with penetrating polychromatic radiation to a niobium crystal grown by a critical strain and annealing procedure. Schulz [3] has shown that this technique allows for the observation of subgrain boundaries due to the overlapping or separation of adjacent subgrain reflections obtained for different wavelengths within a continuous spectrum of X-rays. Reasonably sharply resolved images are obtained by employing a microfocus source of X-rays at a sufficient X-ray source-to-specimen distance to give a small divergence of X-rays. Of particular interest in Fig. 11 is the lowermost nearly vertical boundary at the crystal edge which is observed to be in a twist orientation. From such reflection topographs for strain and annealed crystals as compared with electron beam zone melted crystals, it was found that fewer subgrain boundaries were present in crystals produced by the method of critical straining followed by annealing.

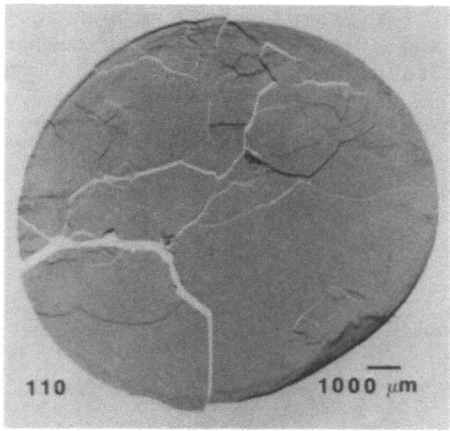

Figure 11. Subgrain structure within a (110) surface reflection obtained by the Schulz technique from a (strained and annealed) niobium crystal [37].

An association of dislocation subgrain boundaries and a surface structure of ledges and steps is shown in Fig. 12(a,b) for an alumina crystal produced by deposition with a vapour transport technique as studied by Farabaugh [6,8]. Fig. 12(a) is a $2\bar{1}\bar{1}6$ reflection principally showing enhancement of intensity within the diffracted beam due to $\{01\bar{1}4\}$ steps leaning away from the incident X-ray beam direction while Fig. 12(b) is a skew reflection obtained from the reverse incident beam direction showing a larger amount of surface shadowing associated with these same steps. Dislocation subgrain boundaries are observed in one or the other of these figures to intersect the prominent corners leading into or away from the parallelogram-shaped ledge structure in the lower left hand corner of Fig. 12(a,b).

Armstrong, Wu and Farabaugh [38] determined an $(01\bar{1}4)$ type of plane normal as the axis of rotation for certain subgrain boundaries observed within the cross-section of an alumina crystal examined by both the Berg-Barrett and Lang X-ray methods. Yip and Brandle [39] have reported that subgrain boundaries within Czochralski-grown white sapphire crystals followed major low index crystallographic planes in a manner seemingly uniquely determined by the crystal growth direction. Arnstein [40] found that subgrain boundaries followed low index planes in cylindrical zinc crystals with an [0001] cylinder axis and, also, he found for the reverse case of crystal decomposition by sublimation that preferential sublimation occurred at positions on the cylinder circumference which were intersected by certain ones of these boundaries. Such observations by X-ray topography that the substructure of dislocation boundaries within crystals follows preferred crystallographic planes and directions carries over to other types of boundary structures, also, as was demonstrated, for example, in a pioneering study by Bousquet, Lambert, Quittet and Guinier [41] on the texture of ferroelectric domains in barium titanate crystals.

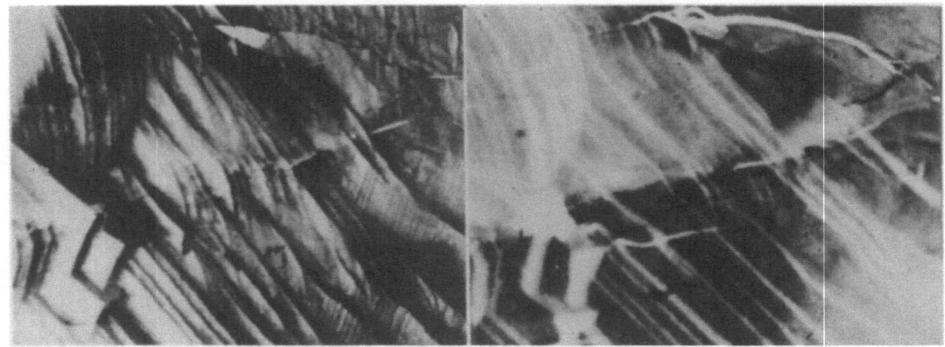

Figure 12. X-ray reflection topographs of the vapour growth surface of an alumina crystal showing the association of dislocation boundaries with the structure of ledges revealed by intensity enhancement or shadowing effects [6,8].

Fig. 13 shows a useful application of Berg-Barrett topography to studying the behaviour of dislocations within a zinc crystal during the growth of an oxide layer on the (0001) crystal surface, as reported by Roessler and Burns [42]. The formation and enlargement of dislocation loops and hexagons due to the condensation of vacancies was followed over a long period of time, say, 529 days, as indicated in Fig. 13. Two or more dislocation loops were observed to coalesce into a single hexagonal shape either in an isolated manner, as at T, or when connected to other dislocation segments, as at N. In the main, the dislocation structure associated with the oxidation process was affected by the prior arrangement of dislocations present within the material say, due to the crystal growth history, crystal cleavage, or subsequent deformation of the crystal. G'Sell and Champier [43] have reported observations on the climb of dislocations during the process of oxidation of zinc crystals. Yoo and Roessler [44] are combining such X-ray observations with optical and electron microscope observations to further investigate the crystal oxidation process.

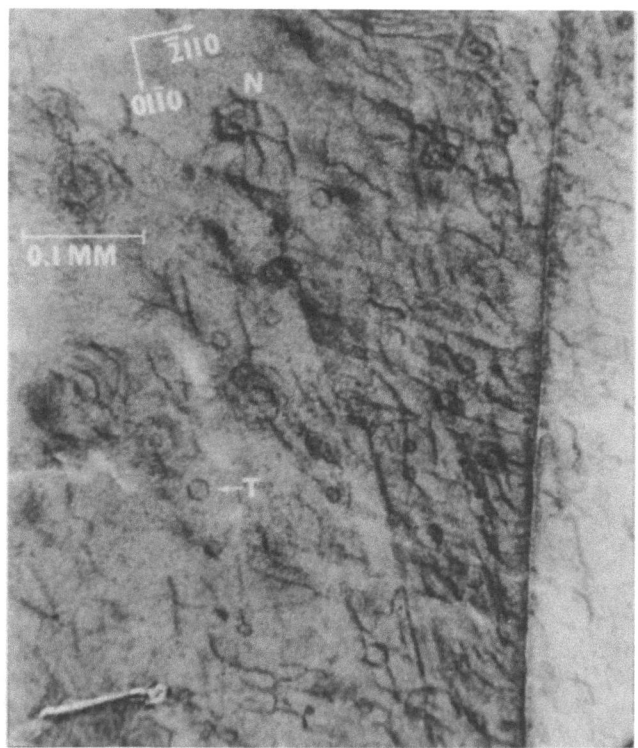

Figure 13. Dislocation loops and hexagons associated with the condensation of vacancies on dislocations during the oxidation of a zinc crystal for 529 days [42].

One further interesting application of X-ray reflection topography, as well as other topographic methods, for which much future research will very probably be accomplished is concerned with the investigation of the growth process and internal defect structure of polymer crystals. Fig. 14 shows a (700) Berg - Barrett reflection topograph obtained by Schultz [45] of a poly-[1,2-bis-(p tolylsulphonyloxmethylene)-1-butane-3-inylene] crystal polymerized from the monomer in the solid state. In this case, the relatively large amount of divergence within the incident X-ray beam for a Berg-Barrett topograph allows for reflection from different crystal regions separated by substantial strains or rotations, as between A and C and D in Fig. 14, while the control of resolution within the recorded image by capturing it at small specimen-to-film distance still allows for the detection of subtle extinction effects possibly associated with individual dislocations, as at D in Fig. 14. The X-ray topography techniques should find increasing application to the study of such polymer crystals and of other molecular crystals, too, as part of the total research effort to produce increasingly perfect crystals involving this type of bonding.

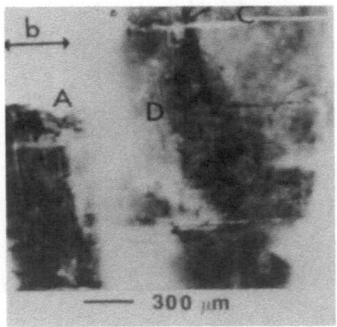

Figure 14. Local extinction contrast at defects within a macroscopic crystal polymerized within the solid-state [45].

REFERENCES

1. W. Berg (1931) Naturwissenschaften 19, 391;Z. Krist. 89, 286
2. C.S. Barrett (1945) Trans AIME 161, 15
3. L.G. Schulz (1954) Trans AIME 200, 1082
4. U. Bonse (1958) Z. Phys. 153, 278 (1962) in Direct Observation of Imperfections in Crystals (eds. J.B. Newkirk and J.H. Wernick), Interscience Publishers, New York, p431; (1964) Z. Phys, 177, 543
5. J.B. Newkirk (1958) Phys. Rev. 110, 1465; (1959) Trans AIME 215, 483
6. E.N. Farabaugh, H.S. Parker and R.W. Armstrong (1973) J. Appl. Cryst 6, 482
7. R.W. Armstrong and C.Cm. Wu (1973) in Microstructural Analysis: Tools and Techniques (eds. J.L. McCall and W.M. Mueller), Plenum Press, New York, p.169
8. E.N. Farabaugh (1977) Ph.D. Thesis, University of Maryland
9. J.M. Schultz and R.W. Armstrong (1966) Acta Met. 14, 436
10. A.P.L. Turner, T. Vreeland, Jr., and D.P Pope (1968) Acta Cryst. A24, 452
11. A.R. Lang (1970) in Modern Diffraction and Imaging Techniques in Material Science (eds. S. Amelinckx, R. Gevers, G. Remaut and J. Van Landuyt), North- Holland Publishing Company Amersterdam, p.407
12. A. Authier, Ibid., p481
13. B.W. Batterman and H. Cole (1964) Rev. Mod Phys 36, 681
14. R.W. James (1963) in Solid State Physics 15 (eds. F. Seitz and D. Turnbull). Academic Press, New York, p.53
15. F.C. Frank and A.R. Lang (1965) in Physical Properties of Diamond (ed. R. Berman), Oxford University Press, p.69
16. A.R. Lang and M. Polcarova (1965) Proc. Roy. Soc.London A285, 297

17. B. Roessler and R.W. Armstrong (1969) in Adv. X-ray Anal. **12**, (eds, C.S. Barrett, J.B. Newkirk and G.R. Mallett), Plenum Press, New York, p. 139
18. International Tables for X-ray Crystallography II Mathematical Tables (eds. J.S. Kasper and K. Lonsdale), [5] Physics of Diffraction Methods (ed. H. Lipson), Intern. Union of Cryst. (1959), Kynoch Press, Birmingham, p.235
19. R.W. James (1954) The Optical Principles of the Diffraction of X-rays, G. Bell and Sons, Ltd. London, p.60
20. J.M. Schultz and R.W. Armstrong (1964) Phil. Mag. **10**, 497
21. T. Vreeland, Jr. (1976) J. Appl. Cryst. **9**, 34
22. M. Kuriyama and T. Miyakawa (1970) Acta Cryst. **A26**, 667
23. S. Weissmann (1956) J. Appl. Phys **27**, 389
24. S. Weissmann, Ibid., 1335
25. W.J. Boettinger, H.E. Burdette, M. Kuriyama and R.E. Green Jr. (1976) Rev. Sci. Instrum **47**, 906
26. K. Kohra, H. Hashizume and J. Yoshimura (1970) Japan J. Appl. Phys **9**, 1029
27. M. Kuriyama, W.J. Boettinger and H.E. Burdette (1978) J. Cryst. Growth **43**, 287
28. R.W. Armstrong, W.J. Boettinger and M. Kuriyama (1979) in preparation.
29. Y. Nakayama, S. Weissmann and T. Imura (1962) in Direct Observation of Imperfections in Crystals (eds. J.B Newkirk and J.H. Wernick), Interscience Publishers New York, p 573
30. A.R. Lang and V.F. Miuscov (1964) Phil. Mag **10**, 263
31. S.J. Burns and O. Birau (1979) unpublished research
32. C.Cm. Wu and R.W. Armstrong (1975) J. Appl. Cryst. **8**, 29
33. W. Bollmann (1964) in Dislocation in Solids, Discussion of the Faraday Society, **38**, p26
34. H. Fehmer and W. Uelhoff (1972) J.Cryst. Growth **13**, 257
35. K.L. Bye (1979) J. Mater, Sci. **14**, 619
36. K.L. Bye and R.S. Cosier, Ibid., 80037.
37. M.R. Achter, C.L. Vold and T.G. Digges, Jr. (1966) Trans TMS-AIME **236**, 1597
38. R.W. Armstrong, C. Cm. Wu and E.N. Farabaugh (1977) in Adv. X-ray Anal. **20** (eds. H.F. McMurdie, C.S. Barrett, J.B. Newkirk and C.O. Ruud),
39. V.F.S. Yip and C.D. Brandle (1978), J. Amer. Ceram. Soc **61**, 8
40. G.M. Arnstein (1972) Ph.D. Thesis, University of Maryland; (1972) G.M. Arnstein, P. Bolsaitis and R.W. Armstrong, Acta Cryst A28, 344
41. C. Bousquet, M. Lambert, A.M. Quittet and A. Guinier (1963) Acta Cryst. **16**, 989
42. B. Roessler and S.J. Burns (1974) Phys. Stat. Sol (a) **24**, 285
43. C.G'Sell and G. Champier (1975) Phil. Mag **32**, 283
44. K. -C. Yoo and B. Roessler (1979) unpublished research.
45. J.M. Schultz (1976) J. Mater. Sci. **11**, 2258

CHAPTER 15

LABORATORY TECHNIQUES FOR TRANSMISSION X-RAY TOPOGRAPHY

J. CHIKAWA

15.1 Introduction

Transmission topography is a powerful tool for surveying defects in crystals, and some useful reviews have been published [1-3] Green [4] and Hartmann [5] provided detailed reviews on direct-viewing techniques of defects which have been developed recently. Therefore it is intended in the present article to describe the author's own experience of the transmission topography rather than a general review, to avoid overlapping with the above review papers.

First, basic photographic techniques will be introduced, and some methods for analysis of defects and special techniques will be described briefly. The second part of this article will be devoted to the direct-viewing technique for rapid characterization of crystals and dynamic observation of defects.

15.2 Basic photographic techniques

Laboratory techniques for transmission topography are divided into two categories. One is the so-called 'section' and 'projection' topography where the incident beam on the specimen crystal should be treated as a spherical wave [6]. The other incorporates special methods using highly collimated beams which can be regarded as plane waves.

15.2.1 Topography by spherical waves

Section topography

Geometry of section topography is shown in Fig.1. X-rays from a focal spot of an X-ray generator enter a specimen crystal through a very narrow slit placed just before the crystal orientated to satisfy a Bragg condition. X-rays emitted from an atom in the focal spot, even though collimated by the slit, have a divergency much larger than the intrinsic angular width of the diffraction peak for

LABORATORY TECHNIQUES: TRANSMISSION TOPOGRAPHY

a perfect crystal and therefore should be treated as a spherical wave which is given by superposition of coherent plane waves, as interpreted by Kato [6]. The spherical wave generates wave fields which spread within the X-ray fan OCD defined by transmitted and diffracted directions (OC) and (OD) in Fig 1 (a). The intensity distribution between C and D at the exit surface of the crystal is recorded with a photographic plate placed behind the crystal. Such a recorded image is called a 'section topograph'. In the X-ray fan, the incident waves are analysed by the degree of their deviations from the Bragg condition and give rise to interference patterns which are informative on crystal perfection. The waves in the incident direction OC are intense and, by intersecting a defect, produce new waves propagating in the diffracted direction which give an image of the defect on the plate as seen from the example of a stacking fault in Fig. 1 (b). The intersection of the sheet of the incident beam and the stacking fault inclined to the crystal surface is imaged as the black line OE in Figs. 1 (c) and (d). As seen from above, section topography gives good sectional information.

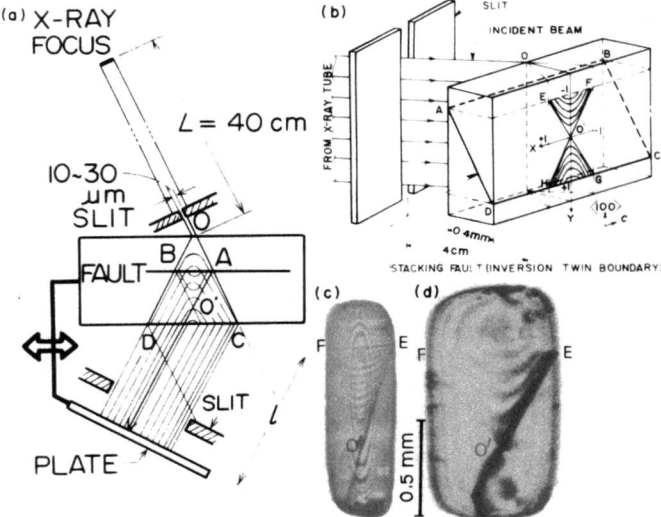

Figure 1. Section topography. (a) Geometry. For traverse topography, the crystal and plate are moved parallel as indicated by the thick arrow with a wider slit width. (b) Schematic illustration for the section topograph (c) and (d) of a stacking fault in a BeO crystal. (c) 002 reflection (d) 004 reflection MoKα_1

Projection topography

In order to obtain overall information on crystals various methods have been devised to get projection topographs of large areas of crystals. The Lang method has been widely used for this purpose. The crystal and plate are moved parallel back and forth, and topographs are called 'transverse topographs'. In this case, some of the information given by section topography is lost. For example, the contrast of the defect images in Fig. 1 (c) is so degraded that the defect is hardly observed in the traverse topograph.

The resolution is given by the size of the image formed by a point of the exit surface of the crystal [7,8]. The vertical resolution (parallel to the slit) is

$$\Lambda_v = f_v (\ell/L) \qquad (1)$$

where f_v is the vertical dimension of the X-ray focal spot used, and L and ℓ are the distances of focus to crystal and of crystal to plate, respectively. To save exposure time, the full spectral width of the characteristic radiation $K\alpha_1$ (wavelength λ) is used by making the slit wider, and the horizontal resolution Λ_h is given by

$$\Lambda_h = \ell (\Delta\lambda/\lambda) \tan\theta_B \qquad (2)$$

where θ_B is the Bragg angle and $\Delta\lambda$ may be taken to be the half height width of the $K\alpha_1$ line. A resolution of a few microns is obtained easily. However, if both the $K\alpha_1$ and $K\alpha_2$ lines are used, the resolution is greatly degraded. In order to obtain images with only $K\alpha_1$ radiation, $f_h/L < \Delta\theta_{12}$, where $\Delta\theta_{12}$ is the difference of the Bragg angles for $K\alpha_1$ and $K\alpha_2$ radiations. In the Lang method, the slit should be narrow enough not to satisfy the Bragg condition for the $K\alpha_2$ line. With the use of a narrow beam one can also make ℓ smaller.

In the transmission Berg-Barrett method [8,9], a large area of a specimen crystal can be imaged by a wide incident beam from a line-focus X-ray tube without the scanning. However, separation of $K\alpha$ lines is difficult. Anderson [10] overcame the disadvantage by placing a Soller slit (divergency 6 min of arc) between an X-ray focus and crystal. Kohra et al [11] used a monochromatized 30 mm wide beam as the incident beam for specimen crystals which was obtained by an asymmetrical reflection of a Ge crystal [non-parallel (+,-) or (+,+) Bragg-Laue double crystal arrangement]. See 2.2. In this method, homogeneous intensity distribution along the line X-ray focus is important, and some devices such as scanning the X-ray source by a small distance were made. The diffraction vectors of the two successive reflections are not parallel and have the opposite senses in the non-parallel (+, -), and have the same sense in the (+, +) setting. The wavelength and angular spreads of the beam diffracted from the specimen crystal are known from the Du Mond diagram.

LABORATORY TECHNIQUES: TRANSMISSION TOPOGRAPHY

Hosoya [12] recorded projection topographs with $AgK\alpha_2$ radiation by absorbing $K\alpha_1$ radiation with a Ru filter. In this case, the homogeneity of X-ray intensities along the focus is important too. Because of widely divergent incident beam, precise alignment of the specimen crystal is not required but sometimes undesired reflections are superimposed on the topographs.

The anomalous transmission is conveniently applied to taking projection topographs for crystals with high absorption coefficients such as Ge, GaAs, etc. X-rays deviated from a Bragg condition are absorbed easily by thick crystals, but X-rays in the Bragg condition make anomalous transmission by propagating between the atomic planes. Distorted regions in the crystal do not allow the anomalous transmission, and defects are imaged as their shadow. (Images of defects near the entrance surface of X-rays become diffuse, and only defects near the exit surface show the sharp shadows). Anomalous transmission topography can be made easily by transmission Berg Barrett methods. Topographs taken by the transmitted beam are free from the effect of the wavelength dispersion [13].

In order to know depths of defects from the crystal surface, stereo pairs are obtained by taking two topographs for hkl and $\bar{h}\bar{k}\bar{l}$ reflections [1] or for the same reflection with two different orientations by rotating the crystal around the diffraction vector [14].

Semiconductor devices are fabricated on one surface or crystal wafers, which are often warped especially for wafers with large diameters. In the methods which employ a collimated incident beam, the Bragg condition is satisfied only for a small region of warped crystal. In order to image entire wafers Schwuttke [15] invented the SOT technique, in which the specimen and plate are oscillated during scanning in the Lang method. Resolution is not degraded if the oscillation range is small enough not to diffract $K\alpha_2$ radiation because the specimen and plate are mounted on the same oscillating holder. This technique was improved by replacing the oscillation with tuning the crystal orientation at the Bragg peak position during the scanning which is made automatically by detecting the X-rays through the plate [16]. For slightly warped wafers, a simple mechanical orientation tuning is effective enough [17]. Orientation tuning methods, however, are not suitable for torsionally distorted crystals. Therefore, Kohra and Takano [18] developed the oscillation method using a monochromatic divergent beam (OMD). The specimen is placed in a divergent monochromatic beam obtained by mounting a slit at the $K\alpha_1$ - focussing position of a curved monochromator and is oscillated so as to scan the position where the Bragg cndition is satisfied. They took projection topographs for crystals containing many sub-grains. In this arrangement, topographs can also be recorded by traversing the crystal and plate as in the Lang method, instead of the oscillation. The present author [17] employed the chromatic-aberration correction method (CAC) in which a divergent beam from a point focus of an X-ray tube is used and the images by X-rays with different wavelengths are superimposed. The method can save exposure time by using both the $K\alpha_1$ and $K\alpha_2$ lines. The CAC technique is effective for recording a high resolution topograph for a higher order reflection with a larger dispersion. In methods using a divergent beam, rough alignment of the specimen crystal is enough, and they are convenient for routine work involving the taking of many topographs.

In characterization of crystals, defects are first surveyed by transmission projection topography, and section topographs are taken at positions of interest. Sometimes, section topographs are recorded in superposition with a traverse topograph to find the position of the observed defects. For taking both traverse and section topographs, an X-ray generator with exchangeable electron guns for line and microfocus is convenient. For full utilization of the advantages of section topography, Kawado and Aoyama [19] developed step-scanning section topography with an automatic Bragg-angle control which allows us to take automatically a series of section topographs at intervals over a large area of crystals.

15.2.2 Topography by the plane waves

Interpretation of images from spherical waves are complicated. Therefore, it is meaningful to make conditions experimentally under which a simple plane-wave theory can be applicable.

Bonse and Kappler [20] first applied the double-crystal technique to topography in a Bragg-Bragg arrangement. In this method, the reference crystal and the specimen are usually made of the same kind of material so that exactly the same spacing of the diffracting planes can be used in both crystals. As a result of this arrangement which is referred to as '(+n, -n) setting' or '(+, -) parallel setting', the dispersion due to the spectral distribution of the radiation used is eliminated. Authier [21] obtained a collimated narrow beam by a (+n, -n) Laue-Laue setting and experimental evidence for existence of four waves in the crystal predicted by the dynamical theory.

Renninger [22] and Kohra [23] found independently that a highly collimated wide beam is obtained with an asymmetric reflection in a Bragg case. In Fig. 2, the angular spread w_o of the incident beam incorporated into the Bragg reflection and that of diffracted beam w_h have a relation

$$w_o = b^{-\frac{1}{2}} w_s \qquad (3)$$
$$w_h = b^{\frac{1}{2}} w_s \qquad (4)$$

where w_s is the angular width for the symmetrical reflection and b is the asymmetry factor given by

$$b = \sin(\theta_B - \alpha)/\sin(\theta_B + \alpha) \qquad (5)$$

where α is the angle between the reflecting plane and the surface of the collimator crystal. For such a collimator, extremely careful polishing should be made for the surface finishing. They also obtained a highly parallel beam by consecutive asymmetric reflections [24] using the technique invented by Bonse and Hart [25]. The collimator system was monolithic, but minor orientation adjustments were made for each asymmetric collimator crystal to correct the Bragg peak shift due to the asymmetric reflection [26]. By n consecutive reflections, the angular spread becomes $w_h = b^{n+\frac{1}{2}} w_s$. They obtained a beam with a divergency of 0.01 sec of arc with $b = 0.1$ and $n = 3$ for the Si 422 reflection and $CuK\alpha_1$. Such a parallel beam can be treated as a single plane wave incident on the specimen crystal. For recording topographs by the beam diffracted from the specimen, only $K\alpha_1$ radiation should be used for a high resolution. The transmitted beam may be recorded if it gives the same information as the diffracted beam because of no dispersion [13].

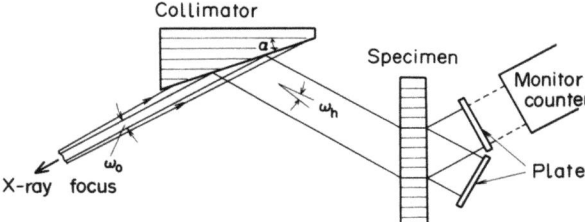

Figure 2. Schematic illustration of plane-wave topography

15.2.3 Analysis of defects

It will be shown by some examples how to analyze defects by topography.

Determination of Burgers vector and fault vectors

Burgers vectors and fault vectors are determined from contrast variations with reflections, using the criteria that the contrast is minimum or disappears for $\underline{h} \cdot \underline{b}$ or $\underline{h} \cdot \underline{f}$ integers, where $\underline{h}$ is the reciprocal lattice vector, and $\underline{b}$ and $\underline{f}$ are the Burgers and fault vectors, respectively. It is convenient for observation of the contrast variations to take Laue patterns of section topographs [27]. In the experimental arrangement of the section topographs, a nuclear plate is placed at a relatively long distance (3 cm) behind the crystal and exposed longer than the exposure time for the usual section topograph taken with the $K\alpha_1$ radiation. An example is shown in Fig. 3, which was taken for the same specimen crystal as that in Fig. 1. The oblique black line due to the stacking fault is seen in each spot (section topograph) except the 1$\bar{1}$2 and $\bar{1}$12 reflections from which the fault vector was determined.

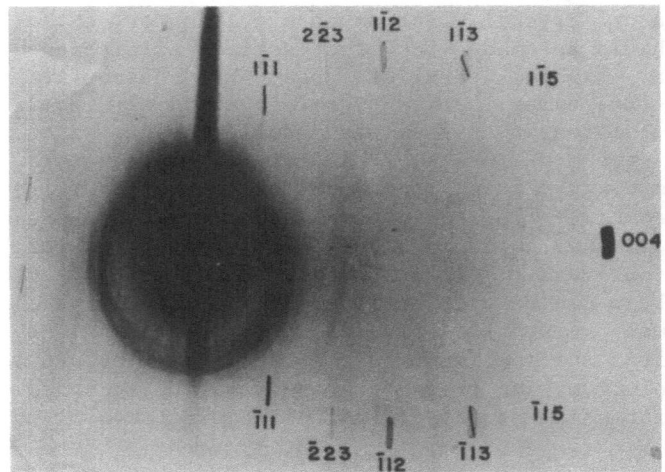

Figure 3. Laue pattern of section topographs. The specimen is the same as in Fig. 1.

Determination of the sense of Burgers vectors

This is sometimes important for dislocation loops and helical dislocations which may give information on the type of dominant point defects in crystals. It is well known that the perfect determination of Burgers vectors (B.V.) can be made by moiré topography [28-30]. Here another method [31] to observe the kinematical images of dislocations will be described briefly.

As shown in Fig. 4, the lattice planes on both sides of a dislocation are strained with the opposite curvatures, and the kinematical images are formed on the opposite side of the dislocation core by the incident X-rays with the opposite deviations from the Bragg condition $\delta\theta = \theta_B - \theta > 0$ and $\delta\theta < 0$. Therefore the sense can be determined by measuring the directions of the X-rays diffracted from both sides of the core. For relatively thin crystals the sense determination can be made with a divergent beam from a point focus, as shown in Fig. 5(a). When a specimen crystal is oriented to satisfy the Bragg condition for the centre of the beam, we have regions with $\delta\theta>0$ and $\delta\theta<0$ on both sides of centre and obtain dislocation images as shown schematically in Fig. 4(c) and (d). depending upon the sign of B.V. Although topographs by this arrangement are a kind of section topograph, we placed a wide slit (more than 200 μm) just before the specimen, unlike the usual section topography described in 2.1. For convenience, this method is

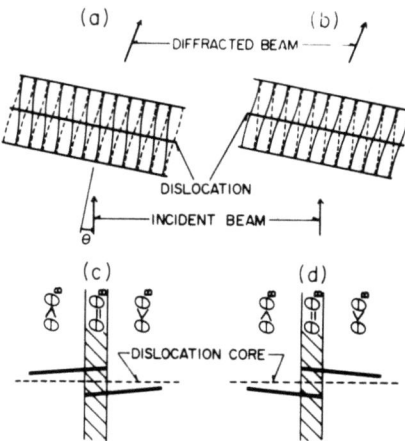

Figure 4. Schematic diagram of screw dislocations with the Burgers vectors having opposite signs. Images by the method shown in Fig. 5. In (a) and (b), the reflecting planes on the lower and upper sides of the dislocation core are indicated by the dotted and solid lines respectively. (c) and (d) are the images (solid lines) for the dislocations shown in (a) and (b), respectively.

referred to as 'divergent beam section topography'. For the geometry in Fig. 5(a), the incident X-rays with different wavelengths satisfy the Bragg condition at different positions on the entrance surface for the X-rays. For simplicity, we consider the two wavelength λ_1 and λ_2 ($\lambda_1 > \lambda_2$) at the half-height of the $K\alpha_1$ spectral line, which satisfy the Bragg condition at the entrance positions O_1 and O_2 in Fig. 5(a), respectively. Suppose a narrow slit is placed at O_1 and O_2. The $K\alpha_1$ radiation spreads onto the X-ray fans $\Delta O_1 A_1 B_1$ and $\Delta O_2 A_2 B_2$. In the former fan, waves with $\theta > \theta_B$ are much stronger than ones with $\theta < \theta_B$, and in the latter fan, vice versa. For our purpose, the two fans should not overlap within the crystal, i.e. the crystal thickness D should be

$$D < (L/2 \cos \theta_B)(\Delta\lambda/\lambda) \qquad (6)$$

where $\Delta\lambda$ is the half-height width of the $K\alpha_1$ line. For L = 40 cm and $MoK\alpha_1$, D ≲ 100 μm. Under this condition, when the crystal is oriented to satisfy the Bragg condition for the centre of the divergent beam, we have $\theta > \theta_B$ and $\theta < \theta_B$ on its sides. An example is shown with a 70 μm thick CdS crystal. Fig. 6(a) is a traverse topograph of dislocations with a B.V. = c_o (6.7A). Single line images are seen. Fig. 6(b) is a divergent beam topograph taken at the position indicated by the dotted line in Fig. 6(a). A vertical black band is due to the central region in the Bragg condition. At its intersection with the looping dislocation, the images shown in Fig. 4(c) and (d) are seen clearly. (The contrast of the images $\theta > \theta_B$ and $\theta < \theta_B$ depends on the depth of the dislocation). Note the

375

opposite signs for the upper and lower sides of the loop. For smaller B.V. (a_0 = 4.1A in CdS), this method is applicable too, and helical dislocations in CdS were found to be formed by absorbing interstitials.

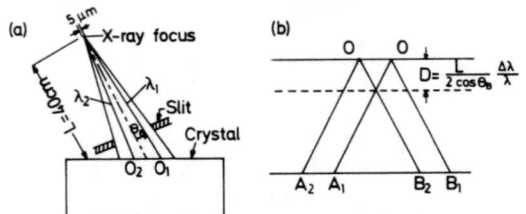

Figure 5. Divergent-beam section topography. (a) Schematic diagram. (b) X-ray fans for the points O_1 and O_2 on the X-ray entrance surface where the wavelengths λ_1 and λ_2 ($\lambda_1 > \lambda_2$) at the half-height of the $K\alpha_1$ spectral line satisfy the Bragg condition, respectively.

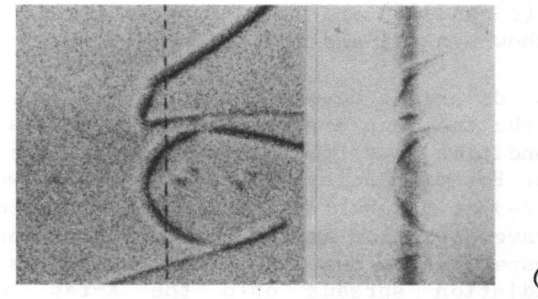

Figure 6. Determination of the signs of Burgers vectors (a) Traverse topograph of a CdS crystal (b) Divergent-beam section topography (see Figs. 4(c) and (d).

For thick crystals, L should be made longer in the above method, or one should employ the plane-wave topography described in 2.2, (Ishida et al [32]). Also, they found for a wedge shaped crystal that the number of plane-wave Pendellosung fringes differs between both sides of a dislocation, and an extra fringe line was observed at the end of the dislocation like a moire dislocation image. From this pattern, both the size and sense of B.V. can be determined [33].

Effect of elastic strains on defect contrast

This effect is often observed as a contrast inversion and varies with the depth of the defect from the crystal surface. Since image formation for dislocations and stacking faults is similar, we can examine the contrast variations of images of a stacking fault which inclines to the crystal surface. This is made in Fig. 7. Fig.

7(a) and ($\bar{a}$) are topographs of a Si crystal for 111 and $\bar{1}\bar{1}\bar{1}$ reflections with the geometry as shown in Fig. 8. Slightly curved black bands are due to growth bands. Stacking faults and dislocations are seen.

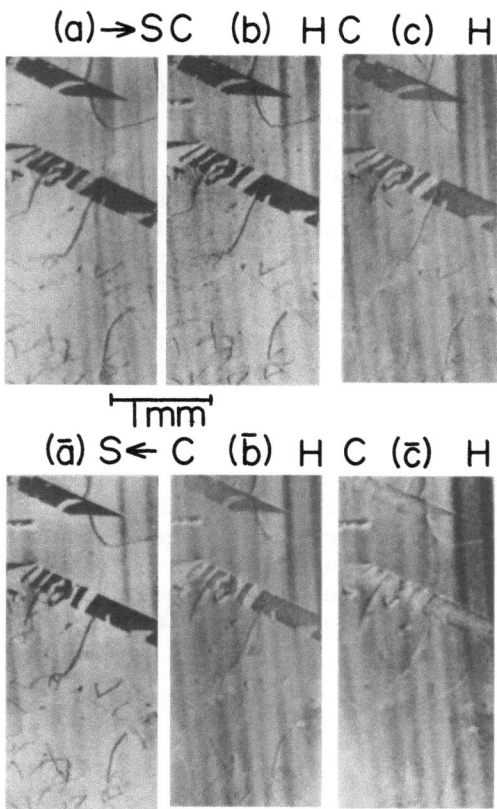

Figure 7. Effect of elastic strains on defect images in traverse topographs (Si. MoKα_1, crystal thickness D = 0.5 mm). (a) (b) (c) for the 111 and ($\bar{a}$) ($\bar{b}$) ($\bar{c}$) for the $\bar{1}\bar{1}\bar{1}$ reflection, as shown schematically in Fig. 8(a) and (b), respectively. (a) and ($\bar{a}$) were taken without elastic strains. (b) (c) and ($\bar{b}$) ($\bar{c}$) were obtained by applying the temperature gradients as shown in Fig. 8(a) and (b), respectively. The hot and cold sides are indicated by H and C, respectively. The temperature gradient is medium for (b) and ($\bar{b}$), and larger for (c) and ($\bar{c}$).

To observe the effect of elastic strains, Figs. 7(b) (c) were taken by applying temperature gradients shown schematically in Fig. 8, as has been done by Hart [34]. The hot and cold sides are indicated by H and C, respectively. The temperature gradients are medium for Figs. 7(b) and ($\bar{b}$), and large for Figs. 7(c) and ($\bar{c}$). For the large temperature gradient, the contrast of the stacking faults and dislocations decreases. Especially in Fig. 7($\bar{c}$), their contrast becomes inverse, i.e. the diffracted intensity from the image of the stacking fault decreases with the depth from the entrance surface for X-rays (the upper edge of the stacking faults). According to the theory for a slightly distorted crystal [6] the paths of the X-ray wave field having lower absorption (branch 1 of the dispersion surface) in defect-free regions is bent towards the transmission direction in Figs. 7(b) and (c), and towards the diffracted direction in Figs. 7($\bar{b}$) and ($\bar{c}$). In Figs. 7(b) and (c), the wavefield produces a new wavefield at the stacking fault which propagates towards the diffracted direction, and therefore the image has an intensity higher than the background intensity (defect-free region). However the background intensity becomes higher by the strain, owing to the wavefield belonging to branch 2. In Figs. 7($\bar{b}$) and ($\bar{c}$) the wavefield (branch 1) which produced the new wavefield at the stacking fault propagates toward the transmitted direction. If it is located near the X-ray exit surface, its image has an intensity lower than the background intensity. For the stacking fault near the entrance surface, however, the wavefield belonging to branch 2 also produces the new wavefield on branch 1 which propagates toward the diffracted direction with a low absorption. Consequently, the intensity of the image becomes similar to the background intensity . Since elastic strains sometimes remain in real crystals, one should keep this effect in mind, especially in observations of moving dislocations which are accompanied with some strain fields.

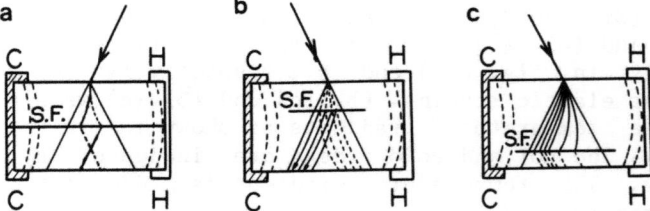

Figure 8. Schematic illustration of explanation for the contrast changes of defect images (stacking fault). H: heater, C: cold heat sink. The propagation of the wavefields belonging to the branches 1 and 2 are shown by the solid and dotted lines.

15.2.4 Special techniques

Weak-beam method in topography

In electron microscopy, the 'weak-beam' technique has been used for obtaining sharp narrow dislocation images and applied to various investigations such as analysis of composite dislocations. As seen from the previous section, the images of dislocations become narrow with the incident beam deviated greatly from the Bragg condition [31]. An example is shown in Fig. 9. Fig. 9(a) is a traverse topograph of dislocations (B.V = 4.1 A) in a CdS crystal. Fig. 9(b) is a traverse topograph which was taken with slit-collimated beam with the deviation of 8 sec of arc. The dislocation images are sharp. For Si crystals, such narrow images of dislocations were obtained by the double-crystal plane- wave topography [32].

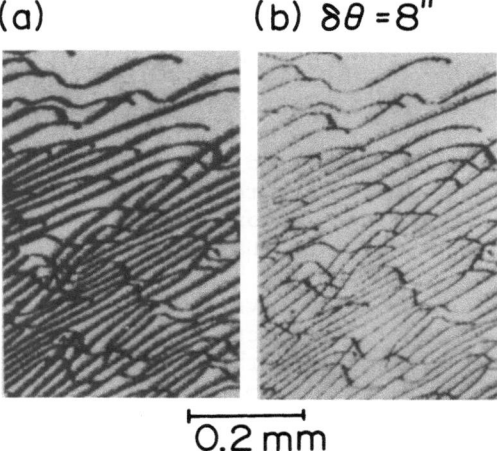

Figure 9. Weak-beam method. (a) Usual traverse topograph of a CdS crystal. (b) Traverse topograph taken with a slit-collimated beam deviated from the Bragg condition by 8 second of arc.

Detection of small defects

Even if a very high resolution is attained, detection of small defects is impossible because their strain fields are very small compared with the extinction distances and do not produce an appreciable contrast against the background (perfect region). However, if the contrast of their images were high, defects could be observed even with a poor resolution, as we can see dust illuminated by a beam of light in a dark room.

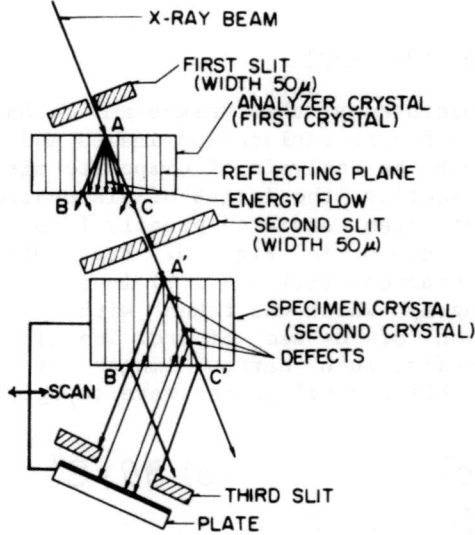

Figure 10. Kinematical image technique.

The present author developed the 'Kinematical Image Technique' (KIT) in which defect images are recorded by separating them from diffraction by the perfect regions [35]. In Fig. 10, the first and specimen crystals are placed so that their reflecting planes are parallel . The slit-collimated incident beam with a divergency of about a minute of arc produces wave fields which spread within the X-ray fan ABC in the first crystal. All the X-rays through the second slit placed in the incident direction are deviated from the Bragg condition and take the two paths A'C' and B'C' in the specimen crystals. If the specimen is a perfect crystal, the diffracted intensity between B' and C' is zero in principle. When defects are present on the path A'C' their kinematical images are formed between B' and C'. An example is shown in Fig. 11(a). A float-zone dislocation-free crystal having swirl defects was observed by the method. Two black bands correspond to B' and C' in Fig. 10. Many black spots due to the defects are seen. Figs. 11(b) and (c) are usual section topographs of the same crystal for 220 and 440 reflections, respectively. In Fig. 11(b), the defect images are not clear, but some relatively large defects are seen in the higher order reflection, Fig. 11(c) . In the KIT topograph, Fig 11(a) , more defects are seen. For much smaller defects such as point defects, individual defects cannot be observed, but they are detected as an intensity enhancement (diffuse scattering) between B and C' in Fig 10, if their density is high enough. For example, such an intensity enhacement was observed by neutron irradiation of 10^6 nvt in a preliminary experiment. A disadvantage of the KIT is the 10 times longer exposure time than that for usual section topographs.

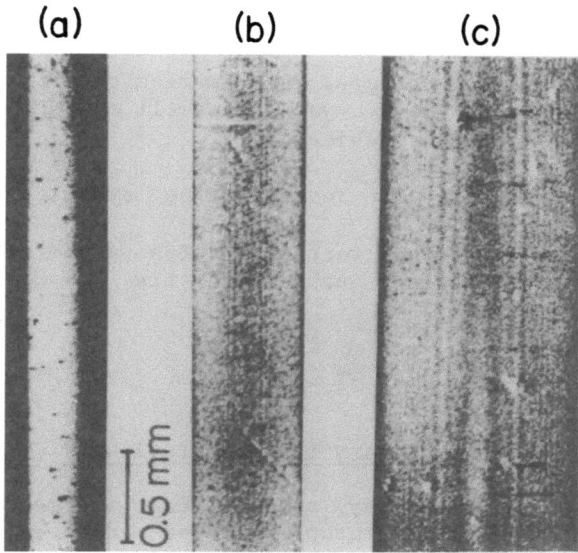

Figure 11. Section topograph of a Si crystal by the kinematical image technique (a) and usual section topographs (b) and (c) which were taken for 220 and 440 reflections. MoKα_1.

In order to save the exposure time, Renninger [36] used the (+, -) double-crystal Bragg-Laue arrangement (the plane-wave topography) and observed swirl defects by positioning the specimen orientation at a deviation from the Bragg peak (1/10 height of the Bragg peak). Microdefects at a high enough density in a random distribution can be detected as an intensity enhancement of diffuse scattering observed at the foot of a Bragg peak, and identification of defect types (vacancy or interstitial) can be made [37].

Small defects at a high density in a random distribution are also detected as a decrease of anomalous transmission intensity. Oxygen precipitates in Si [38], radiation damage in Si, Ge[39] and Cu[40], dislocation loops in GaAs [41] have been detected by the method. In this method, if the defects are distributed inhomogeneously and cause long range strains, the net effect of defects is hardly detected.

De Kock has observed microdefects in Si crystals by a combination of decoration and topography [42]. Strain fields of microdefects are expanded by Cu-precipitation on them to be large enough for topographic observation. Although this method requires a heat treatment which may change the defect structure, it is an advantage to observe without any special techniques.

15.3 Video display of topographic images

The video display has been made for two purposes.

(1) Rapid survey of defects. This is required for selection of suitable specimens from many crystals as well as for nondestructive inspection of semiconductor devices.

(2) Live topography for investigating dynamic behaviour of defects.
In this section, a video technique which was developed for both these purposes by the present author and his co-worker will be described in some detail.

15.3.1 Instrumentation

Diffracted intensities necessary for direct viewing

Video displays of X-ray and optical images have different features. Although X-ray photon energies are very large, the intensities available in X-ray diffraction are extremely low compared with optical images. Therefore, photon noise due to the statistical fluctuation of the number of photons incident upon an image system gives a detection limit of the image.

Consider the case of viewing defects in a crystal by an imaging system. ν_p photons sec^{-1} mm^{-2} are diffracted from the perfect region, and $\eta_o \nu_p$ ($\eta_o < 1$) are absorbed by the X-ray sensing layer of the system. An absorbed photon produces a mean number η_1 of electron or visible photons, each of which may be rescattered to produce a mean number η_2 of electrons or photons. By repeating S such processes, the mean signal height

$$S_p = \eta_o \nu_p \eta_1 \eta_2 \ldots \ldots \eta_s \varepsilon^2 t \tag{7}$$

is obtained from each square-shaped element $\varepsilon \times \varepsilon$ mm^2 for t sec. The value of ε is determined by the resolution of the system. Since $\eta_1 \gtrsim 100$ due to the large photon energy and the values of $\eta_2, \eta_3 \ldots$, η_s are considered to be less than 100, the photon noise N as a standard deviation of S_p is given by [43]

$$N_p = \eta_s \eta_{s-1} \ldots \ldots \eta_1 (\eta_o \nu_p \varepsilon^2)^{\frac{1}{2}} \tag{8}$$

In order to recognise the defect image in the perfect region, the signal-to-noise ratio (SNR R_I of the image must be

$$R_I \equiv (S_d - S_p)/N_p = \varepsilon C(\eta_o \nu_p t)^{\frac{1}{2}} \geq k \tag{9}$$

where S_d is the signal height of the defect image, C is the contrast of the defect image $C \equiv (S_d-S_p)/S_p$, and k is the threshold SNR of 1~5. By taking t = 0.2 sec as the integration time of human eyes [44], the necessary intensity diffracted from the perfect region is $\nu_p = 10^4 \sim 10^5$ photons sec^{-1} mm^{-2} for the case of C=1, $\varepsilon = 30\mu$m, and $\eta_o=1$. Since dislocation images in transmission topography are commonly caused by the distorted region with a width of about 10μm around the dislocation core, the contrast is increased by improving the resolution of the imaging system, and the value of εC is considered to be constant for the range of resolution $\varepsilon > 10\mu$m As described later, the contrast was found to be $C \sim 0.5$ with $\varepsilon \simeq 30\mu$m for dislocation images in Si crystals which were obtained with the reflecting plane perpendicular to their Burgers vectors. Furthermore, $\eta_o < 1$ and the noise of the imaging system is superposed. Therefore, more intensity is required.

The difference in meaning between sensitivity and quantum detection efficiency should be noted. Sensitivity is expressed by the intensity required for a certain signal height to the noise height of the imaging system and is proportional to the value of $\eta_o \eta_1 \ldots \eta_s$. As seen from equation (9), the ratio of the signal to the photon noise depends only on the quantum detection efficiency η_o, which may be considered to be the absorption rate of the X-ray sensing layer because each absorbed photon may contribute to the signal owing to its large energy. For example, 100 μm-thick silicon diode array and 15 μm - thick PbO vidicon camera tubes have similar sensitivities for MoKα radiation but $\eta_o = 0.15$ and $\eta_o = 0.6$, respectively. The former obtains the sensitivity with a higher conversion efficiency $(\eta_1 \eta_2 \ldots \eta_s)$ of absorbed photon energy and gives a $(0.15/0.6)^{\frac{1}{2}} = 0.5$ times lower signal-to-noise ratio.

It is clear from the above discussion that imaging systems for topography should be evaluated by three factors: resolution, integration time for observation, and image quality (SNR).

Geometry for direct viewing

In the standard photographic methods, a small focus size such as 0.1 X 0.1 mm has been used to secure a high resolution. A fairly high brightness can be obtained for a small focus of the usual X-ray tubes. For example, the diffracted intensity for a 220 reflection of a relatively thin perfect Si crystal is about $\nu_p \simeq 10^4$ MoKα_1, photons sec^{-1} mm^{-2}. However, direct viewing with this intensity is considered to be somewhat difficult (see Fig. 14). Also the viewing area is limited to a width of about 100 μm. Therefore the possibility of a high power X-ray generator with a larger focus size will be considered. For a large focus size L should be large enough to separate Kα_1, and Kα_2 radiations. However, the diffracted intensity is proportional to $1/L$. By increasing L, a wider focus can be used without resolution loss and makes the viewing area wider. L should be determined by compromise between the intensity decrease and widening the viewing area with increase of L. In this experiment, the focus size of $f_v = 0.5$ mm and $f_h = 1$ mm was used, as shown in Fig. 12 and the intensity was increased with a rotating target. For the resolution of the television camera (20~25μm, as described later), the geometrical resolution of about 10 μm is enough. Since ℓ is required to be, at least, 3 cm for the dynamic observation with a specimen furnace, L was taken as L=120 mm, so that $\Lambda_v \simeq 10$ μm and $\Lambda_h \simeq 3$ μm by separating the $K\alpha_1$ image from the $K\alpha_2$ image.

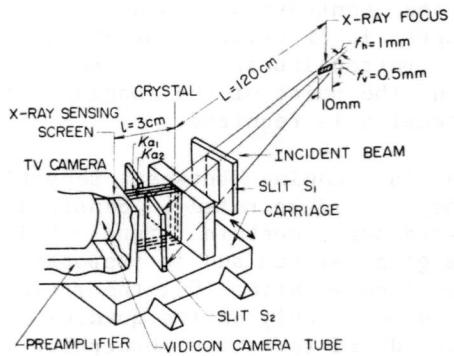

Figure 12. Geometry for x-ray transmission topography using a TV camera.

High power x-ray generator

As seen from the SNR given by equation (9), target materials of the X-ray generator should be chosen to make $C(\eta_o \nu_p)^{\frac{1}{2}}$ as large as possible for the resolution of the camera tube and specimens observed. The thermal limit of X-ray output of a rotating-target generator is proportional to $(T_M - T_o)(ch)^{\frac{1}{2}}$ where T_M=melting point of the target material, T_o=temperature of the cooling water, c=thermal conductivity, and h=heat capacity [45]. By bringing these points into consideration, a Mo-or Nb- rotating target was used for observation of Si wafers with thickness of 0.2 to 0.5 mm. The generator was operated at 60 kV_p and 0.5A with a focus size of 0.5 x 10 mm. For the geometry mentioned above, we obtained about 40 times higher intensity than that by the conventional small focus X-ray generator.

Television system

A block diagram of the imaging system is shown in Fig.13. The TV system consists of a TV camera unit with an X-ray sensing camera tube video amplifier, camera control unit, and picture monitor, as operating a usual closed-circuit TV system [46]. A standard 525 line, interlaced, 30 frame per second scanning system is employed. For weak images, an intermittent scanning unit is used. The TV camera unit and specimen crystal are placed on the carriage of the topographic camera. By orienting the crystal so as to satisfy the Bragg condition for the slightly divergent X-ray incident beam (Kα), two images due to the diffracted $K\alpha_1$ and $K\alpha_2$ beams each with a width of 1 mm are received by the camera tube, as shown schematically in Fig. 12 Two band-shaped regions of the crystal are imaged simultaneously at about 30 diameter-enlargement on the picture monitor. Such images are called 'direct-view images' for convenience. To image traverse topographs, the video signal due to the $K\alpha_1$ image is selected by the electric slit unit and stored in the digital image processor [47], while the carriage is moved for 3 seconds, to several minutes depending upon the diffracted intensities. At the end of the carriage motion, a traverse topograph with the imaging area of the camera tube (9 x 13) is displayed on the picture monitor. Hereafter, such images are referred to as 'synthesized images.' Next the characteristics and performance of each component in the system will be described.

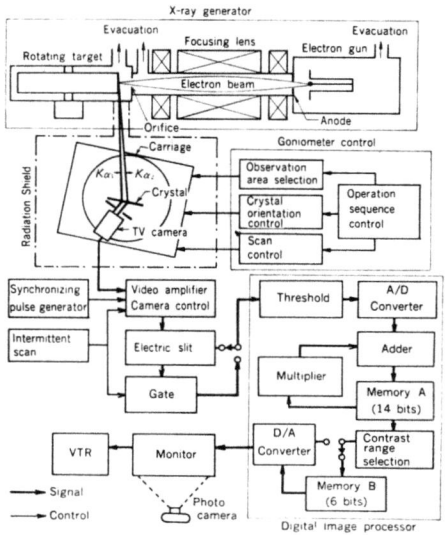

Figure 13. Block diagram of the imaging system for X-ray topography.

(1) TV camera unit

The unit consists of the camera tube, coil assembly for the electron beam focussing, alignment and deflection, and preamplifier which were mounted in a magnetic shielded box with a size of 8 x 8 x 20 cm (total weight 1.5 kg).

(2) PbO camera tube

The construction and operation of the tube are similar to a

light-sensing vidicon except for an X-ray responsive target layer and a beryllium window. The camera tube was developed by Nishida and Okamoto [48] who have reported on the characteristics in detail. The tubes have been manufactured by Hamamatsu TV Co. Ltd. Sensitivity and resolution for characteristic radiation will be described here.

To obtain a high sensitivity the thickness of the PbO target layer should be thick enough to absorb the characteristic radiation used. However, thick targets result in a low resolution due to lateral diffusion of the X-ray induced carriers. Taking into account both the resolution and sensitivity for MoKα thickness of the PbO layer was made about 15µm. For this thickness, the absorption efficiencies are 35, 60 and 78% for AgKα (22 keV), MoKα (17 keV), and CuKα(8 keV), respectively. Fig 14 shows sensitivity measured for MoKα and CuKα radiations. Each photon, if absorbed, produces about 400 and 200 electrons in the video signal, respectively. The CuK radiations are suitable because of a very high photon generation efficiency of X-ray tubes, in spite of the low gain per photon. The X-ray input intensities and amplifier noise levels for bandwidths of 3, 5 and 7 MHz in the standard scanning mode are compared in Fig. 14.

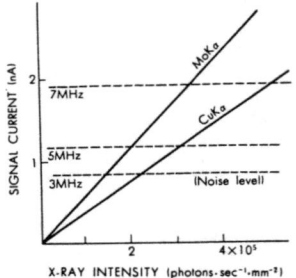

Figure 14. The signal current due to X-ray input and noise levels of the amplifier.

The resolution of the camera tube is shown in Fig.15 by the square-wave modulation transfer function (MTF) measured with an amplifier having a bandwidth of 7 MHz. The limiting resolution (at MTF=5%)is 20 to 25 µm.

(3) Intermittent scan unit

The inner surface of the target layer in the camera tube is scanned repetitively by an electron beam, which simultaneously reads out and erases the image charge built up on the surface during the scan interval. The amount of the charge can be increased by making the interval longer using the intermittent scan unit, and higher signal currents are produced. The resolution loss due to charge diffusion is minor up to a 1 sec interval because of the very low dark current of the camera tube.

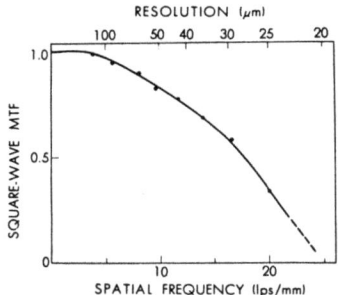

Figure 15. Square-wave modulation transfer function (MTF) measured for the PbO-vidicon camera tube. This function shows the magnitude of brightness modulation in the output image obtained for an input image with a square-wave intensity distribution and is defined by

$$\text{MTF} = \frac{I_o(\max) - I_o(\min)}{I_o(\max) + I_o(\min)} \bigg/ \frac{I_i(\max) - I_i(\min)}{I_i(\max) + I_i(\min)}$$

where $I_i(\max)$ and $I_i(\min)$ are the maximum and minimum intensities of the input image, respectively, and $I_o(\max)$ and $I_o(\min)$ are those of the output.

(4) Electric slit unit

To obtain good-quality synthesized images, the electric slit is used so as not to store the $K\alpha_2$ image and amplifier noise outside of the $K\alpha_1$ image. The TV camera is placed so that the horizontal scanning lines may be parallel to the reflecting plane used, and blanking is made outside of the slit region. The position of the electric slit is moved in synchronization with the carriage motion, and the speed, intitial position and slit width are adjusted with viewing topographic images on the picture monitor. For uniformly warped crystals, as often seen in device wafers, synthesized images can be obtained easily by adjusting the speed of the electric slit. The direct viewing is made for $K\alpha_1$ and $K\alpha_2$ radiations, but the synthesized images are obtained with the 1 mm width $K\alpha_1$ images.

(5) Digital image processor

Digital image processing [47] has become popular, and techniques such as noise reducers [49, 50] are very useful for video topography. A digital image processor was designed for the following three basic modes to improve quality of images. (1) Sliding summation of successive frames with the weighting of new to old data adjustable from the control panel. (2) Image integration for a time interval adjustable from the control panel. Moving objects are seen as images changing discretely like animation pictures. (3) Synthesization of traverse topographs from the $K\alpha_1$ images through the electric slit.

A block diagram of the digital image processor is shown in Fig. 13. The video is sampled and digitized by the A/D converter (6 bits per pixel) and the digital video is sent to the adder and thence to Memory A. Image information in the memory is continually sent both to the adder through the multiplier for combination with incoming data and to Memory B through the contrast-range selection. The images in Memory A or B are converted to analog video which can be displayed or recorded with conventional video accessories by adding sync and blanking information. The weighting of new to old data is made by changing the factor k of the multiplier in the range of $0 \leqslant k \leqslant 1$. For k=0, the original input image is displayed. When the intermittent scanning is used for low-brightness images, each image is continued to be displayed until the next image comes. Therefore, if the interval of the intermittent scanning is long, the display is like animation pictures. In a range of $0 < k < 1$, a sliding summation of successive frames is displayed by sending image information in Memory A directly to the D/A converter through the contrast range selection. The output image in continuous operation is given by summation of a geometric series $S_\infty = S_0/(1-k)$, where S_0 is the input signal, and the SNR is improved by a factor of $[(1+k)/(1-k)]^{\frac{1}{2}}$. For k=1, incoming and stored image data are combined on a pixel-sequential basis, i.e. the memory acts as an image integrator, allowing improvement of SNR through the addition of a selectable number of successive frames. With a memory depth of 14 bits and an initial A/D encoding of 6 bits per pixel, it is possible to sum 256 full-amplitude images in the memory before overflow takes place. The 6-bit contrast range of (64 grey levels) image data in Memory A is selected and stored in Memory B for display. In the synthesized image mode, the $K\alpha_1$ images are stored in Memory A during a traverse of the carriage, and the synthesized image is sent to Memory B and displayed. Simultaneously, the next traverse and image storage in Memory A start, and at the end of the traverse, the old image on display is replaced wih the new one.

Acquisition of extremely low intensity images, dramatic improvements in signal-to-noise ratios via frame integration, and isolation and enhancement of selected-contrast ranges are possible by digital image processing and intermittent scanning.

15.3.2 Rapid survey of defects

The TV system can be applied to rapid inspection of crystals without photographic development. It is desired to attain this purpose with an X-ray power as low as possible. As an example, high quality topographs are displayed with an X-ray power as low as 10 kW (65 kV$_p$, 150 mA NbKα). Fig. 16 shows photographic images of dislocations in a Si wafer. The video amplifier was used with a bandwidth of 3 MHz which was found to be wide enough to obtain sharp dislocation images. Fig. 16(a) is a direct-view image displayed by the standard scanning code (real-time basis). Since the photograph was taken with an exposure time of 1/4 seconds, (the integration time for the human eye is 0.2 sec) , the image is similar to the one seen instantanously on the picture monitor. The two band-shaped images are due to the NbKα_1 and Kα_2 radiation. The black lines are dislocation images. (The images are shown in negative contrast). As seen from the video waveform in Fig. 16(b), the SNR of the image is extremely poor. The image quality can be improved by both the intermittent scanning and digital image integration. Fig. 16(c) is an improved image in which intermittent scanning was made with 30/8 frames per second and 125 frames were superimposed by the digital memory. The video waveform in Fig. 16(d) shows how the image quality is improved. Fig. 16(e) is a synthesized image obtained for about 1 minute by scanning the carriage at 0.17 mm/sec. The intensiy heterogeneities in the background are due to thickness variations of the wafer (Pendellosung fringes). The images compare favourably with the photographically recorded topograph.

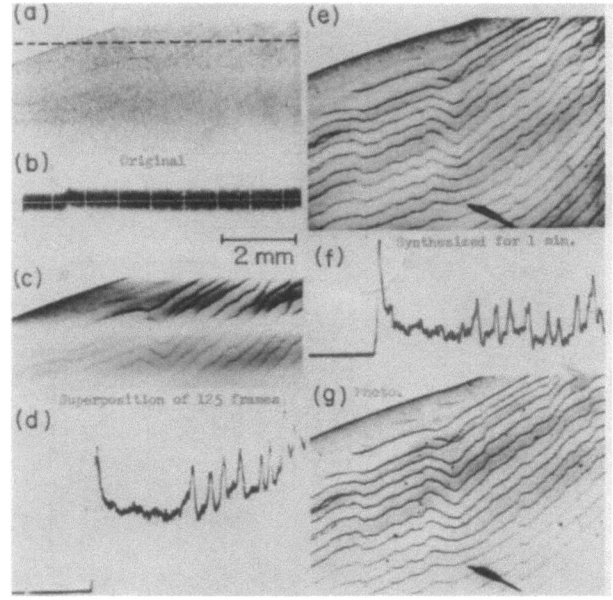

Figure 16. Video display of a topograph for a Si crystal using a 10 kW (65 kV, 150mA) X-ray generator. The black lines are dislocation images. The Burgers vectors are perpendicular to the 220 reflecting plane. (a) Original image by standard scanning (30 frames/sec). (b) Video waveform obtained along the dotted line in (a). (c) Image improved by intermittent scanning and digital image storage (total integration time 33 sec). (d) Video waveform for (c). (e) Synthesized image for 1 min. (f) Video waveform for (e). (g) Topograph by the photographic method for comparison with the video images.

15.3.3 Live topography

For the image-quality improvement of moving defects, the digital image processor should be used with the optimum integration time. In general, rates of image changes cannot be estimated beforehand. Therefore, the original images should be tape-recorded and afterwards the reproduced images should be improved by the processor. In this case, the quality of the original images is important, which depends upon the diffracted intensity. Fig. 17 shows original images of the dislocations in the Si crystal (almost the same area) obtained by operating the X-ray generator at 60 kV and 500 mA. Fig. 17(a) is a direct-view image by the standard scanning. The two band-shaped images are due to the Mo$K\alpha_1$ and $K\alpha_2$ radiations. The dislocations can be seen even in the $K\alpha_2$ image.

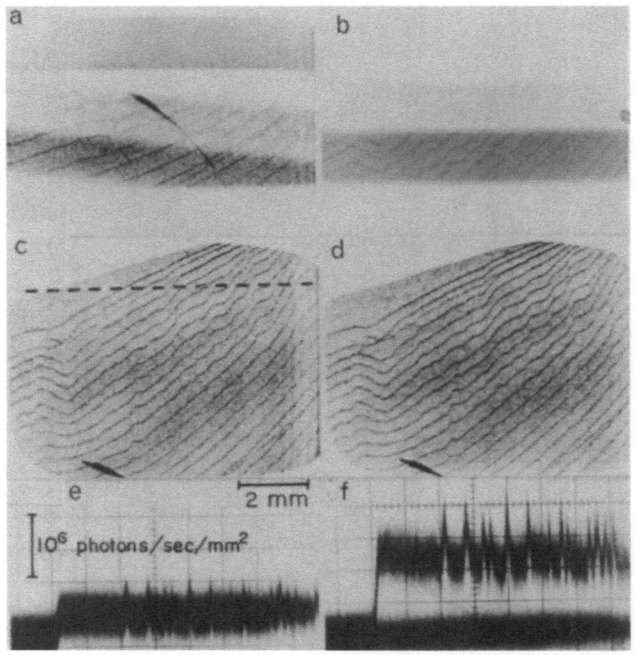

Figure 17. Video display of topographs of the same crystal in Fig. 16. (a) Image as seen instantaneously on the picture monitor. (b) Superposition of the electric slit and the $K\alpha_1$ image. (c) Synthesized image obtained in about 3 sec (standard scan). (d) Synthesized image with the intermittent scan. (e) Output signal of the video amplifier for the position indicated by the dotted line in (c). (f) Output sgnal increased by using intermittent scan.

After the electric slit is positioned on the $K\alpha_1$ image, as seen in Fig. 17(b), a synthesized image can be obtained by moving the carriage, as shown in Fig. 17(c). Fig. 17(e) shows a video waveform of the original image by the standard scanning measured at the position of the dotted line in Fig. 17(c). The peaks due to the dislocations are visible. The intensity of the perfect region is about 4×10^5 MoKα_1 photons sec^{-1} mm^{-2}. The signal has a high noise level. Fig. 17(d) was synthesized by using the intermittent scanning of 5 frames per sec. In this case, the video signal is higher as seen in Fig. 17(f). Although both Figs. 17(c) and (d) were synthesized with the carriage motion of 4mm/sec, the image in Fig. 17(d) has a better quality due to the intermittent scanning. The quality of the images is inferior to those in Figs. 16(c) and (e), becuse the total photon numbers forming the images are smaller for Fig. 17. However, the images quality is enough for observation of dislocations moving at about 1 mm/sec.

Next, some applications of the imaging system will be described.

Measurement of dislocation mobilities

Dislocation mobilities for various crystals have been measured by various techniques such as etching [51], X-ray topography [52]. Dislocation velocities were obtained by the comparison of dislocation configurations observed before and after the application of a stress pulse. Transmission electron microscopy has been the only technique to observe moving dislocations in thin foils. Recently, high-voltage electron microscopes have been used for observing dislocation behaviour similar to that in bulky states. For example, Si crystals as thick as 9 μm can be observed by a 1 MeV electron microscope [53]. To measure dislocation mobilities precisely, the specimen seems to be still too thin to apply stresses uniformly. Since highly perfect Si crystals have been obtained easily, many studies on dislocation motion were carried out by the etching method, X-ray topography [52] and electron microscopy [54]. However, any <u>in situ</u> observations of moving dislocations in Si crystals had not been reported, and the present imaging system was applied to the measurement using the specimen furnace shown schematically in Fig. 18. A Si specimen crystal was placed between two carbon - plate heaters and was observed through the carbon heaters by the TV camera. The observed velocity of the dislocations results in a very large mobility, compared to those obtained from the mean velocities in the duration of a stress pulse. This result may be due to non-uniform stress distribution in the crystal. Recently, Sumino and Harada [55] measured dislocation mobilities with caution for stress uniformity using the same imaging system. Their result was similar to those obtained by many investigators using stress pulses.

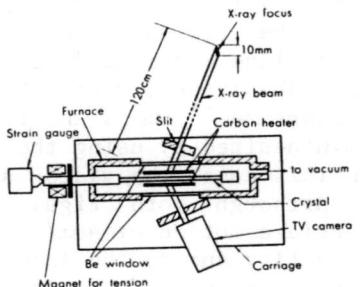

Figure 18. Schematic drawing of the furnace for the X-ray observation of moving dislocations.

Dislocation generation

Dislocation generation in dislocation-free crystals occurs under large stresses, and generated dislocations move very fast. Such dislocations are usually accompanied by long range strains and sometimes their images are buried under the strong contrast due to the strains. For example, cross-slip of dislocations takes place under relatively large stresses, and we can see only images of long-range strains by the imaging system. In order to observe the dislocation configuration, we have to quench the crystal from the processes and observe the dislocations by the usual photographic method. The topographs in Fig. 19 were obtained in this way.

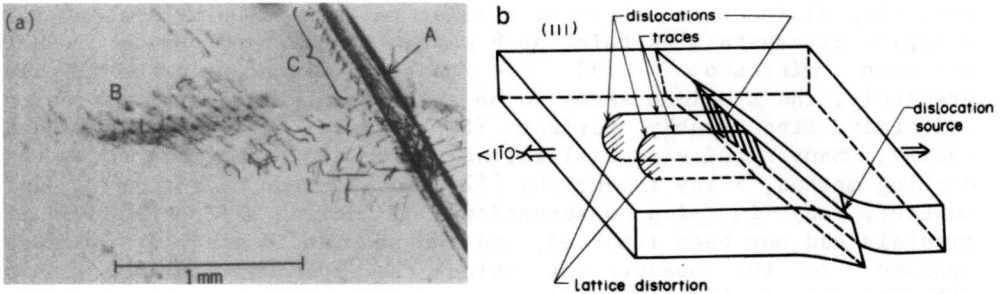

Figure 19. Dislocations introduced into a Si crystal. Individual moving dislocations were not observed owing to the elastic strains accompanied with them. The photographic topograph in (a) was taken after quenching the crystal during the dislocation motion. The cross slip as shown in (b) is seen.

The TV observation shows that the dislocations are introduced as shown schematically in Fig. 19(b). First a slip band marked 'A' is introduced and then dislocations ('B') moving towards the left-hand side are formed by a cross slip. In the edge of the slip band ('C'), dislocations just leaving from the slip band are seen. In such cases, the imaging system is an important tool as a monitor to catch a phenomenon.

Observation of growing crystals

Electron microscopy provides very high resolution and until recently has been the only technique used to observe defect structures during the growth process. For electron microscopy, however, the following disadvantages have been pointed out [56].

(1) The extreme thinness of specimens used may result in an unusual situation different from any common macroscopic growth process.

(2) Its application is limited to slow growth rates (nearly equilibrium states) because of the large magnification, and to the specimens with sufficiently low vapour pressure.

If *in situ* X-ray observation is possible, the growth processes fairly similar to common macroscopic ones can be investigated owing to the high penetrating power of X-rays and low-magnification observation.

As an example, Si crystals melting and growing were observed by the imaging system [57]. A dislocation-free crystal with a thickness of 0.2 ~ 0.3 mm placed between two carbon heaters on a furnace similar to that in Fig. 18, and in the middle part of the crystal was melted in an argon gas flow. To observe growth processes, the floating zone method was applied by moving both the crystal and TV camera.

Melting process

Different melting behaviour was observed between dislocated and dislocation-free crystals. Dislocated crystals melt homogeneously from their surfaces as seen from the direct-view images in Fig. 20. The black parts are due to many dislocations which were generated by thermal stresses during temperature increase to the melting point ($1412^\circ C$). Figs. 20(a) and (b) were obtained before the melting began. The central region with the highest temperature becomes of a low dislocation density, as seen in Figs. 20(b) and (c). The continuous observation showed that the dislocations moved rapidly away from the central region. In Figs. 20(c) and (d), the surface regions were melted, and melting through caused a white region on the image Fig. 20(e) , because the melt does not diffract the X-rays. Some round fringes appear in Figs. 20(c) and (e). They are 'Pendellosung fringes' (equal-thickness fringes) due to wedge-shaped interfaces. This observation shows that the crystal becomes highly perfect just before melting.

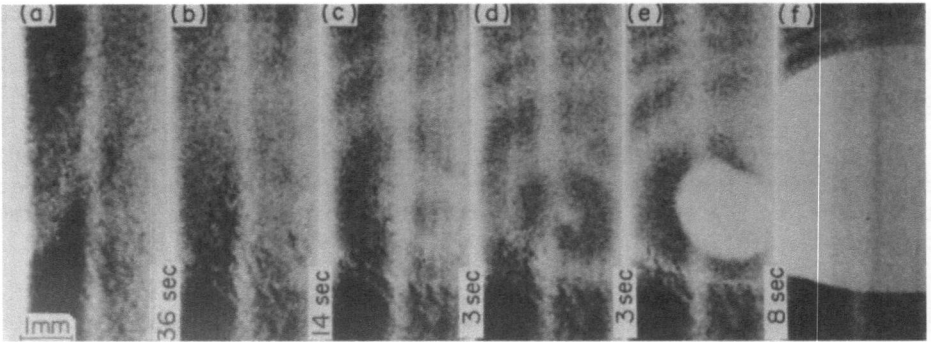

Figure 20. Direct-view images showing the melting sequence of a dislocated crystal. 220 reflection. Each image consists of two bands due to the MoKα_1 and Kα_2; their intensity difference was adjusted in printing from the movie film, in order to make the Kα_2 image useful. Time intervals are indicated between the photographs.

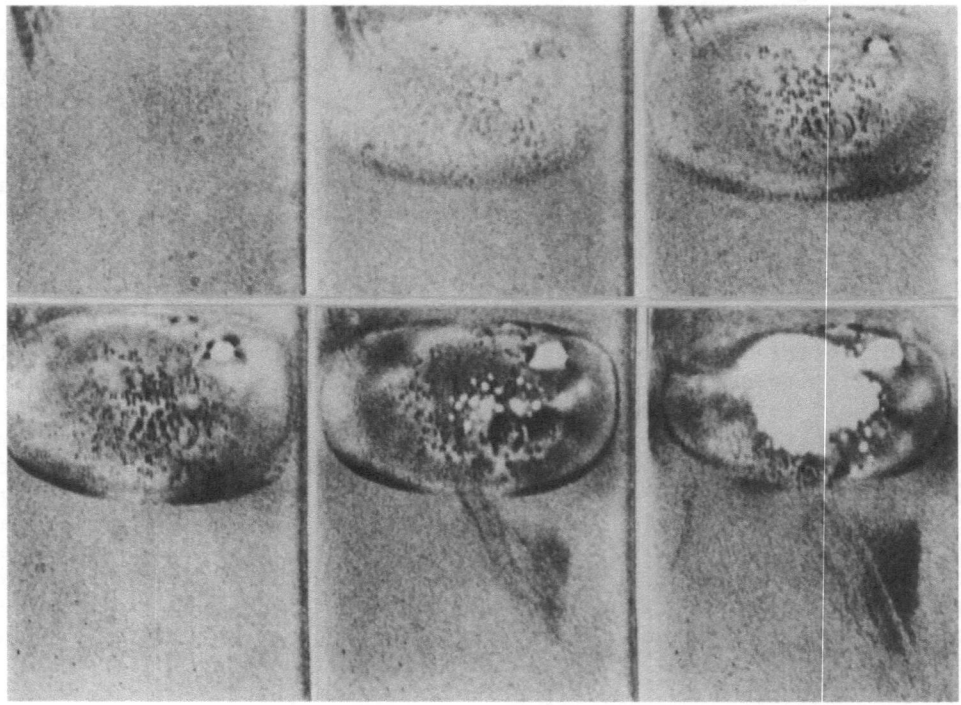

Figure 21. Synthesized images showing the melting sequence of a dislocation-free crystal, 220 reflection, 10 sec intervals.

On the other hand, droplets of the liquid are formed inside dislocation-free crystals, simultaneously with melting from their surfaces. Fig. 21(a) shows the state just before the melting began from the surfaces. In Figs. 21 (b) - (e), there is an ellipse-shaped region which is covered with the melt. There are many black spots. In Fig 21 (e), the crystal was very thin at the centre of the ellipse, and many white holes appear. In Fig. 21 (f), a large hole is seen due to perforating with the melt. Comparison of Figs 21(d) and (e) shows that melting proceeds from the black spots. They were found to be due to droplets of the liquid silicon inside the crystal. Many dislocations were generated from the region and propagated downwards, as shown in Figs. 21(e) and (f). Such dislocations were almost always generated just before the region was melted completely.

It was found that microdefects are formed by solidification of the droplets and grow larger in the cooling process so that they can be detected by topography. Although the above observation was made for thin crystals, it can be expected that such droplets, though small, are formed in the growth process of dislocation-free bulk crystals under the condition of periodic remelt or temperature fluctuations. The droplets are considered to be an origin of microdefects 'swirls' [58].

Growth process

An example of the growth is shown in Figs. 22(a) - (f) which are direct-view images. The lower edges of each image is the interface between the crystal and melt, and the crystal is growing downwards. Fig. 22(a) is the state of the seed crystal just before the growth starts. The crystal is seen black due to many dislocations generated during the temperature increase. In Fig. 22(b), a newly grown region is seen in where many hairpin-shaped dislocations are inherited from the seed crystal, as shown schematically in Fig. 22(g). These dislocations follow the interface motion, touching the interface with their points, Figs. 22(b) and (c). In Fig. 22(e), however, the majority of the dislocations are left behind the interface. They were found to have the common type of Burgers vectors 1/2 $\langle 110 \rangle$. In Figs. 22(e) and (f), two dislocations intersect nearly perpendicular to the interface. They were found to be of composite type; each of them consists of three dislocations and is stable enough to intersect with the interface. As seen from the above observation, common dislocations cannot intersect with the growth interface, i.e. it was confirmed by some other observations that a dislocation-free region first grows and then dislocations are driven in the newly grown region by the thermal gradients.

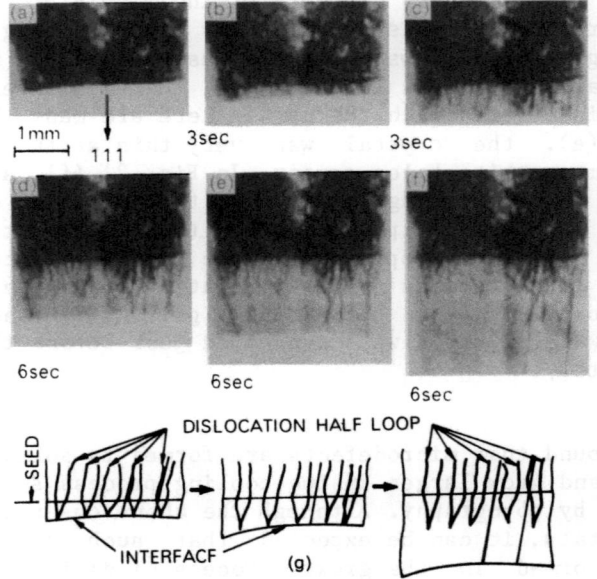

Figure 22. Direct-view images showing a growth process of a Si crystal.

It may be concluded from the above examples that the imaging system is useful for <u>in situ</u> observation of defect structures during deformation and phase transformations.

15.3.4 Future trends

Possible developments on imaging systems will be discussed by taking into account three characteristics; resolution, integration time for observation, and image quality (SNR).

<u>Resolution</u>

Resolution is always a compromise between the two opposing factors.

(1) Suppression of lateral diffusion of the image signal by making X-ray sensing layers thin.

(2) Obtaining the layer thickness required for high enough detection efficiency η_o in equation (9).

Especially for video display of large areas with a high resolution, a wide bandwidth of the video amplifier is required so that the amplifier noise is increased. Therefore, the conversion efficiency $\eta_1 \ldots \eta_s$ (equation (7)) from X-rays to signal may also be required to be high enough to overcome the amplifier noise. Hartmann et al [59] are opening a promising technique with thin rare-earth phosphor screens which have high enough detection and emission efficiencies. Another trend is the technique of isolation between small picture elements. For silicon vidicon camera tubes, such isolation seems to be possible even for thick photodiode targets [60]. Also Smith [61] tried to secure a high absorption efficiency by the glass capillary array plates which the phosphor was fused into. It is desired that each picture element is made of thick transparent single crystal which has a high emission efficiency. In general, resolution is limited by the interval of the isolated elements or grain size of phosphor screens. Amorphous photoconductive layers may give a very high resolution. For example, the camera tube with an amorphous alloy Se-As-Te has been used for TV broadcasting. The layer thickness of a few microns is enough for visible light, and a resolution of 6μm was attained by optimizing sizes of the electron beam which scan the photo conductive layer. For X-ray topography, a preliminary experiment with a 10 μm thick Se-As on a Be window showed an absorption efficiency of 31% for $MoK\alpha_1$, radiation and a 1/4 times smaller sensitivity than that of the PbO camera tube. Further improvement is in progress.

X-ray source

In the $K\alpha_1$ image in Fig. 17 (Si 220 ref. X-ray power 60kV 0.5A), the number of photons received per picture element (30 x 30) per frame (1/30 sec) is only 12 for the perfect region, .i.e., the photon noise is important for the image quality. Especially by improving the resolution (smaller ϵ), the SNR decreases as seen from equation (9), and we need more powerful X-ray generators. As shown in Fig. 13, an X-ray generator has been tried in which the electron gun and rotating target are separated by placing a focussing electron lens between them thus making the operation at a high power stable and the focus size adjustable. Another possibility in the laboratory technique may be a laser-produced plasma pulsed X-ray source, which already enables nanosecond X-ray diffraction from biological samples with 4.45 A radiation, although the pulse intervals are too long for topography at present [62].

Improvement of image quality

In 3.1, the basic digital technique for improvement by image integration was described. We can also improve images by a digital two-dimensional filter. For example, images of Pendellosung and Moire fringes are improved by averaging locally parallel to the fringes.

By future improvements as described above, it is desired to materialize real-time observation by more informative means such as section topography.

15.4 Conclusion

Both interpretation of topographic images and techniques for defect analysis have been well established. Recent developments in video display techniques are overcoming the disadvantages of the photographic technique, in particular long exposure times and the tedious procedure of photographic development. By fully utilizing its advantages; primarily that it is non-destructive, that thick specimens, and a large viewing area may now be used, wide applications will open from fundamental research to industrial purposes.

REFERENCES

1. A.R. Lang (1970) in Modern Diffraction and Imaging Techniques in Materials Science (ed S. Amelinckx R. Gevers, G. Remaut, J. van Landuyt, North Holland Amsterdam) p. 407
2. A. Authier (1978) in X-ray Optics (ed. H.-J. Queisser) Springer-Verlag p. 145
3. B.K. Tanner (1976) X-ray Diffraction Topography, Pergamon Press, Oxford
4. R.E. Green Jnr. (1971) Advances in X-ray Analysis $\underline{14}$, 311
5. W. Hartmann in Ref. 2 p. 191
6. N. Kato (1974) In X-ray Diffraction (ed. L.V. Azaroff R. Kaplov, N. Kato, R.J. Weiss, A.J.C. Wilson, R.A. Young) McGraw-Hill, New York p.222
7. M. Yoshimatsu and K. Kohra (1960) J.Phys.Soc. Japan $\underline{15}$, 1760
8. S.B. Austerman and J.B. Newkirk (1968) Advances in X-ray Analysis $\underline{10}$, 134
9. H. Barth and R. Hosemann (1958) Z. Naturforsch $\underline{13a}$, 792
10. A.L. Anderson (1965) Rev. Sci. Instr. $\underline{36}$, 1888. Also, S. Oki and K. Futagami (1969) Japan. J. Appl. Phys. $\underline{8}$, 1574
11. K. Kohra, H. Hashizume and J. Yoshimura (1970), Japan J. Appl Phys. $\underline{9}$, 1029 Also Ref 13
12. S. Hosoya (1968) Japan J. Appl. Phys. $\underline{7}$, 1
13. W.J. Boettinger, H.E. Burdette and M. Kuriyama (1976) Rev. Sci. Instr $\underline{47}$, 906
14. K. Haruta (1965) J. Appl. Phys. $\underline{36}$, 1789
15. G.H. Schwuttke, (1965) J. Appl. Phys. $\underline{36}$, 2712
16. L.J. van Mellaert and G.H. Schwuttke (1970) Phys. Stat. Sol. (a) $\underline{3}$, 687
17. J. Chikawa, I. Fujimoto and Y. Asaeda (1971), J. Appl. Phys. $\underline{42}$, 4731
18. K. Kohra and Y. Takano (1979) Japan J. Appl. Phys. $\underline{7}$, 982
19. S. Kawado and J. Aoyama (1979) Appl. Phys. Lett. $\underline{34}$, 428
20. U. Bonse and E. Kappler (1958) Z Naturforsch $\underline{13a}$, 348
21. A. Authier (1966) J. Phys. Radium, 57

22. M. Renninger (1961) Z. Naturforsch, 16, 1110
23. K. Kohra (1962) J. Phys. Soc. Japan 17, 589
24. S. Kikuta (1971) J. Phys. Soc. Japan 30, 222
25. U. Bonse and M. Hart (1965) Appl. Phys. Lett. 7, 238; (1966) Z. Phys. 189, 151
26. T. Matsushita, S. Kikuta and K. Kohra (1971) J. Phys. Soc. Japan 30, 1136
27. J. Chikawa and S. B. Austerman (1968) J. Appl. Cryst. 1, 165
28. U. Bonse and M. Hart (1965) Appl. Phys. Lett. 6, 155; (1965) 7, 99; (1965) Z. Phys. 188, 154; (1966) Z. Phys. 190, 455; (1966) Z. Phys. 194; (1968) Acta Cryst. A24, 290
29. A.R. Lang (1968) Nature 220, 652 J. Bradler and A.R. Lang (1968) Acta Cryst. A24, 246
30. J. Chikawa (1965) Appl. Phys. Lett 7, 193; (1967) Proceeding of International Conference of Crystal Growth 817
31. J. Chikawa (1965) J. Appl. Phys. 36, 3496
32. H. Ishida, N. Miyamoto and K. Kohra (1976) J. Appl. Cryst. 9, 240
33. H. Ishida (1980) J. Appl. Cryst. 13, 58.
34. M. Hart (1966) Z. Phys. 189, 269
35. J. Chikawa, Y. Asaeda and I. Fujimoto (1970) J. Appl. Phy. 41, 1922
36. M. Renninger (1976) J. Appl. Cryst. 9, 178
37. For examples, see J.R. Patel (1975) J. Appl. Cryst. 8, 186, K. Lal, B.P. Singh and A.R. Verma (1979) Acta Cryst. A35, 286
38. J.R. Patel and B.W. Batterman (1963) J. Appl. Phys. 34, 2716; J.R. Patel (1973) J. Appl. Phys. 44, 3903
39. R. Collela and A. Merlini (1966) Phys. Stat. Sol. 14, 81; T.O. Baldwin and J.E. Thomas (1968) J. Appl. Phys. 39, 4391
40. T.O. Baldwin, F.A. Sherrill and F.W. Young Jnr. (1968) J. Appl. Phys. 39, 1541; F.W.Young Jnr., T.O. Baldwin and P.H. Dederichs (1970) in Vacancies and Interstitials in Metals (Eds. A. Seeger et al.) North Holland, Amsterdam p. 619; B.C. Larson and F.W. Young Jnr. (1971), Phys. Rev. B4, 1709
41. L.I. Datsenko, A.N. Guerrv, and M.I. Strarchik (1975), Phys. Stat. Sol. (a) 32, 549
42. A.J.R. de Kock (1973) Philips Res. Rept. Suppl. No. 1
43. A. Arcese (1964), Appl. Opt. 3, 435
44. A. Rose (1948) J. Opt. Soc. Am. 38, 196
45. J. Chikawa and I. Fujimoto (1974) NHK Technical Monograph No. 23 (available by request)
46. For example, see B. Shackel and G.R. Watson (1968) Photography for the Scientist, (ed. C.E. Engel) Academic Press, Inc. London, p 553
47. For example, see G.D. Heynes (1977) SMPTE Journal 86, 6.
48. R. Nishida and S. Okamoto (1966) Rept. Res. Inst. Electron. Shizuoka Univ. 1, 21, 185
49. R.H. McMann, S. Kreinik, J.K. Moore, A. Kaiser and J. Rossi (1978) SMPTE Journal 87, 129

50. J. Rossi (1978) SMPTE Journal 87, 134
51. W.G. Johnston and J.J. Gilman (1959) J. Appl. Phys. 30, 129
52. For example, see A. George, C. Escaravage, G. Champier and W. Schroter (1972) Phys. Stat. Sol. (b) 53, 483
53. G. Thomas, (1968) Phil. Mag. 17, 1097
54. For example, see V.G. Eremenko and V.I. Nikitenko, (1972) Phys. Stat. Sol. (a) 14, 317
55. K. Sumino and J. Harada to be published
56. M.E. Glicksman (1970) in Solidification, (eds. J.J. Hughel, and G.F. Bolling) American Society for Metals, Metals Park, Ohio p.155
57. J. Chikawa and S. Shirai, J. Crystal Growth 39, 328
58. J. Chikawa and S. Shirai, (1978) Proc. 10th Conf. Solid State Devices, Tokyo, Japan. J. Appl. Phys. 18, (1979) Suppl. 1. 153
59. W. Hartmann, G. Markewitz, U. Rettenmaier and H.J. Queisser (1975) Appl. Phys. Lett. 27, 308
60. For example, see W.J. Beyen and O.B. Cecil, (1973)
61. R.W. Smith RCA Review 36 (1975) 632 in Semiconductor Silicon (1973) (eds. H.R. Huff and RR. Burgess) Electrochemical Society, Princeton, p.1.

CHAPTER 16

WHITE BEAM SYNCHROTRON RADIATION TOPOGRAPHY

J. MILTAT

16.1 Introduction

White beam X-ray topography (Laue Technique) is probably the simplest available X-ray imaging technique. Pioneering work by Guinier and Tennevin [1] and Schultz [2] was however followed with little applications due to the lack of intense, low divergence white beam X-ray sources. Synchrotron sources possess both of these qualities. They therefore appear as ideal sources for such experiments.

Source characteristics [3] may be schematically listed as follows.

1. Synchrotron sources deliver a continuous spectrum with emission power $P(\lambda)$, where $P(\lambda)$ is here understood as the photon flux reaching the sample. It is useful to consider that ring geometry, operating energy, beam line absorption ... define a spectral range $\lambda_I \lambda_{II}$ within which the largest part of the beam intensity is concentrated.

2. Beam divergence is small, being typically of the order of 10^{-5} to $3\ 10^{-4}$ on existing machines [4]

3. The X-ray beam is highly polarized in the plane of the orbit (the horizontal plane).

4. The beam is characterized by its time structure. It consists of pulses with a high frequency repetition rate.

16.2 Experimental set-up

Any crystalline material imbedded in the beam will diffract along given directions provided:

1. the reflection is not structure factor forbidden

2. the angle θ between lattice planes and impinging beam corresponds through Bragg's law

$$2d_{hk\ell} \sin\theta = n\lambda \qquad (1)$$

to wavelengths comprised between λ_I and λ_{II}. Each Laue spot may consist of the superposition of several harmonic reflections (see equ. 1). In the transmission geometry (Fig. 1a), diffraction spots due to reflection on lattice planes belonging to a crystallographic zone lie on ellipses, one major axis of which passes through the centre of the diagram, whereas, in the reflection geometry (Fig. 1b) they lie on hyperbole branches [5,6].

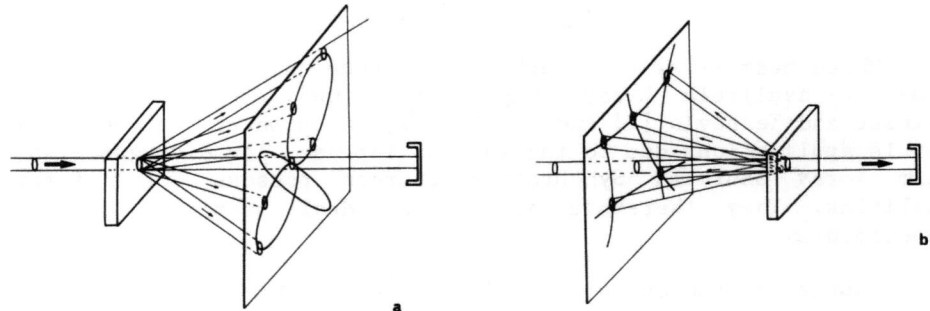

Figure 1. Experimental geometry for the Laue experiment a) transmission geometry b) reflection geometry

Assuming the source to be a point source, it appears that different parts of the same single crystal reflect slightly different wavelengths. The effect is however negligible since specimen to source distances are usually large with respect to sample dimensions (typically 20 m or more versus 1 cm).

Fig. 2 shows Laue spots from a single crystal of olivine (Mg, Fe)$_2$ SiO$_4$ orthorhombic; a = 4,8A, b= 10.5 A, c= 6.1 A) in the transmission geometry. The crystal was just inserted in the beam without any particular precaution. It suggests the following remarks.

1. Signal to noise ratio must be as good as possible; in other words, background noise from fluorescence, diffusion .... must be strongly attenuated.

2. Laue spots must not superimpose, as in the case of Fig. 2. This condition imposes, according to crystal parameters and symmetry, a minimum distance between sample and detector, possibly at the expense of resolution.

3. Only those spots with comparable recorded intensities may be handled simultaneously if similar image qualities are sought for.

4. The geometry of the experiment may have to be carefully worked out if severe geometrical image distortion is to be avoided.

Figure 2. Diffraction spots from an olivine single crystal in the transmission geometry. Vermicular lines visible in some of the spots are dislocations.

An elegant solution to the fulfilment of condition 1. may be found in ref. [7]. Condition 4. is specific to the particular experiment involved. Conditions 2. and 3. have a general character. They are discussed in the following section.

16.3 The perfect crystal: general properties of Laue topographs

16.3.1 Geometrical resolution

Let us consider the source to be emitting X-rays over an area of diameter S and let us call D and d the distances from specimen to source and detector, respectively (Fig. 3) It is a property of synchrotron sources that each point of the source emits X-rays in a cone of apex angle 2ϕ, which is a function of wavelength. Therefore, the geometrical resolution is given by:

$$R = S\, d/D \text{ if } 2\phi D > S$$
$$R = 2\phi d \text{ if } 2\phi D < S \qquad (2)$$

provided the detector is perpendicular to the diffracted beam (Fig. 3a). Equation (2) therefore provides optimum figures for geometrical resolution. It is a direct consequence of the broad spectral range of the beam.

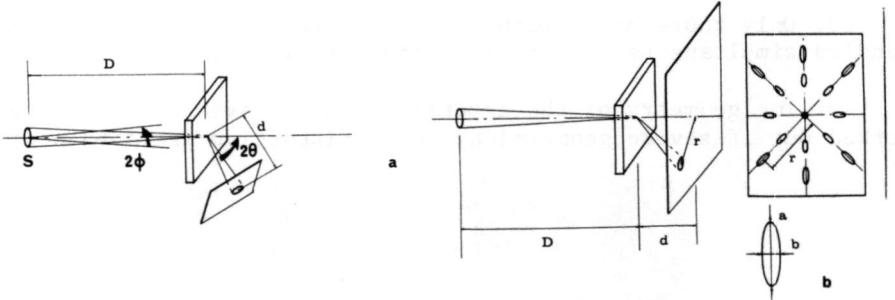

Figure 3. a) Definition of experimental parameters contributing to geometrical resolution. b) Behaviour of geometrical resolution as a function of 2θ when the detector is perpendicular to the impinging beam. $r = d\,\mathrm{tg}\,2\theta$; $a = S\,\dfrac{d}{D}\,\dfrac{1}{\cos 2\theta}$; $b = S\,\dfrac{d}{D}$

For an arbitrary orientation of the detector, geometrical resolution depends on Bragg angle and orientation of the detector. It is schematically represented in Fig. 3b when the detector is perpendicular to the incident beam. Resolution loss is only important at high Bragg angles. High resolution experiments require geometrical resolutions of the order of a few microns (say, less than 3). With $D \sim 20$ to 30 m, $d = 5$ cm for experimental comfort, S should be less than ~ 1.5 mm. This condition is not realized in machines such as LURE-DCI or DESY. Therefore, as in the laboratory, care should be taken when working with such machines, to place the detector as close to the sample as possible. Further, source areas are generally ellipses with their largest dimension in the horizontal plane. Geometrical resolution is therefore expected to be a complex function of diffracting plane and detector orientations.

16.3.2 Integrated diffracted intensity

In a white beam experiment, θ is fixed by positioning the crystal in the beam. As already stated, due to the broad spectral range of the beam, several harmonic reflections may superpose in a given Laue spot. This is reflected in λ, θ (or Du Mond) diagram as a set of curves (Fig. 4a).

It is already known from the previous paragraph that each point of the sample reflects X-rays within the angular range

$$\Delta\theta_S = S'_i/D \qquad (3)$$

where S'_i is the effective source dimension in the incidence plane (which strictly speaking is a function of wavelength). Further, for each harmonics ($\lambda, \lambda', \lambda''\ldots$ in Fig. 4a), the crystal is characterized by an angular width of reflection $\delta(\lambda)$ [8].

$$\delta(\lambda) = \frac{2|C|}{\sin 2\theta} \sqrt{\frac{\gamma_0}{|\gamma_h|}} F_h \frac{e^2}{mc^2} \lambda^2 \frac{1}{\pi V} \qquad (4)$$

where C is the polarization factor, V is the unit cell volume, F_h the structure factor $\gamma_0 = \cos\phi_0$; $\gamma_h = \cos\phi_h$; $\phi_0 = (\underline{n}, \underline{s}_0)$; $\phi_h = (\underline{n}, \underline{s}_h)$ $\underline{n}$ is the normal to the entrance surface, $\underline{s}_0$, $\underline{s}_h$ the incident and reflected directions, respectively.

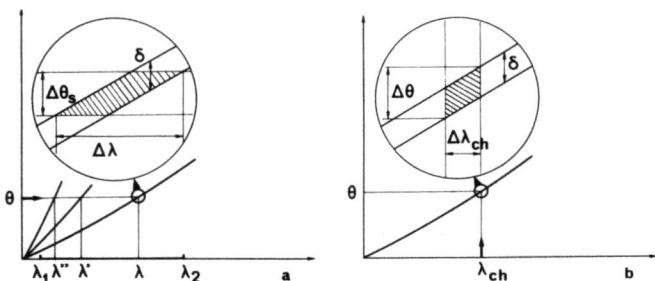

Figure 4. a) Du Mond diagram for white beam X-ray topography b) Du Mond diagram for characteristic radiation X-ray topography.

To the angles $\Delta\theta_s$ and δ corresponds a wavelength spread which may be deduced from the Du Mond diagram (Fig. 4a). The Du Mond diagram pertaining to a characteristic line topography experiment is shown in Fig. 4b. Here on the contrary, wavelength spread is fixed by characteristic line width $\Delta\lambda_c$. The angular range $\Delta\theta$ may be deduced from the Du Mond diagram.

Due to wavelength spread, the diffracted intensity for a given harmonic is an integrated intensity. It is proportional to source power $P(\lambda)$ and to the hatched area in Fig. 4a ($\sim\delta\Delta\lambda$). Since derivation of Bragg's law yields:

$$\Delta\lambda = \lambda \cot g\theta \Delta\theta_s \qquad (5)$$

the integrated diffracted intensity is proportional to:

$$I_h \alpha \, C \, \lambda^3 \, P(\lambda) \, F_h \, \frac{1}{\sin^2\theta} \, \frac{S_i'}{D} \sqrt{\frac{\gamma_0}{|\gamma_h|}} \qquad (6)$$

This relation has been derived in a close form by Tuomi et al [9] as well as by Hart [7]. It enables one to compute the contributions of various harmonics to a Laue image as well as to compare integrated diffracted intensities of different Laue images, provided the following corrections are made:

i) in the transmission case, absorption may be taken into account by multiplying (6) by $\exp - \mu(\lambda)t$ where $\mu(\lambda)$ is the true absorption coefficient, including Bormann transmission [10]

ii) since, one is ultimately interested in comparing recorded intensities, one should also take into account;
- the absorption by air which affects mostly long wavelengths [7, 9].
- the detector response as a function of wavelength [7]. This is particularly important for photographic emulsions since AgBr has two absorptions edges at 0.484 A and 0.920 A in the spectral range usually employed for topography [11,12].

Expression (6) suggests the following remarks: a) Due to the $\lambda^3 P(\lambda)$ term, a strong decrease in I_h at short wavelengths is expected. This means that Laue spots will be composed of a moderate number of harmonics with appreciable intensity [7]
b) According to reflection geometry, the $C\, S_i^!/D$ term may take quite different values, especially if the source is markedly elliptic. If the plane of incidence is vertical, $C = 1$ but the effective source size is small. On the contrary, if the plane of incidence is horizontal, $C = \cos 2\theta$, but the effective source size is large, which may in fact more than compensate polarization losses, at the expense, however of resolution.
c) Finally, although highly improbably, multiple scattering may occur (several Laue spots extract exactly the same wavelengths out of the beam). Equation (6) obviously ceases to be valid in such a case.

16.3.3 Effects of polarization properties on Pendellosung fringes

If a non polarized X-ray source is used, Pendellosung fringes, such as observed in wedge-shaped crystals exhibits a periodical fading due to the incoherent superposition of the σ and π polarization components [13 - 15], each polarization state giving rise to a Pendellosung length Λ_σ and Λ_π.

If a plane polarized X-ray source is ultilized, it has been shown that no fading occurs if the polarization is either normal or parallel to the incidence plane [13]. However if the polarization is neither normal or parallel to the incidence, Skalicky and Malgrange [16] have shown that a periodical fading may still be observed. They interpreted the result as a coherent splitting of the incident wave into σ and π components accompanied by a modification of the wavefield polarization state during propagation.

These results have been confirmed directly by fringe observations in a wedge-shaped silicon crystal by Sauvage, Petroff and Skalicky [17] in the case of reflections with vanishingly small harmonic content. Fig. 5a shows that no periodical fading is observed when the polarization is normal to the plane of incidence. The result may be interpreted as a confirmation of the highly defined polarization state of the beam. With polarization at 45° to the plane of incidence, periodical fading is observed (Fig. 5b).

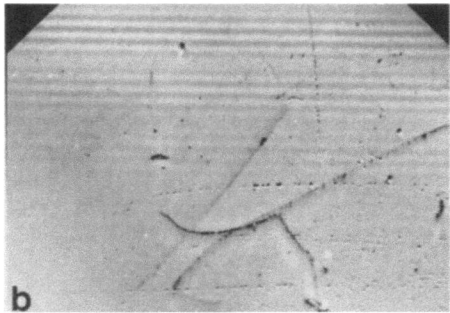
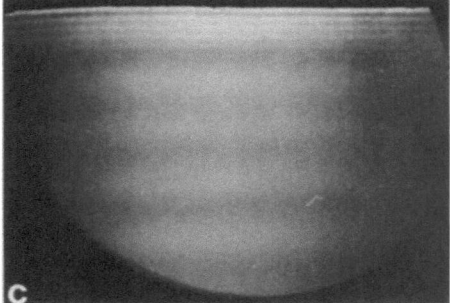

Figure 5. Fringes in wedge-shaped silicon single crystals: a) polarization normal to the plane of incidence b) polarization at 45° to the plane of incidence 220 reflection $\lambda = 0.9A$. Vanishingly small harmonic contamination [17] c) Complex behaviour of fringes due to the presence of harmonics. See text for details [20]

Finally, when the recorded harmonic content in a Laue spot is high, fringes undergo major modifications. In Fig. 5c, from the work of Tuomi et al [18], rapid oscillations close to the specimen edge were interpreted as due to interferences between wavefields of the fundamental (111) reflection ($\lambda = 2.14$ A) whereas weaker broader fringes were due to the superposition of fringes pertaining to the (333) ($\lambda = 0.71A$) and (444) ($\lambda = 0.54A$) reflections. Such behaviour which is specific of white beam X-ray topography, should be borne in mind when dealing for instance with planar defect contrast.

16.4 The imperfect crystal

16.4.1 Sensitivity to entrance surface misorientations

Let us consider two regions with identical lattice parameters, the reflecting planes in region 2 being rotated with respect to those in region 1 by the angle ω around the normal to the incidence plane (Fig. 6a). In a white beam experiment, regions 1 and 2 will select the wavelengths λ_1 and λ_2, respectively, and possibly associated harmonics. Regions 1 and 2 will yield diffracted intensities provided λ_1 and λ_2 lie within the $\lambda_I \lambda_{II}$ wavelength range (Fig. 4a). Relative integrated intensities from regions 1 and 2 depend on λ and eventually, in the case of large ω's on harmonic content.

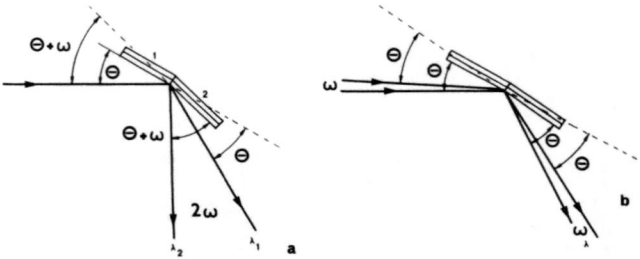

Figure 6. Influence of entrance surface misorientations on topographs: a) white beam X-ray topography b) characteristic line X-ray topography.

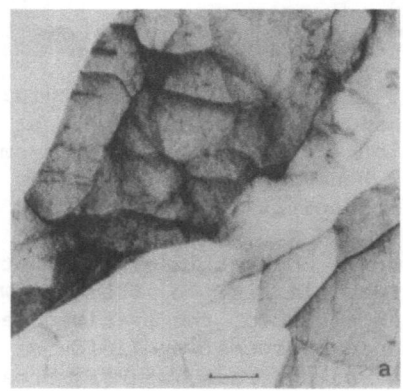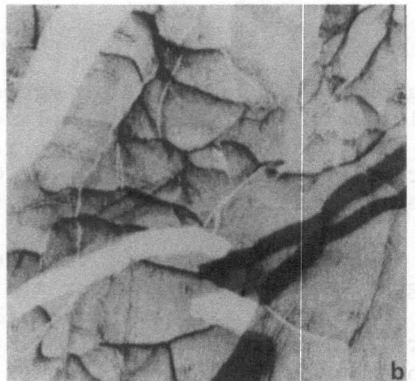

Figure 7. Subgrains in (001) LiF single crystals. Reflection geometry. a) 004 Bragg reflection of $CuK\alpha_1$ radiation b) 002 Bragg reflexion, mainly, at $\lambda = 2,845\text{A}$ Scale Mark : 500 μm. [7]

WHITE BEAM SYNCHROTRON RADIATION TOPOGRAPHY

The corresponding situation in a conventional Lang topography experiment is illustrated in Fig. 6b Here, regions 1 and 2 diffract simultaneously only if ω is smaller than the beam divergence. Beam divergence is typically of the order of $5 \ 10^{-4}$ Rd, which limits observable disorientations to rotations of about ±50 sec. No such limitation is imposed to white beam X-ray topography since the $\lambda_I \lambda_{II}$ range may be as large as 2A. White beam X-ray topography appears therefore as a very useful tool to investigate crystals with, for instance subgrain boundaries, such as emphasized in Fig. 7 (reflection geometry), or Fig. 12 (transmission geometry) or, even, polycrystals [19,20]. It may be further noticed that:

1) the angle between exit beams is 2ω in the case of white beam radiation whereas it is ω for characteristic radiation.

2) according to the sign of ω, images of regions 1 and 2 may either be separated by a gap (white contrast) or overlap (black contrast). Along the trace of the incidence plane, the width of the gap or of the overlap stripe is for white beam radiation:

$$W = 2\omega.d \qquad (8)$$

Mapping of lattice rotations from one subgrain to the next may therefore easily be deduced from experimental observations provided at least three independent reflections are used.

16.4.2 Defect contrast

Very little attention has, up to now, been paid to the study of defect contrast in white beam X-ray topographs. This section will therefore be reduced to a schematic discussion of dislocation direct images and to a few general comments. Let us, in a first step, neglect harmonic contamination.

16.4.2.1 Dislocation direct images

In the case of conventional Lang X-ray topography, it is supposed [21], that misoriented regions around the dislocation (Fig. 8a) are able to reflect rays which do not participate in the diffraction from the perfect crystal. This is possible because beam divergence is much larger than the angular reflection width of the perfect crystal. Those rays, while travelling in the perfect crystal, do not suffer primary extinction, but only normal photoelectric absorption. Direct images will therefore be strong only if absorption is low ($\mu d < 1$).

The effective misorientations around a defect may be defined as

$$\delta(\Delta\theta) = \frac{1}{k\sin 2\theta} \frac{\partial}{\partial s_h} (\vec{g}.\vec{u}) \qquad (9)$$

[21]; where $k = 1/\lambda$, <u>s</u> is the reflected beam direction, <u>g</u> the reflection vector, u the displacement vector.

It was shown [22,24] that direct image width (in Fig. 8a) is roughly proportional to the width of the region where

$$\delta(\Delta\theta) > \alpha\delta \text{ with } 1 < \alpha < 2 \qquad (10)$$

Further, Chikawa [25] has shown that opposite Burgers vector dislocations yield bimodal or unimodel profile images as sketched in Fig. 8b. Turning to a white beam experiment, let us first assume that the beam is divergence-less (Fig. 8d) (source at infinity). It is easy to conceive that misoriented regions are able to pump out of the beam wavelengths which are not participating in the diffraction from the perfect crystal. Corresponding rays do not suffer primary extinction while travelling in the perfect crystal.

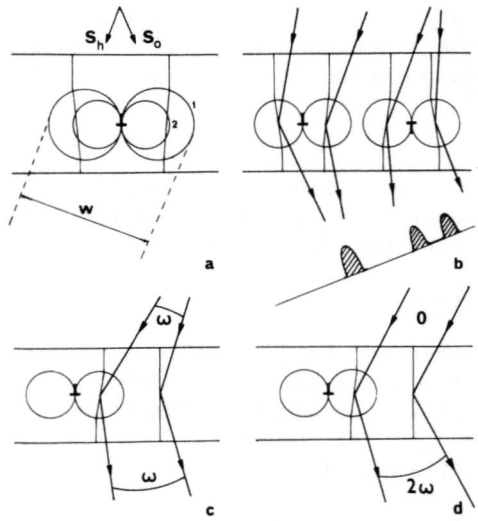

Figure 8. Dislocation direct image formation: a) regions (schematic) around an edge dislocation where the effective misorientation is greater than δ (curve 1) or 2δ (curve 2). b) Influence of the sign of the Burgers vector on dislocation image profile. c) Ray geometry in a characteristic line topographic experiment d) Ray geometry in a zero divergence white beam topographic experiment.

It appears that zero divergence, or say very small divergence ($\leq \delta$) in a white beam experiment is equivalent to a zero, or very small wavelength spread in a monochromatic experiment. Such a situation was approached at the late NINA synchrotron source, since source dimensions were small (0.5x0.5 mm^2) and specimen to source distance large (47 m). Fig. 9 shows dislocation direct images in a silicon single crystal obtained at NINA by Tanner et al. [26]. As distance from specimen to detector is increased, dislocations marked by arrows exhibit marked bimodal profile images. There is very little qualitative difference, indeed, between those images and corresponding images, obtained in a characteristic line experiment [26].

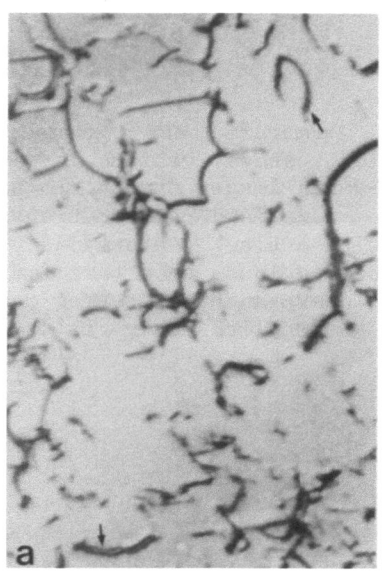 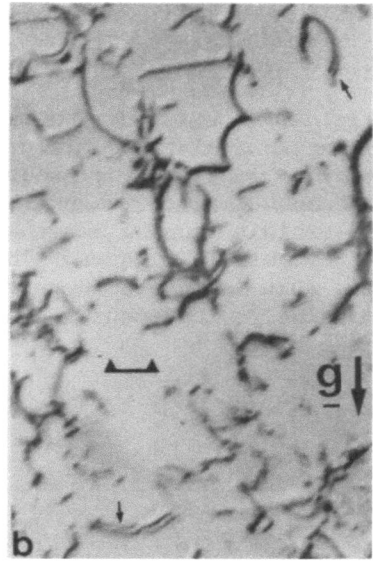

Figure 9. Direct images of dislocations in a silicon crystal recorded with white beam X-ray topography; $2\bar{2}0$ reflection at $\lambda = 0.95 A$. Harmonic contamination of about 5%. Transmission geometry. Specimen to plate distance: a) 50 mm b) 100 mm Scale Mark: 100 μm [26]

It is interesting to notice that, as in the case of entrance surface misorientations, the angle between beams reflected in the perfect crystal and the misoriented regions around the dislocation is $2\delta(\Delta\theta)$ in a white beam experiment (Fig. 8d) whereas it is only $\delta(\Delta\theta)$ in a monochromatic experiment (Fig. 8c) [26].

In intermediate situations, i.e. when beam divergence is greater than, say, δ, direct image formation may be accounted for by both of the two mechanisms described earlier. However, misoriented regions around the dislocation line, will be able to pump, out of the beam, wavelengths which do not participate in the reflection from the perfect crystal only if

$$\delta(\Delta\theta) > S_i'/D \qquad (11)$$

Assuming $|\underline{b}| = 3A$, it is easy to show from equation 8 that $\delta(\Delta\theta)$ is only of the order of 8.10^{-5} Rd at 1 μm from the dislocation core. Therefore, for sources such as LURE-DCI for which $S_i/D \sim 10^{-4}$ Rd, only regions very close to the core may fulfill inequality (10), neglecting the fact that the model is probably inadequate for such small regions very close to the core. It would therefore appear that, for such sources, direct images arise mostly from beam divergence. Source characteristics are essential parameters in defining the contrast mechanism.

16.4.2.2 Other type of contrast

Fig. 10. from the work of Petroff and Sauvage on heterojunctions misfit dislocations [10], shows various types of dislocation images. One sees black-white images from dislocations parallel and close to the exit surface. Inclined dislocations are seen to exhibit fanning image widths with a black (intermediary) and white (dynamical) images. These observations show that familiar concepts such as wavefield curvature, interbranch scattering remain, unsurprisingly, fundamental concepts. No detailed study is however yet available.

Figure 10. Epitaxial and substrate dislocations in a $(Ga_{1-x}A\ell_xAs_{1-y}P_y)$/GaAs heterojunction. 422 reflection at $\lambda=1.2A$. Negligible harmonics contamination. Transmission geometry. Scale mark: 100 μm [10]

16.4.2.3 Influence of harmonics superposition

In order to discuss the influence of harmonics superposition on defect contrast, it will be schematically assumed [21,23,27,28] that regions around the defect participating in direct image formation are defined by equation (9) whereas regions in which interbranch scattering occurs may be defined by

$$G = \frac{\partial}{\partial s_0} [\delta(\Delta\theta)] > \beta \frac{\delta}{\Lambda} \quad \text{with } \beta \sim 1. \tag{12}$$

where s_0 is the incident beam direction and Λ the Pendellosung periodicity. For G values less than δ/Λ, wavefield curvature is assumed to occur and appreciable contrast may be observed down to G values of the order of $10^{-2}\delta/\Lambda$. Now, both quantities δ and δ/λ are decreasing functions if λ decreases. This means that the width of various types of topographic images are expected to increase if harmonic content is appreciable. In different terms, it may be said that strain sensitivity is increased. This behaviour is illustrated in Fig. 11 showing oxide window edge images in a silicon crystal [7].

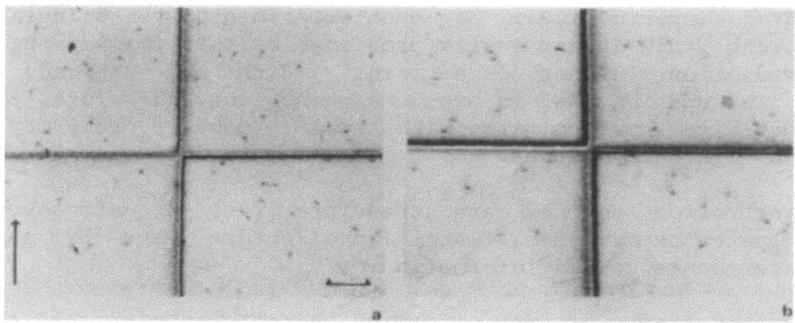

Figure 11. Contrast due to oxide windows on a silicon crystal. Reflection geometry. The arrow indicates the direction of incidence a) 333 Bragg reflection taken with CuKα radiation (λ = 1.54A) b) superposition of 62% 333 (λ = 1.478A) and 22% 444 (λ = 1.019A) Bragg reflections with synchrotron radiation. Scale Mark: 200 μm [7]

Fig. 11a is a 333 reflection recorded with CuKα characteristic radiation (λ = 1.54A) whereas Fig. 11b is a synchrotron white beam topograph in which reflection 333 (λ = 1.478A) and reflection 444 (λ = 1.019A) contribute to 62% and 22% of the recorded intensity, respectively. Increased image width is obvious. Harmonics superposition is also expected to perturb fringe patterns in planar defect images, as well as in intermediary images fringes. Finally, polarization properties of the source should be taken into account.

16.5 Applications

White beam X-ray topography has been already utilized in a number of fields of scientific activity. On top of publications of a rather technical concern [20,29 to 32], published work includes studies of recrystallization [19,33 to 35], defects associated with crystal growth [10,36,37], phase transitions [38], polytypes [39], magnetic domains in various materials [40 to 46] and plastic deformation [47]. Most studies rely on high source intensity as well as on the fact that recording several Laue spots simultaneously may allow for quick defect identification, as shown by Tuomi et al [9].

A fair review of all these studies is almost impossible in a limited space. Therefore, although their choice may appear arbitrary, four experimental studies only, will be briefly reviewed here, stressing in each case, the major inputs of the technique.

16.5.1 Recrystallization of Al polycrystals [19,33,34]

Zone refined Aluminium polycrystals (grain size 2-3 mm) strained a few percent have been recrystallised in situ, the aim of the experiment being to visualize growing grains as well as associated imperfections. In such experiments, the orientation of the growing grain is not easily predictable; this imposes the use of white radiation. Further, once grain growth has started, it is highly interesting not to stop the phenomenon. Therefore, exposure time should be short compared with grain boundary velocity. This requires high X-ray fluxes.

Synchrotron sources are therefore very adequate sources for these type of experiments. Gastaldi and Jourdan have by means of such experiments gained information on

1) the shape of growing grains whose boundaries were shown to follow low index crystallographic planes at the onset of grain growth,

2) the shape of moving boundaries,

3) the interaction of moving boundaries with obstacles and subsequent dislocation generation,

4) the evolution of growth defects upon cooling the sample.

Fig. 12 shows two (111) annealing twins (fringe contrast) which grew during the primary recrystallization of a weakly strained sample. Defects labelled 1 are steps in the twin boundary, those labelled 2, matrix dislocations. Defects labelled 3 and 4 are dislocations in the twin plane or very close to it, since they are seen in the images of crystals 1 and 2, which bound the twin plane.

Since their Burgers vector is $1/6\ [1\bar{2}1]$, they would be twinning dislocations if their position exactly in the twin plane could be ascertained. The figure shows that those dislocations move as a function of time, presumably indicating that a closer fit between crystals 1 and 2 is gradually achieved. Such experiments obviously provide numerous informations on crystal growth from a solid phase. No other technique really allows for comparable observations to be made.

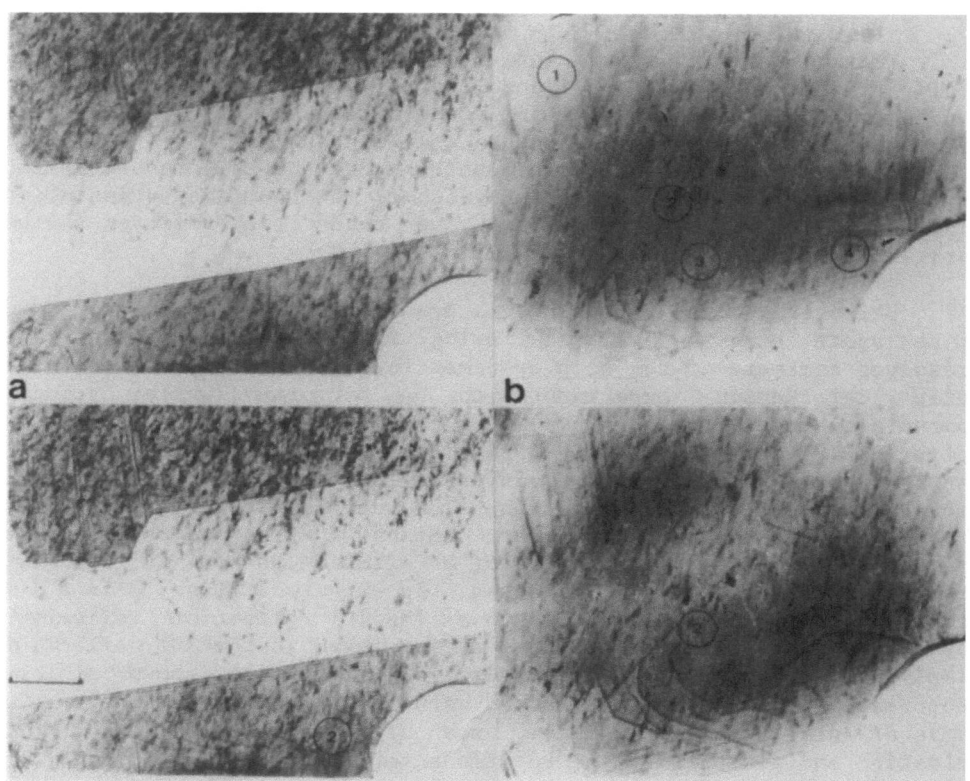

Figure 12. Annealing twins in aluminium: evolution of defects in the twin plane a) crystal I: reflection 111 mainly top: 44 m ; bottom 68 m b) crystal II: reflection 113 mainly top: 44 m ; bottom 68 m Scale mark: 1mm [19]

16.5.2 Misfit dislocations [10]

Petroff and Sauvage have studied misfit dislocations in $(Ga_{1-x}Al_xAs_{1-y}P_y)/GaAs$ epilayers. They made use of the tunability of the source to select a wavelength just above the K-absorption edge of Ga. Absorption could thus be reduced to a value comparable to that obtained with AgKα characteristic radiation. Due to the longer wavelength however, defect image width was reduced and high resolution pictures could be obtained. Besides, wafer elastic curvature did not, as in the case of a conventional experiment, limit drastically the field of view.

Fig. 10 shows one of their topographs. Burgers vector analysis could be performed and dislocation generation linked to an elastic model. The origin of misfit dislocations were related to substrate dislocations, surface or sample's edge defects.

16.5.3 Plastic deformation of Fe-Si single crystals [47]

Miltat and Bowen have deformed *in situ* Fe-Si single crystals. Their aim was to put into evidence the role of surface orientation upon the choice of slip systems in a material in which the velocity of edge dislocations is larger than that of screws. Two types of samples were therefore deformed, both having the same tensile axis, but different surface normals. Evidence for surface orientation influence is rather subtle and the reader is referred to the original paper for details.

White beam X-ray topography is used here to record selected Laue spots simultaneously, allowing for at least partial time resolved analysis of the slip systems. Low exposure times allow for step deformation experiments to be performed while avoiding parasitic creep. Fig. 13(a,b,c) (101 reflection) shows the onset of plastic deformation in one type of sample. Slip bands are seen to nucleate at surface imperfections as well as at specimen edges (Fig. 1.). Broad vertical lines in Fig. 13a delineate magnetic domain boundaries belonging to a rather complex Dijkstra and Martius type structure [48]. Under the action of an elastic stress (Fig. 12b), the domain periodicity decreases, as expected. Fig. 13d is a $\bar{1}01$ recorded on the same nuclear plate as the 101 reflection reflection of Fig. 13c. Slip bands are extinguished in the $\bar{1}01$ reflection implying that the Burgers vector is [111]. Fig. 13e shows a Lang topograph of a deformed specimen with the same surface and tensile axis orientations. The limited area of image formation due to elastic and plastic lattice rotations implies a loss of information which may be very severe.

16.5.4 Wall movements in the antiferromagnet KCoF3 [44]

In domain studies, X-ray topography is a unique tool for a number of reasons. One of the most prominent features of the technique is that it enables one to observe both domain walls in magnetostrictive (or electrostrictive) materials and lattice imperfections in relatively massive samples. It is therefore particularly suited for the study of interactions between domain walls and lattice imperfections. Safa and Tanner have observed the movement of walls in $KCoF_3$, an antiferromagnet. Fig. 14 shows a sequence of topographs recorded with increasing applied magnetic field.

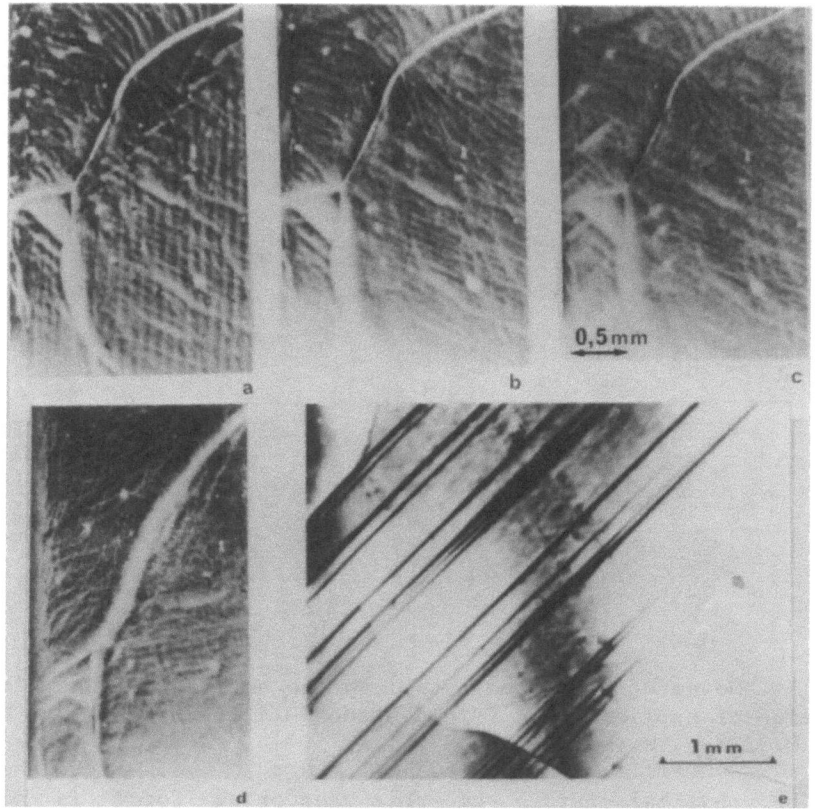

Figure 13. Onset of plastic deformation in a Fe-Si single crystal a) $\sigma = 0$ b) $\sigma = 134$ MPa c) $\sigma = 208$ MPa a,b,c: 101 reflection mainly at σ $\lambda \sim 1$Å d) $\sigma = 208$ MPa. 101 reflection mainly at ~ 1Å e) Lang topograph of another deformed crystal with the same tensile axis and surface orientations. Note the restricted area of image formation. 200 reflection.

Domains I and II have spins parallel and antiparallel to [010] and [001], respectively. The boundary between domains I and II is the broad oblique black stripe on the topographs. When a field is applied along [001], the boundary sweeps to the right since spins tend to align perpendicularly to the field. It was shown that wall displacement as a function of field agrees well with an energy model in which it is assumed that

1) wall displacement may only occur if the field exceeds a given field, the coercive field,

2) wall displacement is resisted by a restoring force proportional to displacement.

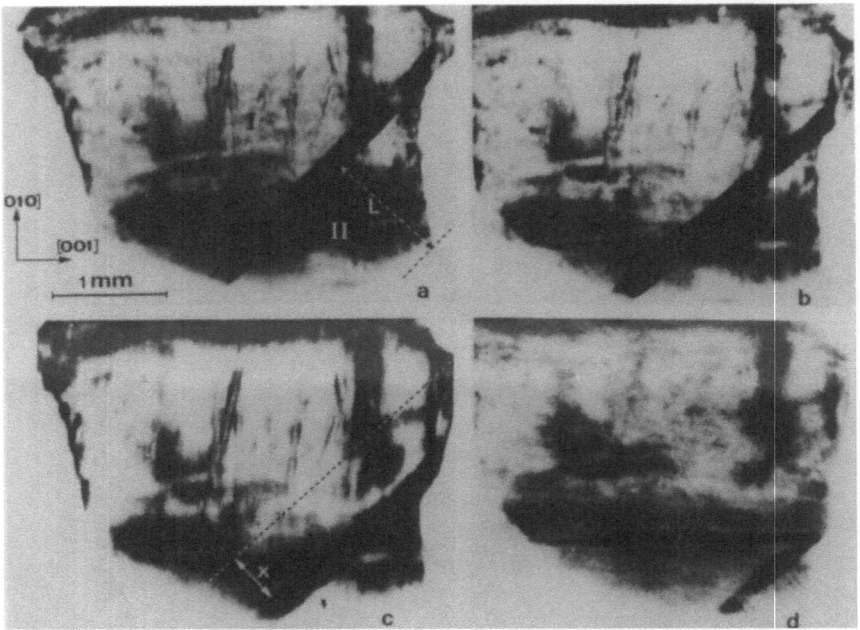

Figure 14. Movement of a domain wall in the antiferromagnet $KCoF_3$ as a function of applied field (field along 001) T = 77K a) H = 0.22T; b) H = 0.72T; c) H = 0.94T; d) H = 1.4T [44]

The physical nature of this restoring force is not yet entirely elucidated. Magnetoelastic coupling may, at least partially, account for it. Accordingly, wall motion should be reversible upon decreasing the field. Such a behaviour has been observed by Safa and Tanner. In experiments of this type, it is rather essential to avoid long term rearrangements of wall geometry. Therefore, short exposure times are necessary. Synchrotron based topography is, here again, a very valuable tool.

16.6 Conclusion

Synchrotron white beam X-ray topography is a now well established topographic technique. More should be however known on defect contrast properties. Applications up to now have neglected the time structure of the beam. Direct stroboscopic experiments are certainly feasible, provided some kind of lattice imperfections may be moved at frequencies in the MHz range. Magnetic domain walls are obvious candidates.

WHITE BEAM SYNCHROTRON RADIATION TOPOGRAPHY

REFERENCES

1. A. Guinier and J. Tennevin (1949), Acta Cryst. $\underline{2}$, 133
2. L.G. Schultz (1954), Trans. AIME $\underline{200}$, 1082 3.
3. U. Bonse, this volume
4. M. Sauvage in Synchrotron Radiation Research, to be published
5. B. Cullity (1956) in Elements of X-ray Diffraction Addison Wesley, Reading Mass.
6. J.B. Cohen (1966) in Diffraction Methods in Materials Science Macmillan, New York
7. M. Hart (1975), J. Appl. Cryst. $\underline{8}$, 436
8. M. Hart, this volume
9. T. Tuomi, K. Naukkarinen and P. Rabe (1974), Phys. Stat. Sol. (a) $\underline{25}$, 93
10. J.F. Petroff and M. Sauvage (1978), J. Crystal Growth $\underline{43}$, 628
A.R. Lang (1978), in Modern Diffraction and Imaging Techniques in Material Science (ed. S. Amelinckx et al) North Holland - Amsterdam - London, 2nd Edition
12. J. Miltat in Proceedings of the workshop on Imaging Processes and Coherence in Physics Les Houches - March 1979. Springer-Verlag, Berlin
13. M. Hart and A.R. Lang (1965), Acta Cryst. $\underline{19}$, 73
14. M. Hattori, H. Kuriyama and N. Kato (1965), J. Phys. Soc. Jap. $\underline{20}$, 1047
15. N. Kato, this volume
16. P. Skalicky and C. Malgrange (1972), Acta Cryst. A28, 501
17. M. Sauvage, J.F. Petroff and P. Skalicky (1977), Phys. Stat. Sol. (a) $\underline{43}$, 473
18. T. Tuomi, M. Tilli, V. Kelha and J.D. Stephenson (1978), Phys. Stat. Soli (a) $\underline{50}$, 427
19. J. Gastaldi and G. Jourdan (1978), Phys. Stat. Sol. (a) $\underline{49}$, 529
20. J.D. Stephenson, V. Kelha, M. Tilli and T. Tuomi (1978) Nucl. Inst. Methods $\underline{152}$, 319
21. A. Authier (1967), Adv. X-ray Analysis $\underline{10}$, 9
22. C. Willaime and A. Authier (1969), Bull. Soc. Fr. Miner. Gryst. 89 269
23. J. Miltat and D.K. Bowen (1975), J. Appl. Cryst. $\underline{8}$, 657
24. H. Klapper (1976), J. Appl. Cryst. $\underline{9}$, 310
25. J.-I. Chikawa (1965), J. Appl. Phys. $\underline{36}$, 3496
26. B.K. Tanner, D. Midgley and M. Safa (1977), J. Appl. Cryst. $\underline{10}$, 281
27. A. Authier and F. Balibar (1970), Acta Cryst. A26, 647
28. Y. Ando and N. Kato (1970), J. Appl. Cryst. $\underline{3}$, 74
29. M. Sauvage (1978), Nucl. Inst. Methods $\underline{152}$, 313
30. J. Miltat (1978), Nucl. Inst. Meth. $\underline{152}$, 323
31. B.K. Tanner, M. Safa and D. Midgley (1977), J. Appl. Cryst. $\underline{10}$, 91
32. I.M. Buckley-Golder, B.K. Tanner and G.F. Clark (1977), J. Appl.

Cryst. 10, 502
33. C. Jourdan and J. Gastaldi (1979), Scripta Met. 13, 55
34. J. Gastaldi and C. Jourdan (1979), Phys. Stat. Sol. (a) 52, 139
35. I.B. MacCormack and B.K. Tanner (1978), J. Appl. Cryst. 11, 40
36. M. Safa, B.K. Tanner, H. Klapper and B.M. Wanklyn (1977), Phil. Mag. 35, 811
37. B.K. Tanner (1977), Progress in Crystal Growth and Assessment 1, 23
38. J. Bordas, A.M. Glazer and H. Hauser (1975), Phil. Mag. 32, 471
39. I.T. Steinberger, J. Bordas, Z.H. Kalman (1977), Phil Mag. 35, 1257
40. B.K. Tanner, M. Safa, D. Midgley and J. Bordas (1976), J. Magn. Mat. 1, 337
41. R.S. Sery, H.T. Savage, B.K. Tanner and G.F. Clark (1978), J. Appl. Phys. 49, 2010
42. G.F. Clark, B.K. Tanner, R.S. Sery and H.T. Savage (1979), J. de Physique 40, C5-183
43. Y. Chikaura and B.K. Tanner (1979), Jap. J. Appl. Phys. 18, 1389
44. M. Safa and B.K. Tanner (1978), Phil. Mag. B37, 739
45. J.D. Stephenson, V. Kelha, M. Tilli and T. Tuomi (1979), Phys. Stat. Sol. (a) 51, 93
46. T. Tuomi, J.D. Stephenson, M. Tilli and V. Kelha (1979), Phys. Stat. Sol. (a) 53, 571
47. J. Miltat and D.K. Bowen (1979), J. de Physique 40, 389
49. L.J. Dijkstra and U.M. Martius (1953), Rev. Mod. Phys. 25-1, 146

CHAPTER 17

CONTROL OF WAVELENGTH, POLARIZATION, TIME-STRUCTURE AND

DIVERGENCE FOR SYNCHROTRON RADIATION TOPOGRAPHY

MICHAEL HART

17.1 Introduction

In the wavelength range near $\lambda = 1\text{Å}$ Bragg reflection perfect crystals are the most useful optical components available. Modern commercially available single crystals of silicon and germanium are inexpensive, almost undamaged by radiation and, from the diffraction point, ideally perfect.

Although perfect crystals Bragg reflect X-rays only over a very small range of incident angles, they do so with very high efficiency so that multiple crystal systems and multiple Bragg reflections can be used without a serious loss of intensity. Multiple Bragg reflections allow great freedom in X-ray optical design and we will explore here the control and measurement of spectral profile, polarization, time structure and beam divergence.

17.2 Summary of results from dynamical diffraction theory

The theory of diffraction in perfect crystals is fully developed in a number of standard texts [1-9]. Two important cases are distinguished; in the Bragg-case (reflection) the diffracted wave $\underline{K}_h$ emerges from the same (entrance) surface as the incident wave $\underline{K}_o$ while in the Laue-case (transmission) the diffracted waves emerge from the lower surface. Only in the symmetric Laue-case when the Bragg planes are normal to the crystal surface does the Bragg law give the exact Bragg angle θ_L, so that

$$2d \sin \theta_L = n\lambda \qquad (1)$$

In all other cases the Bragg angle is shifted slightly by refraction by an amount which depends on the optical parameters and on the relative orientations of the crystal surface and the wave vectors. The principal features of multiple reflection systems can be understood in terms of the examples shown in Fig. 1 which are drawn for the symmetric Bragg-case.

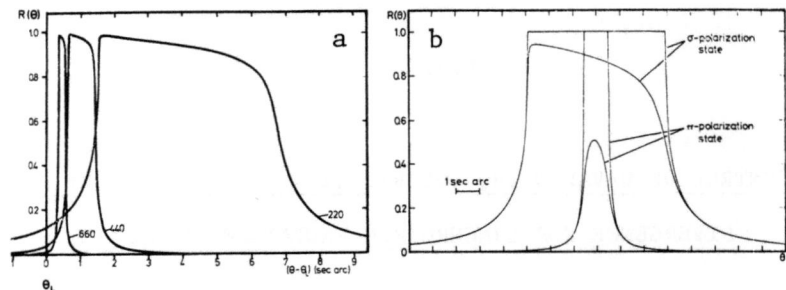

Figure 1. (a) Graph of the reflectivity $R(\theta)$ for the fundamental (n = 1) 220 Bragg reflection from silicon at λ = 1.54A and for the harmonics n = 2 (440 at λ = 0.77A) and n = 3 (660 at λ = 0.385A). The σ-state of polarization is assumed. (b) $R(\theta)$ for the two polarization states in the 440 Bragg reflection from germanium at λ = 1.54 A. The flat-topped curves show the result when absorption is ignored.

The principal effect of refraction is to shift the position of the Bragg peak and its harmonics away from θ_L. The shift of the centre of the Bragg peak is given by

$$\theta_o - \theta_L = \frac{|\chi_{ro}|}{2\sin 2\theta} \left[1 + \frac{\sin(\theta - \phi)}{\sin(\theta + \phi)} \right] \quad (2)$$

where ϕ is the angle between the crystal surface and the Bragg planes and $\chi_{ro} = -8 r_e \lambda^2 (Z + f')/\pi a^3$ in diamond structure materials. For the harmonics the shift is proportional to λ^2 or n^{-2}.

The peak widths $\Delta\theta_o$ are given by

$$\Delta\theta_o = \frac{2 |C| |\chi_{rh}|}{\sin 2\theta} \sqrt{\frac{\sin(\theta - \phi)}{\sin(\theta + \phi)}} \quad (3)$$

where C is 1 for the σ- polarized state and $\cos 2\theta$ for the π-polarized state. $\chi_{rh} = r_e \lambda^2 F_h/\pi a^3$ where F_h is the structure amplitude for the active Bragg reflection. The peak widths are therefore proportional to $F_h n^{-2}$ for the harmonics.

Equation (3) and Fig. 1(b) also show that the width of the Bragg peak varies as the polarization state of the beam is changed. In the zero absorption approximation the ratio of the widths for π-state and σ-state polarization is $\cos 2\theta$.

The width of the Bragg reflection, given by equation (3) is simply related to the extinction length of Pendellosung length Δ_o. At the centre of the Bragg range in the symmetric Laue-case ($\phi = 0$) we have

$$\Delta_o = \lambda\cos\theta / |C| |\chi_{rh}| \tag{4}$$

17.3 Control of wavelength

There are only two generic designs for X-ray spectroscopes [10,11]. They are sketched in Fig. 2 in the Bragg-case, though Laue-case and mixed Laue-Bragg-case systems can also be used.

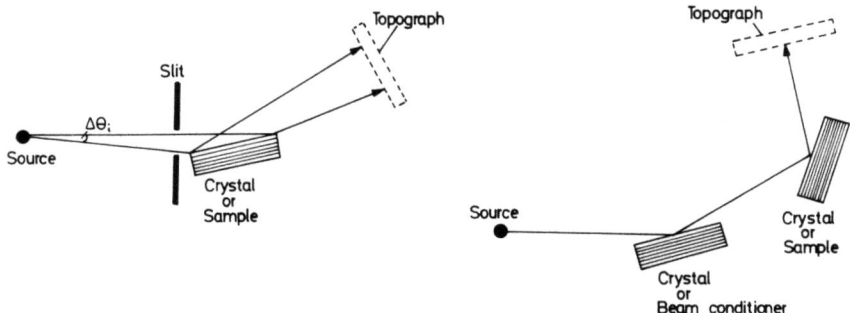

Figure 2. (a) Bragg spectrometer or white radiation topography arrangement, (b) Double crystal spectrometer or double crystal topography arrangement. Laue-case and mixed Laue-Bragg case arrangements can also be used.

In the Bragg spectrometer the wavelength resolution is determined by the divergence of the incident beam $\Delta\theta_i$ which is fixed in practice by the geometry of the source and collimator slit. $\Delta\theta_i > 10^{-4}$ can be achieved in the laboratory with reasonable intensity and at synchrotron radiation sources it might be possible to work with $\Delta\theta_i < 10^{-5}$. The corresponding energy resolution follows from Bragg's law

$$\Delta E/E = \cot\theta \cdot \Delta\theta_i$$

Higher resolution can be achieved in the double crystal arrangement because the angular pass band is determined by the Bragg width $\Delta\theta_o$. Thus large sources can be used to gain intensity. In practice $\Delta\theta_o$ lies within the range $10^{-8} < \Delta\theta_o < 10^{-4}$ and the corresponding energy or momentum resolution is determined by the width of the convolution of the two crystal profiles. It is approximately

$$\Delta E/E \simeq \cot\theta \cdot \Delta\theta_o$$

The Bragg spectrometer (Fig. 2a) becomes a white beam topography camera if the crystal is the sample under investigation. Similarly, the double crystal spectrometer (Fig. 2b) becomes a double crystal topography camera if we regard the second crystal as the sample and the first crystal as an X-ray beam conditioner. We are concerned here with the problems of beam conditioning to achieve single wavelength topographs, plane wave topographs, polarized radiation topographs and time resolved topographs. In practice these objectives will almost invariably require double or multiple crystal systems.

17.3.1 Multiple Bragg reflections

Since the peak reflectivity is almost 1 (Fig. 1) multiple Bragg reflections can be made with little intensity loss. This design freedom allows one to control the shape of the Bragg reflection profile, to split and recombine beams and to control their paths through a system. For example, multiple reflections lead to the elimination of the tails of the Bragg reflection curve [12]. In practice most of the multiple reflection diffraction optical systems are monolithic perfect crystals, but a new design freedom is added by the realisation that elastic adjustments are practicable if the crystal is suitably designed. In this way it has been possible to make ultra-stable scanning interferometers and fast interferometric choppers [13,14], harmonic-free monochromators [15] and tuneable polarizers [16].

17.3.2 Harmonic-free systems for topography

When the hkl Bragg reflection chosen for a topograph is the lowest order (n=1 in equation (1)) the white beam topograph will consist of a series of superimposed images with the intensity ratios proportional to F_h n^{-2} (equation (3)). Thus, about 2/3 of the total intensity is expected from the hkl reflection and the remainder is due to higher order Bragg reflections [17]. In practice, there are other important effects. The reduced intensity at high energies in the source will lead to a preferential suppression of the harmonics as too will a reduced structure amplitude for high order reflections. On the other hand, the reduced absorption on most materials at high X-ray energies often leads to preferential enhancement of the higher order contributions to the topograph.
If it is important to ensure single wavelength imaging then X-ray beam conditioning can be achieved with any of the methods illustrated in Fig. 3. Although specular reflection at a mirror (usually metal-coated quartz, bent to focus the beam) has been widely used to reduce crystal monochromator harmonics at synchrotron radiation sources, there appears to be little information on their actual performance. It is straight-forward (e.g. from James [1]) to show that the harmonic reflectivity at twice the critical angle is 0.4%. In practice, if we include the influence of absorption, the harmonic content will be higher.

Three of the harmonic reducing optical systems rely on shifting the relative positions of fundamental and harmonics by refraction (Fig. 1). Bonse, Materlik and Schroder [18] found that the Laue-Bragg combination did not reduce the harmonic content below a few percent of the fundamental intensity. However, when the Laue-case device is an interferometer and the Bragg-case device is a double-reflection grooved crystal as it was in those experiments, harmonics can at least be conveniently reduced by this method. The refractive index shift can also be exploited by missing oblique and symmetric Bragg reflections [19]. To exploit this system in X-ray diffraction topography would require a large number of different oblique cuts. By mixing crystals with different refractive indices, for example, a double 220 reflection from silicon and a single 220 reflection from germanium, harmonic suppression to 0.3% can be achieved [18].

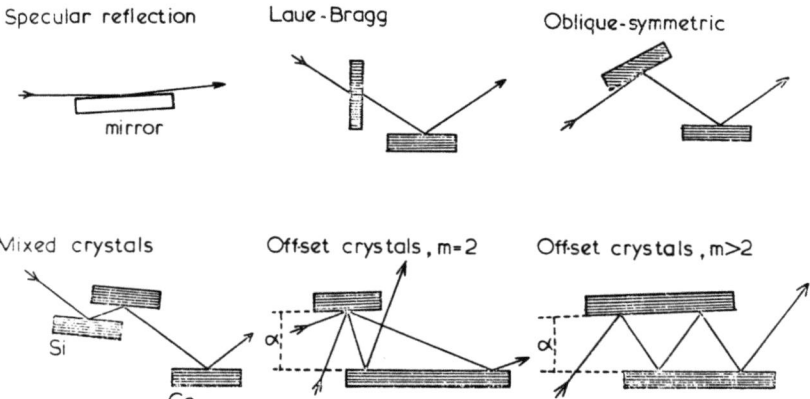

Figure 3. Experimental arrangements which have been used to control the intensity of harmonics.

The off-set grooved crystal systems suffer from none of the disadvantages of the refraction systems. Their operation depends not on the refraction effect but on the fact that the Bragg reflection widths are different for the fundamental and the harmonics. The off-set angle α is larger than the width of the harmonic peak but much less than the width of the fundamental (F. 1). With $m = 2$ Bragg reflections there is no restriction on the wavelength at which topographs can be obtained compared with the single Bragg reflection shown in Fig. 2, but the harmonic intensity is reduced to well below 0.5% [15]. As more Bragg reflections are added the harmonic ratio is reduced substantially and can be chosen in the range 10^{-2} to (say) 10^{-10}.

In practice, the problems caused by harmonic contamination in X-ray diffraction topographs will become more severe as storage ring sources are enhanced by the installation of wigglers etc. which are designed to increase the intensity available. In experiments on weak scattering phenomena harmonic control is already essential.

17.4 Polarizing optical systems

Sources of synchrotron radiation are linearly polarized in the plane of the ideal electron orbit. Fig. 4 shows the angular distribution of intensity calculated for the UK Storage Ring Source at 2 GeV energy. Theoretically, the photon emission into the two principal polarization states is coherent with the phases in quadrature. Consequently the radiation between 0.1 and 0.2 m rad above and below the mean orbit plane is expected to be elliptically polarized with an intensity per m rad approximately 15 times lower than the central beam of horizontally polarized synchrotron radiation. The ellipticity is between 2 and 3 over the angular range giving maximum intensity of elliptically polarized X-rays. In practice, the polarization is less complete than indicated in Fig. 4 because the source size is finite and the radiation pattern is smeared, typically over several tenths of an m rad by betatron oscillations.

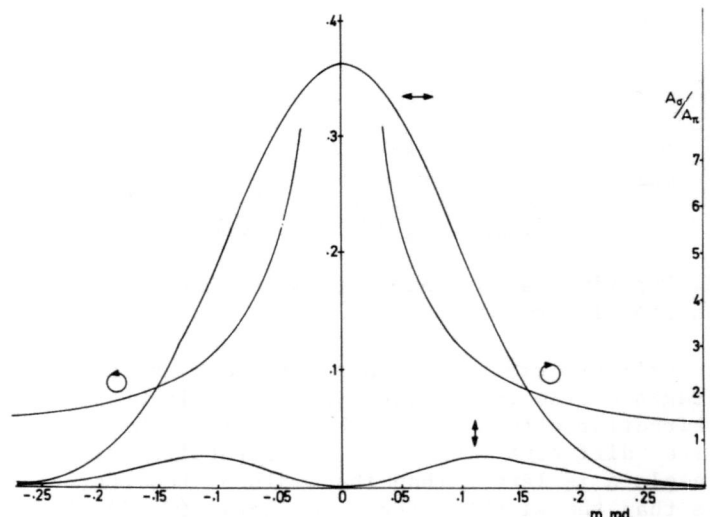

Figure 4. Angular emission of radiation from the SRS at Daresbury

17.4.1 Linear polarization

Prior to the 1960's very little work had been done with polarized X-rays [20]. Simple Rayleigh scattering and coherent Bragg scattering at 90° were the only available methods for producing polarized beams. Cole, Chambers and Wood [21] showed that the

Borrmann effect could also be used to produce polarized beams but only at long wavelengths. $\lambda \gtrsim 1A$, and with low intensities [20].

Recently a tunable X-ray polarizer has been developed [16]. As Fig. 1 shows, the peak reflectivity of a crystal is lower for the π - polarization state than for the σ - state. By making multiple Bragg reflections in a groove cut into a perfect crystal we can produce polarized beams. Fig. 5 shows calculatons of the polarizing ratio I_π/I_σ for $1 < m < 10$ in the case of the 440 Bragg reflection from germanium for the wavelength range near the polarizing Bragg angle.

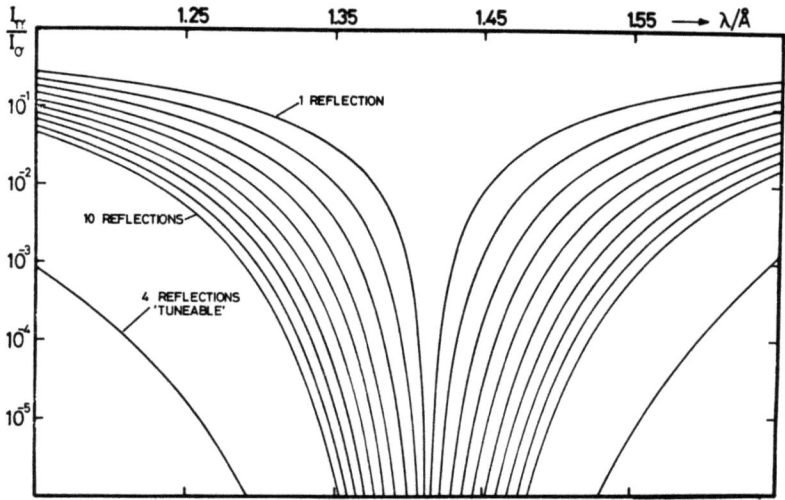

Figure 5. Calculated polarization ratio in a multiple reflection tunable polarizer as a function of wavelength [16]. Germanium 440 reflection.

As the π-state reflection curve is much narrower than the σ-state reflection curve (Fig. 1), the polarization ratio can be further improved if one side of the groove is offset by a small angle α with respect to the other as in the arrangement for harmonic elimination. The result, for a four-fold Bragg reflection system, is shown as the outermost curve in Fig. 5. Detailed calculations in Table 1 below (for the silicon 422 Bragg reflection with m = 4 Bragg reflections at λ = 1.38A) show how the same offset grooved crystal acts both as a harmonic-free monochromator and as a polarizer.

Table 1

α/sec.arc	$I_\sigma(\alpha)/I_\sigma(0)$	$I_\pi(\alpha)/I_\sigma(\alpha)$	I_{844}/I_{422}
0.0	1.00	5.8×10^{-2}	9.6×10^{-2}
0.1	0.95	3.6×10^{-2}	1.0×10^{-3}
0.2	0.85	6.3×10^{-3}	5.0×10^{-6}
0.3	0.73	5.4×10^{-4}	1.5×10^{-7}
0.4	0.60	7.9×10^{-5}	2.5×10^{-8}
0.5	0.47	2.0×10^{-5}	7.5×10^{-9}

17.4.2 Circular polarization

Elliptically polarized white radiation is available above and below the mean orbit plane but, as Fig. 4 shows, the <u>integrated</u> intensity is only about 1% of the total intensity if the ellipticity is between 2 and 3.

Fortunately [20] three-quarter-wave X-ray phase plates can be made from Bragg reflecting crystals which permit the conversion of linearly polarized X-rays to elliptically polarized X-rays with an efficiency of about 30%. For example, for the 220 Bragg reflection from silicon at 1.5A wavelength, equation (4) gives Δ_o^σ = 15.6μm and Δ_o^π = 23.0μm so that a crystal 24.2μm thick acts as a quarter-wave plate. Although higher monochromatic intensities should be obtained with crystal quarter-wave plates they, of course, can only provide monochromatic beams.

17.5 Stroboscopy, modulation and timing

Fig. 6 shows a range of methods by which time can be introduced as an experimental parameter in X-ray diffraction topography. Stroboscopic techniques [22] permit, for example, the "freezing" of the motion of domain walls under the influence of alternating magnetic fields. Modulation of the X-ray beam wave-length or direction permits the measurement of derivatives of the diffraction contrast and integration techniques can be used to enhance or otherwise process the images obtained. Precise timing is essential if transient phenomena are to be investigated.

By far the simplest system consists of a rotating vane which interrupts the X-ray beam (Fig. 6a). Ultimately speeds are limited by the strength of materials, as in ultracentrifuges, but in practice frequencies in the range 1Hz - 1kHz can be achieved conveniently in air and 10 kHz in vacuum. Equivalent rotating crystal systems can conveniently provide monochromatic beams at up to 10 kHz in air as Fig. 6b and Fig. 6c show.

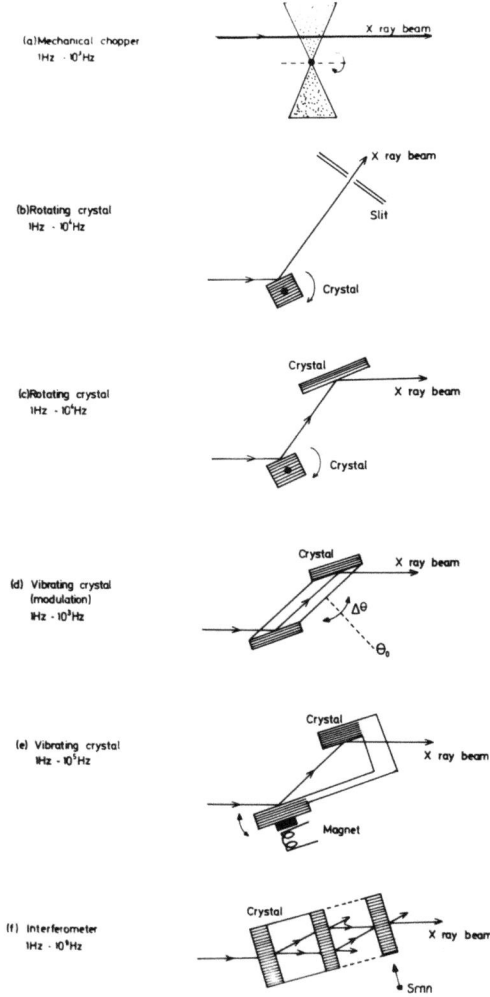

Figure 6. Experimental arrangements suitable for frequency modulation and chopping of X-ray beams: (a) mechanical chopper, (b) rotating crystal, (c) rotating crystal, (d) vibrating crystal (modulation), (e) vibrating crystal, (f) interferometer.

The improvement in frequency response can be expected because the size and moment of inertia of the rotating element has been reduced compared with the mechancial chopper. If wavelength modulation rather than beam chopping is required then oscillation rather than rotation can be used to obtain higher duty cycles as Fig. 6d shows. The mean wavelength $\lambda = 2 d \sin\theta_o$ and the modulation range is $\Delta\lambda/\lambda = \cot \theta_o \Delta\theta$. Modulation frequencies up to 1 kHz should be straightforward. By using a monolithic elastic double reflection system (Fig. 6e) it should be possible to work at up to 100 kHz. Much higher frequencies can be achieved, at the expense of complexity, by using a scanning X-ray interferometer as a beam chopper. Chopping frequencies of 2 MHz have already been achieved [13,14] and straightforward extrapolation suggests that operation at 1 GHz is possible. At very high frequencies electronic techniques, based on the source characteristics, provide simple methods of timing and modulation. At the largest storage ring VEPP 4 at Novosibirsk and CESR at Cornell University, the basic source frequency is between 0.5 and 1 MHz, rising to 4 MHz at the smallest storage rings. In multiple electron bunch modes the modulation frequency of the source is increased in proportion to the number of electron bunches. For example, the fundamental single bunch frequency of the UK Storage Ring Source is 3 MHz (320 ns cycle time) and with the maximum of 160 electron bunches in the ring source the modulation frequency would be 0.5 GHz. Use of these high modulation frequencies will only be possible if the frequency response of detector systems is sufficiently good.

17.6 Control of beam divergence

Equation (3) gives peak widths in the range of 10^{-3} m rad to 1 mrad for typical combination of structure amplitude and wavelength used in x-ray diffraction topography. On the other hand, the source divergence (Fig. 4) is approximately 1/4 mrad. It is sometimes desirable to increase the divergence of the ray bundle accepted by the crystal by using oblique reflections with $\phi \to -\theta$ or to increase the selectivity of the Bragg reflection by working with $\phi \to +\theta$ [23] as shown in Fig. 7. In practice it is not convenient to work with $|\theta-\phi| < 1°$ since surface roughness becomes very important and external total specular reflection can occur. A greater degree of divergence control can be achieved with curved crystal systems at the cost of increasing complexity.

If we write $b = \sin(\theta - \phi)/\sin(\theta + \phi)$ then the width of Bragg reflection is $\omega_s/\sqrt{b}$ in the primary beam and $\omega_s\sqrt{b}$ in the diffracted beam where ω_s is the Bragg width in the case of symmetric reflection, when $\phi = 0$. Thus, an oblique Bragg reflection can be used to increase the beam acceptance angle on the primary beam side so as to match the source divergence and incidentally to provide an exit beam which is expanded in width as in Fig. 7(b) or to compress the exit beam and accept reduced divergence beams so as to obtain high angular resolution as in Fig. 7(a). Two Bragg reflections in a monolithic monochromator can be used to provide either higher or low angular resolution without altering the spatial width, as in Fig. 7(c) and Fig. 7(d).

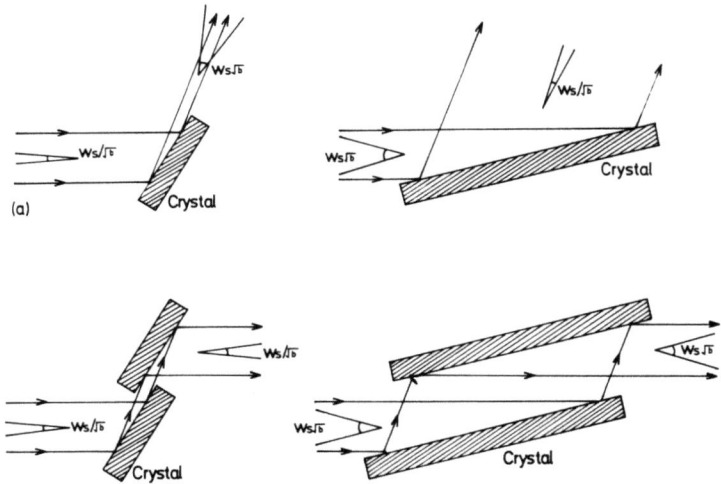

Figure 7. Oblique Bragg reflections which serve to control beam shape and divergence.

REFERENCES

1. R.W. James (1948). The optical principles of the diffraction of X-rays, Bell, London
2. W.H. Zachariasen (1945) Theory of X-ray diffraction in crystals, Wiley, New York.
3. J.C. Slater (1958) Rev. Mod.Phys. $\underline{30}$, 197
4. G. Borrmann, (1959) Rontgenwellenfelder in Beitrage zur Physik und Chemie des 20 Jahrhunderts, Vieweg und Sohn, Brunswick
5. M. von Laue (1960) Rontgenstrahl - Interferenzen, Akademische Verlag, Frankfurt
6. R.W. James (1963) Solid State Phys. $\underline{15}$, 55
7. B. W. Batterman and H. Cole (1964) Rev. Mod. Phys. $\underline{36}$, 681
8. L.V. Azaroff, K. Kaplow, N. Kato, R.J. Weiss, A.J.C. Wilson and R.A. Young (1974) X-ray diffraction, Mcgraw-Hill, New York
9. Z.G. Pinsker (1978) Dynamical scattering of X-rays in crystals. Springer, Berlin
10. A.H. Compton and S.K. Allison (1935) X-rays in theory and experiment, Van Nostrand, New York.
11. L.V. Azaroff (1974) X-ray Spectroscopy, McGraw-Hill, New York
12. U. Bonse and M. Hart (1965) App. Phys. Lett. $\underline{7}$, 238
13. M. Hart and D.P. Siddons (1978) Nature, London $\underline{275}$, 45
14. M. Hart and D.P. Siddons (1978) Workshop on X-ray and neutron interferometry. Ed. U. Bonse and H. Rauch. (OUP)

15. M. Hart and A.R.D. Rodrigues (1978) J. Appl Cryst. **11**, 248
16. M. Hart and A.R.D. Rodrigues (1979) Phil. Mag **40**, 149
17. M. Hart (1975) J. Appl. Cryst. **8**, 436
18. U. Bonse, G. Materlik and W. Schroder (1976) J. Appl. Cryst. **9**, 223.
19. S. Kikuta and K. Kohra (1970) J. Phys. Soc. Japan **29**, 1322
20. M. Hart (1978) Phil. Mag **38**, 41
21. H. Cole, F.W. Chambers and C.G. Wood (1961) J. Appl. Phys. **32**, 1942
22. J. Miltat (1979) Imaging Processes and coherence in physics, Les Houches, Springer
23. K. Kohra, M. Ando and T. Matsushita (1978) Nucl. Instrum & Meth. **152**, 161

CHAPTER 18

MONOCHROMATIC SYNCHROTRON RADIATION TOPOGRAPHY

MICHELE SAUVAGE

18.1 Introduction

Synchrotron radiation has been used as an X-ray source for topography since 1974. A significant number of results have already been obtained by means of white beam topography as reported by Miltat [1]. However, since 1978, topographic experiments where a monochromatized beam was first extracted from the white spectrum prior to reflection in the sample have also been performed. When thinking of physical studies where the observation technique is topography, one has thus to decide which among the following imaging methods is the most appropriate : white beam topography, Lang or double-crystal topography using laboratory generators or monochromatized synchrotron radiation topography. The purpose of the present chapter is first to produce some guide lines which will help in promoting monochromatized synchrotron radiation topography and secondly to give typical examples of applications.

18.2 Selection of the experimental conditions

Two different situations occur depending on whether integrated or plane-wave imaging techniques are considered.

18.2.1 Integrated images

Let us review some particular circumstances in which white beam topography has to be ruled out. Among these, one finds crystal constraints such as high sensitivity to radiation damage or thermal stresses, high surface reactivity to the oxidizing ozone atmosphere surrounding the white beam, or the presence of an oxide layer that introduces important modifications in the magnetic domain pattern close to the surface of metallic materials. The diffraction and absorption properties of the sample may also be unfavourable. For example, in the case of $Gd_2(MoO_4)_3$, which presents an interesting phase transition at 159°C, the absorption coefficient is so high that due to the lack of strong Borrmann effect, only short wavelengths are efficient in forming topographs. Since the lattice parameters are large, Bragg angles are very small for such wavelengths. Moreover numerous Bragg spots occur due to the low symmetry of the crystal space group, both effects leading to spot overlap (Fig. 1). Such a disadvantage can be avoided either by reducing drastically the field of observation, which is not suitable when interaction effects are sought, or by increasing the sample-film distance, which impairs the resolution. A monochromatic experiment seems more promising in such cases.

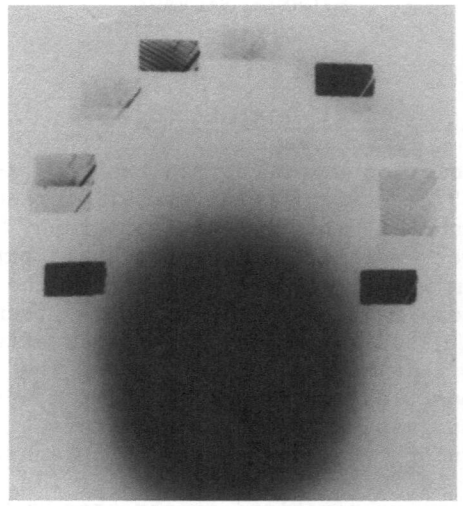

Figure 1. White beam diffraction pattern of $Gd_2(MoO_4)_3$, spot width: 4.3mm (courtesy of C. Malgrange and B. Capelle).

A more fundamental reason which would prevent the use of white beam topography is the difficulty in interpreting quantitatively the contrast features, due to harmonic superposition on the one hand and to the lack of theoretical studies concerning image formation in the case of a divergent and white incident beam on the other hand, cf. [1]. Besides the above arguments, white beam topography may also be prohibited on the basis of practical constraints such as the lack of available space around the sample when the topography station is not an end-of-line port.

Let us consider now a few points which enable one to make a choice between laboratory Lang topography and monochromatic synchrotron radiation topography. The first one is the expected gain in intensity by a factor of approximately 100 in comparable band widths [2]. For example, the stroboscopic experiment described in section 5.1 could not be realized in the laboratory even with a rotating anode generator. Secondly, the tunability of the synchrotron radiation source offers a possible optimization of the absorption conditions by working just above absorption edges.

MONOCHROMATIC SYNCHROTRON RADIATION TOPOGRAPHY

Finally in synchrotron radiation techniques, the instantaneous images are two-dimensional which is a major advantage when compared to Lang topography, especially when real-time detection is achieved by means of an X-ray sensing TV camera [3].

18.2.2 Plane-wave imaging techniques

Such techniques are particularly useful to measure minute distortions in crystals. A standard laboratory set-up is the two crystal arrangement in the (+,-) non dispersive parallel setting introduced by Bonse [4] (Fig.2). However such experiments can only be performed if a perfect crystal identical to the sample is available. Moreover, the sensitivity of the method is preserved only if the width of the reflected beam R_I is sufficiently small, which may require a highly asymmetric cut, valid for a given reflexion only.

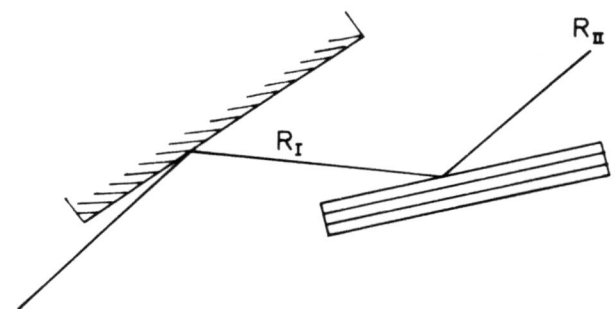

Figure 2. Highly accurate (+,-) parallel setting

Besides silicon, germanium and quartz, almost no single crystal materials present the required properties of perfection and shaping facility which in addition, have to be coupled with a good stability in the high flux delivered by modern X-ray generators. Such requirements have prevented double-crystal studies on many materials that are very interesting from the point of view of crystal growth, such as those presented by Klapper [5]. A possible compromise would be to look for a matched reflection in a silicon crystal (the first example presented in section 5.2 relies on such a procedure, which is valid also for a synchrotron source). However, each reflection in each single crystal is a new problem. On the contrary, by inserting a suitable 'beam conditioner' in the synchrotron radiation path [6], it is possible to extract a wave sufficiently narrow in angle and wave-length spread to be considered as a plane wave in most experimental situations. Due to the high photon flux, the output beam is intense enough to obtain topographs of a sample crystal within reasonable exposure times. Several applications for such tailored waves are given in section 5.2.

18.3 Available and future facilities

Monochromatic synchrotron radiation topography requires at least a two-axis spectrometer. Such an experimental set-up has been on line at LURE-DCI (Orsay-France) since July 1977 [7]. The block diagram of the unit is represented in Fig. 3. The fine rotations are driven by piezoelectric translators and profiles as narrow as 0.6" have been recorded. In addition the mechanical and thermal stability of the set-up have proved to be quite satisfactory and no regulation has been attempted. A two-crystal spectrometer designed for SSRL (Stanford - USA) by Parrish and his coworkers from IBM San Jose [8] has been operated in 1978. A new set up accessible to SSRL users is being built [9]. Similarly, CESR (Cornell - USA) and DORIS (Hamburg-GFR) will be equipped with two-axis spectrometers. At the SRS (Daresbury-UK) one station dedicated to interferometry and 'beam-conditioned' synchrotron radiation topography plus another station shared between white radiation and beam-conditioned topography are planned. the experimental set-up is being built by Hart, Siddons and Bowen and full operation is expected for 1981. In the long term, it can be foreseen that BNL (Brookhaven - USA) and the Photon Factory (Tokyo-Japan) will provide similar facilities.

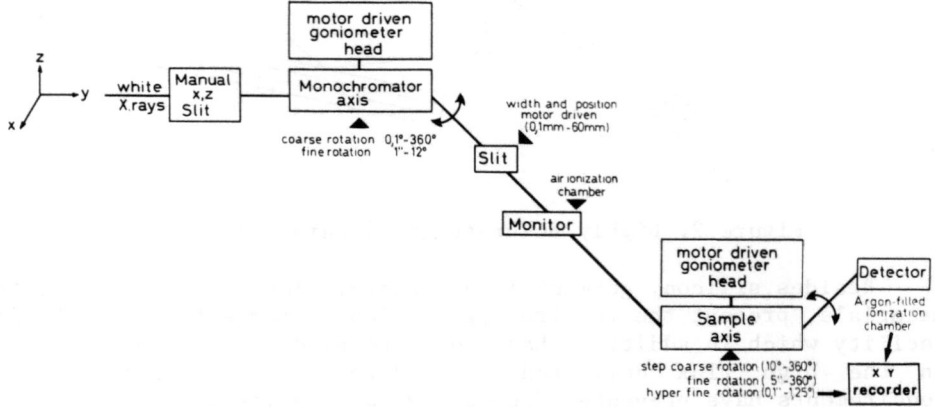

Figure 3. Block diagram of the LURE-DCI two axis spectrometer.

18.4 Analysis of an experiment

In order to take full advantage of the technique, an estimation as accurate as possible of the angular divergence and spectral width which are involved in the image formation process must be obtained. This is very adequately performed by means of the Du Mond diagram representation [10] which will be briefly recalled.

18.4.1 Du Mond diagram

Such diagram represents the correlation between wavelength

and angle provided by Bragg reflection in a single crystal. Actually it is a sine band representing Bragg's Law whose width on the angular scale is taken equal to the reflection width of the crystal (Fig. 4). A significant intensity output in the reflected beam is only obtained when the representative point in λ and θ coordinates belongs to this correlation band. With synchrotron radiation sources, the primary beam incident on a first crystal presents a finite angular divergence $\Delta\theta_0$ mostly governed by the source size, and an infinite wavelength spread. The distribution in λ and θ for the reflected beam R_I (fundamental only) is thus represented by the shaded parallelogram in the blow-up in Fig. 4.

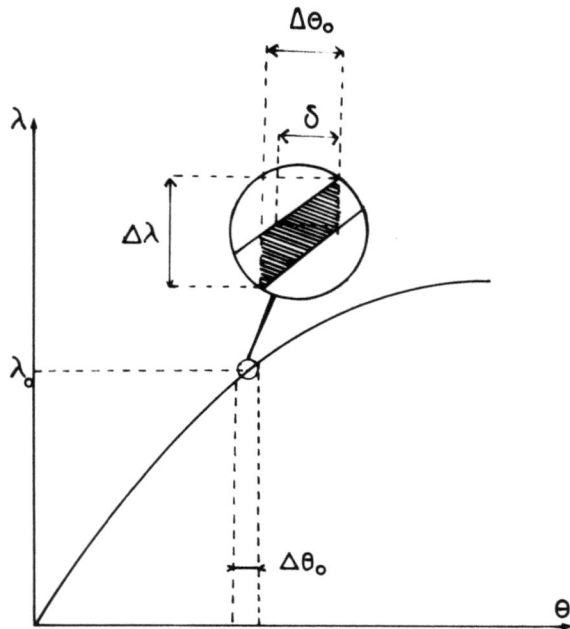

Figure 4. Du Mond diagram representation for a single Bragg reflection

For multiple-crystal arrangements, it is adequate to represent successive reflexions in the same diagram. Two geometries must be distinguished : first the so-called (+,+) setting where deviations of the beam at successive reflexions add up (Fig. 5a). If ω denotes the dihedral angle between the reflecting plane in crystals I and II the following relations hold :

$$\theta_I + \theta_{II} = \omega$$
$$d\theta_I = - d\theta_{II}$$

when ω is kept constant as in a monolithic device. In order to use a common set of axes for both correlation curves, the θ_I and θ_{II} axes should be oriented in opposite sense and their origins separated by a quantity ω (Fig. 5b). In the case of independent crystals, a relative rotation $\Delta\omega$ appears as an additional shift between the origins which actually means a translation $\Delta\omega$ of curve II along the angle axis, curve I being kept fixed (Fig. 5b).

In the other geometry classically denoted as the (+,-) setting successive deviations take place in opposite sense (Fig. 6a) and the relations to be used are now :

$$|\theta_I - \theta_{II}| = \omega$$

$$d\theta_I = d\theta_{II}$$

The origins are still shifted by an amount equal to ω but both θ_I and θ_{II} axes have the same orientation (Fig. 6b)

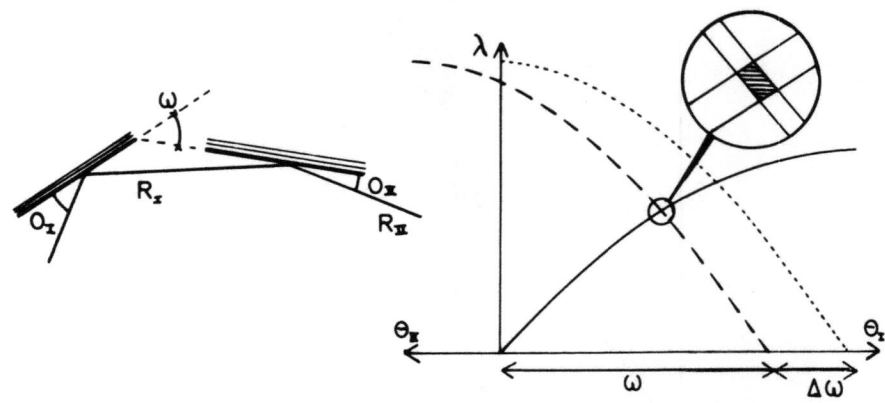

Figure 5. a) the (+,+) setting b) Du Mond diagram representation: full curve : crystal I, dashed curve : crystal II , dotted curve : crystal II after a rotation $\Delta\omega$

In multiple-crystal arrangements, the λ and θ distribution in the double reflected beam R_{II} is represented by the dashed overlap region of both bands. When keeping all other parameters unchanged, the intensity is roughly proportional to the area of the overlap parallelogram. It thus appears clearly that more intensity is collected from R_I in a (+,-) than in a (+,+) geometry. The particular case of the (+,-) parallel setting when both reflection bands are identical is the most efficient in that sense but the overlap is only ensured on a very narrow range of ω values.

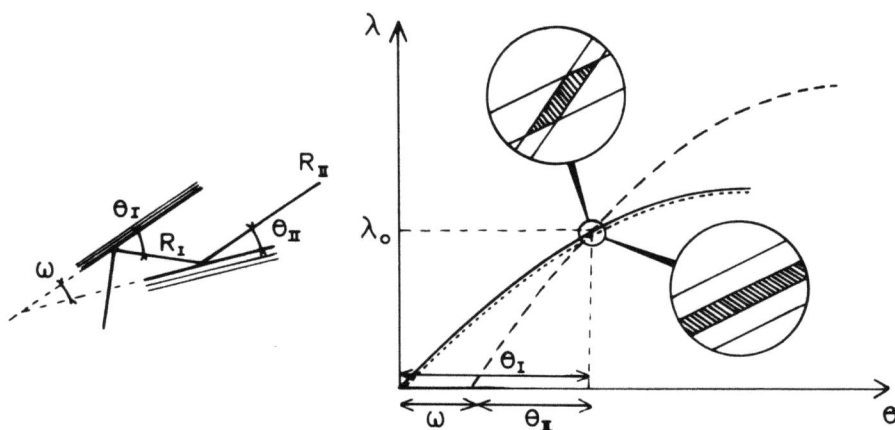

Figure 6. a) the (+,-) setting b) Du Mond diagram : full curve : crystal I dashed curve : crystal II, dotted curve : crystal II in a parallel setting.

18.4.2 Numerical application

Let us consider a particular example, the coupling of a silicon monochromator adjusted for the 333 reflection close to 0.7 A with a Fe-3.5% Si single crystal adjusted for the 110 reflection. The (+,+) and (+,-) configurations are respectively displayed in Fig. 7a and b. By assuming rectangular distributions in the single crystal bands, the intensity 'profiles' obtained in both cases by rotating the sample crystal may be sketched as shown in Fig. 8. Two points should be emphasized:

(a) 75% of the available intensity delivered by the silicon crystal is collected by the iron-silicon crystal at the apex of the (+,-) 'rocking curve' whereas 25% only is collected in the (+,+) case;

(b) the (+,-) curve is much steeper, which means a higher strain sensitivity. In the case of crystals with subgrain boundaries, such as metallic materials only small areas will be imaged simultaneously.

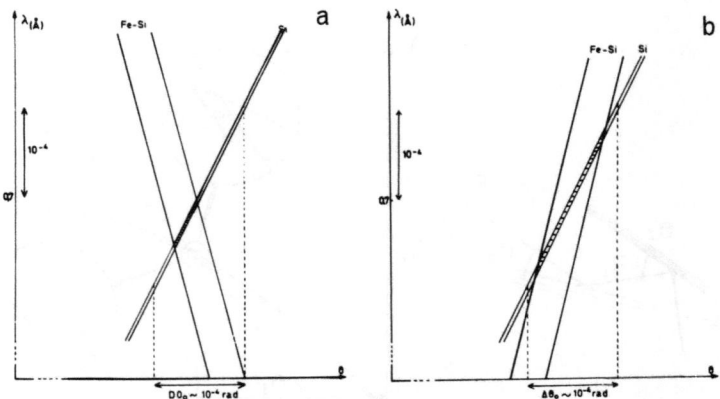

Figure 7. Du Mond diagram representation for : Si-reflection 333 Fe-3.5%Si reflection 110 $\lambda = 0.7$ A a) (+,+) setting b) (+,-) setting.

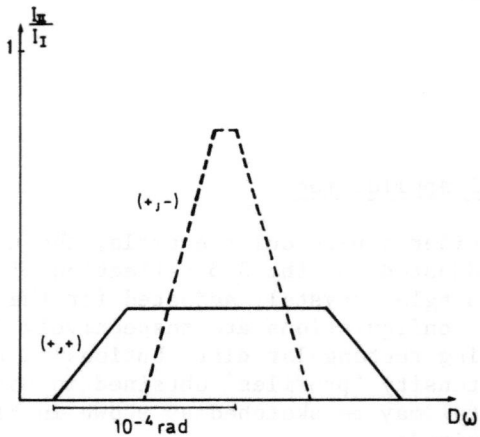

Figure 8. Schematic 'rocking curves', for Fe-3.5% Si in the situations described in Fig. 7a and b.

The two topographs presented in Fig. 9 illustrate these remarks. When designing an experiment one has to decide whether it is more important to obtain higher local intensity and strain sensitivity or a larger, homogeneous field of observation. By increasing the signal to noise ratio, the first solution is more favourable for TV detection where a permanent control of the imaged areas is possible. [3]. For photographic detection the second solution is safer especially when interaction effects are sought as in the stroboscopic experiment described in the next section.

Figure 9. Synchrotron radiation topographs ($\lambda = 0.7$ A). Monochromator Si-reflection 333. Sample Fe-3.5% Si - reflection 110 a) (+,-) setting, b) (+,+) setting; scalemark 1mm

Whenever the width of the pseudo-rocking curve is much broader than the sample reflection width, the image formation can be interpreted in terms of the spherical wave theory (integrated images). In the other extreme case where the incident beam only probes a small fraction of the sample reflection domain, the plane-wave theory of contrast should be used. The examples given in the next section all belong to these 'simple' situations. However, any intermediary case may occur and some theoretical work is still needed to have a full understanding of the experimental results.

18.5 Examples of applications

18.5.1 Stroboscopic observation of magnetic domain wall motion

The aim of the experiment [11], in Fe-3.5% Si, was to observe interactions between crystal defects and magnetic domain walls moving under an alternative magnetic field phase-locked to a mechancial beam chopper. Since 1/60th only of the reflected intensity is allowed to reach the nuclear plate, such an experiment could not be performed on a laboratory generator. Moreover, the first synchrotron radiation trial analyzed in section 4.2 where a Si 333 was used as a monochromator did not provide a sufficient flux to collect the necessary data (about 30 exposures) during one single experimental session.

In the final set-up a Ge 220 adjusted for 0.85A was used. The sample µt value is 3.5 which is a favourable situation to image long range strain fields. The Du Mond diagram is presented in Fig. 10. By comparing with Fig. 7a, it appears that the overlap area is multiplied by a factor 20, moreover the emission power of the source is approximately twice as large at 0.85A as at 0.7A. On the other hand, due to its high absorption, the reflection profile for Ge is far from being rectangular and shows a pronounced decrease in maximum reflectivity and similarly the absorption in the sample is enhanced. These various effects account quite satisfactorily for the gain by a factor 10 observed experimentally. As can be derived from Fig. 10, the divergence of the beam Δ is about five times the reflection width of the sample δ which means that integrated images are expected.

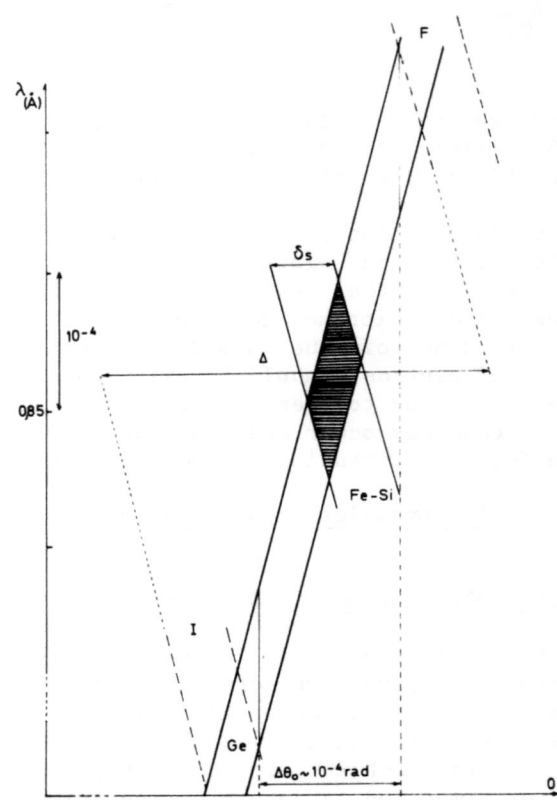

Figure 10. Du Mond diagram : monochromator Ge reflexion 220. Sample Fe-3.5% Si, reflection 110, λ = 0.85 A. Δ : angular range (2.7 10^{-4} rad) between the initial position I and the final position F of the sample for which an overlap region exists. δ_s : sample reflection width : 4.6×10^{-5} rad.

MONOCHROMATIC SYNCHROTRON RADIATION TOPOGRAPHY

Examples of interactions between walls and precipitates or dislocations are shown in Fig. 11a and b where the images of some 90° walls are broadened due to local instabilities. The magnetic domain configuration is sketched in Fig. 12, the extreme positions of the moving wall (the 'vertical' 180° wall for this direction of the alternating applied field) are denoted by heavy lines while the alternately created and annihilated sections of 90° walls are denoted by dotted lines.

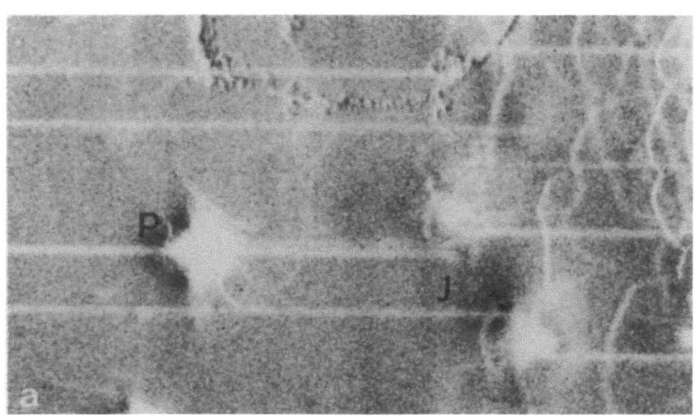

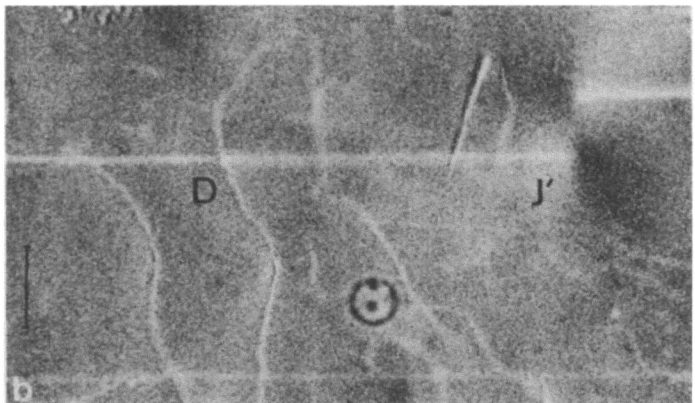

Figure 11. Stroboscopic topography. a) broadening of a 90° wall due to the interaction with a precipitate : the broadening appears between the precipitate P and the moving junction J. b) broadening due to the interaction with a dislocation detected between junction J and dislocation D. Scale mark : 100 μm (after [11]).

443

A sequence of topographs recorded for a series of different field chopper phase relationships is displayed in Fig. 13. Such images represent the first truly dynamic experiment performed with synchrotron radiation, the observation of wall images as narrow as in a static situation implies a remarkable periodicity of the displacements apart from the defect-induced instabilities mentioned previously. The principle of stroboscopic topography is bound to find many applications to other periodic changes in crystals. Extension to higher frequencies (KHz or even more) might be feasible by use of a crystal chopper [6] provided the resulting intensity is not too drastically reduced.

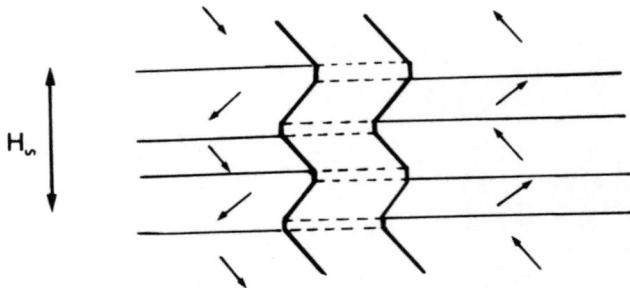

Figure 12. Schematic drawing of the magnetic domain pattern.

18.5.2 Plane-wave topography

One of the interests of plane-wave topography is to provide a very high sensitivity to local variation of the lattice plane spacing and orientation. This enables a detailed study of defect strain fields in otherwise perfect crystals and gives the possibility of separately imaging crystal areas where the reflection conditions are not simultaneously fulfilled. A major application with technological spin-off is the separate imaging of epilayer and substrate in heteroepitaxial systems such as garnets (bubble devices) or III-V compounds (I-R laser diodes). Another interesting field of investigation is the quantitative study of impurity induced lattice parameter variation in crystals.

(a) Heteroepitaxial systems

Fig. 14 shows pseudo plane-wave 444 topographs of a thin magnetic garnet layer (2µm) deposited on a thick GGG substrate [8]. A silicon 220 adjusted for λ = 2.25 Å was used as a first crystal (reflection width 8") and the angular departure from a true parallel setting was quite small ($\theta_{GGG} - \theta_{Si}$ = 3.15°). Such a set-up is not very accurate but suffices in the present example since the Bragg angle peak shift between layer and substrate is large (255"). It can be derived from the topographs that the striation pattern characteristic of the Czochralski grown substrate is not transmitted in the epilayer.

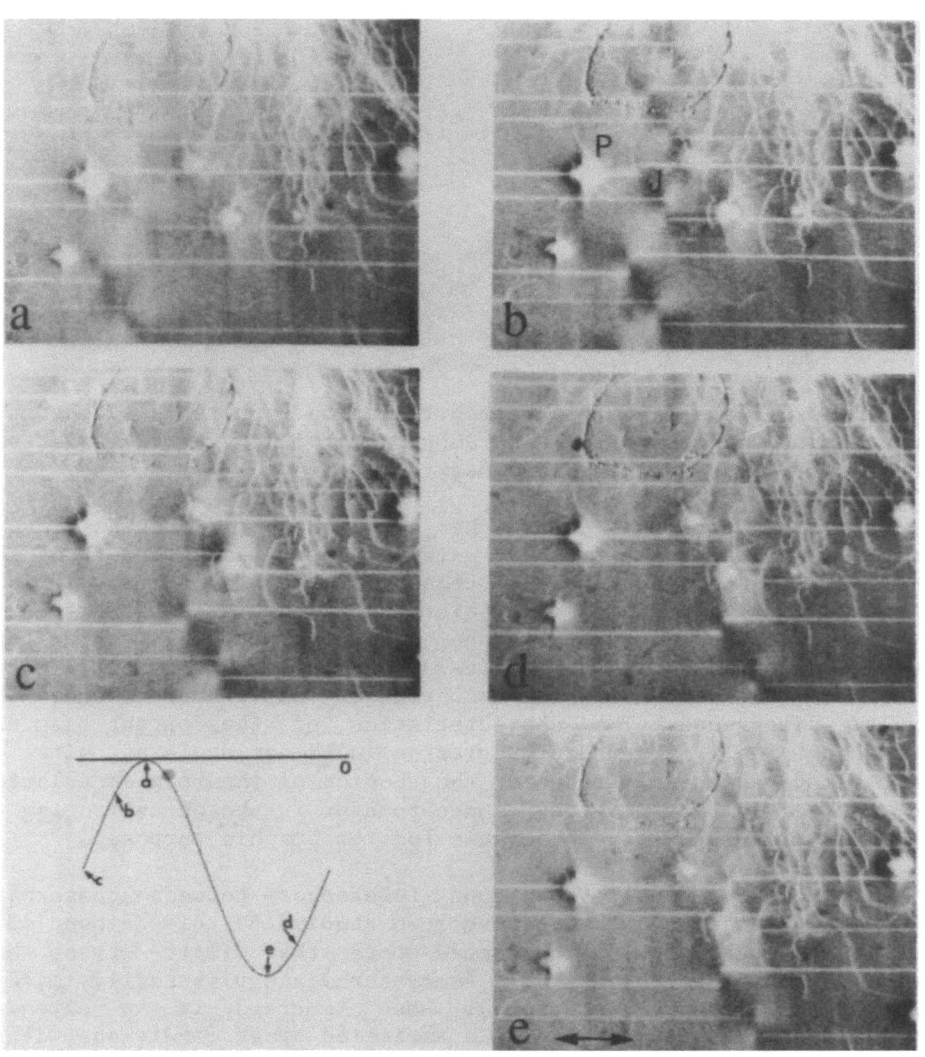

Figure 13. Sequence of stroboscopic topographs showing the displacement of the 'vertical' 180° domain wall. Scale mark 200 μm (after [11]).

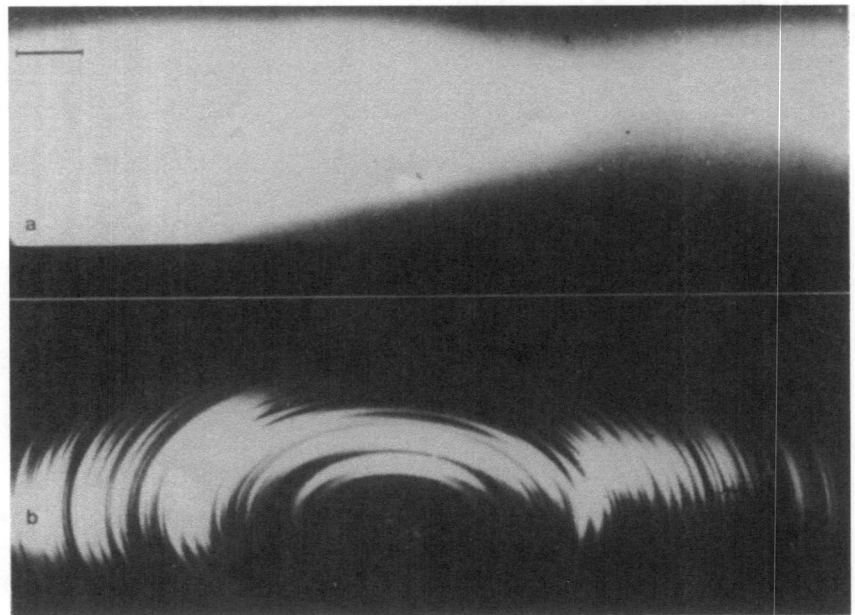

Figure 14. Plane wave synchrotron radiation topographs a) magnetic garnet layer defect free b) GGG substrate with growth striations (courtesy of W. Parrish). Scale mark 2mm

At LURE-DCI, the choice has been somewhat different since a monochromator able to extract an extended quasi plane wave from the white divergent spectrum has been designed [12]. Fig. 15 is a schematic drawing of this monolithic device. The reflections 1 and 2 in the (+,+) setting defines the wavelength average value and bandwidth, whereas 2 (symmetrical) and 2' (asymmetrical) in the (+,-) parallel setting eliminates harmonics and further reduces the angular divergence. The characteristics of the output beam are listed in table 1. By use of a grazing incidence angle at site 1, the final divergence is reduced, the section of the beam is enlarged and the efficiency of the monochromator in photon collecting is enhanced which is mostly favourable for topographic purposes.

Dislocations in III-V compound interfaces between quaternary layers and binary substrates have been studied by this method [13]. Fig. 16 is a schematic of the sample where the misfit stress has been relaxed both elastically (curvature) and plastically (misfit dislocations at the interface). When immersed in an extended parallel beam such a warped sample satisfied Bragg conditions within very limited areas only. Two reflection contours are recorded on the photographic plate, one corresponding to the substrate and one to the epilayer. By rotating the sample, any dislocation can be brought to any desired value of departure from the Bragg angle and a systematic contrast study can be performed.

Table 1

CHARACTERISTICS OF LURE-DCI MONOCHROMATOR

Reflection 1	333
Reflection 2	131
Reflection 2'	131
Output wavelength	1.2378Å
Spectral window	$\Delta\lambda/\lambda = 7.10^{-6}$
Angular divergence	$\Delta\theta = 2.10^{-6}$
Number of photons	$N = 4.10^7$ ph/sec/cm^2
(at 1.72 GeV and 200 mA)	

Figure 15. Schematic of ray paths in the harmonic free narrow band pass monochromator : larger dimension 5 cm.

Fig. 17 shows an overall view of the epilayer contour. Subsidiary maxima characteristic of thin crystal profiles are observed and the focussing of the contours towards the sample edge denotes a progresssive thickening of the layer, typical of the liquid phase epitaxy growth process. The estimation of the local value of the departure from Bragg angle is readily obtained by comparing the plate densitometer track (Fig. 17) to a computed reflection profile. The dislocation images observed in the contour of the thick substrate show the well-known features of double crystal images [14]:

(a) white images at the peak apex,

(b) black-white images on the flanks,

(c) black, thin images on the remote wings.

However, in addition to these main contrasts, Bedynska [15] has predicted subsidiary maxima due to interbranch scattering. Fig. 18 shows a densitometer track across dislocation images recorded for local departures from Bragg incidence $\Delta\theta$ increasing from left to right, the secondary peak is clearly detected and its distance to the main peak, as a function of $\Delta\theta$, follows the predicted behaviour. Fig. 19 gives a display of images in the epilayer contour. Two families of dislocations are present and respectively labelled by full or dotted arrows. It can be derived very rapidly that both families give opposite contrast, except at the peak apex and that a given dislocation shows a reverse behaviour on both sides of the reflection curve. Quite unusual double black and black-white-black broad images are also observed.

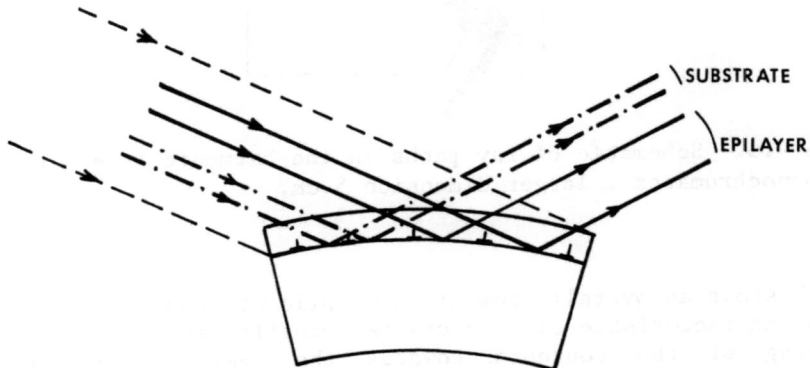

Figure 16. Reflection of an extended plane wave by a curved heterostructure.

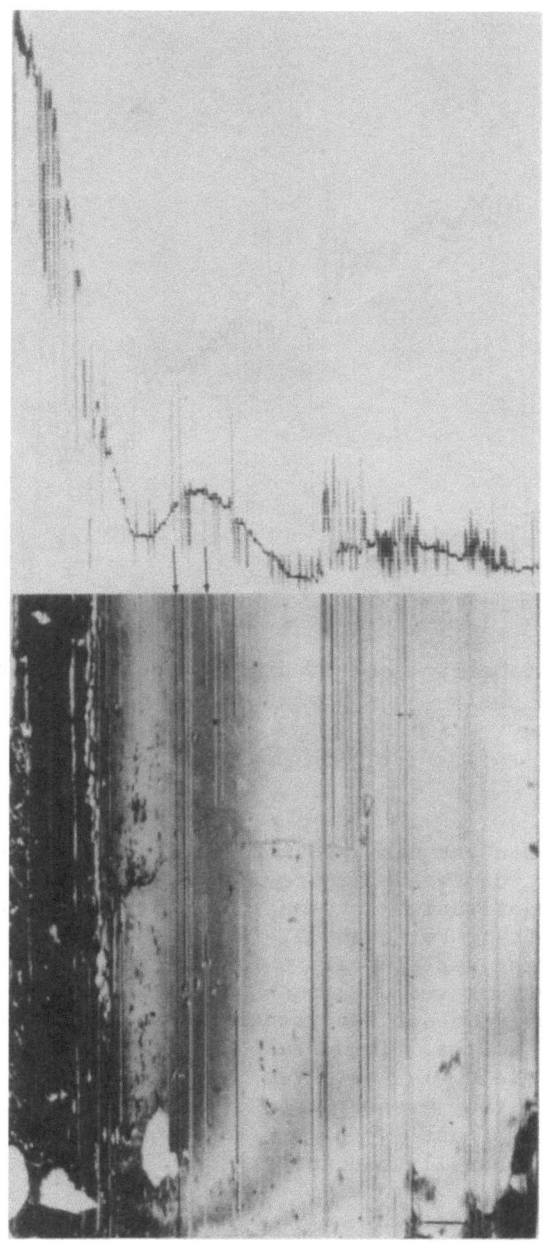

Figure 17. Main and subsidiary maxima in the epilayer contour. Evaluation of the local departure from peak incidence $\Delta\theta$ at a given dislocation by use of the densitometer track (after [13]).

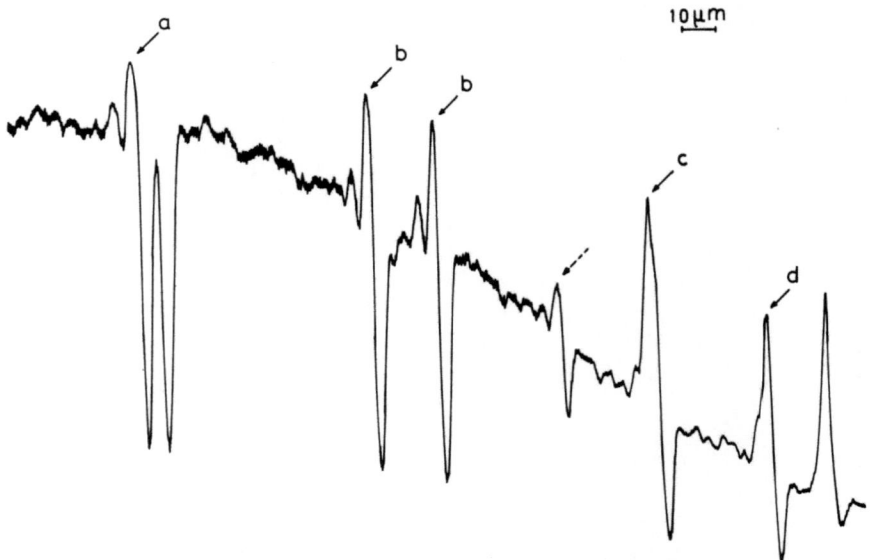

Figure 18. Densitometer track of dislocation images in the substrate contour, around half peak height - variation of spacing for the subsidiary maximum.

The comparison between experimental and computed images for selected values of the departure from Bragg incidence [16] enables one to obtain quantitative information on the strain distribution around the defects. For example, in the epilayer images calculated without introducing surface relaxation are much too thin and the fit is significantly improved when such an effect is taken into account although a factor of about two is still present between observed and computed image widths. This residual discrepancy shows that the actual strain field is not yet properly described in the computation. In the present case of shallow dislocations, surface relaxation effects must then be introduced with more accurate functions than is usually necessary.

Plane wave topography may also be used to provide a high <u>spatial resolution</u>. This is a 'weak beam' technique somewhat similar to one of the methods used in transmission electron microscopy, where the sample is adjusted far from the reflection condition. In the present example, images as narrow as one micron have been obtained. (Fig. 20).

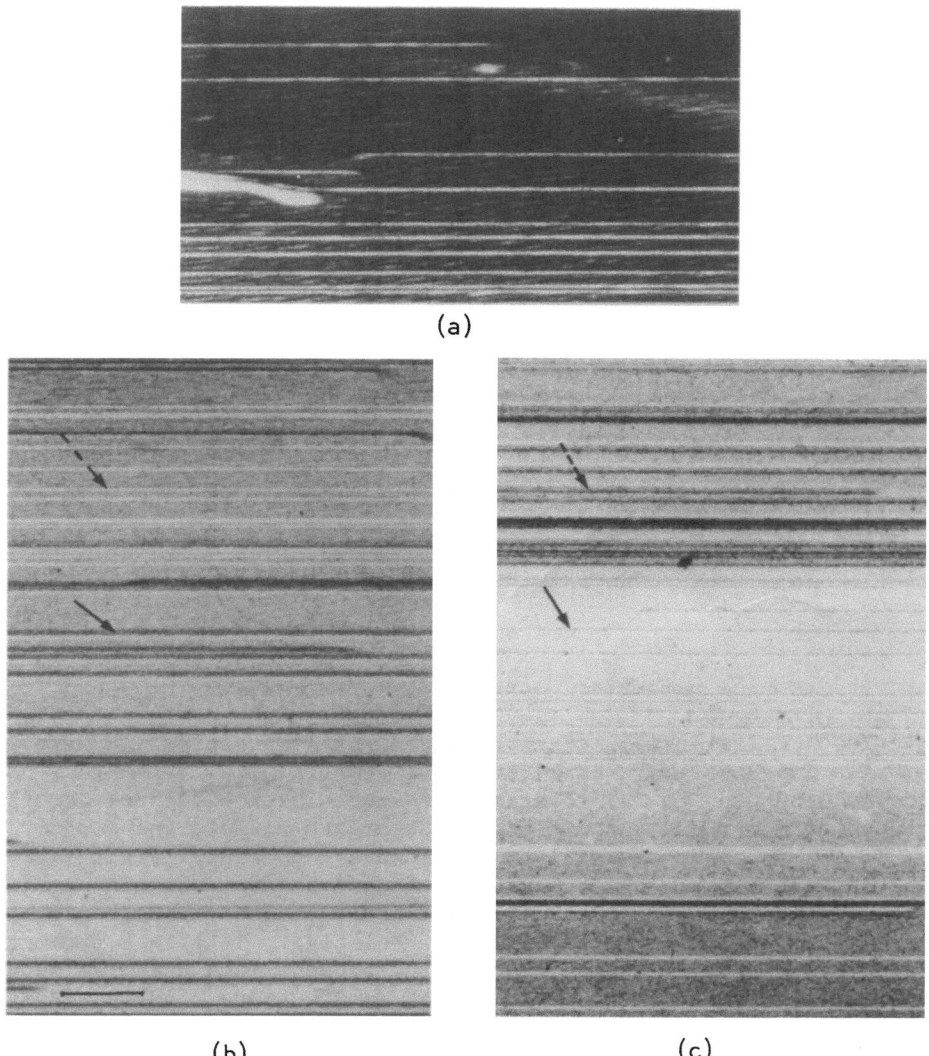

Figure 19. Display of dislocation images on the epilayer contour a) peak apex ; b) negative flank ; c) positive flank. Notice the complex images for single dislocations. Scale mark 100 μm (after [13]).

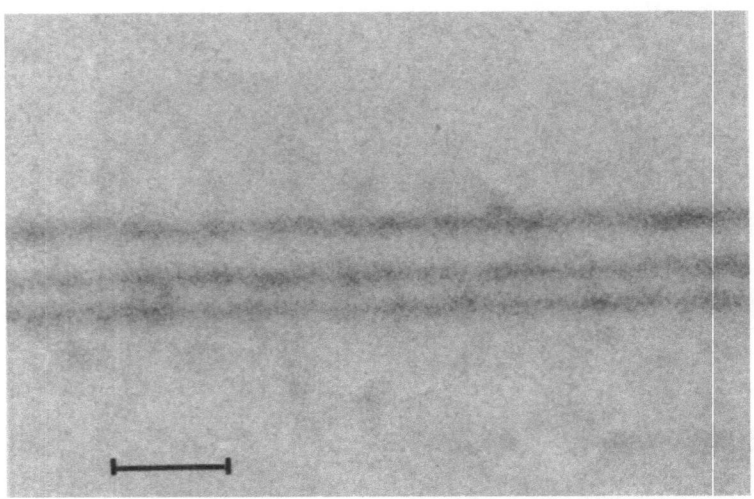

Figure 20. High resolution 'weak beam' images. Scale mark 10 μm.

(b) Lattice parameter mapping

The differential incorporation of impurities in growing crystals induces small variations of the lattice parameter $\Delta d/d$ usually coupled to a rotation of the lattice planes. Accurate measurement of these quantities, which might be of the order of 10^{-6}, is possible by means of plane wave topography. Fig. 21 shows two topographs taken on both sides of an $Sr(NO_3)_2$ sample profile [17]. A contrast reversal between the cube and octahedron growth sectors is observed, denoting a relative lattice parameter difference $\Delta d/d$ of the order of 10^{-5}. between both sectors. Such a crystal is very sensitive to radiation damage and a double crystal experiment in the laboratory would be difficult to achieve. Fig. 22 shows a series of quartz topographs, the angular separation between successive images is about 0.8". Quartz crystals may indeed be studied in the laboratory [18], however the higher flux delivered in the synchrotron radiation set-up enables us to use fine-grain emulsions and more detailed information may be expected particularly when growth sector boundaries are concerned.

Figure 21. Contrast reversal in Sr $(NO_3)_2$ cube and octahedron growth sectors at half peak height on both sides of the profile (experimental FWHM : 5") Scale mark : 5 mm. (courtesy of M. Ribet, F. Lefaucheux and M.C. Robert).

18.6 Conclusion

Synchrotron radiation white beam topography has been very popular from the beginning due to the extreme simplicity and rapidity of data collection. The second generation of experiments, using more or less sophisticated monochromators is less straightforward and might seem repellent to would-be users. The purpose of the present chapter was to point out that such techniques deserve the effort since they provide an unique access to original studies either on fine interaction effects or on high precision distortion evaluation in all kinds of single crystal materials.

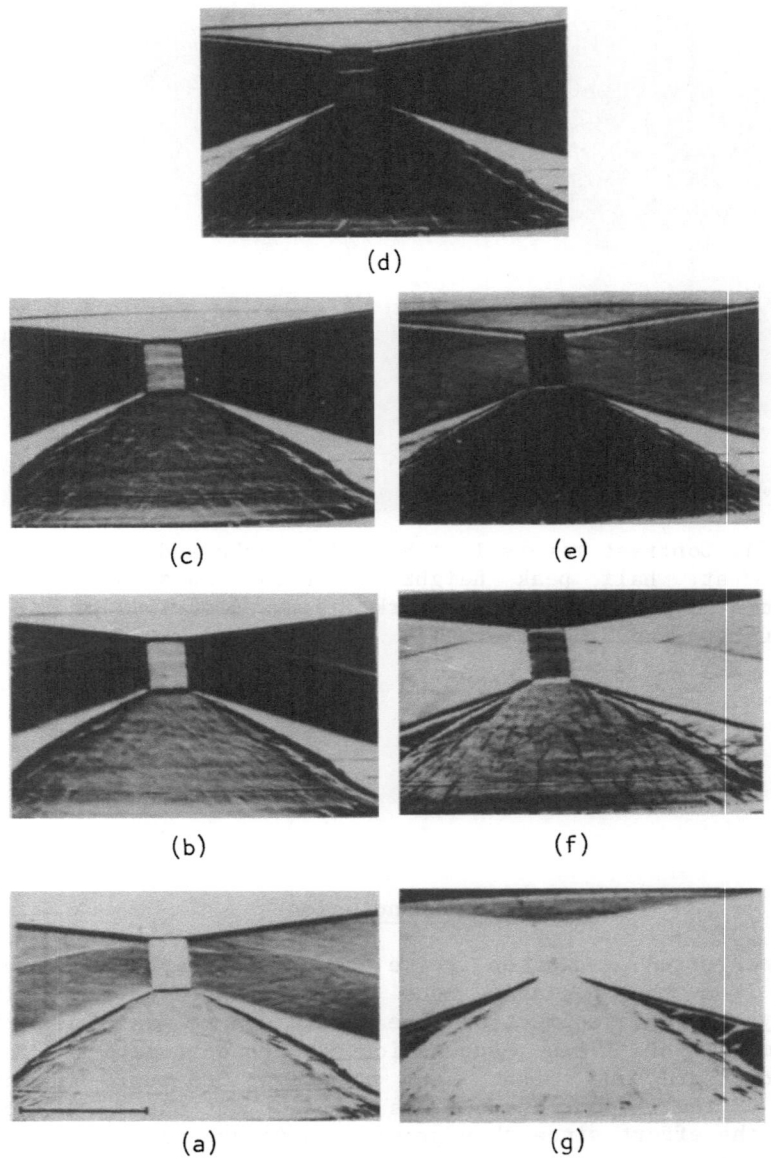

Figure 22. Series of quartz topographs recorded along the profile 0.8" between successive images. Scale mark : 5 mm (courtesy of A. Zarka and T. Soledade).

REFERENCES

1. J. Miltat, this volume, chapter 16
2. U. Bonse, this volume, chapter 11
3. W. Hartmann and J. Miltat, submitted to Appl. Phys. Lett.
4. U. Bonse, (1958), Z. Phyz. 153, 278
5. H. Klapper, this volume, chapter 6
6. M. Hart, this volume, chapter 17
7. M. Sauvage (1978) Nucl. Instr. Meth. 152, 313
8. W. Parrish and C.G. Erickson (1978) Acta Cryst. A34, S4, 331
9. G. Brown and P. Eisenberger (1979) SSRL reports proposal no. 217
10. J.W.M. Du Mond (1937) Phys. Rev. 52, 871
11. J. Miltat and M. Kleman (1980) J. Appl. Phys., in press
12. H. Hashizume, M. Sauvage, J. F. Petroff, B. Capelle, P. Riglet and T. Matsushita, submitted to Jap. J. Appl. Phys.
13. J.F. Petroff, M. Sauvage and P. Riglet (1980) Phil. Mag. (in press)
14. U. Bonse (1959) Z. Naturforsch 149, 1079
15. T. Bedynska (1979) private communication
16. P. Riglet, M. Sauvage, J.F. Petroff and Y. Epelboin (1980) Phil. Mag. (in press)
17. M. Ribet, J.C. Ribet, F. Lefaucheux and M.C. Robert, submitted to J. Cryst. Growth
18. J. Yoshimura, T. Miyazaki, T. Wada, K. Kohra, M. Hosaka, T. Ogawa and S. Taki (1979) J. Cryst. Growth 46, 691.

CHAPTER 19

ENVIRONMENTAL STAGES AND DYNAMIC EXPERIMENTS

B.K. TANNER

19.1 Introduction

In studies involving X-ray topography there is, as with all research, a subtle temptation to allow the study of the technique to obliterate the original scientific problem. There is, of course, a very proper place for carefully conceived model experiments to test theoretical predictions of defect contrast and one must not under-emphasize the importance of detailed examination of image contrast in reaching conclusions about the origin and significance of crystal defects. However, X-ray topography is only an analytic tool, akin to electron and optical microscopy, to be used in conjunction with other measurements. In Lang's studies of diamond [1,2] for example the crystals were examined by X-ray topography, visible light and ultra-violet microscopy, cathodoluminescence topography and ultra-microscopy. It is when X-ray topographic results, which provide data on lattice distortions, are combined with measurements of totally different parameters that some of the most valuable insights are obtained.

Very many X-ray topographic experiments involve the study of lattice defects introduced during or after some well controlled physical or chemical process. It is in relatively few cases, for example in the study of minerals, that one attempts to deduce the growth conditions from the defects. Usually the conditions are specified and the origin of the defects is sought. Two approaches may be adopted.

(a) The same specimen is examined by X-ray topography after each step in the process.

(b) Many specimens are examined, samples being taken from the batch at each stage in the treatment.
The former procedure is preferable in a number of ways as the latter relies on a statistically significant number of samples being taken and suffers the disadvantage that the evolution of individual defects cannot be traced. However, it is less convenient to the collaborating crystal grower or device engineer and there exists a

real risk that removal of the specimen may introduce spurious effects. The ideal experiment is one in which X-ray topographs are taken in situ as the changes take place.

Although many experiments fall into both classes, in situ experiments will, for convenience, be divided into two classes.

1. Experiments in which simultaneous measurements of another physical parameter of the crystal is required.

2. Dynamic experiments during which time-dependent changes in micro-structure are followed simultaneously or in rapid sequence. The first type of experiment is accessable to all topographers with a little ingenuity, the second is restricted to those with access to a high brightness X-ray source.

19.2 Simultaneous measurements

As discussed in several reviews [3,4] the vertical resolution δ of a topograph taken with Bragg planes vertical is given by

$$\delta = V L/D \qquad (1)$$

where V is the projected height of the source, L is the specimen to plate distance and D is the specimen to source distance. In order to minimize the exposure time for a traverse topograph, we select the horizontal width of the beam slits to be equal to the horizontal length H of the source and the distance D such that the $K\alpha_1 - K\alpha_2$ doublet is just resolved in the rocking curve. Thus

$$H = D \Delta \theta_{\alpha_1 \alpha_2} \qquad (2)$$

where $\Delta \theta_{\alpha_1 \alpha_2}$ is the angular separation of the doublet.

Combining (1) and (2) yields

$$\delta = L \Delta \theta_{\alpha_1 \alpha_2} (V/H) \qquad (3)$$

The geometry of most X-ray tubes is such that when a take off angle of a few degrees is used, $V \simeq H$. $\Delta \theta_{\alpha_1 \alpha_2}$ is typically 2×10^{-4} and hence $\delta = 2 \mu m$ only if L = 1 cm. Unless one is prepared to sacrifice exposure time by increasing D, it is thus essential to place the photographic plate within one or two centimetres of the crystal if detail is not to be lost. A very severe spatial constraint is therefore imposed on any in situ experiment. In addition, diffracted beam slits must be used in the transmission geometry and these provide an additional obstacle to the design of a special stage. Finally, if the popular Lang technique is used, the specimen must still be traversed across the beam.

In general the approach in the past has been to design stages for in situ experiments which will fit onto the scanning stage of a conventional Lang camera. An exception is the major modification reported by Argemi et al. [5]. In order to keep the specimen vertical for creep experiments using a weight attached directly to the sample, they designed a scanning camera in which the incident beam is not generally horizontal. As design of special stages cannot be divorced from the specific problem to be investigated a number of papers will be reviewed in which information on stage design is given. These will be grouped according to the parameter simultaneously measured.

19.2.1 Electric field

This is perhaps the easiest of parameters to monitor as electrodes can be evaporated directly onto the specimen surface. If commercial quartz oscillators are used, no preparation is needed as the electrodes are already present on the thin plates. Under an oscillatory a.c. field the standing waves in the quartz oscillator can be revealed on the topographs and the presence of unwanted modes detected [6,7]. Application of a d.c. field in situ results in curious radiation damage effects in the quartz [8,9] which appears to be related to changes in structure factor. Niizeki and Hasegawa [10] were able to follow polarization reversal due to movements of $180°$ ferroelectric domain walls in barium titanate revealed by anomalous dispersion contrast. In a pioneering experiment, Itagaki [11] demonstrated that dislocations in ice were charged as they were seen to vibrate in an a.c. electric field applied parallel to the specimen surface. In view of their simplicity it is surprising how few reports of electric field experiments exist in the literature.

19.2.2 Magnetic field

While it proves easy to apply intense electric fields to the sample, it is extremely difficult to apply magnetic fields of more than 50 Oe to a crystal mounted in a Lang camera. Fields up to 50 Oe can be obtained by use of a large Helmholtz pair of coils mounted directly on the traversing stage. Many field directions are, however, precluded by the difficulty of accommodating the diffracted beam slits and plate holder close to the sample. A very limited length of traverse is usually only permitted. Although in fields of this strength substantial magnetic domain evolution occurs in soft magnetic materials, there are serious problems associated with stray fields from the X-ray generator and camera components and a real risk of accidental changes in meta-stable configurations induced by unsuspecting colleagues. Only one report of conventional X-ray topography studies of domains in soft ferromagnets under systematically applied fields is apparent in the literature, the work on iron whiskers by Nagakura and Chikaura [12].

However, as early as 1966, Yamada, Saito and Shimomura [13,14] had succeeded in observing the evolution of S domain walls in antiferromagnetic NiO in fields up to 9.5 KOe. A double crystal arrangement was used employing an asymmetric reflection from the specimen crystal. As in the single crystal Berg-Barrett method the photographic plate was located very close to the crystal and a region of specimen 2 mm wide imaged without scanning. By placing the photographic plate parallel to the electromagnet pole pieces (Fig. 1) a small pole gap and hence high fields could be obtained.

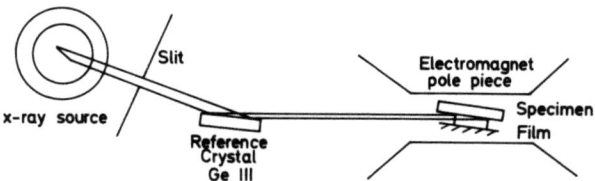

Figure 1. Double crystal arrangement for studying antiferromagnetic domains in NiO (Courtesy of S. Saito and colleagues [13])

19.2.3 Low temperature

Many phenomena of interest to solid state physicists occur at low temperatures where the effects of lattice vibration are reduced. Low temperature topography is much easier to perform than is realised, provided that temperatures below about 80K are not required. If a competent glass-blower is available a very useful cryostat can be fabricated for very little cost. Fig. 2 shows the cryostat made at Durham capable of reaching about 100K and used in studies of domains in $KNiF_3$ [15]. Below the liquid nitrogen reservoir a copper cold finger extends to the specimen holder. This is covered by a glass cap with Mylar X-ray windows and the specimen area can be evacuated through the Dewar walls. Temperature measurement is best made using a silicon diode thermometer as the close proximity of the electrical lead-throughs introduces severe thermal e.m.f. problems when using thermo-couples. (At a constant current of 100 μA a silicon diode develops about ½ volt at room temperature, rising to about 2 volts at liquid helium temperature). A heater on the cold finger permits temperatures above 100K to be maintained.

A disadvantage, though not serious if small specimens are used, is that no adjustment about the X-ray beam axis is permitted. In the commercial cryostat, designed for Lang topography used by Mathiot and Petroff [16,17] and the Durham group [47] for study of magnetic domains, a standard liquid helium reservoir cryostat is supported above the Lang camera traverse unit by three legs. The specimen is contained in a thin tail (Fig. 3) made from waveguide section extending some 10 cm below the reservoir base and cooled by helium gas flow. A double beryllium window allows passage of X-rays and Bragg angles in symmetric reflection up to 45° can be used. Thermal insulation of the tail is improved by the presence of a copper shield in the vacuum space cooled by the shielding liquid nitrogen reservoir. A worm drive mechanism is incorporated to rotate the specimen about the beam direction and the system has proved extremely stable. A single fill of the 2 ℓ helium reservoir lasts for 10 hours at 4.2K, quite sufficient for topography of small crystals. Temperature control can be better than 0.1K using a standard commercial controller.

Figure 2. Inexpensive glass cryostat for use at temperatures down to liquid nitrogen temperature.

Figure 3. Tail of the liquid helium cryostat designed for Lang topography.

19.2.4 High temperature

One of the earliest high temperature studies reported was that into the dislocation density in aluminium during annealing [18-20]. A small resistance furnace was constructed by Nøst et al which performed well up to 510°C but construction details have not been published.

An extremely elegant little furnace (Fig. 4) has been constructed for work up to 600°C by Kume and Kato [21] and used in the study of phase transitions in quartz. Full construction details are given in the above reference. The spatial temperature gradient was less than 0.2°C/mm across the specimen and the temperature variation over a 10 hour exposure was only 1°C. In operation the specimen, on its independent rotation axis, is set for a Bragg reflection and the furnace independently rotated so the diffracted beam passes normally through the aluminium window A_2. Good spatial resolution can be obtained as the photographic plate can be placed very close to this window. Further reflections from a zone can be taken without reducing the temperature.

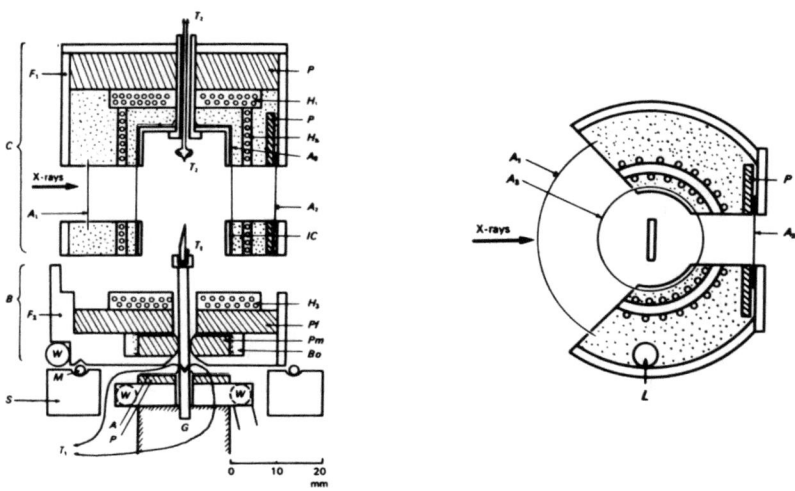

Figure 4. Furnace for Lang topography up to 600°C (Courtesy of S. Kume and N. Kato [21]

19.2.5 Stress

Nøst and Nes [22] continued the elevated temperature work on aluminium by a series of section topography studies of dislocation density during plastic deformation. Unfortunately few details of apparatus were given either there or in reports of subsequent room temperature studies [23].

Oki and Futagami [24] have published the design of a small tensile stage capable of taking topographs under loads of up to 4 Kg and temperatures up to 800°C. With this device (Fig. 5) they were able to follow dislocation movements during creep deformation silicon specimens 1.5 mm thick using the scanning Soller slit technique [25]. (These experiments are almost classed as 'dynamic' as exposure times were only minutes). The chamber was made from stainless steel and a load applied by a steel wire leading outside the chamber. Coiled thermocoax heaters were supplemented by nichrome wire above and below the specimen and the chamber was sufficiently thin that specimen to plate distance did not exceed 5 cm.

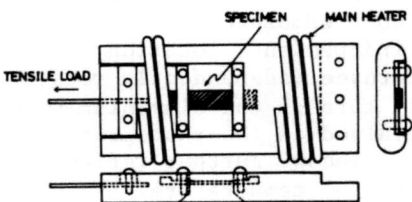

Figure 5. High temperature tensile stage. (Courtesy of S. Oki and K. Futagami [24]

There is a serious problem associated with applying a true axial load at low stresses when using very thin specimens. For iron, to obtain µt~3 with MoKα radiation a specimen thickness of 100 µm is required. Bowen and Miltat [26] have developed a room temperature tensile stage incorporating precision linear roller bearings to carry one of the specimen mounts and maintain this in the same plane as the other fixed mount (Fig. 6). Stress is applied by pushing a micrometer against a steel spring cantilever, and measurement of the cantilever deflection using a capacitive transducer enables the applied force to be maintained. Forces up to 30 Kg weight could be applied, with a minimum load, when the specimen is vertical, of 75 g due to the weight of the stage.

This stage is not suitable for very soft materials such as copper. A lighter stage had earlier been constructed for anomalous transmission studies on 1 mm thick copper crystals by Young and Sherrill [27]. Loads up to 50 g could be applied. Both these stages fitted directly onto a standard goniometer head.

Minari, Pichaud and Capella [28] have reported an ingenious device in which stress was applied by the weight of a float whose immersion could be controlled by a micrometer (Fig. 7). As the apparatus must be vertical, a camera modification such as described by Argemi et al [5] was necessary. Although the device introduced little bending stress, subsequent experiments using a bending stage [29] showed that nucleation of dislocation sources in copper was associated with slight bending stresses. The latter experiments were performed on a stage in which the copper crystal was cemented by one edge to a silicon single crystal wafer 200μm thick and the silicon bent by deflection of the free end by a micrometer. Topographs were recorded through the silicon plate, and the stress derived from the local radius of curvature which could be measured from the shift in Bragg peak across the crystal.

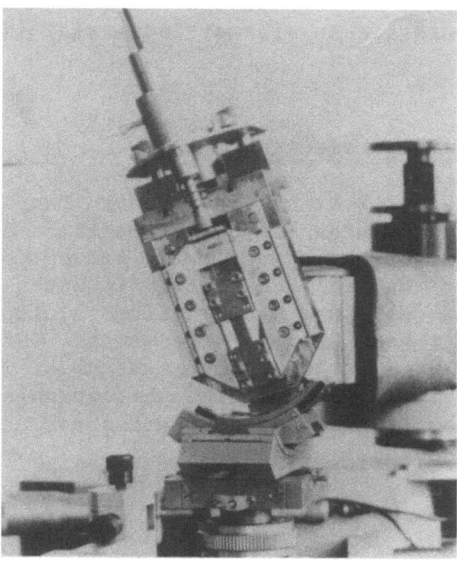

Figure 6. Room temperature tensile stage designed to apply axial loads to thin specimens (Courtesy of J. Miltat and D.K. Bowen [26])

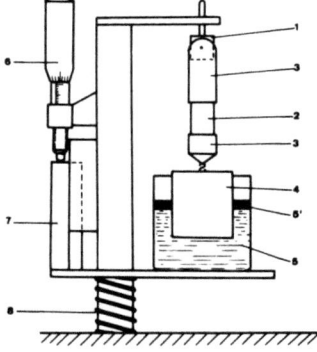

Figure 7. Tensile stage incorporating a float mechanism for very smooth loading. (Courtesy of F. Minari and colleagues [28])

Other deformation stages have been described by Nittono [30] Fukuda and Higashi [31] and Jourdan and Sauvage [32].

19.2.6 Time lapse in a corrosive environment

A number of studies have involved the action of atmospheric oxygen on reactive crystals such as tin [33], cadmium and zinc [34,35] where no special stage design has been necessary. Studies of chemical attack provide an area of industrial importance where few have ventured.

19.2.7 Crystal growth parameters

In situ crystal growth of ice on a Lang camera has been performed by Oguro and Higashi [36]. A modified Bridgman method was used (Fig. 8) in which the crystal was grown downwards from the seed D into the growth cell E immersed in water kept at 4°C by the heater H. The growth cell was slowly raised by the motor M and as the crystal grew, heat was extracted at the surface of the growth cell where the ambient air temperature was -5°C. The growth cell and crucible were thin (7 mm) so that the photographic plate could be brought near to the crystal and topographs taken of the dislocation arrangement during crystal growth.

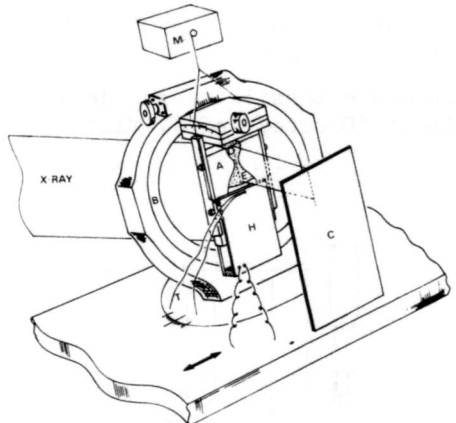

Figure 8. Stage for in situ crystal growth experiments on ice. (Courtesy of M. Oguro and A. Higashi)

A further in-situ experiment worth mentioning here is the study of melting of gallium by Wenzl and Mair [37,38]. Here the interface temperature was cleverly measured to an accuracy of 10^{-4}K by growing a thermistor into the crystal.

ENVIRONMENTAL STAGES AND DYNAMIC EXPERIMENTS

19.3 Dynamic experiments

19.3.1 Conventional source experiments

The only difference between these experiments and those discussed above is the availability of a much higher X-ray flux and hence a substantial reduction in exposure time. Developments in rotating anode targets have led to sources of loading over 30kW on 5 mm^2 target area. Hagen and Queisser [39] exploited such a high intensity in a Berg-Barrett topography study of the growth of epitaxial Ge layers on GaAs substrates. Use of an asymmetric reflection enabled the whole 5 x 5 mm surface to be imaged without scanning and the reflection geometry is such that the epitaxial reactor need not be extremely thin (Fig. 9). The photographic plate was moved step-wise for each exposure and an array of 10 x 10 topographs obtained on one plate. Exposure times on Ilford G5 nuclear emulsions were between 2 and 20 sec depending on the thickness of the epilayer. Generation and evolution of misfit dislocations was followed.

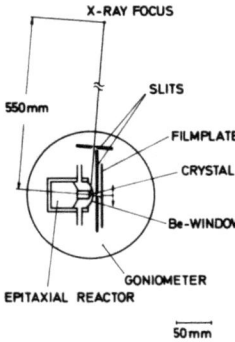

Figure 9. Reactor for _in situ_ Berg-Barrett studies of epitaxial growth (Courtesy of W. Hagen and H. Queisser [39])

Chikawa and his co-workers have exploited their T.V. imaging system [40] in pioneering dynamic experiments Using a small high temperature straining stage they observed directly dislocations moving at up to 0.3 mm sec^{-1} during plastic deformation. Although the dislocation sources could not be identified, new dislocations were introduced as half loops from the crystal surfaces. Perhaps the most ambitious experiments have been their dynamic studies of growth and melting [41,43]. These studies have led to a significant advance in the understanding of swirl defects [44]. The heating stage was essentially very simple (Fig. 10). Melting of the crystal (at 1412°C) was achieved by placing it between carbon heaters in a double walled chamber, with the wall space evacuated. Argon gas was used in the chamber and the X-ray beam passed through the carbon heaters surrounding the specimen. Beryllium windows were incorporated into the vacuum chamber.

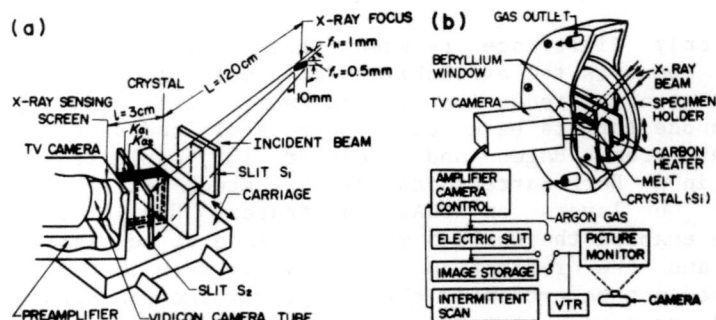

Figure 10. Heating stage for in situ Lang topography studies of melting and solidification of silicon. (courtesy of J. Chikawa [43])

19.3.2 Synchrotron radiation experiments

There are three properties associated with synchrotron radiation (SR) which makes it ideal for dynamic and in-situ experiments.

(a) The extremely high flux means that exposure time on nuclear plates are short, of the order of seconds, and the video imaging can be employed to perform truly dynamic experiments.

(b) With a small source typically 50 to 100 m from the specimen the photographic plate or TV detector may be placed several cm from the crystal without impairing resolution. At NINA, a geometrical resolution of 1μm could be obtained at a specimen to plate distance at 10 cm [45,46]. In addition, with specimens of less than about 1 to 2 cm^2 superficial area there is no need to scan. Consequently the geometrical constraints on special stages are relaxed and it becomes straightforward to incorporate, for example, electro-magnets and cryostats around the sample [47].

(c) As the source is spectrally continuous several reflections can be imaged simultaneously, although not with optimum resolution. This is often extremely valuable when destructive experiments are being performed on a sample. Further, each point on the crystal selects its own wavelength for Bragg reflection and bent crystals yield good topographs. This is particularly important as crystals often distort during in situ experimentation. If more than one grain or sub-grain is present, all grains are imaged simultaneously, and the misorientation between grains deduced from the image separation. In recrystallization studies this feature is most important.

ENVIRONMENTAL STAGES AND DYNAMIC EXPERIMENTS

The pioneering quasi-dynamic experiment using SR was performed by Bordas et al [48]. This was a preliminary study of the cubic to tetragonal ferroelectric phase transition in BaTiO$_3$. The crystal was mounted with alumina cement on the tip of a thermocouple and both crystal and thermocouple placed inside a copper block with a hole for the passage of X-rays such that the angular aperture for diffracted beams was 60°. Soldering iron elements were used to heat the block to around 130°C. As exposure times on nuclear emulsions were only 30 sec, the thermal inertia was sufficient to maintain the temperature constant to 0.25°C over each exposure without the need for any electronic feedback control.

The above experiment, and a subsequent one to study changes in ZnS polytypes on heating [49] acted as a spur to utilize SR for magnetic domain experiments. In a programme to study antifer-romagnetic domains in the cubic perovskites KNiF$_3$ and KCoF$_3$ [50,51], it was shown that quasi-dynamic X-ray topography at temperatures down to 4.2K and in fields up to 15 KOe was straightforward [47]. Most of the experiments were performed with the cryostats illustrated in Figs. 2 and 3 mounted in an Oxford Instruments 4 inch water cooled electromagnet. However for experiments involving the highest available fields and a fixed temperature of 77K the samples were immersed in a glass Dewar with a narrow (25mm) tail and no X-ray window. Because of the extremely high flux the 20 fold increase in exposure times due to the absorption in the glass (20 sec to 6 min) was quite acceptable. Work down to liquid nitrogen temperature is extremely easy if no magnetic fields are required and simple immersion of the sample in an expanded polystyrene trough resulted in good quality topographs being obtained. This simple cryostat, with X-rays penetrating the polystyrene wall and liquid nitrogen has been used in experiments to study the effect of stress on anti- ferromagntic domain walls in KCoF$_3$. The crystal was mounted in a slotted perspex holder in the polystyrene trough and weights applied directly [53]

At the other extreme of temperature, MacCormack and Tanner [54] studied the recrystallization of iron 3.5% silicon at 1000°C directly. Prior to recrystallization no Laue images were visible and as each grain recrystallized its individual Lane images could be detected on polaroid film. In these preliminary experiments a standard resistance heated furnace was hoisted onto the X-ray beam line and diffracted X-rays emerged through the wall of the alumina insert tube (Fig 11a). Polaroid films were exposed for 30 sec and exposure taken every 2 minutes. Individual grain growth could be followed and growth rates measured (Fig 11).

Subsequently Gastaldi and Jourdan [55] have studied the recrystallization of aluminium at LURE. They used a vacuum insulate furnace built from a high vacuum copper gasket with large area beryllium windows on each end. With Inconel coated resistance heaters mounted in the chamber walls, 600°C could be reached. Recrystallizations were performed at 300°C and it proved possible to observe individual growth defects evolve [55,56] as Kodak M films were exposed every 3 minutes. Kalman et al [57] have demonstrated that information on very small crystallites can be obtained from streaking of Laue spots during recrystallization of Xe and Kr.

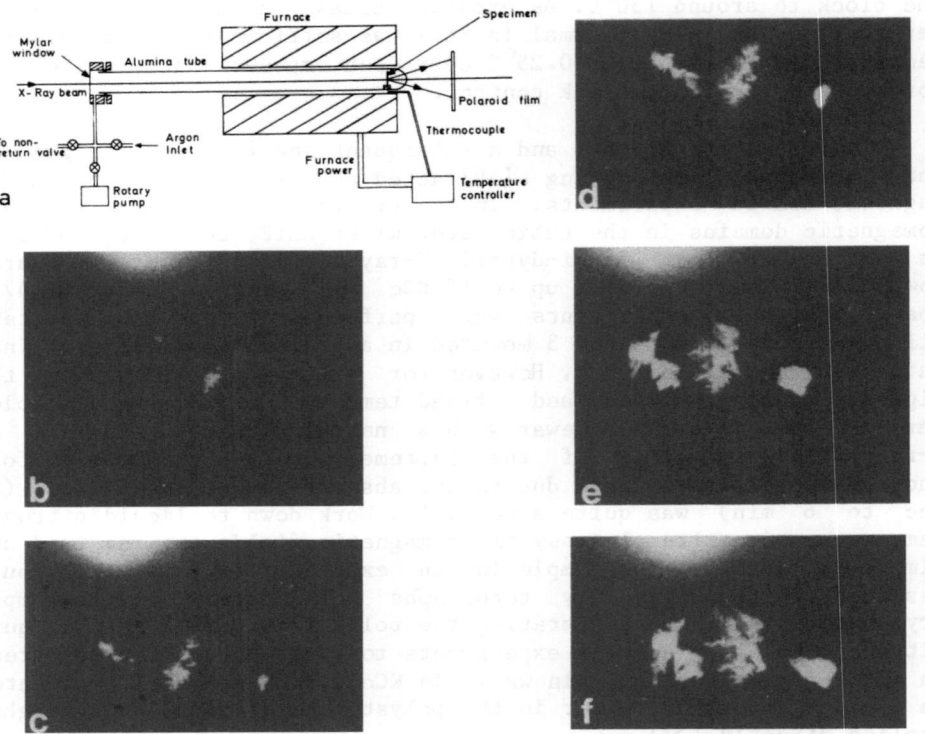

Figure 11. (a) Apparatus used for <u>in situ</u> recrystallization studies of iron-silicon using synchrotron radiation, (b) - (f) sequence of synchrotron topographs showing grain growth.

Increase in the space available round the specimen means that SR experiments on soft magnetic materials at room temperature are easy to instrument and a number of quasi dynamic studies of ferromagnetic domains in iron-silicon [58,59] iron whiskers [60] and rare earth iron alloys [61] have been reported.

Miltat and Bowen [62,56] have reported plastic deformation experiments on iron-silicon using the straining stage shown in Fig 6. Useful data was obtained at high strains where lattice rotation is severe. A similar quasi dynamic deformation experiment was performed on copper by Buckley-Golder and Tanner [63].

ENVIRONMENTAL STAGES AND DYNAMIC EXPERIMENTS

19.4 Instrumentation developments at SR sources

There can be no doubt that the availability of synchrotron radiation has revolutionized the scale and scope of X-ray topographic experiments. We have now passed beyond the 'sealing wax and string' attitude of the pioneers and it is vital at this time for sophisticated, high precision instruments to be installed at all the major synchrotron radiation facilities. As only one or, at most, two topography stations will be accommodated on each ring, it is important that a very close liaison exists between topography groups as the instruments will be communally used and hence must fulfill a wide range of needs. In the U.K, the topography groups have collaborated on specifying the outline requirements for instruments at the new synchrotron radiation source (SRS) under construction at Daresbury. Three communal instruments have so far been funded. Designs of these have been effectively frozen and construction is in progress. While it would be improper at this stage to report detailed designs, it is important to publicize the outline specifications so that experimenters can plan their own projects in the light of the new facilities.

19.4.1 White radiation camera

A white radiation camera is to be installed on the first X-ray beam line at the Daresbury SRS, and will be ready for the commissioning of the SRS in late 1980. Design and construction are the responsibilities of Dr. D.K Bowen and colleagues at Warwick University. The instrument is designed for loads up to 25 Kg in particular the environmental stage (section 19.4.4). The geometry is novel. The specimen arm is provided with two permanent rotation axes, one _coincident_ with the X-ray beam and the other perpendicular to it. These axes intersect at the specimen and no translation of the specimen accompanies these rotations. A translation stage is also incorporated for positioning small crystals, scanning large wafers or step scanning. Other translation or tilting axes may be added. A single vertical axis of rotation will be available for mounting heavy magnets or exceptionally heavy specimen stages such as a crystal puller. The detector arm has two orthogonal rotation axes such that the detector can assume almost any orientation and will always point at the specimen.

All camera motions are driven by stepping motors under complete minicomputer control. The control programme will include simple means of orienting camera and detector, automatic selection of any diffraction conditions for crystals whose orientation is specified, complete programming of dynamic experiments (e.g. temperature ramps), general housekeeping facilities and, of course, manual override. Topography experiments will be controlled from a teletype keyboard and a certain readjustment in the philosophy of the experimenter may initially be necessary.

19.4.2 Double crystal camera

A precision double axis goniometer is also being constructed by the Warwick group to Prof. Hart's design (as is being used for the interferometry station), for use in the double crystal camera. The latter has two novel features. First, the whole goniometer can be rotated around the X-ray beam so that the axes can point horizontally for normal use, vertically downwards for immersing the specimen in a large magnet, or vertically upwards for mounting special stages such as cryostats. Secondly, the station itself will be mounted ahead of the white radiation camera and in 'tandem' with it, so that with channel-cut or channel-mounted crystals on each axis the whole camera can act as a beam conditioner for the 'white' radiation camera. This should give great versatility in the selection of optimum conditions for any given experiment. Both cameras will be controlled through the same minicomputer system.

19.4.3 X-ray TV detector

This is being constructed under the direction of the author by R.T. Laboratories, Wells. An indirect conversion system is being used, as this is less susceptible to radiation damage and inadvertent pointing into the direct beam. The image formed on a fluorescent screen will be optionally magnified optically and intensified by a high resolution single stage electrostatic image intensifier incorporating an electronic zoom facility This output will be viewed by a SIT TV camera. Recording will initially be on video-tape. The unit is small, robust and ideally suited to the complex motions, required by the X-ray camera.

19.4.4 Environmental stage

The responsibility for this item is vested with Professor Sherwood's group at Strathclyde University. Built from standard UHV components for convenience the stage will consist of an evacuable cylindrical chamber 30 cm diameter and 10 cm deep. X-rays will enter and emerge through beryllium windows and blanked off ports will allow insertion of windows for observation or sample irradiation, for example with U.V. light. Controlled admission of gases will enable gas - solid reactions to be studied. Various resistance heating elements will enable $1000^{\circ}C$ to be reached. Temperature control will be via the microcomputer interface controlling the white radiation cameras.

19.5 Future developments

Particularly in metallurgy and solid state chemistry, there are a number of dynamic X-ray topography experiments simply waiting to be performed. Probably the most widely publicized X-ray topography experiments of the next decade will be such dynamic studies.

ENVIRONMENTAL STAGES AND DYNAMIC EXPERIMENTS

Certainly a very large part of the case for synchrotron radiation facilities rests on this type of work.

Unfortunately, use of synchrotron radiation and communal equipment will mean a drift from the home based laboratory work to the central facilities. It is vital to remember that specimen selection and characterization is best done steadily on a conventional generator away from the constraints of 'beam time' present at the SR facilities. When the experiment is set up and working, then and only then should it be transferred to the synchrotron radiation facility. The beam lines must not be clogged up by a few people setting up elaborate dynamic experiments.

Further, the impression must never be given that the only good results are those which come from synchroton radiation experiments. Adequate funds must be available for home based back-up experiments and the funding bodies must be made aware of this situation and pressed for fair spread of resources. While dynamic sychrotron radiation experiments will certainly feature as some of the highlights of research in the 1980s they must not be at the expense of all else.

REFERENCES

1. A.R. Lang (1974) Proc. Roy. Soc. Lond A 340, 233
2. A.R. Lang (1977) J. Crystal Growth 42 625
3. A.R. Lang (1970) in Modern Diffraction and Imaging Techniques in Material Science (ed. S. Amelinckx, R. Gevers, G. Remaut and J. Van Landuyt) N. Holland p. 407
4. B.K. Tanner (1976) X-ray Diffraction Topography, Pergamon Press
5. R. Argemi, C. G'Sell and B. Baudelet (1971) Rev. Sci. Inst. 42 1711
6. A.C. Greenham, B.J. Isherwood and C.A. Wallace (1965) Brit. J. Appl. Phys 16, 1759
7. B.J. Isherwood and C.A. Wallace (1975) J. Phys. D. 8, 1827
8. S. Yamashita and N. Kato (1975) J. Appl. Cryst. 8, 623
9. K. Yasuda and N. Kato (1978) J. Appl. Cryst. 11, 705
10. N. Niizeki and M. Hasegawa (1964) J. Phys. Soc. Japan 19, 550
11. K. Itagaki (1970) Adv. X-ray Analysis 13, 526
12. S. Nagakura and Y. Chikaura (1971) J. Phys. Soc. Japan 30, 495
13. T. Yamada, S. Saito and Y. Shimomura (1966) J. Phys. Soc. Japan 21, 672
14. K. Nakahigashi, N. Fukuoka and Y. Shimomura (1975) J. Phys. Soc. Japan 38, 1634
15. M. Safa, D. Midgley and B.K. Tanner (1975) Phys. Stat. Sol (a) 28, K 89
16. A. Mathiot and J.F. Petroff (1974) Mat. Res. Bull. 9, 319.

17. A. Mathiot and J.F. Petroff (1976) J. Appl. Phys. **47**, 1639.
18. B. Nøst and G. Sørensen (1966) Phil. Mag. **13**, 1075.
19. B. Nøst, G. Sørensen and E. Nes (1967) in Crystal Growth (ed. H. Steffen Peiser) Pergamon Press p. 801
20. B. Nøst, G. Sørensen and E. Nes (1967) J. Crystal Growth **1**, 149.
21. S. Kume and N. Kato (1974) J. Appl. Cryst. **7**, 427
22. B. Nøst and E. Nes (1969) Acta. Met. **17**, 13.
23. O. Rustad and O. Lohne (1971) Phys. Stat. Sol. (a) **6**, 153
24. S. Oki and K. Futagami (1974) Japan J. Appl. Phys. **13**, 605.
25. S. Oki and K. Futagami (1969) Japan J. Appl. Phys. **8**, 1574.
26. D.K. Bowen and J. Miltat (1976) J. Phys. E. **9**, 868.
27. F.W. Young Jr. and F.A. Sherrill (1971) J. Appl. Phys. **42**, 230.
28. F. Minari, B. Pichaud and L. Capella (1975) Phil. Mag. **31**, 275.
29. J. Kellerhals, F. Minari and B. Pichaud (1979) Phil. Mag. A. **39**, 341.
30. O. Nittono (1971) Japan J. Appl. Phys. **10**, 188.
31. A. Fukuda and A. Higashi (1973) Crystal Lattice Defects **4**, 203.
32. C. Jourdan and M. Sauvage (1970) Phys. Stat. Sol. (a) **3**, 343.
33. R. Fiedler and A.R. Lang (1972) J. Mater. Sci. **7**, 531.
34. C. G' Sell and G. Champier (1975) Phil. Mag. **32**, 283.
35. B. Roessler and S.J. Burns (1978) Phys. Stat. Sol. (a) **24**, 285.
36. A. Higashi (1974) J. Crystal Growth **24/25**, 102.
37. G. Mair and H. Wenzl (1976) Kristall und Technik **11**, 1059.
38. H. Wenzl and G. Mair (1975) Z. Phys. B. **21** 95.
39. W. Hagen and H.J. Queisser (1978) Appl. Phys. Lett **32**, 269.
40. J. Chikawa and I. Fujimoto (1968) Appl. Phys. Lett. **13**, 387.
41. J. Chikawa (1974) J. Crystal Growth **24/25**, 61.
42. J. Chikawa and S. Shirai (1977) J. Crystal Growth **39**, 328.
43. J. Chikawa (1978) J. Japan. Assoc. Crystal Growth **5**, 141.
44. J. Chikawa and S. Shirai (1979) Japan. J. Appl. Phys. **18**, (suppl. 18 - 1) 153.
45. M. Hart (1975) J. Appl. Cryst. **8**, 436.
46. B.K. Tanner, D. Midgley and M. Safa (1977) J. Appl. Cryst. **10**, 281
47. B.K. Tanner, M. Safa and D. Midgley (1977) J. Appl. Cryst. **10**, 91.
48. J. Bordas, A.M. Glazer and H. Hauser (1975) Phil. Mag. **32**, 471.
49. I.T. Steinberger, J. Bordas and Z.S. Kalman (1977). Phil. Mag. **35**, 1257.
50. B.K. Tanner, M. Safa, D. Midgley and J. Bordas (1976) J. Magn. Mag. Materials **1** 337.
51. M. Safa and B.K. Tanner (1977) Physica **86 - 88 B**, 347.
52. M. Safa and B.K. Tanner (1978) Phil. Mag. B **37** 739.
53. G.F. Clark and B.K. Tanner (1980) Phys. Stat. Sol.(a) **59**, 241

54. I.B. MacCormack and B.K. Tanner (1978) J. Appl. Cryst. 11 40.
55. J. Gastaldi and C. Jourdan (1978) Phys. Stat. Sol. (a) 49 529.
56. J. Miltat (1978) Nuclear Inst. and Methods 152, 323.
57. Z.H. Kalman I.T. Steinberger and S.S. Hasnain (1979) J. Appl. Cryst. (in press).
58. J.D. Stephenson, V. Kelha, M. Tilli and T. Tuomi (1979) Phys. Stat. Sol (a) 51 93.
59. J.D. Stephenson, T. Tuomi, V. Kelha and M. Tilli (1979) Phys. Stat. Sol. (a) 53, 271.
60. Y. Chikaura and B.K. Tanner (1978) Jap. J. Appl. Phys. 18, 1389.
61. G.F. Clark, B.K. Tanner, R.S. Sery and H.T. Savage (1979) J. de. Phys. 40, C5 - 183.
62. J. Miltat and D.K. Bowen (1979) J. de Phys. 40, 389.
63. I.M. Buckley - Golder and B.K. Tanner. Daresbury Preprint DL/SCI/P207E (Phil. Mag. in press)

CHAPTER 20

TECHNOLOGY AND COSTS OF X-RAY DIFFRACTION TOPOGRAPHY

MICHAEL HART

20.1 Introduction

Having determined that X-ray diffraction topography is relevant to your needs you have the task of obtaining the necessary equipment. This, of necessity the closing talk of the NATO Advanced Study Institute, is concerned with essential apparatus - in brief what you need and how much it costs.

Fig. 1 illustrates the problem faced by any newcomer to the field. All the topographs shown were recorded on the same Ilford type L-4 nuclear emulsion to eliminate differences in recording technique from the discussion. The topographs of dislocations in ice [1] were obtained in 2 minutes at almost zero cost whereas the Lang topograph of dislocations in germanium [2] required three days exposure with a Lang camera, microfocus generator and the associated detector electronics. The double crystal topographs [3] required two days exposure on a servo-controlled diffractometer using a stabilised X-ray generator while the synchrotron radiation topograph of oxide films on silicon [4] was obtained in a few seconds using an extremely simple (and inexpensive) goniometer with the NINA synchrotron. Obviously the complexity of the apparatus employed and its cost vary widely over these four examples. In practice the choice of technique and apparatus should be determined by the demands of the job in hand, but that is rarely possible since apparatus must usually be applicable in a wide range of situations. These four examples have been deliberately chosen to demonstrate the wide range of specifications which are possible and to emphasise the point that one needs to form a clear view of the results required before the apparatus can be selected. Visits and discussions with manufacturers and with practising topographers are of course invaluable.

We will look separately at each component of the system i.e. the X-ray source, the goniometer and the detector used for setting-up. In each case we shall try to identify the basic minimum requirement, the commonly used arrangements and features which might be required in demanding applications. Two experimental requirements, for Lang topography and for double crystal topography and rocking curve studies, will be considered in the laboratory environment. Synchrotron radiation sources will be considered separately.

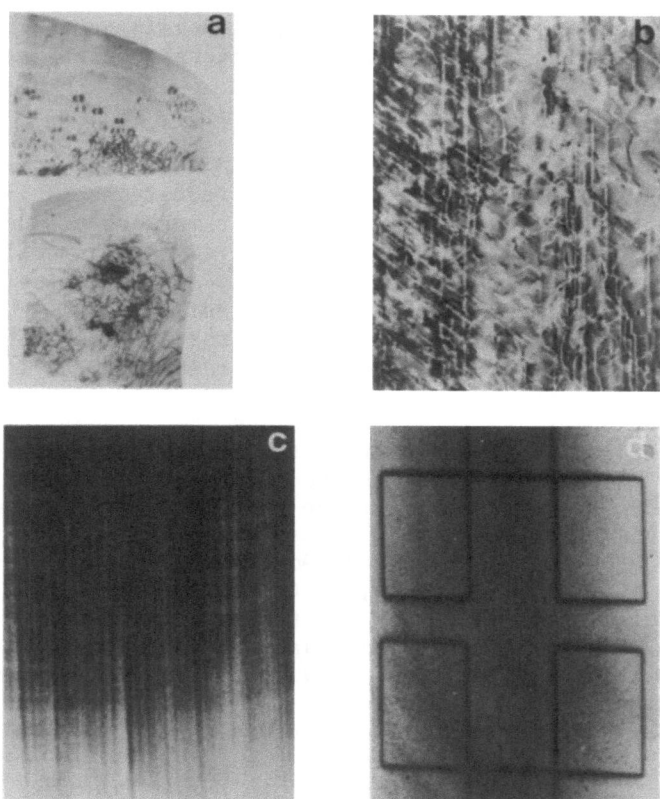

Figure 1. A selection of diffraction topographs illustrating the wide variety of requirements. (a) Dislocations in ice, wide beam topograph, field 3mm x 3mm (b) Dislocations in germanium. Lang topograph, field 3 mm x 2mm. (c) Banding in silicon. Double crystal topograph, field 3.5 mm x 2.5 mm (d) Oxide films on silicon synchrotron radiation topograph field 8 mm x 5mm

20.1.1 Cost estimates

It is quite impossible to give accurate costs and prices without detailing specifications, exchange rates and local conditions. The objective of this paper is to provide the newcomer with an economic framework within which he can plan. The figures given are based on pricelists from manufacturers and cost estimates based on experience and they are probably reliable to one significant figure. Exchange rates have been used, where necessary, as £1 = $2 = DM4 = FF8.

20.2 Experimental arrangements for transmission topography

During the last twenty years many topographers have designed and built their own cameras. By far the largest number have probably been built to Lang's design and several designs have been made commercially.

20.2.1 The X-ray source

A general discussion of source requirements is given in Tanner's book [5]. In common with other workers he adopts the view that single imaging is essential (i.e. the collimation is sufficiently fine to separate the $K\alpha_1$ and $K\alpha_2$ lines at, say, 6° Bragg angle). This matter of double images is often confused with the question of resolution. Fig. 1a, for example, required such a short exposure because it is not a Lang topograph but it is a wide beam topograph which contains both copper $K\alpha_1$ and $K\alpha_2$ images - but the copper doublet is closely spaced and the kinematic dislocation image width is rather larger than the spatial separation of the two images on the film. By contrast the germanium topograph formed part of a study of dislocation image structure and therefore demanded single imaging with 1 μm spatial resolution - a situation in which a microfocus generator is required. An important point, not so far discussed in the literature, is that with very large crystals and straight collimation slits one has to work at a larger distance from the source than is strictly necessary for adequate separation of the $K\alpha_1$ and $K\alpha_2$ lines. Under those conditions the microfocus generator has no significant advantage but the total power available in the point source is the important parameter.

Table 1 shows a selection of X-ray sources from those available. A fundamental decision has to be taken between sealed-off X-ray sources and continuously pumped X-ray sources. Since sealed-off sources require far less servicing than pumped sources one may need compelling reasons, for example the extra brightness and power available from a rotating anode or the possibility of fabricating non-standard targets for special purposes, before buying one. But Table 2 shows that those who count servicing as an important cost are probably not aware of total costs. Fortunately for the customer there are several manufacturers of X-ray generators and sealed-off X-ray tubes which are compatible.

Sealed-off X-ray tubes are available with a wide range of target materials and cost approximately $2000. The choice of target in rotating anode generators is also very wide, but in practise it is rather more costly to purchase (and certainly more difficult and time consuming to install) alternative anodes than to purchase several sealed-off X-ray tubes. Running costs can be estimated as the sum of the costs of electricity, water and maintenance manpower. For 4000 hrs, that is for 45% utilisation over one year the costs are given in Table 2.

Table 1

X-RAY SOURCES: A SELECTION FROM THOSE AVAILABLE

Type	Manufacturer	Source mm^2	Loading kW(Mo)	Cr	Ni	Co	Fe	Cu	W	Au	Mo	Ag
Sealed-off tubes	AEG	0.15x8	1.2	✓	x	✓	✓	✓	✓	✓	✓	✓
	Siemens	0.15x8	0.8	✓	x	✓	✓	✓	✓	x	✓	✓
	AEG	0.4 x8	2.0	✓	x	✓	✓	✓	✓	✓	✓	✓
	Elliott	0.4 x8	1.2	✓	x	✓	✓	✓	✓	✓	✓	✓
	Philips	0.4 x8	2.0	✓	x	✓	✓	✓	✓	✓	✓	✓
	Siemens	0.4 x8	2.0	✓	x	✓	✓	✓	✓	x	✓	✓
Pumped Rotating anode	Elliott	0.1 x1	1.2	✓	x	x	✓	✓	x	✓	✓	✓
	Rigaku	0.1 x1	1.2	✓	✓	✓	✓	✓	x	✓	✓	✓
	Elliott	0.3 x3	5.4	✓	x	✓	✓	✓	x	✓	✓	x
	Rigaku	0.3 x3	5.4	✓	✓	✓	✓	✓	x	✓	✓	✓
	Rigaku	0.5 x10	12	✓	✓	✓	✓	✓	x	✓	✓	✓
	Elliott	0.5 x10	15	✓	x	x	✓	✓	x	✓	✓	✓
	Rigaku	1 x10	30	✓	✓	✓	✓	✓	x	✓	✓	✓
	Rigaku	1 x10	90	x	x	x	x	✓	x	x	✓	✓

In many places one is obliged to use closed circuit cooling. It is straightforward to calculate the total cost in terms of capital and recurrent costs including electricity. Small systems are totally uneconomic. As a rough guide, a cooling system rated at 6 kW which will comfortably handle two high power X-ray generators consumes $480/yr of electricity at the lower price in Table 2, saving $420/yr water costs. The capital cost of the installation is about $5000. Even at the higher prices one would require ten years trouble free service from a closed cooling system to break even.

It is much more difficult to give capital cost guidelines for X-ray generators since the packages purchased vary considerably. For example, the smallest 1.5 kW generators may be either rectified and smoothed with a current stabiliser or fully stabilised. While 3 kW generators are only running at half-power with the single tubes listed in Table 1, some provide for the possibility of running two tubes simultaneously. In some cases one need only purchase shutter assemblies and accessories for those windows which are actually used. For transmission topography Tanner [5] found that the cost of a generator was nearly proportional to its power and this is still the case if one concentrates on one series of generators, for example, the Rigaku rotating anode generators.

Table 2

RECURRENT COSTS OF X-RAY GENERATORS

Tube kW	Power Gen.kVA	MWh/yr	Water L/min	m³/yr	Service Man yr		Costs $ Power	Water	Service	Total
1	2.5	10	3.5	840	0.01	*	400	210	75	685
						+	600	840	150	1590
3	6	24	3.5	840	0.01	*	960	210	75	1245
						+	1440	840	150	2430
3	6	24	8	1920	0.2	*	960	480	1500	2940
						+	1440	1920	3000	6360
10	20	80	20	4800	0.2	*	3200	1200	1500	5900
						+	4800	4800	3000	12600
30	60	240	40	9600	0.2	*	9600	2400	1500	13500
						+	14400	9600	3000	27000
100	200	800	100	24000	0.2	*	32000	6000	1500	39500
						+	48000	24000	3000	75000

*Electricity costs 0.04 $/kWh, water costs 0.25 $/m^3, salary $7500/yr.
+Electricity costs 0.06 $/kWh, water costs 1 $/m^3, salary $15000/yr.
Service time is based on a small sample of users' experience of the range 1 - 20 kW in tube power. Consumption figures are taken from manufactures' catalogues with averaging and interpolation where necessary. There are inconsistencies between the manufacturers' literature and the article by Yoshimatsu and Kozaki [6].

Table 3 gives approximate costs, including X-ray tubes in the case of the high power rotating anode generators. The prices, excluding value-added or sales tax, are taken from the current catalogues of Elliott, Enraf-Nonius, Philips, Rigaku and Siemens and relate to both stabilised and unstabilised generators and, in the case of two-tube operation, both simultaneous and sequential operation of the X-ray tubes.

TECHNOLOGY AND COSTS OF X-RAY TOPOGRAPHY

Table 3

CAPITAL COST OF X-RAY GENERATORS

	Tube kW	Price range k$
Sealed off	1.5	11-16
	3-4	13-31
	3-4 (2 tubes)	30-40
Rotating anode	3-4	28-74
	3-4 (2 tubes)	89
	12-15	70-90
	30	350
	60	700
	90	1000

20.2.2 Cameras for transmission topography

Without doubt, the preferred camera for serious work is the Lang camera. It has two main applications; for section topographs [7] and for projection topographs [8]. The fundamental requirements are the same for the two techniques viz. crystal angle setting to 1 sec.arc, counter angle setting to $1/2°$ and a sufficiently smooth calibrated linear traverse. Collimator slit design is important and slits of 10 μm spacing are required for section topography when source brightness is also very important. Many cameras have been built by individuals to Lang's original design and at least eight companies have marketed Lang-type cameras, though less than half of them are now available. Two commercial cameras are illustrated in Fig. 2. During the last 15 years design criteria have changed. Whereas a 25 mm x 25 mm maximum topograph area was the norm, some Lang cameras can now accommodate the 'industry-standard' 4 inch diameter silicon wafers. Generally speaking there is no better test of a traverse platform than to take a topograph with a high single crystal background intensity and low dislocation image contrast, for example of germanium. If one happens to be interested in high speed recording then fast traverse speeds must be available.

One minor problem, particularly with the large cameras which are designed for collimator lengths of at least 1 m, is that the X-ray generator table provided by the manufacturer will seldom accommodate one camera and never two. Camera prices depend upon size and upon the complexity of automation. Usually only the traversing system is remotely controlled. Prices range from k$ 14-63. If crystals contaning long range strains are to be studied then the crystal axis must have a motor which can be driven by a peak servo controller. A suitable system [5] is marketed by Elliott as an attachment for their Lang camera.

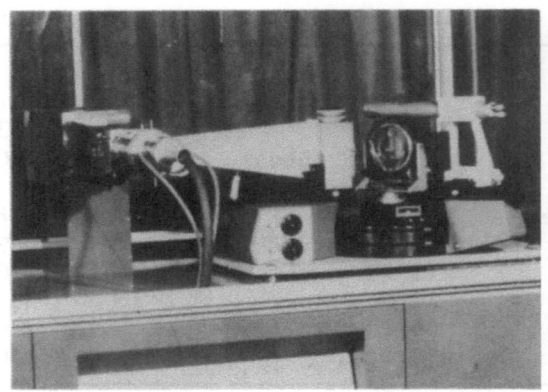

(a)

(b)

Figure 2. (a) Elliott model. Lang type camera for wafers up to 100 mm diameter, (b) Rigaku model 1515A1 Lang-type camera for wafers up to 102 mm diameter. It is fitted with four stepping motors and has built in Bragg angle control system for topographs of bent crystals.

20.2.3 Detectors for transmission topography

Transmission diffraction topography often requires the use of hard radiation such as MoKα and AgKα and, in light materials CuKα. Neither proportional counters nor Geiger counters are suitable as general purpose detectors. Scintillation counters, which provide for large area, uniformly sensitive, efficient X-ray detection are preferred.

TECHNOLOGY AND COSTS OF X-RAY TOPOGRAPHY

Suitable electronics is available from a very large number of commercial sources and falls broadly into three classes. Scintillation-type Radiation Monitors are inexpensive. They should be already available in any X-ray laboratory and they are entirely adequate for finding Bragg peaks and for setting up Lang topographs of single crystals. Fig. 3 shows a block diagram of the counting electronics which is required for Lang topography. Although a pulse height analyser is essential when working with fluorescent samples, in practice a discriminator is often adequate. The chart recorder is a great convenience when setting up and is regarded as essential for work with inhomogeneous samples when one obtains only partial topographs. The electronics is available as a Single Unit from several manufacturers. It is known under various names, for example, 'ratemeter' or 'measuring electronics' and may also have a scaler/timer included. Finally one can assemble a tailor-made system from the internationally standardised NIM modules. One starts with a 'bin' and power supplies and simply chooses the necessary modular components from the range available. Although this route is by far the most expensive it provides the possibility of later modification, enhancement and expansion (with computer control if necessary). Approximate costs are shown in Table 4.

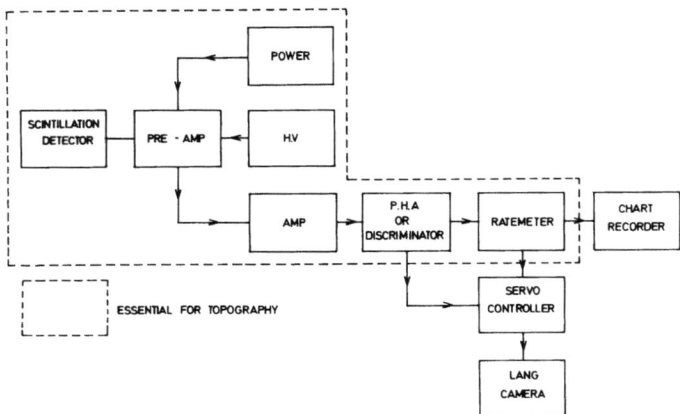

Figure 3. Block diagram showing detector systems which are used for single crystal methods

Chart recorder prices depend very much on the type. At the higher end of the price range one might be dealing with two-pen recorders or X-Y plotters for correlating X-ray intensity with other experimental parameters, while the least expensive chart recorder is suitable for setting up Lang topographs. It is convenient if the chart recorder has, in addition, a very slow speed so that it can be left running during the exposure of a topograph so that drift can be detected and corrected in situ. Speeds of 2-5 cm/hr are suitable for that purpose and speeds around 1-5 cm/min are convenient for setting-up.

Table 4

COSTS OF THE SCINTILLATION DETECTOR SYSTEMS

type	$
Radiation monitor	400 - 500
Single unit	1000 - 3000
NIM bin & modules	>3000
Chart recorder	200 - 2000

Specialised detectors such as position-sensitive detectors and television imaging systems are outside the present discussion. We assume that the final recording will invariably be done on either nuclear emulsion or high speed X-ray film and provision for processing and microscopic examination must therefore be made.

20.3 Double crystal experimental arrangements

For studies of defects the double crystal diffractometer can operate in one of two modes. It should be possible to measure small area rocking curves and to obtain diffraction topographs [3,9,10] from large areas of crystal.

20.3.1 X-ray sources

All of the X-ray sources listed in Table 1 are quite suitable for measuring rocking curves. Ideally, one would use the point projected source collimated with a pinhole equal in size to the source so that the probe area ranges from 0.4 mm x 0.4 mm with the sealed-off X-ray tubes to 0.1 mm x 0.1 mm with a microfocus rotating anode. There is no advantage (in routine tasks) in high power generators because the diffracted intensity is usually very high. For work with very small collimators or on weakly diffracting structures, such as epitaxial layers [11], then the brightness of the source is very important and rotating anode generators may be necessary.

In double crystal topography (where the line source can be employed) larger X-ray sources than those listed in Table 1 can also be used and total power may thereby be increased. A wide range of compatible sealed-off X-ray tubes is available at approximately the same price as the fine focus tubes. Most of the manufacturers listed in Table 1 also supply compatible tubes with 2 mm x 12 mm foci and Cr, Cu and Mo targets. These high power tubes may present a minor cost disadvantage in some circumstances. Whereas (Table 3) one can run two 2 kW fine focus tubes from one 4 kW generator, only one high power tube can be run. Thus, two generators would be required to run two high power tubes.

It is usually necessary to use servocontrol during the exposure of double crystal topographs so that a stabilised X-ray generator is considered essential. The detailed analysis of rocking curve profiles too can only be contemplated with constant intensity sources.

20.3.2 Double crystal cameras

The Bragg reflections which have been used for studies of defects in materials range in width from 0.2 sec arc to 20 sec arc. To cope with this range the camera must be set with a sensitivity of a few millisec arc. Drift rates in excess of 1 sec arc/hr are at best inconvenient and render high resolution work impossible. Fortunately, the absolute angles need not be determined to better than $1°$ or so and the high precision axes need only have ranges of between $5°$ and $10°$. Thus, the design and construction of a double crystal machine for defect studies is cheaper and more simple than the manufacture of a spectroscopic double crystal instrument.

Double crystal diffractometers constructed during the last twenty years were almost invariably built by the individuals who used them. On this basis, more than twenty copies have been made of the instrument, shown in Fig. 4, which was designed by the author about ten years ago.

The diffractometer has four axes which are arranged as two coincident pairs. The two high precision spindles are driven by levers, micrometer screws, gearboxes and stepping motors arranged to give steps of between 0.2 sec arc and 2 sec arc (determined by design parameters such as lever lengths and gearbox ratios). The steps are reliable to better than 0.01 sec arc and interpolation over approximately 30 sec arc can be achieved with the piezo-electric ceramics which are built into each lever system. The whole casting can be rotated about the first axis through $360°$ on the rotary table and the detector can be rotated about the second axis on a crude bearing.

Commercial cameras have been offered from time to time but, as far as can be determined, none has achieved a wide distribution. At present Rigaku appear to be the only major company offering a double crystal camera, at a cost of about k$ 27.

Figure 4. (a) Top view of a double crystal camera built at King's College. Set for rocking curve measurements. With a larger collimator the same system can be used for double crystal topography, (b) Double crystal camera showing the drive mechanism which consists of step-motor driven tangent screws and piezo-electric ceramics for fine motion and servo control, (c) Rigaku double crystal camera.

For large samples Rigaku also manufacture a scanning type double crystal topographic camera. Our experience has been that after discussion and minor use of one of our diffractometers, new users have been able to supervise the construction of their own copy from our sketches, at a cost of k$ 3 for parts and 200 man-hours of skilled machine shop time. [Bede Scientific Instruments, Coxhoe, Co. Durham, U.K. now offer a similar designed camera and PET interface for about k$ 16. (Ed.)]

20.3.3 Detector systems for double crystal diffraction

Double crystal topographs can be set up with the more advanced detection systems outlined in Fig. 3. In practice more systematic adjustments are required than with a Lang camera and a chart recorder is essential. Fig. 5 gives the schematic layout of the detector electronics which is required for double crystal topography and also indicates additional parts which may be required. For work which requires thermostatic enclosures an end-window scintillation detector is physically inconvenient and the more compact proportional counter may be used. By far the most convenient system (for displaying rocking curves) measures the crystal angles with a high resolution electronic micrometer (arranged as an angle transducer) and uses the output from that and from the ratemeter to drive an X-Y plotter. It is unrealistic to expect than even a well-made instrument will not drift off the rocking curve flank during the exposure and the provision of a servo-control system is necessary. That can be conveniently implemented by replacing the rate-meter ammeter with a photoelectric meter relay whose output contacts can be used to control the piezoelectric ceramic. If the diffractomer has motor steps which are much smaller than the rocking curve width then the feedback can be directly applied to the motor instead.

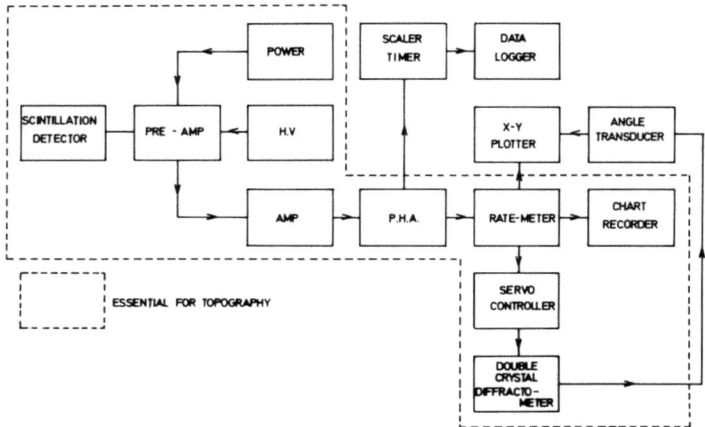

Figure 5. Block diagram showing the detector systems which are used for double crystal methods.

When rocking curve profile analysis is performed, the X-Y plotter system is best replaced by a scaler-timer and data-logger so that computer aided data analysis can be undertaken. In this context it may be useful to remark that a very wide range of equal crystal rocking curves fit very well to a linear combination of a Gaussian and a coincident Lorentzian curve. Only five independent parameters are necessary; the two widths and two heights and one peak position. With care it is possible to resolve angle changes of a few millisec arc with these diffractometers.

Double crystal topographs are recorded on film and on nuclear emulsions and the same considerations apply as for transmission Lang topography. When high resolution is required long exposure times will be necessary but with MoKα and harder radiations the necessary servo control can be done with the beam which is transmitted through the plate. The attenuation factor for Ilford glass plates is 10x for MoKα. When using servo-control with softer radiations it is necessary to use film rather than plates.

20.4 Building your own apparatus

There are two main reasons for building your own apparatus; either because no equivalent commercial apparatus is available or because it is cheaper. In practice the apparent cost is determined more by local accounting procedures than by the true costs.

20.4.1 X-ray sources

There appears to be no case at present, for X-ray topographic purposes, for research into new X-ray sources since the capital costs are likely to be high. An excellent review of high power generators is given in reference [6] (but see the caveat at Table 2).

20.4.2 Cameras for defect studies

As already indicated, a large number of workers have successfully constructed their own Lang cameras and double crystal diffractometers. Most laboratories welcome visitors. Before building your own cameras you can visit several laboratories and might even arrange an extended visit to gain first hand experience, sketches and advice.

In constructing cameras we have come to two important conclusions. First, that wherever possible commercial parts should be used and secondly that, partly because we are mainly concerned with precise angular control, there are advantages in building large instruments. The Lang camera shown in Fig. 6 illustrates these points. The basic axis is a large (30 cm diameter) workshop-model standard rotary table driven through a commercial 100:1 worm and wheel gearbox. Thus, one rotation of the handle corresponds to 108 sec arc and setting to 1 sec arc is easily achieved with a stability of 5 sec arc per day and a backlash of 1 min arc. The loading capacity is very large indeed so that furnaces, magnets and cryostats can easily be handled. The traverse is constructed from a Schneeberger linear bearing and a specially made precision ground lead screw giving a range of 120 mm under stepping motor control. For our simplified needs we did not build an axis for either the detector or the diffracted beam slits which are clamped to the large rotating circular ring. On this large scale it is comparatively easy to construct slit systems which are convenient and well-labyrinthed for safety. As Fig. 6 shows, the camera is rudimentary and not as easy to work with as commercial models but it can perform all the necessary functions. It is robust, inexpensive and requires no precision in-house machining.

Figure 6. Simple Lang-camera for wafers up to 105 mm diameter using precision commercial parts wherever possible to minimise the need for in-house workshop time.

20.4.3 Electronics for X-ray diffraction topography

The integrated circuit revolution and the development of microprocessors have dramatically changed our methods of working at King's College in the last two years. Several suitable microprocessor systems are available but we have based our system on the Commodore PET [12] using standard circuit hardware of modular design. A typical experimental station is shown in Fig. 7; 8 k-bytes of memory is entirely adequate for all of the tasks which have been outlined in sections 3 and 4. Taking account of the <u>parts cost only</u> these systems are very inexpensive as Table 5 shows.

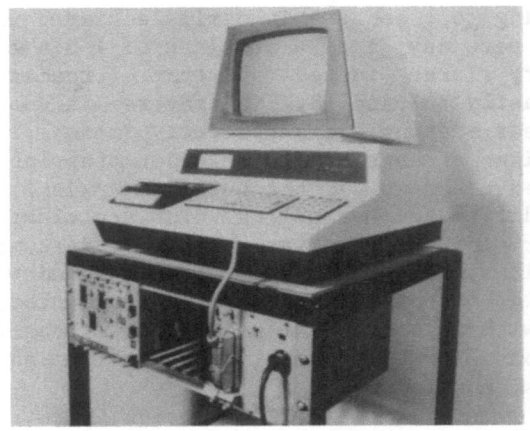

Figure 7. Microprocessor control system [12] capable of controlling a wide variety of instruments. From left to right there are 3 step-motor modules, a dual scaler, dual analogue output and dual analogue input modules.

Table 5

PARTS COST FOR MICROPROCESSOR CONTROL SYSTEMS

Method	System	Modules	Cost $
Lang; section and traverse	PET 8 k bytes	–	1000*
	Power supplies & interface	–	400
	2 x scaler	1	
	2 x analogue output	1	400
	2 x motor drive	2	
		total	1800
double crystal	PET 8k bytes	–	1000*
	Power supplies & interface	–	400
	2 x scaler	1	
	2 x analogue output	1	
	2 x motor drive	2	500
	1 x analogue input	1	
		total	1900

* $500 in U.S.A.

TECHNOLOGY AND COSTS OF X-RAY TOPOGRAPHY

Compared with Fig. 3 and Fig. 5 the detector systems themselves are much simpler because only the components up to the pulse height analyser are required. Substantial savings are made and a new flexibility is introduced. A few examples of regular tasks will indicate the features of the microprocessor control system outlined in Table 5.

detector setting. With one scaler and one analogue output to the pulse height analyser, the pulse height distribution can be measured and displayed and the pulse height analyser can be set properly.

Measure and display rocking curve. For many purposes chart recorders are no longer required since the screen can be used to display data. The PET provides control of motor-steps, scaler and recording facilities. Timing is under programme control.

Rocking curves. Rocking curves, for example as a function of tilt when setting up double crystal rocking curves, can be measured, recorded on the internal magnetic tape, displayed on the screen and plotted (through the analogue outputs) on an X-Y plotter under software control.

Traverse. No control system is required for the Lang-camera traverse since the traverse stepping motor is under programme control. Setting up may not be necessary since the programme can specify 'traverse until the intensity is lower than I, then reverse' in addition either peak or flank control of the Bragg angle can be readily implemented.

Peak and flank control. Software is easily implemented to control on any feature of the Bragg peak so that the servo systems shown in Fig. 3 and 5 are not usually required.

Peak search. Especially in double crystal settings, automatic peak searching techniques can be implemented. A factor of at least ten in speed over manual techniques is obtained in practice.

Ratemeter. The PET can act as a simple ratemeter displaying counting-rate graphically and numerically with any desired scale factors and time constants.

Sequential operations. Sequential operations such as those needed to create multiple exposure Bragg angle maps [13,14] are easily controlled by means of software. Additional parameters such as force, field or temperature can also be controlled.

Many other variations are possible, for example about eight modules would be required to make a completely automatic Lang or double crystal camera for routine production inspection and the experimenter is free to alter the experimental strategy through software. In our particular installation, subroutines are permanently loaded in EPROM so that the user software is extremely simple. For example, programmes to fulfill the first two functions are listed below:

```
10   REM   DETECTOR SETTING
20   POKE1,0:POKE2,148:A%=0:N%=0:T%=0
30   PRINT" "
40   DIM IN(79)
50   FOR V=0 TO 7.9 STEP .1
60   A%=2:N%=V:A=USR(4)
70   A%=22:T%=10:IN(10*V)=USR(2)
80   N%=V:T%=IN(10*V):A=USR(6)
90   NEXTV
```

```
10 REM MEASURE AND DISPLAY ROCKING CURVE
20 POKE1,0:POKE2,148:A%=0:N%=0:T%=0
30 PRINT" "
40 DIM IN(79)
50 FORI=0TO79
60 A%=22:T%=10:IN(I)=USR(2)
70 A%=56:N%=2:T%=1:A=USR(1)
80 N%=I:T%=IN(I):A=USR(6)
90 NEXTI
```

Line 20 locates the interface subroutines which are recorded in EPROM and assigns zero values to the variables. Line 30 clears the screen of the PET and line 40 reserves storage space for an 80 element data array of measured intensities.

To measure and display a pulse height spectrum, line 60 sets the analogue output (module address A%=2) to the value V (output voltage N%=V) using subroutine USR(4). The lower level of the pulse height analyser is connected to the analogue output module. In line 70 a counter module (address A%=22) accumulates counts from the detector for a time T%=10 and subroutine USR (2) stores the total number of counts in the data array IN. Line 80 uses subroutine USR(6) to plot the spectrum with ordinate N% and abscissa T% on the PET screen.

Variables can be reassigned very rapidly using the screen editing capacity of the PET and simple programmes such as those listed can be easily developed. In fact we find that users do not keep short programme libraries but prefer to write short programmes as they are needed.

Even the most complicated systems which we have so far constructed, for example a completely automatic scanning X-ray interferometric spectrometer [15] or an X-ray polarimeter [16] are well within the capabilities of the system.

20.5 Synchrotron radiation sources

At present X-ray beam time is quite scarce at all of the available sources of synchrotron radiation. Presently active sources, which all operate in a parasitic mode, are located at Stanford (USA), Paris (France) and Hamburg (W. Germany) and the first dedicated source will open next year at Daresbury (UK). All of these sources have (or will have soon) provision for both single crystal (white beam) topography and double crystal techniques. There are few restrictions on access to these facilities other than the requirement that the project should demand high priority in the view of the assessors.

The sources and facilities are listed below. In all cases it is essential that prospective users should discuss the feasibility and timing of experiments with the resident staff <u>before writing proposals</u>.

20.5.1 Stanford Synchrotron Radiation Laboratory (SSRL)

White beam topography is available and has been set up by Parrish's group from IBM San José. SSRL supplies no special equipment for topography though several users have made double crystal systems, both for spectroscopy and topography, which could have diffraction topographical applications. The possibility of providing an 'end-of-line' station for topography is being discussed.

Two developments will be of particular interest to X-ray diffraction topographers. Wiggler magnets have recently been installed which have considerably enhanced the output of X-rays below 1Å wavelengths and within the next year the source will be dedicated to synchrotron radiation work for 50% of the time. This will result in approximately a tenfold increase in the available X-ray beam time. By 1981 the number of available beam lines will have been doubled.

A booklet giving advice about 'Proposal guidelines and general information' is available on request and initial contact should be made with either the Director, Prof. A. Bienenstock or the Deputy Director Dr. H. Winick at

Stanford Synchrotron Radiation Laboratory
SLAC Bin 69 P.O. Box 4349
Stanford CALIFORNIA 94305

All proposals are considered on the basis of scientific merit and there are no restrictions based on the nationality or residence of the proposer.

20.5.2 Laboratoire pour l'Utilization de Rayonnement Electromagnetique (LURE)

A great deal of X-ray diffraction topography is undertaken using the DCI storage ring at Orsay. The first station dedicated to topography has facilities for double crystal methods as well as white beam topography and television detectors for real time experimentation [17]. At present approximately 40 shifts of 24 hours of dedicated beam time are available each year. Within the next two years a second beam line will be installed and there are plans for a second topography station (with a white beam) at the end-of-line position. Although no detailed camera plans are yet available it is expected that provision will be made for work in difficult environments, and computer control is envisaged.

Proposals are dealt with twice per year and initial contact for topographic work should be made with

Dr. M. Sauvage,
LURE
Laboratoire de l'accelerateure lineare
Batiment 209c, Universite Paris-Sud
91405 - Orsay, FRANCE

The CNRS has various bilateral international agreements which may cover users proposals for beam time but the CNRS does not normally support other costs of foreign users.

20.5.3 Hamburg, Deutsches Electron-Synchrotron (DESY, DORIS)

Although the first synchrotron radiation transmission topography was done with the DESY source [18,19], interest now centres on the related storage ring DORIS. Two organisations have facilities on DORIS, the Hamburger Synchrotron strahlungslabor (HASYLAB) and the European Molecular Biology Organisation (EMBO) which has an outstation (EMBL) there.

HASYLAB has at present one white beam topography camera available at DORIS which will be converted into a general topography station next year in the new experimental hall. At the new station it will be possible to perform both single crystal (white beam) topography and double crystal (monochromatic) topography. In the first instance prospective users should contact Prof. Dr. C. Kunz at

TECHNOLOGY AND COSTS OF X-RAY TOPOGRAPHY

Deutsches Electronen-Synchrotron DESY,
Notkestrasse 85,
2000 Hamburg 52,
W. GERMANY

At present there are no restrictions on potential users by nationality or place of work.

The EMBL outstation has no facilities for X-ray diffraction topography at present. Initial contact can be made through EMBO itself or with the EMBL outstation director Prof. Dr. H. Stuhrmann at

European Molecular Biology Laboratory,
c/o DESY
2000 Hamburg 52,
Notkestieg 1
W. GERMANY

20.5.4 Cornell University (CHESS)

The storage ring source produced X-rays during the present year and synchrotron radiation experiments will start at the end of 1979. For the foreseeable future CHESS will be the most intense hard X-ray source available to synchrotron radiation users. Three beam lines have been built and a topography station is planned.

There are no restrictions based upon a users place of work or nationality except in the case of industrial use for proprietary research. Potential users should contact the Director.

Prof B.W. Batterman,
School of Applied Engineering Physics
Cornell University,
Ithaca,
New York 14853
U.S.A.

20.5.5 Daresbury, storage ring source (SRS)

A great deal of diffraction topography was done in the U.K. with the 5 GeV synchrotron NINA [20-22]. That facility is now closed. It will be replaced within a year by the first dedicated storage ring source SRS.

In late 1980 two in-line topography experimental stations will become available. Equipment will comprise a universal white radiation camera with an environmental chamber and an X-ray TV imaging system at the end of an 80 m beam line, and a double crystal camera on the same line at 60 m. With this arrangement the double-crystal camera can also be set up as a beam-conditioner for the universal camera, to perform for example beam-modulation experiments. The camera design and construction will be done by Dr. D.K. Bowen, the computer control software by Dr. S.T. Davies, the environmental stage by Prof. J.N. Sherwood and Dr. K.J. Roberts while the TV imaging systems are the responsibility of Dr. B.K. Tanner. A wide range of more specialised stages are planned for in situ studies of crystal growth and for the application of stresses, fields and temperature changes as well as the facilities of the environmental chamber for high and low temperature work in inert or corrosive atmospheres. More details are given in chapter 19 and ref. [23].

Initial contacts for prospective users should be made with the Synchrotron Radiation Experiments Group Leader, Dr. P.J. Duke at

Daresbury Laboratory
Daresbury,
Warrington WA4 4AD
UK.

Science Research Council (SRC) funds and facilities are provided for approved research programmes originating from the UK universities and polytechnics. Scientists from other institutes, charitable foundations and industrial research laboratories are also welcome (though in some cases a charge may have to be made for beam time). While UK users had no source, a number of fruitful collaborations were established abroad. To continue that experience the SRC will try to accommodate users from universities abroad for up to about 10% of the total beam time on the SRS. Formal agreements also exist between, for example, Daresbury Laboratory and Frascati and there is an Anglo-Soviet agreement on synchrotron radiation.

20.5.6 National Synchrotron Light Source (Brookhaven)

The dedicated 2.5 GeV storage ring is exected to produce X-rays in late 1981. In collaboration with a group headed by Professor J. Bilello from the State University of New York at Stony Brook a diffraction topography beam line is being planned. Applications for funding are being made. As at other sources, proposals will be reviewed and the time allocated will be based on the review. Proprietary research will be charged so as to recover costs.

The co-ordinator for the diffraction topography program is

Dr. W. Thomlinson
National Synchrotron Light Source
Brookhaven National Laboratory,
Upton,
New York 11973, USA

TECHNOLOGY AND COSTS OF X-RAY TOPOGRAPHY

20.5.7 The Photon Factory (Japan)

An extensive program of X-ray topography is planned for the Photon Factory. First X-ray beams are planned in April 1982 and the topography program will start in July 1982. Four X-ray beam lines, including one from a 6T wiggler magnet, are planned together with experimental stations for white beam topography, plane wave topography, X-ray interferometry and 'new techniques in X-ray diffraction topography'. It is expected that overseas and industrial collaborations will be possible. Contact in the first instance should be with the Director.

>Prof. K. Kohra
>KEK (National Laboratory for High Energy Physics)
>Oho-Machi, Tsukuba-Gun,
>Ibaraki-Ken,
>305 - Japan

20.5.8 Costs and funding

By their very nature, synchrotron radiation sources are funded nationally and, perhaps in the future, internationally. It follows that their use and funding will follow more closely the pattern of existing neutron facilities than the pattern which has been traditional in laboratory-based X-ray diffraction topography. Nevertheless it will become possible to purchase beam-time on a commercial basis and it is therefore instructive to compare the cost of topographs obtained with conventional Lang apparatus or double crystal methods and with a white beam synchrotron radiation camera. Let us consider taking a 50 mm x 50 mm topograph.

Roughly speaking, we know from many peoples' experience, that comparable topographs can be taken in 24 hours with a 1 kW conventional source, in 1/4 hour with a 100 kW conventional source and in 100 s with the best storage ring sources (10 s for a field 50 mm x 5 mm). From Tables 2 and 3 we find, therefore, an X-ray source cost of about $25 per topograph including both recurrent and amortised capital costs while the synchrotron radiation source would, on the same basis, cost about ten times more per exposure. Whereas, in the laboratory, a source could be shared by two users, the dedicated storage ring source might have fifty simultaneous users and several hundred sequential users. Thus, it appears that a multi-user storage ring source could actually produce less expensive diffraction topographs than convential systems. We have implicitly assumed that camera and recording medium costs remain the same in the two cases.

However, the principal advantages of synchrotron radiation topography lie in other directions and they will determine the main research effort for the foreseeable future.

REFERENCES

1. Undergraduate project (1965) University of Bristol Unpublished.
2. M. Hart (1963) Ph.D. thesis University of Bristol
3. M. Hart (1968) Science Progress (Oxford) 56, 429
4. M.Hart (1975) J. Appl. Cryst 8, 436
5. B.K. Tanner (1976) X-ray diffraction topography, Pergamon, Oxford.
6. M. Yoshimatsu and S. Kozaki (1977) in X-ray Optics (ed. H-J Queisser), Springer, Berlin
7. A.R. Lang (1958) J. Appl. Phys. 29, 597
8. A. R. Lang (1959) Acta Cryst. 12, 249
9. U. Bonse (1962) in Direct observations of imperfections in crystals, Wiley, New York, p. 431.
10. U. Bonse, (1965) Z. Phys 184, 71
11. M. Hart and K.H. Lloyd (1975) J. Appl Cryst 8 42
12. A.R.D. Rodrigues and D. P. Siddons (1979) J. Phys E. 12, 403
13. M. Renninger (1965) Z. Angew. Phys 19, 20
14. L. Jacobs and M. Hart (1977) Nucl. Instrum. and Meth.
15. M. Hart and D.P. Siddons (1978) Nature 275, 45
16. M. Hart (1978) Phil. Mag B28, 41
17. M. Sauvage (1978) Nucl. Instrum Meth. 152 313
18. T. Tuomi, K. Naukkarinen and R. Rabe (1974) Phys. Stat. Sol (a) 25, 93
19. T. Tuomi, K. Naukkarinen, E. Laurila and P. Rabe (1973) Acta Poltech. Scand Ph100 1
20. B.K. Tanner, M. Safa, D. Midgley, and J. Bordas (1976) J. Mag. and Mag.Mat. 1 337
21. M. Safa and B.K. Tanner (1978) Phil. Mag B. 37, 739
22. B.K. Tanner, M. Safa and D. Midgley (1977) J. Appl. Cryst. 10 91
23. D.K. Bowen (1980) Annals of the New York Academy of Sciences (in press)

CHAPTER 21

X-RAY TV IMAGING AND REAL-TIME EXPERIMENTS

W. HARTMANN

21.1 Introduction

Real-time X-ray topography is a powerful modern tool to study directly the dynamic behaviour of strain fields and defects in crystals under various influences. Real-time methods may be grouped into two broad categories depending on the general principle used to permit rapid viewing and recording of topographic images, namely the single-stage and the multiple-stage imaging methods. In the single-stage imaging method, an X-ray sensitive vidicon tube directly converts the X-ray topograph into an electronic charge pattern. This charge pattern is read out by a scanning electron beam and displayed as a visible image on a television (TV) monitor. Chikawa and collaborators [1,2] have developed this technique which is presently, however, limited to a spatial resolution of 30µm. On the other hand, we prefer the multiple-stage imaging method where the X-ray image is first converted into a visible pattern by a fluorescent screen [3]. The visible light pattern is then optically coupled either by a lens or a fiber-optic plate to the input photocathode of a light-sensitive electro-optical device. The output image is displayed on a TV monitor. This method is inherently capable of higher resolution (presently ~ 10 µm), which is a prerequisite for detection of individual dislocations. A survey and discussion about X-ray TV systems is given by Tanner [4] Green [5] and the present author [6].

21.2 Experimental arrangement

Our arrangement of a real-time X-ray topography system is seen schematically in Fig. 1. The collimated X-ray beam enters the specimen at a distance of about 380 mm from the source. The specimen is set in the Bragg condition. A diaphragm stops the primary beam behind the crystal, and only the secondary beam which contains the desired information about the lattice defects reaches the fluorescent screen. This 5µm thick Gd_2O_2S: Tb fluorescent screen converts the X-ray pattern into an optical image. The optical picture is imaged onto the target of a low-light-level TV camera through a magnifying optical system. The optical magnification is usually adjusted to about 10 times. The required resolution of about 10 μm for the detector system makes this optical magnification indispensable because the TV camera tube has a limited resolution of about 25-30 μm. The electronic magnification between the target of the TV camera and the TV monitor is about 17. Therefore, we obtain an image of a crystal area of about 0.4 x 1.5 mm^2 in an overall magnification of 170 times on the TV monitor. In Fig. 1 one finds the principle of a furnace to heat samples for high temperature experiments up to 1500°C. The crystal is placed between two graphite slices. These slices can be heated directly by passage of a current of a few amperes. The furnace has two Be-windows to ensure low X-ray absorption. Temperatures up to 1500°C can thus easily be achieved, which enables us even to reach the melting point of Si.

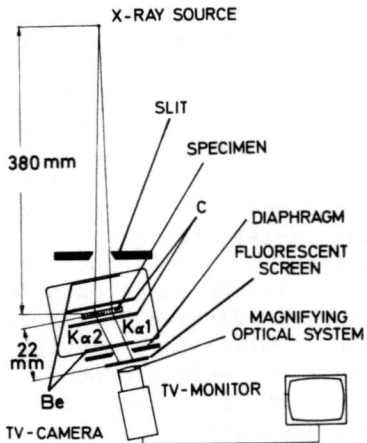

Figure 1. Schematic diagram of a real-time X-ray topography arrangement for high temperature experiments in transmission geometry.

The TV-pictures directly imaged on the monitor are mostly rather noisy. In collaboration with the Institute of Physical Electronics of the University of Stuttgart we have developed digital image processing equipment to enhance the signal-to-noise ratio by means of a trade-off against time resolution.

Fig. 2 shows the arrangement of this processing system. The signal levels of the analog electronic picture from the TV-camera are adjusted in a video-processor unit. An analog-to-digital converter (ADC) transforms the analog signal into a digital one and then the first picture is stored in a storage unit consisting of semiconductor memories. The consecutive TV-picture is also digitized and subsequently added to the previous one and stored in a second storage unit. This addition process can be continued up to 128 pictures. Time resolution can be thus varied between 20 ms and 2.5 s in order to achieve a better signal-to-noise ratio and therefore a better spatial resolution. The trade-off between time and spatial resolution is important because many dynamical processes of defect configuration pass slowly, but it is necessary to have a high spatial resolution to detect fine details of defects during the dynamical process. The added and stored picture is next transferred into another storage unit after the integration time in order to visualize the previous stored picture during the addition process of the next one. The TV-signal is converted into an analog one. We insert then additional experimental data, for example time and temperature, into the TV-picture. The picture is rendered visible on a black-and-white monitor and further stored on a video tape.

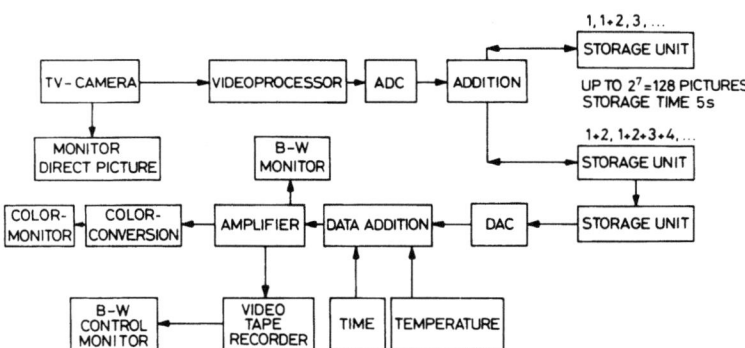

Figure 2. Schematic drawing of the electronic image system with digital integration.

21.3 Experiments

A movie showing some examples of real-time experiments was presented.

a) Movement of dislocations (W.Hartmann)
The sample was indented in [110] direction, then heated up to 1280°C. A crystal area of about 0.8×1.5 mm^2 was imaged in (220) reflection and MoKα radiation. Dislocations were created by the strain field of the indentation above 1000°C. The dislocation velocities were in order of several tens of micrometers per second. Around 1280°C, two dislocation lines separate and move away one from another. The dislocation velocities are in the first storage period of 2.5 s 32 µm/s in the second period 8 µm/s and in the third 5 µm/s.

b) Magnetic domains influenced by an oscillating magnetic field (W. Hartmann, W. Hagen, J. Miltat [7])
We studied the movement of ferromagnetic domains in Fe - 3,5% Si crystals at the synchrotron laboratory LURE in Orsay. The

experimental set-up was a double crystal arrangement which works in (+,-) geometry. The domains were influenced by an oscillatory magnetic field. The applied field varied between ± 1080 Oe in a saw-tooth curve with an oscillation frequency of 0.1 Hz. The time resolution was 20 ms. Fig. 3 shows, as an example, some prints of this part of the movie. A domain wall configuration starts at an arbitrary time (t) zero in Fig. 3a. The white area indicates the Bragg contour where the crystal fulfills the Bragg condition. This Bragg contour moves up and down together with the domain walls under the influence of the applied magnetic field. The upper horizontal line is the crystal edge. Fig. 3b shows the same crystal area 20ms later. The domain walls on the left side have been moved down with a velocity of about 11 mm/s. The configuration has been changed after 140 ms into the pattern seen in Fig. 3c. The domain speed is now reduced to about 5.5 mm/s. It is obvious that pairs of domain walls exist with a distance of about 100μm. The distance between the pairs is measured to 250 μm. Fig. 3d shows the development after 240 ms. The velocity falls down to about 4 mm/s.

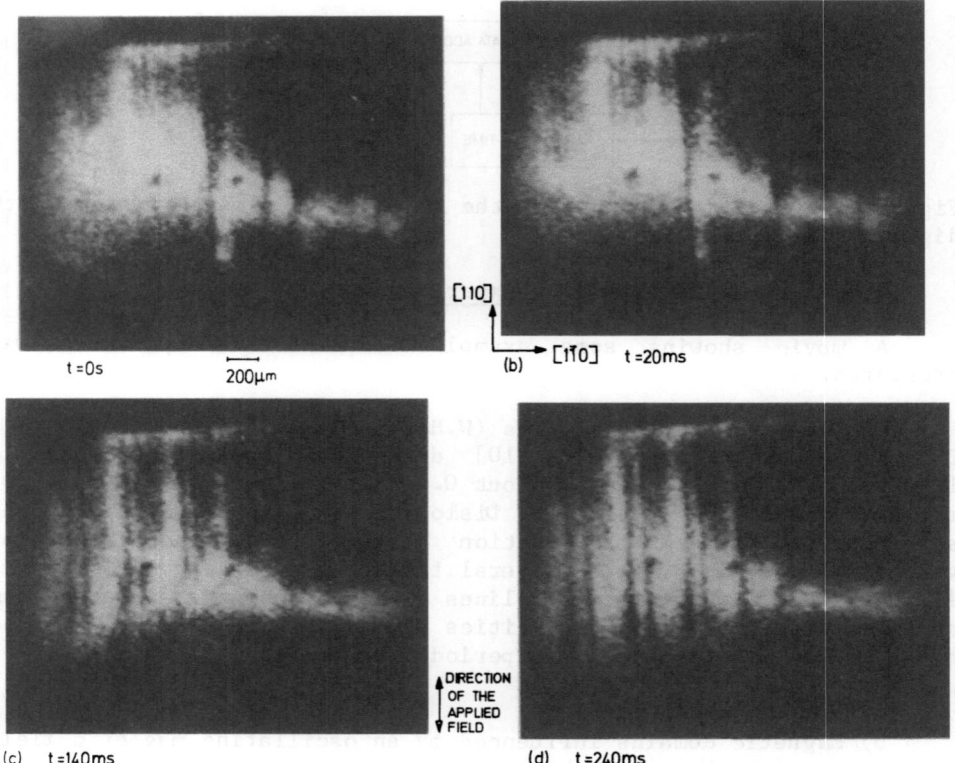

Figure 3. Examples of magnetic domain wall motion taken from the movie.

c) High temperature behaviour of strains along edges of oxide windows. (W. Hartmann, G. Franz [8])
The movie showed the heating up of the sample to about 1040°C. The strain contrast vanishes at about 1020°C because the oxide was grown and adapted to the Si-lattice at this particular temperature. The contrasts exist above and below 1020°C because the oxide layer is not adapted to the lattice.

d) Misfit dislocations created during the epitaxial growth of Ge on a GaAs substrate. (W. Hagen, H.J. Queisser, W. Hartmann [9,10])
This last part of the movie showed misfit dislocations at the interface Ge/GaAs. The Ge layer was grown at 635°C on the GaAs substrate. Misfit dislocations arise from the strain at the interface due to the different lattice constants of Ge and GaAs. The dislocations were observed in a (+,-) double crystal arrangement and an integration time of about 9s of the video processing equipment.

21.4 Summary

Our real-time topography system allows us to observe the dynamics of defects and strain fields with a spatial resolution up to 10μm and a time resolution up to 20 ms. The X-ray pattern is converted to an electronic picture by a multiple-stage method. A digital integration system allows an addition of up to 128 TV-pictures.

REFERENCES

1. J. Chikawa, J. Fujimoto and T. Abe (1972) Appl. Phys. Lett. $\underline{21}$, 295
2. J. Chikawa (1974) J. Cryst. Growth 24/25, 61
3. W. Hartmann, G. Markewitz, U. Rettenmaier and H.J. Queisser (1975) Appl. Phys. Lett. $\underline{27}$, 308
4. B.K. Tanner (1976) X-ray Diffraction Topography (Pergamon Press, Oxford)
5. R.E. Green Jr. in Advances in X-ray Analysis (ed. by H.F.McMurdie, C.S. Barrett, J.B. Newkirk and C.O. Rund) vol. 20,Plenum Press,New York p.221.
6. W. Hartmann (1977) in Topics in Applied Physics, Vol. 22 'X-ray Optics' (ed. by H.J. Queisser) Springer Verlag Berlin, Heidelberg, New York p.191
7. W. Hartmann, W. Hagen and J. Miltat, to be published
8. W. Hartman and G. Franz, to be published
9. W. Hagen and H.J. Queisser (1978) Appl. Phys. Lett. $\underline{32}$, 269
10. W. Hagen and W. Hartmann, to be published

21.5 Discussion

1) Is the growth temperature exactly correlated to the temperature where the topograph shows no strain contrast? The growth temperature was 1020°C. The oxide can be adapted strain free above 900°C. It depends on time. We learned from the experiment that the adaptation time at temperatures around 1000°C is about 30-40 min. Therefore, the cooling rate after growth was much faster than the adaptation rate and the growth temperature is exactly correlated to the temperature seen in the topographs.

2) Did you use the full power of digital image processing? Yes, due to the special properties and claims of real-time X-ray topography. We found only conceptions about the solution of the problem to have no time and a noisy picture.

3) Do you expect major improvements and innovations in this field for the future? Yes, there is a fast progress for random access and other semiconductor memories. Great efforts are made, especially in the medical and military fields, to develop the soft and hardware for such applications.

4) Did you have the fulltime resolution of 20 ms during the synchrotron experiments? Yes, the topographs are taken without digital integration.

5) Participants discussed the capacity of semiconductor storage elements and computer systems which are able to store and evaluate the immense data rate from a TV system. The computer systems start to be good enough, but this field is under progress.

6) The analysis of real-time experiments can be very difficult due to the high density of information. That is definitely true. Each experiment should, therefore be carefully selected and prepared.

CHAPTER 22

COMPUTER MODELLING OF CRYSTAL GROWTH AND DISSOLUTION

MORETON MOORE AND IAN O. ANGELL

22.1 Crystal growth

Computer programs have been written in FORTRAN to display perspective views of various solids on an IMLAC PDS1G interactive graphics terminal linked to the University of London CDC 6400 computer (Fig. 1). The solids may be rotated a chosen number of times about any desired [hkl] axis. Rear edges are shown as dashed lines (also in perspective).

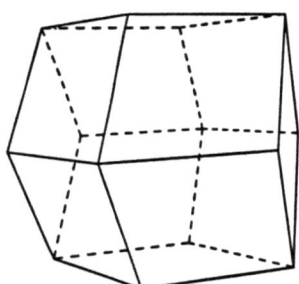

Figure 1. Perspective view of a rhombic dodecahedron

Films have been made from computer drawings of crystals growing with a variety of habits, showing changes of habit as a result of variations in the relative growth rates of different faces.

22.2 Growth horizons

Parallel (hkl) slices through a crystal, considered as an assembly of concentric growth horizons, may be displayed to show the growth sectors in a variety of habits, including mixtures of habit. (Figs. 2 and 3). After choosing the habit and (hkl), the computer gives the maximum distance from the centre for such a plane to intersect the solid, so that suitable choices of distance may be made. If desired, the picture may also be foreshortened (by twice the Bragg angle) for direct comparison with section X-ray topographs.

Figure 2. Cuboctahedron with vertices (1,1,0) etc. (100) section distant 0.4 from the centre. Number of growth horizons = 24, (not all seen in a non-central section). Note the centre-cross pattern [1].

Figure 3. (321) section, 0.4 distant from the centre of the same cuboctahedron.

22.3 Crystal dissolution

Dissolved crystals often exhibit rounded facets bounded by curved edges, but vertices may still be relatively sharp [2]. In the cylindrical approximation [3] each edge of the polyhedral growth form is considered to be blunted by a cylindrical surface. As dissolution proceeds these rounded surfaces eventually meet up with other such surfaces belonging to the same zone, giving the solid a circular cross-section. The dissolution body may then be considered as the intersection of symmetrically oriented cylinders. These shapes have been drawn on the computer and compared with dissolved cubic crystals (Fig. 4).

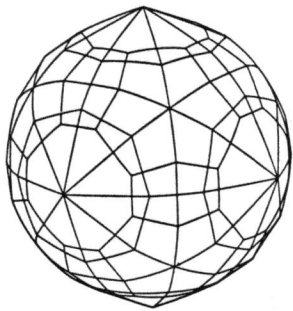

Figure 4. The <211> dissolution body, formed by twelve cylinders.

REFERENCES

1. M. Moore (1979) in The Properties of Diamond (ed. J.E. Field) Academic Press, p 263. A. R. Lang (1979) ibid p 443
2. F.C. Frank (1972) Z. Phys. Chem. (Neue Folge) 77, 84
3. M. Moore (1974) Mathl. Gaz. 58, 181

CHAPTER 23

MICRORADIOGRAPHY AND ABSORPTION MICROSCOPY

D. KEITH BOWEN

23.1 Introduction

The techniques of radiography are as old as the knowledge of X-rays, and of course are widely used in medicine, biological research and inspection of engineering components. A lack of high-intensity tunable sources has meant that its use on a microstructural level for materials has been very limited so far. However, we now have synchrotron radiation sources. As Lang has pointed out [1], almost any camera that will take diffraction topographs will take high-resolution radiographs, so the instrumentation will be available. The purposes of this short chapter are to summarise the experimental techniques for obtaining microradiographs, to state the conditions determining contrast and resolution and to speculate on possible applications in the study of crystals and crystal growth.

23.2 Experimental methods for microradiography

These are governed mainly by the choice of radiation, which may be white, white but with wavelength selected by the specimen, beam conditioned or beam conditioned with wavelength modulation.

23.2.1 White radiation method

This scarcely justifies the title 'method' since all one does is to put the specimen in the X-ray beam and put a film or other image detector behind it. As in topography, the specimen-film distance should be as short as possible to demagnify the source size (see chapter 7); for radiography there are no slits getting in the way and the specimen-film distance will vary from ~ 0 at the exit surface to the specimen thickness at the entrance surface. As will be seen below, this does limit the resolution on thick samples, but in principle it is usually very much better than for diffraction topography. Detectors are also discussed below.

The drawbacks of this simple method are that the contrast is determined by a rather random means; it will depend upon the density and atomic number variation in the specimen and on the particular spectral distribution of the source. Moreover, the contrast is often low. These factors make it rather unsuitable for microstructural work, but of course it is quite useful when only density variations are to be expected in a specimen of uniform chemical composition since the beam-conditioned methods are then of very limited value. Its main use is therefore in the traditional areas of low-resolution (by topographer's standards) radiography of people, animals and engineering components to look for relatively large-scale voids and density variations.

23.2.2 White radiation: self-monochromatisation

This ingenious method, due to Aleshko-Ozhevsky [2], uses the specimen as its own beam-conditioner and is very convenient for synchrotron radiation sources. With a white incident beam in a topographic set-up, the Bragg angle is chosen so as to select a wavelength on the high-absorption side of an absorption edge of an element. Variation in the concentration of this element then gives variation in contrast, which can be detected by comparing topographs taken either side of the edge. Although there is no control over the spectral resolution, this is a most effective means of adding chemical contrast to diffraction contrast, and should be very useful for studying, for example, segregation in crystal growth with relatively simple apparatus.

23.2.3 Beam-conditioned radiography: difference maps

Some improvement in contrast over the white-radiation method can be obtained by using an external beam-conditioner to take a single exposure on the high-absorption side of an edge; this is more flexible (and more expensive) than using the specimen itself, and the contrast is calculable (see section 23.4). However, the most sensitive method is that of taking radiographs with wavelengths either side of the absorption edge. This was first suggested over 50 years ago [3] but is only now properly exploitable; without a synchrotron radiation source one has to use the continuous spectrum of a standard set, or use the nearest K or L lines either side of the edge in question. The latter is not a very effective technique either in practice or in theoretical performance. As shown below, this technique gives a large increase in contrast and hence in noise-limited resolution, and gives specific indication of chemical element distribution. It should be possible to make the chemical analysis quite accurate once the appropriate corrections are worked out.

This method has recently been applied to various mineralogical samples by Polack et al. at LURE, Orsay [4]. The X-ray images give striking contrast either side of the edges, and it is noticeable that they are much less noisy than usual electron microprobe images.

This is expected from the statistics involved, and given also that the method is a transmission one, it could be highly competitive or superior to the microprobe for certain applications. For microanalysis, Polack et al. estimate a detection limit of ~ 100 ppm in a 1 μm^3 volume, which is certainly very interesting for many materials studies. It is logical to combine this technique with bulk X-ray spectrometry, for which phenomenally low concentrations (in the 10^{-8} to 10^{-12} g/g range) should be detectable by beam-conditioned fluorescent analysis with a storage ring source.

23.2.4 Scanning methods

It is possible to make X-ray apertures down to $\sim 1\,\mu m$ diameter (including the effects of diffraction). For this resolution then, one may scan the specimen in a two-dimensional raster with a precise mechanical drive (or scan the aperture - it makes little difference with a synchrotron radiation source but does with an X-ray set), and measure the transmitted radiation with any detector. Use of a solid-state detector has obvious advantages in that fluorescent emission can be detected together with the transmitted. A prototype instrument of the fluorescent type has been constructed by Horowitz and Howell [5]. Although the resolution of this method is not the highest ever attainable (see below), it could actually be superior to standard imaging techniques for very hard radiation, for which all imaging detectors will give much worse than 1 μm resolution because of photoelectron tracking (see Chapters 7 and 12). Its real advantage, however, is in the suitability of the final signal for data processing: for application of correction factors for quantitative analysis, for automatic production of 'difference maps' in the beam-conditioned case, or for any variety of image processing.

23.3 Resolution

The resolution in microradiography is governed by the same rules as in topography, but some of the conditions are now much more favourable; and since one does not have intrinsically broad images (as one does for diffraction topographs of dislocations, for example) it is interesting to discover what the limits are. The resolution d is given by

$$d^2 = (hb/a)^2 + r^2 + m^2 b^2 + b\lambda$$

where h is the maximum X-ray source dimension, b the specimen-detector distance, a the source-specimen distance, m the acceptance angle of any monochromator placed <u>after</u> the specimen and λ is the wavelength. The first term is ordinary geometric resolution in topography (Chapter 7), the second is the detector resolution, the third is the monochromator blurring and the last is the diffraction blurring. It is worth noting (since the point caused considerable confusion when this seminar was given at the Study Institute!) that the diffraction limitation is for Fresnel diffraction, which is appropriate in microradiography when considering ultimate resolution since the distance b is then very small, perhaps only microns for the exit surface of the specimen in contact with a plate or film. This diffraction blurring increases with the square root of wavelength. When considering diffraction from the edges of slits or apertures a few mm or cm away from the specimen - as when trying to make a 1 μm aperture for example - the Fraunhofer regime is entered, where the blurring increases linearly with distance.

How good a resolution can one get? In the sub-micron range there are certain detectors available (see below) which do not in practice limit the resolution at least with softer X-rays. The main limitation is then diffraction. Using the above equation, if b=1 mm, λ=2 A then d=0.45 μm; if b=10 μm, λ=1 A then d=0.03 μm. Thus for ordinary topographic specimens one may expect a resolution in the same region as that of a very good diffraction topograph, but for thin specimens in 'contact' microradiography, or for regions near the exit surface of a thick specimen in contact with the detector, one can expect an order of magnitude better. This is good news!

23.4 Contrast in microradiography

One can only realise such high resolutions if one has enough contrast, as emphasised by Lang in Chapter 12. The equations governing contrast in the various methods are therefore of great interest. Defining the contrast C as the fractional change in signal with respect to background level, the contrast in a 'one-shot' monochromatic radiation exposure of a particle of diameter d embedded in a matrix is [6]

$$C = \exp\{-(\mu_s - \mu_b)d\} - 1$$

which if C < 0.2 is to a good approximation

$$C = (\mu_s - \mu_b)d$$

where μ_s and μ_b are the linear absorption coefficients of the particle and matrix respectively. The equation for contrast in the difference method, assuming that the intensities at corresponding image points in the two exposures are subtracted, is more complex [7]. It is

$$C = \frac{\exp(-\mu_b^1 t)\left[\exp-(\mu_s^1-\mu_b^1)d-1\right]-\exp(-\mu_b^2 t)\left[\exp(-\mu_s^2-\mu_b^2)d-1\right]}{\exp-(\mu_b^1 t)-\exp(-\mu_b^2 t)}$$

Superscripts 1 and 2 refer to exposures (wavelengths) 1 and 2, and the thickness of the specimen t must now be included. The contrast depends as expected on the height of the absorption edge (numerator) and on how close together the two wavelengths can be (denominator), since the variation of μ with λ in between absorption edges depends roughly on λ^3. An approximation of the contrast amplification factor over the single-exposure method is

$$1/(1 - \exp\{-3\lambda^2 \delta\lambda p t\})$$

where the approximation $\mu = p\lambda^3$ has been used to describe the variation of absorption between edges (for an extremely small variation of λ). $\lambda_1 = \lambda$ and $\lambda_2 = \lambda + \delta\lambda$.

These equations were derived and discussed in [7]. It was shown there that the contrast amplification using the differential method is limited in practice by the natural width of the absorption edge, but that contrast amplifications of 6 - 12 x should be typical. Combining this with the criteria for noise- limited resolution it was shown that typically an order of magnitude improvement in resolution could be obtained, and as a general guide elements of comparable atomic number should give enough mutual contrast for visibility down to at least 0.1 µm in specimens 1 mm thick, and the situation improves as the matrix atomic number decreases. The method could be particularly important in polymers and light organic and inorganic crystals or minerals.

23.5 Detectors

The same considerations apply as in topography (Chapter 12), but we are clearly interested in the 'grainless' media: PMMA, other photoresists and possibly lead iodide. There is rumoured to be considerable industrial development work going on on photoresists doped with heavy elements to increase their speed and stopping power and these would be most useful. Sayre [8] has demonstrated resolution around 10 nm in PMMA (on a biological cell placed directly on the film, using $\lambda=4$ nm), and ~ 2 nm should be possible. Very little use has been made of resists in applications to crystals so far, no doubt because their extreme slowness to X-rays makes them hopeless for conventional sources. As mentioned above, the solid-state detector should be a powerful tool in scanning techniques.

23.6 Applications and Speculations

This section is necessarily speculative since hardly any work relevant to crystal growth has yet been performed. However, the following topics are worth considering:
    (a) The study of segregation at growth horizons and growth sector boundaries,

(b) The study of phase transformations involving diffusion,
(c) The study of grain boundary segregation,
(d) Imaging and measurement of segregation in the liquid ahead of a growing interface, and its correlation with (a). This could apply to both melts and solutions; the highest resolution would not be needed, and the information is not usually attainable by other means.
(e) Investigation of segregation produced by chemical reactions, especially in the solid state, and correlation with both chemical kinetics and microstructure revealed by topography. Chemical shifts in the absorption edge may yield extra information with a good beam-conditioner.
(f) The ultimate technique will be to use the fine structure of the absorption edge to obtain microstructural EXAFS information. With the best storage ring sources (dedicated!) the counting times may not be too prohibitive.

Many technological developments can be expected in this field. One may hope for beam concentration with X-ray mirrors to aid the spot-scanning methods, and one may obtain very small spots by asymmetric diffraction as has recently been demonstrated for section topography [9]. The universal camera/beam conditioner that we are building for the Daresbury SRS should be a powerful tool for this type of experiment and we are planning to begin such investigations in the near future.

REFERENCES

1. A.R. Lang (1970) in Modern Diffraction and Imaging Techniques in Materials Science (ed. S. Amelinckx, R. Gevers, G.Remaut and J. van Landuyt), North Holland, Amsterdam p 407
2. O. Aleshko-Ozhevsky (1979) Moscow Institute of Crystallography, unpublished work.
3. R. Glocker and W. Frohnmayer (1925) Ann. Phys. Leipzig 76, 369
4. F. Polack, S. Lowenthal, Y. Petroff and Y. Farge (1977) App. Phys. Letts. 31, 785
5. J. Horowitz and P. Howell (1972) Science 178, 608
6. V.E. Coslett and W.C. Nixon (1960) X-Ray Microscopy, C.U.P. Cambridge
7. D.K. Bowen (1979) in Applications of Synchrotron Radiation to the Study of Large Molecules of Chemical and Biological Interest (eds. R. B.Cundall and I.N. Munro) Daresbury Laboratory Proceedings DL/SCI/R13, p 37
8. D. Sayre (1980) in Imaging Processes and Coherence in Physics, Springer-Verlag,Berlin Heidelberg New York, p229; J.Kirz and D.Sayre (1980) in Synchrotron Radiation Research (eds. S. Doniach and H. Winick), Plenum Press, New York (in press)
9. Mai Zhen-Hong, S. Mardix and A.R. Lang (1980) J. Appl. Cryst. 13, 180

CHAPTER 24

RECIPROCAL LATTICE SPIKE TOPOGRAPHY

A.R. LANG

This topic 'spike topography' for short, involves measuring scattered X-ray intensity as a function of position in two spaces: in real space as a function of location of the scattering volume element within the specimen crystal and in reciprocal space as a function of position of the scattering vector relative to a reciprocal lattice point (relp) of the perfect crystal. This chapter touches upon (1) background theory, (2) history of study of the anomalous 'spike' diffuse reflexions from diamond, (3) the findings of 'spike' topography and (4) likely future developments.

(1) We are concerned with the kinematical diffraction theory of the bounded and/or imperfect crystal domain [1-2], and the keys to understanding are the Fourier transform and the convolution theorem [3]. The latter tell us, for example, that if a perfect crystal is subdivided into domains with plane faceted surfaces, there being a break in perfect lattice periodicity and/or change in scattering amplitude at such surfaces, then the lattice interference function maxima at relps will in general exhibit rod-like extensions ('spikes') perpendicular to the above-mentioned facets. The greater the facet area (in the crystal), the smaller the diameter of the rod (in reciprocal space).

(2) The diamond story begins with the discovery by Raman and Nilakantan [4] of strong diffuse reflections from diamonds. Lonsdale and Smith [5] and Lonsdale [6] then demonstrated that diffuse reflections from diamonds were of two sorts: temperature-dependent, produced similarly by all specimens, due to thermal vibrations; and temperature - independent, relatively sharp, generated by spike-like, <100> oriented extensions of reflecting power from relps, with intensity varying from specimen to specimen. Indeed, the latter sort were produced only by the common type I, ultra-violet absorbing diamonds, and were absent from the rare, ultra-violet transparent, Type II (using the type classification introduced by Robertson, Fox and Martin [7].) Guinier [8] pointed out that these 'spikes' indicated lattice displacement faults on {100}, of lateral extent not less that ~ 0.1 µm to account for the spike sharpness; but many years elapsed before this idea was developed further. Major

progress experimentally was made by Hoerni and Wooster [9] who measured spike intensities and extracted therefrom the 'spike magnitude' associated with a given cube direction at each relp. They found (1) spike magnitudes at a relp hkl in, say, the [100] direction depended only on the corresponding index h; (2) spike intensity at reciprocal lattice distance R from a relp decays as $R^{-2.2}$ and (3) the variation of spike magnitude with order h gave the sequence of magnitude 0, 100, 75, 5 and 30 for orders 0, 1, 2, 3 and 4 respectively. Frank [10] developed Guinier's idea of planar defects into a silicon platelet precipitate model, and a rigorous one-dimensional kinematic diffraction theory for a random sequence of occasional anomalous interplanar spacings was worked out by Caticha-Ellis and Cochran [11]. Then in 1959 Kaiser and Bond [12] discovered that nitrogen was a major impurity in Type I diamonds, in 1962 Evans and Phaal [13] directly observed platelets on {100} in natural Type I diamonds by transmission electron microscopy, and in 1964 Lang [14] proposed a structurally plausible nitrogen platelet model consonant with both X-ray spike data and electron microscope contrast observations. A direct probing, by transmission electron microscopic techniques, of the composition and structure of individual platelets is an extremely difficult task. Perhaps it may be accomplished by the end of the 1980's.

(3) Takagi and Lang [15] developed the spike topography technique to overcome difficulties arising from the great inhomogeneity of natural diamonds. The high spatial and angular resolution inherent in the topographic technique enable estimates of average lateral dimensions of platelets to be made from the sharpness of spike topography images [16,17]. The topographic technique also discloses pitfalls in the conventional 'spike' recording methods [18]. When the spike topographic technique is pushed to greater resolution interesting anomalies manifest themselves [19], involving departures from the results of earlier experiment [9] and theory [11].

(4) When using characteristic X-ray sources, the study of diffuse reflections is much handicapped by the limited range of orientations of sections of reciprocal space that can be cut in the vicinity of a given relp, given the fixed radius of the Ewald sphere. A tunable wavelength source would be highly desirable. So too would be a well-defined bandwidth. The latter need not be very narrow but should be significantly less than the usual $K\alpha_1$ to $K\alpha_2$ separation when studying regions close to relps (where scattered intensity is a rapidly varying function of position of scattering vector). Synchrotron radiation sources, joined with appropriate monochromatising devices, will be of great benefit; and a monochromator geometry that would enable high spatial resolution section topographs to be taken of both Bragg and diffuse reflections has been described [20].

REFERENCES

1. S.G. Lipson and H. Lipson (1969) Optical Physics, Cambridge University Press.
2. R.W. James (1954) The Optical Principles of the Diffraction of X-rays, G. Bell and Sons Ltd. London.
3. P.P. Ewald (1940) Proc, Phys. Soc. Lond. $\underline{52}$, 167
4. C.V. Raman and P. Nilakantan (1940) Proc Indian Acad. Sci., $\underline{A}$, 389
5. K. Lonsdale and H. Smith (1941) Nature, Lond. $\underline{148}$, 112
6. K. Lonsdale (1942) Proc. R. Soc. Lond. $\underline{A179}$, 315
7. R. Robertson, J.J. Fox and A.F. Martin (1934) Phil. Trans. R. Soc. Lond. $\underline{A\ 232}$, 482
8. A. Guinier (1942) Comptes Rend. Acad. Sci. Paris $\underline{215}$, 114
9. J.A. Hoerni and W.A. Wooster (1955) Acta Cryst. $\underline{8}$, 187
10. F.C. Frank (1956) Proc. R. Soc. Lond. $\underline{A237}$, 168
11. S. Caticha-Ellis and W. Cochran (1958) Acta Cryst. $\underline{11}$, 245
12. W. Kaiser and W.L. Bond (1959) Phys. Rev. $\underline{115}$, 857
13. T. Evans and C. Phaal (1962) Proc. R. Soc. Lond. $\underline{A270}$, 535
14. A.R. Lang (1964) Proc. Phys. Soc. Lond. $\underline{84}$, 871
15. M. Takagi and A.R. Lang (1964) Proc. R. Soc. Lond. $\underline{A\ 281}$, 310
16. M. Moore and A.R. Lang (1972) Phil. Mag $\underline{25}$, 219
17. M. Moore and A.R. Lang (1977) J. Appl. Cryst. $\underline{10}$, 422
18. S. Suzuki and A.R. Lang (1976) J. Appl. Cryst. $\underline{9}$, 95
19. S. Suzuki and A.R. Lang (1975) Acta Cryst. $\underline{A\ 31}$, S260
20. Z-H. Mai, S. Mardix and A.R. Lang (1980) J. Appl. Cryst. $\underline{13}$, 180

CHAPTER 25

REFLECTION TOPOGRAPHY: PANEL DISCUSSION

R.W. ARMSTRONG

25.1 Introduction

In order to stimulate interest in the technique and applications of reflection topography and to discuss problems of contrast in this method, a general panel discussion was held at the Advanced Study Institute. R.W. Armstrong was the chairman and D.K. Bowen assisted in the preparation of this report. The panel members included A.Authier, U.Bonse, D.K.Bowen, J.Chikawa, J.Gronkowski, M.Hart, W.Hartmann, N.Kato, A.R.Lang, E.S.Meieran, J.R.Patel, M.Sauvage and B.K.Tanner. The discussion was carried on with questions and comments from individual members of the audience also.

25.2 Discussion

Bowen began by asking, on the one hand, whether the relative importance of dynamical versus kinematical considerations was established for understanding image formation in reflection topographs and, on the other hand, whether it was clear how one should match the geometrical divergence of the experimental set-up with the perfection of the crystal being examined so as to obtain optimum information in any case? Authier gave his opinion that a complete understanding of image contrast for reflection topographs must require use of the dynamical theory as had been demonstrated, for example, for the Lang transmission method. D.Y.Parpia asked whether stacking faults inclined at a small angle to the crystal surface might be studied either experimentally or theoretically for the reflection case as had been done in transmission. Patel mentioned that this had not been done to his knowledge and might be worthwhile to do.

Kato pointed out that the plane wave and spherical wave theories for the reflection case are well-established for perfect crystals, say, as recently described by Chukhovski and colleagues. One reason for the reflection case not being easily understood is that the possible wave vectors are complex even in the non-absorbing case if the incident wave falls in the region of total reflection. Then, one cannot visualise the wave fields in terms of ray concepts.

Hart suggested that a fuller understanding of the reflection method or other reflection methods, too, might have been obtained by now if more than one topography technique had been applied to any single problem. Armstrong agreed and stated that this was now being done at a number of laboratories. Tanner asked whether reflection experiments were easily done with the standard Lang camera and Lang mentioned that with appropriate positioning of slits, etc., no trouble was encountered at all. Furthermore, Lang suggested that there might be something like a topographer's uncertainty principle to be kept in mind for image contrast in that the product of the spatial and angular resolution within images might be nearly constant.

J. Miltat said that it should be very useful to apply both reflection and transmission methods to the complete characterisation of boundary structures. The directions of dislocation lines and the crystal rotations can be determined in reflection topographs taken with controlled small divergences. Adjustment of the beam absorption can be effected in transmission to obtain information through different depths. Armstrong pointed out that the resolution of individual dislocations within subgrain boundaries in lithium fluoride crystals by Burns and Birau had been accomplished with reflection topography utilising a geometrical divergence of 100 secs of arc. Vreeland, Jr., also has obtained topographs of individual dislocations within annealed subgrain boundaries in zinc crystals.

Bonse supplied a fuller description of the double crystal reflection case. Resolution with the double crystal technique depends mainly on two conditions: (a) normal to the plane of diffraction (the ray plane), the image smear is hd/D where h is the source size normal to the ray plane, d is the sample-to-film distance, and D is the source-to-sample distance; and (b) in the ray plane, the image smear is proportional to $d(\Delta\lambda)\tan\theta_B/\lambda$, where $\Delta\lambda$ is the wavelength range effectively contributing to a particular image point and θ_B is the Bragg angle. $\Delta\lambda$ can be $(\lambda_{\alpha 2} - \lambda_{\alpha 1})$ if the images of the α_2 and α_1 lines overlap in that region or just the natural width of one of the lines or even only a fraction of this if the source width, w, in the ray plane and its divergence are so limited by slits that a fraction of a line is cut-out.

From the previous work for a double crystal topograph taken so that the overall intensity is maintained in the linear range on one of the slopes of the rocking curve, it is known that the intensity change corresponding to the long reaching strain field of a dislocation is proportional to the total distortion at the crystal surface. This is true provided the overall distortion is small enough that contributions to the intensity from crystal regions below the surface can be neglected. The two contributions to the intensity from lattice parameter changes and lattice rotations, respectively, can be separately measured. The conventional direct image of the dislocation may be measured alone in an exposure during which the sample is rocked over the reflecting curve width in which case the linear contributions to the intensity cancel because of the opposite sign of the rocking curve slopes. This rocking curve method is useful for avoiding the overlap of dislocation images because they are narrower in this case. This points to a strength of the reflection method where the integration depth can be made relatively small compared to that employed in transmission and, hence, larger dislocation densities give less overlaps in reflection.

A last consideration from Bonse involved the determination of dislocation Burgers vectors from the linear model interpretation of the far reaching strain fields measured at the crystal surface by double crystal method. It appears that the linear model has proved sufficient for this purpose mainly because there is usually only a limited choice of possible Burgers vectors for a given dislocation line. The complete wave optical simulation of dislocation images for the reflection case, say, as employed by Gronkowski and by Chukhovski and colleagues, gives very valuable information in cases where the simpler interpretation may leave doubt about the type of dislocation being characterised.

Meieran commented that neither the reflection topography methods nor the other topographical methods have kept pace with other imaging techniques as far as concerns the use of these techniques in technology. For example, optical and scanning electron microscopy (SEM) methods are widely used in industry even though they are governed by reasonably complex physical laws that determine contrast, have equally difficult sample preparation procedures, and are equally expensive. One part of the problem may be the lack of interface persons or companies, between universities and factories, involved with the manufacture of instruments in large quantities to be used by relatively unskilled operators. Also, the benefits of topography might have been oversold for solving device properties before the technology of devices had reached the present stage where they are sensitive to the crystal defects which are observed in topographical images. D.C.Miller endorsed the consideration even among the topographical methods that the easiest techniques should be employed first and so on until the observations match the level of sophistication required to solve whatever problem is involved. D.J.Viechnicki pointed out that polycrystalline, heterogeneous materials could be characterised to some extent by optical microscopy, conventional X-ray diffraction, and SEM methods whereas X-ray topography is currently confined to single crystal studies. The restricted applications of single crystals to materials technology makes the broader characterisation techniques more cost effective. M.Moore wondered whether the lack of stronger interest in utilising topographical methods, necessarily involving crystal adjustments in three-dimensional space, was even traceable to the unfashionableness of teaching geometry in schools. Parpia reminded the audience that there were safety considerations, both directly related to the damage of human tissue and indirectly related to separately ionised molecules presenting health hazards, which were not easily dealt with in the various topographical techniques employed until now. Tanner mentioned that this last consideration certainly required that total remote control was involved in doing synchrotron topography.

Hartmann gave his judgement that, thus far, real time topographs taken in Berg-Barrett reflection with a television system show relatively blurred images as compared to those observed as section topographs recorded with nuclear emulsion. The difference in the recording systems might be solely responsible but this was not yet established. Y.Epelboin commented that it would be helpful if the fields of application for the various techniques were more firmly established so that one could match the technique with the problem and switch accordingly ---- even for the consideration of using real time systems.

Sauvage pointed out that the parameter for the formation of kinematical images in reflection is not the extinction distance but is the penetration depth, t_A, determined by the photoelectric absorption process as $t_a = \mu^{-1}[\csc|\theta_B + \chi| + \csc(\theta_B - \chi)]^{-1}$ where χ is the negative angle between the reflecting plane and the surface and μ is the photoelectric absorption coefficient. t_A can be changed by changing the wavelength and the reflection. As an axample, the analysis of defects introduced during the successive surface processing of semiconductor materials is easily performed since by changing from grazing to large incidence angles t_A can be adjusted from a small fraction of a micron to a few tens of microns. Additionally, separate imaging of nearly matched thin epitaxial layers over a thick substrate is only possible by use of a low divergence incident wave reflected in the Bragg case.

S.R.Stock inquired whether the images of defects observed in any topograph were influenced by the presence of other defects too deep in the crystal to be seen and, also, whether surface "image forces" might be important to consider in topographs of thin crystal sections? Gronkowski recommended that each of the topographical methods should probably be employed in a complementary way fully to characterise the the defect structure within any crystal. It should not be expected, for example, that reflection topography would give complete information about the total defect content of a bulk sample. For a complete interpretation of defect contrast features in reflection, a transmission method such as the Lang method should be used. From the combined results, the presence of other defects in the vicinity of one defect of interest could be established.

REFLECTION TOPOGRAPHY: PANEL DISCUSSION

Chikawa drew attention to the fact that the incident beam in a Berg-Barrett set-up normally has a divergence of a few minutes of arc. Some of the incident X-rays are reflected in the range of total reflection without entering the crystal. Therefore, they do not contribute to the formation of direct images of those defects appreciably below the crystal surface. These X-rays enhance the background intensity of the perfect crystal except that they are affected by surface roughness. The internal defect images are formed mainly by the beam deviated from the Bragg condition. For the best contrast, the double crystal (+n,-n) setting should be used. In general, however, the background intensity generated by the white X-rays in the Berg-Barrett method is usefully suppressed by employing a crystal monochromator so as to give acceptable contrast of defects.

APPENDIX 1

DESIGNING A TOPOGRAPHIC EXPERIMENT

A.R. LANG

Section topograph exercises

Fig. 1 shows schematically the arrangement for taking section topographs. The rectangle (dimensions in mm) at F represents the X-ray tube focal area which is in the horizontal plane and from which X-rays are taken off at a mean angle of 4° with the horizontal. IBS is the incident beam slit, X is the specimen crystal. DBS is the diffracted beam slit and P is the photographic plate. The diffracting planes make an angle α with the normal to the surfaces of the plate-shaped specimen, α being positive in the same sense as the deviation 2θ of the diffracted beam. Answer the following questions:

A.1. If the distance FX is 0.5 m, what is the maximum distance XP can take if the resolution in the vertical (axial) plane is to be not worse than 2 μm?

A.2. The section topograph in Fig. 2 is the $11\bar{1}$ reflection from diamond (a = 0.3567nm) using $CuK\alpha_1$ radiation. It was not actually taken in symmetric transmission, but to simplify calculation assume that it was. Measure the maximum width of the section on the print (magnification x36), and calculate therefrom the specimen thickness.

A.3. You are given a wafer of silicon (a = 0.5431 nm) polished parallel to (111) and wish to take a section topograph with the incident beam cutting the specimen as nearly perpendicularly as possible. Can you do this using a reflection (possibly higher order) from one of the {111} planes which make 70 1/2° with (111)? You have choice of MoKα or CuKα. Consult Tables of 'd' versus θ and the standard (111) stereographic projection (Fig 3). Sketch the diffraction geometry you would use. (For further thought: what other planes would be suitable?)

A.4. You have taken section topographs of a parallel-sided plate-shaped specimen in symmetrical transmission with $2\theta = 60°$, using the hkl and $h\bar{k}\bar{l}$ reflections. On both topographs you observe the image of a fault surface. It makes 30° with the image margins, in both cases. (See sketches of topographs in Fig. 4. in which the X-ray entrance and exit surfaces are indicated.) Determine the orientation of the fault surface. If you only had <u>one</u> of the sections, what information would that give you about the orientation of the fault surface?

Projection topographs

B.1 You wish to make the IBS as wide as possible without allowing a significant contribution from the α_2 component when the specimen is set on the α_1 peak. Using the dimensions of F and of distance FX($\simeq$F to IBS) given above, what (if any) slit widths would be satisfactory in the following cases: interplanar spacing, d = 0.1nm, CuKα; d = 0.4 nm, CuKα; d = 0.2 nm, MoKα.

B.2 Look at the projection topograph of the diamond (Fig. 5) The X-ray entrance and exit surfaces are ($\bar{1}1\bar{1}$) and (111) respectively, the reflection was $11\bar{1}$ with CuKα_1. The [$\bar{1}$10] axis was pointing vertically upwards. The topograph was enlarged in conventional orientation, i.e. image viewed looking <u>towards</u> the specimen. Set the print in this orientation. The section topograph was taken in the same geometry, also using the $11\bar{1}$ reflection. Can you place the section topograph image in correct location on the projection topograph image?

B.3 If the projection topograph was taken using IBS 140 μm wide and the section topograph with IBS 7 μm wide, estimate the relative durations of suitable exposures to give similar densities in the two experiments.

Stereo pairs

C.1. The pair of topographs of LiF in Fig 6 are the 020 and 0$\bar{2}$0 stereo pair of a thin plate with surfaces parallel to (001) and (00$\bar{1}$), oriented with [00$\bar{1}$] towards the X-ray tube and [100] pointing <u>downwards</u>. CuKα radiation 2θ = 45.1°. From the appearance of the images, confirm that the prints are reproduced in the correct orientation, and in the correct relative position for viewing as a stereo pair. (Note the print with the black dot defect is the 020.). Hint: look for the dynamical images of dislocations, which appear as a 'shadow', broadening towards the exit surface; sketch the Borrmann fan, from which you should see the relationship between the direct image (dark line) and the dynamical image. If you do not, reread chapters 7, 9, and 15!

C.2. By measurements of parallax determine the thickness of the specimen (Question for further thought: what other evidence do the images give concerning specimen thickness.)

C.3. If you wanted to take stereo pairs of the diamond specimen whose projection topograph you have, would you use the hkl, h̄k̄l̄ method, or the Haruta method (rotation about the g-vector). If you chose the latter, what magnitude of rotation would be suitable?

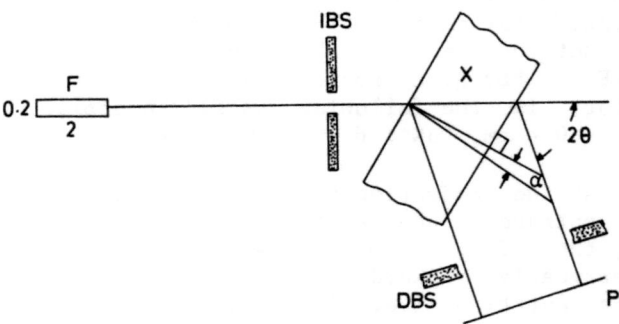

Figure 1. Schematic arrangement for taking section topographs.

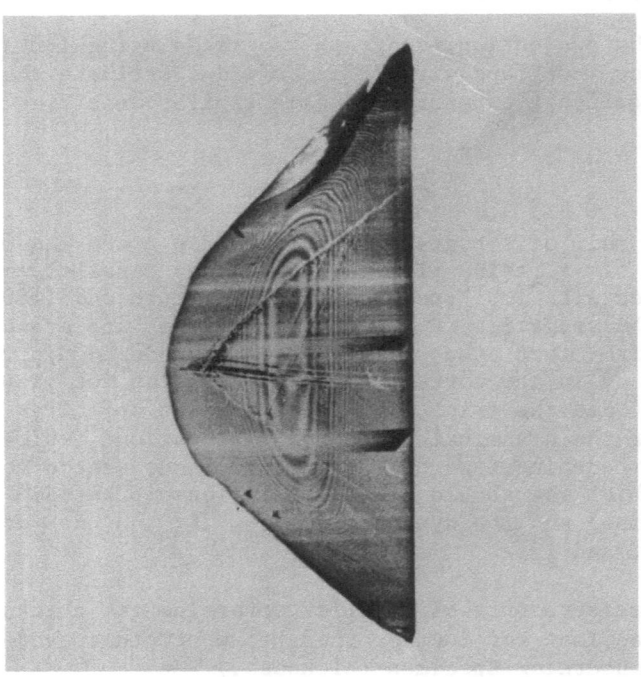

Figure 2. Section topograph of diamond (see question A.2) [110] vertically upwards

DESIGNING A TOPOGRAPHIC EXPERIMENT

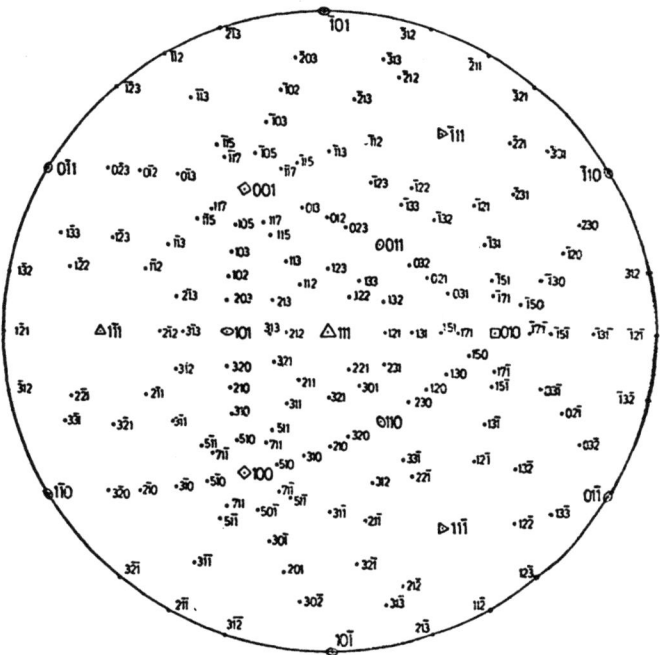

Figure 3. [111] stereographic projection for cubic crystals

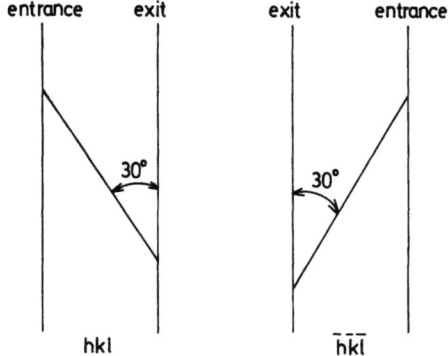

Figure 4. Geometry of topographs for question A.4

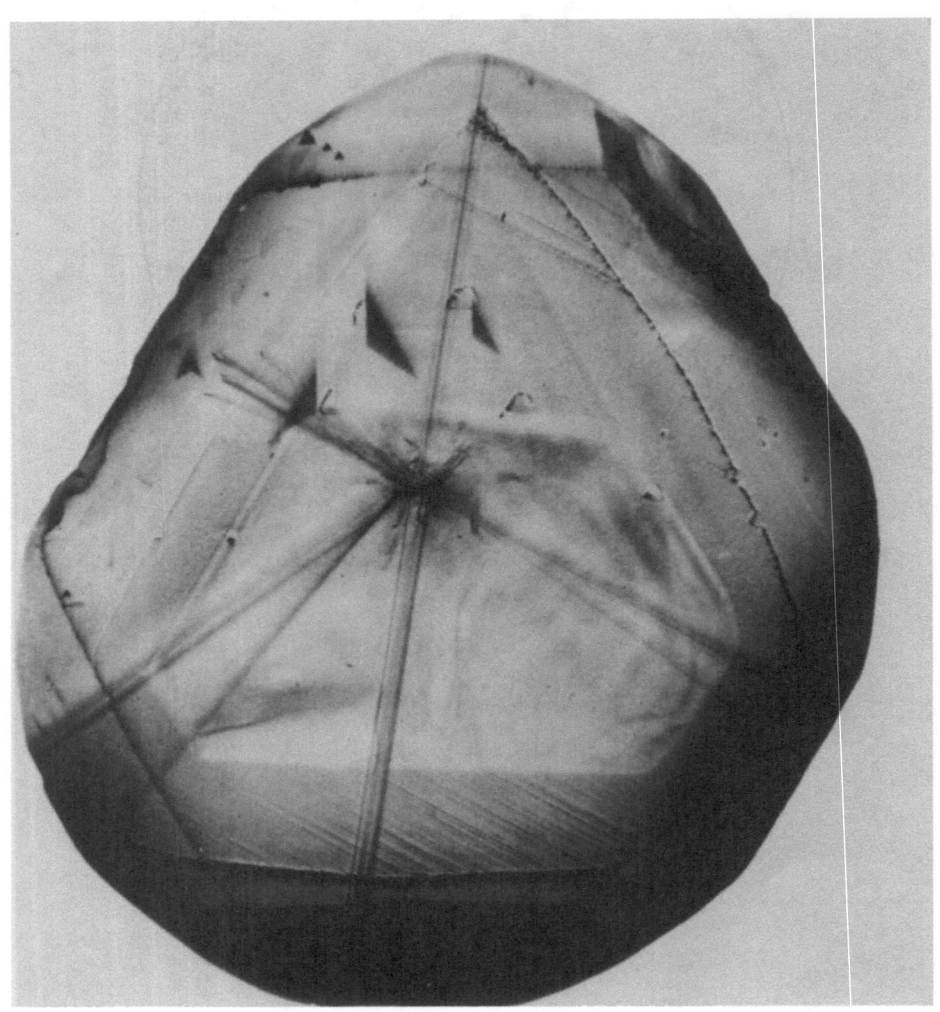

Figure 5. Projection topograph of diamond, $CuK\alpha_1$, g=111
$[\bar{1}10]$ points to right, roughly horizontally

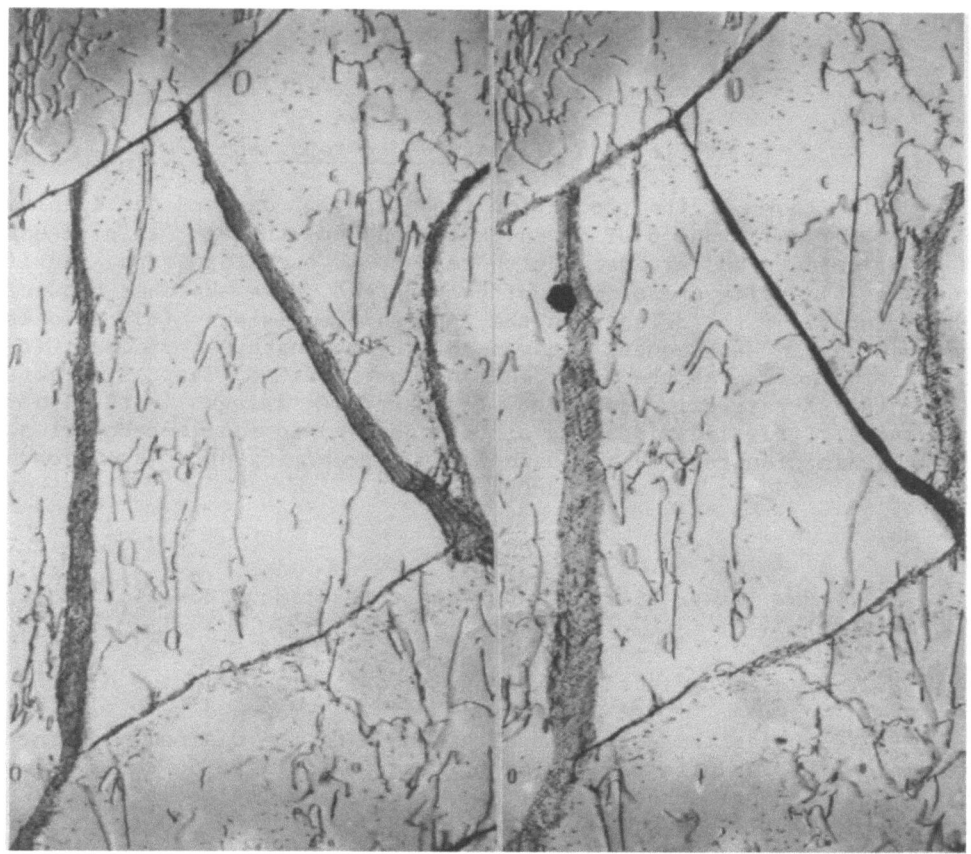

Figure 6. Stereo pair of topographs of lithium fluoride. $0\bar{2}0$ (left) and 020 (right); CuKα_1, $2\theta=45.1°$

APPENDIX 2

DEFECTS AND ARTIFACTS

A.R. LANG

Identification of features on topographs

A.1. Look at the projection topographs of diamond and lithium fluoride (Figs. 5 and 6 of Appendix 1) and go through this check list of defects and artifacts with respect to each topograph. See if you can identify one or more of each defect. Make sketches showing their location. (a) defects in the nuclear emulsion; (b) defects introduced in subsequent stages of the photographic process; (c) lattice defects near the X-ray entrance surface; (d) lattice defects near the X-ray exit surface; (e) surface damage (natural or man-made); (f) individual dislocations; (g) low-angle-boundaries; (h) stacking faults; (i) inclusions; (j) precipitates; (k) growth bands; (m) cracks; (n) slip bands.

Analysis

B.1. How many dislocations does this diamond contain? (Rough estimate only)

B.2. Find the orientation of one stacking fault in the diamond

B.3. Knowing the approximate thickness of the diamond specimen estimate the Pendellosung period by counting fringes on the section topograph. Account quantitatively for the apparent 'beat' in the fringe pattern.

B.4. Do any of the dislocations in the diamond follow low-index directions?

B.5. Do you see any evidence for decoration or dissociation of dislocations in any of the topographs?

B.6. Using the stereo pair of LiF topographs, choose a dislocation whose orientation you think you can determine easily, and do so.

DEFECTS AND ARTIFACTS

Questions for further study and discussion

C.1. Suppose the LiF slice were 200 μm thick and filled with dislocations with Burgers vectors and line directions randomly distributed (i.e. dislocations <u>not</u> concentrated in low-angle boundaries). What is the maximum dislocation density for which you could resolve individuals in (a) a projection topograph (b) a section topograph?

C.2. You are asked to give a figure for dislocaton density in the diamond. What is your answer?

C.3 How would you assess the relative number of dislocations in the LiF sample which are concentrated in low-angle boundaries and which are distributed more or less randomly within the sub-grains (To answer this, you must consider ways available to you for assessing the average density of dislocations in the low-angle boundaries.)

C.4. Some of the dislocations in the diamond are very straight. How would you determine their orientations precisely, and what precision could you achieve?

C.5. Derive an estimate of the width of the reflection curve of a perfect LiF crystal from the observed widths of appropriate dislocation images in the topographs.

APPENDIX 3

EXERCISES IN DIFFRACTION CONTRAST

D.K. BOWEN

Introduction

The purpose of this appendix is to give practice in the determination of dislocation Burgers vectors from topographs taken in different reflections. A realistic situation is chosen, that is, one where there are insufficient reflections for the determination to be quite unambiguous. We shall see how other crystallographic information can be used to help decide the probable Burgers vectors. Two crystals are shown, sodium chlorate and silicon. The former shows straightforward invisibility of dislocations, but 'residual contrast' is often seen in the silicon specimen and this must be recognised. The silicon specimen also shows traces of dynamical images, even though the value of μt is only 0.15.

Sodium chlorate

Fig. 1. Shows topographs of a sodium chlorate crystal 15 mm in longest dimension, taken in 200, 020 and 111 reflections (courtesy Prof. J. Sherwood, Strathclyde University).

1. Dislocations 1 - 11 are marked on one of the three reflections on the figure. For each dislocation, determine whether it is visible or invisible in the other reflections and complete a table with a 1 for visible and 0 for invisible.

2. Where a 0 is entered is there <u>any</u> trace of an image or is it completely invisible?

3. Are there any uncertainties in the table? If so, why?

4. The images should fall into two groups on the basis of invisibility.
Group A contains dislocations 1, 2 and which others?
Group B contains dislocations 5 and which others?
Why is dislocation 4 an oddity in its group?

5. What can be said about the Burgers vectors of group A dislocations?

EXERCISES IN DIFFRACTION CONTRAST

6. What can be said about the Burgers vectors of group B dislocations?

7. Sodium chlorate has a primitive cubic unit cell. What is therefore the likeliest Burgers vector (and why)?

8. What is the likely Burgers vector for group B dislocations?

9. Add to the table the line directions of the dislocations. It is necessary to know that the major crystal edges are parallel to the cube axes. Could the directions be determined from the 200 and 020 topographs alone?

10. Identify group B dislocations as edge, screw or mixed.

11. For the edge/mixed types, confirm that $\underline{g} \cdot \underline{b} \times \underline{u} = 0$ for the reflection in which they are invisible.

12. There are no invisibilities for type A dislocations. Examine the possibility that these are edge dislocations with [100] type vectors, for which $\underline{g}.\underline{b}x\underline{u} \neq 0$ for all three reflections.

13. Is it possible for group A to be mixed dislocations with Burgers vector of type [100]?

14. Give examples of possibilities for A group dislocations. What extra topographs should be taken to identify them positively?

Hints

I am not going to give all the answers, as that would make it too easy, but the following points may help if you get stuck.

You should be able to identify group B dislocations positively. The invisibility is clear. However, you will probably not succeed until you have answered question 9 on line directions. If you are still stuck, remember that the <u>only</u> reflection for which a pure edge dislocation is invisible is that of the plane perpendicular to its line direction. If you can find the line direction and check this point it confirms the analysis.

The group A dislocations are much more difficult, and cannot really be determined from these topographs - but working through them does show the problems. The best guess of the tutors at the A.S.I. was that they are not individual dislocations but bundles, maybe with mixed Burgers vectors.

It should now be obvious that the above order of questions, although it follows the pedagogical approach of systematic examination of invisibility criteria, is not the best way to analyse

Burgers vectors. It is extremely helpful to get the line directions at an early stage, and to use other crystallographic information (such as knowledge of likely Burgers vectors).

Silicon

Figs. 2 and 3 show various reflections in a (111) silicon slice (taken from J. Miltat and D.K. Bowen, 1975, J. Appl. Cryst. **8**, 657). The thickness was 100 μm and with MoKα radiation $\mu t = 0.15$

Analysis of dislocations

You may assume that all dislocations have Burgers vectors parallel to a [111] direction. Identify the dislocations marked a, b, c, d in the line diagram.

Method. First study the stereo pair 4 and 5 to visualise the dislocations; then compile a table showing $\underline{g}\cdot\underline{b}$ values for all the reflections and possible b values; then compile a table showing the strengths of the images (strong, very strong, weak, very weak and invisible) in each reflection and compare with the invisibility table.

You will probably run into difficulties or ambiguities; when you do, go on to the remaining questions and return to the analysis later.

Observation of complex effects

With the low value of μt, kinematic effect (direct images) should predominate. However .....

1. Dislocations 'a' clearly thread the slice from top to bottom. Is the contrast uniform all the way? What could be the cause of the non uniformity?

2. Observe the shape of the image of dislocation 'e' (also labelled 'c' in Fig. 2(c), in several topographs, and deduce which end is at the entrance surface for the X-ray beam (remember appendix 2)

3. Dislocations c and c' in the line diagram (B and A in Fig. 2(c) are clearly of the same type. Are they parallel or inclined to the surface? Which is nearer the surface? Give two possible explanations for the differences in image widths.

4. Dislocations 'a' are least visible in reflections $1\bar{1}\bar{1}$ and $\bar{1}11$. In fact they are pure screw dislocations and $\underline{g}\cdot\underline{b} = 0$ in these cases. However, there is still some residual contrast. Is this uniform across the slice? Suggest possible reasons for it.

EXERCISES IN DIFFRACTION CONTRAST

5. Now that you have an idea of the ambiguities go back and finish the analysis. Do any ambiguities remain, and if so, what topographs should be taken to resolve them?

Hints

This is a complex situation, which would take an experienced topographer some hours to solve, so do not give up too soon. In addition to practice in Burgers vector analysis, these topographs show many features in which dynamical effects blur the simple kinematical images, although the μt value was low. Also, the dislocation images never go absolutely invisible, but there is some 'residual contrast', e.g. dislocations a on fig. 2(a) (remember that the geometric distortion on each topograph is different). Residual contrast may be caused by dynamical effects in certain cases, but a more likely cause is surface relaxation - the stress fields of dislocations are given by the usual expressions only in an infinite medium, and the presence of the surface modifies the field substantially. It is worth re-examining the topographs to look for image differences at the surface from those inside the specimen.

Identification of all these dislocations and a discussion of the various image effects appears in the original publication cited above.

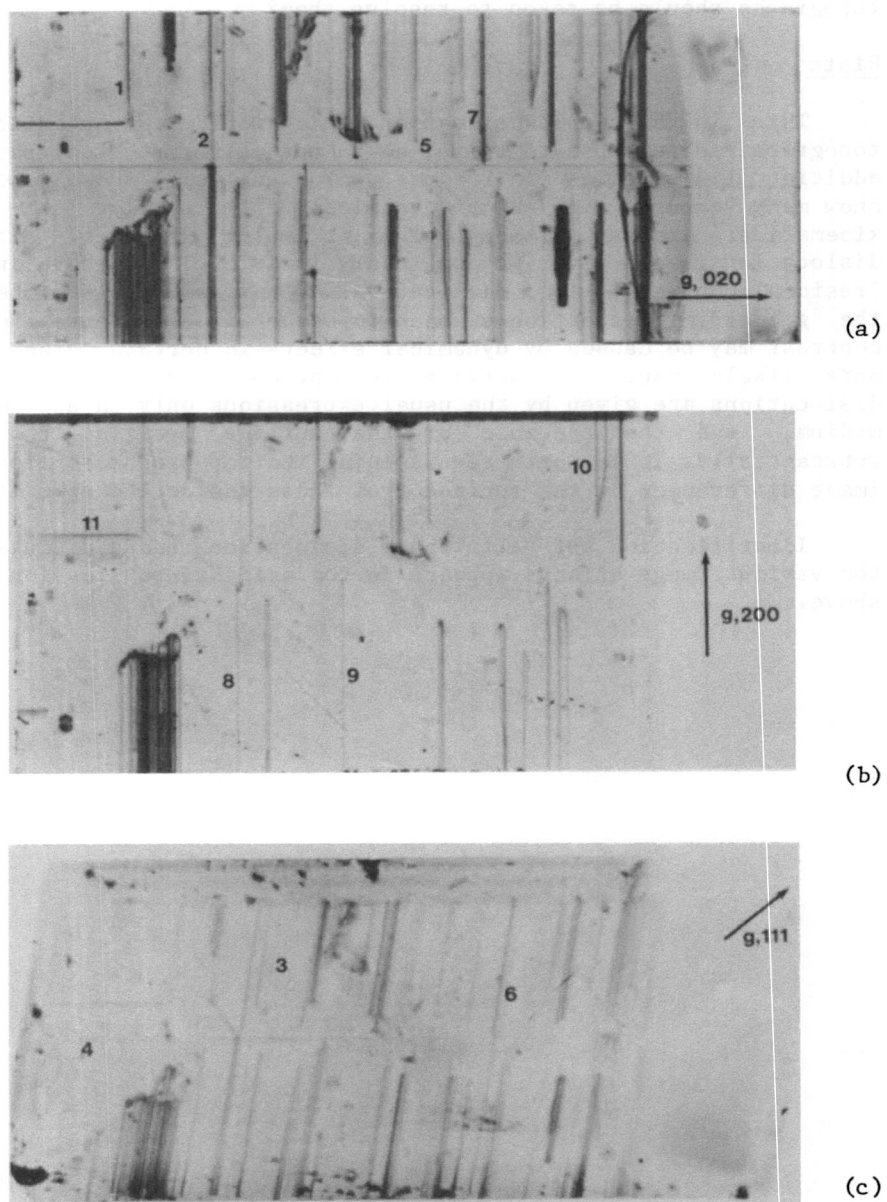

Figure 1. NaClO$_3$ (a) 020 reflection, (b) 200 reflection, (c) 111 reflection.

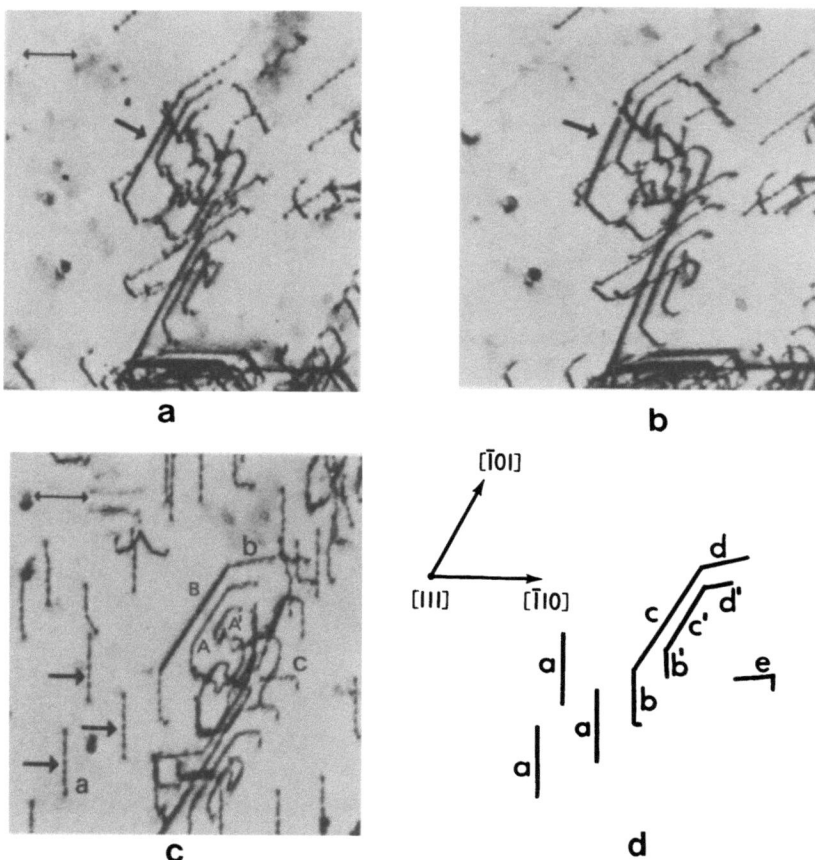

Figure 2. Silicon (a) $1\bar{1}\bar{1}$ reflection, (b) $\bar{1}11$ reflection [stereo pair with Fig. 2(a)], (c) $\bar{1}\bar{1}1$ reflection, (d) line diagram labelling dislocations.

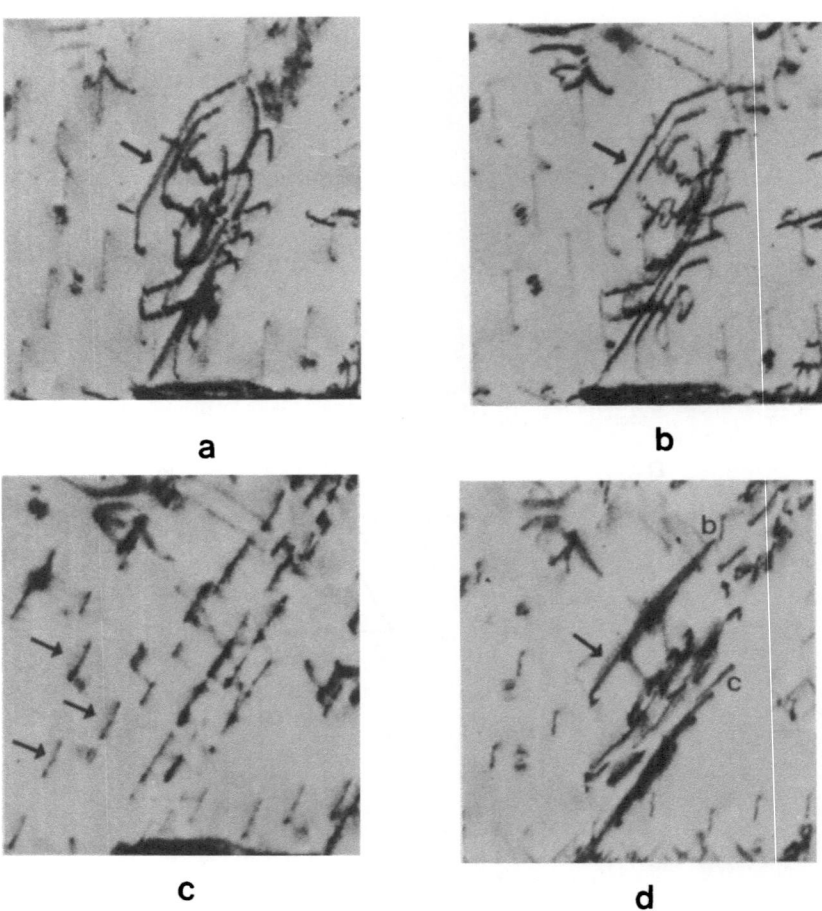

Figure 3. Silicon (a) $20\bar{2}$ reflection (b) $02\bar{2}$ reflection (c) $2\bar{4}2$ reflection (d) $\bar{2}\bar{2}4$ reflection.

APPENDIX 4

STEREOGRAPHIC PROJECTION DESCRIPTION FOR X-RAY TOPOGRAPHY:

SUBGRAIN BOUNDARIES AND STEREO-PAIRS

R.W. ARMSTRONG

The geometrical conditions for obtaining X-ray topographs of crystals and for describing the crystallography of their structures are advantageously described with a stereographic projection method of analysis.

Adjacent subgrains with misorientations, say, on the order of, or greater than, $0.1°$ are easily recognised in reflection topographs because of the relative displacement of the subgrain images. The ORIENTATION width of such boundaries is normally observed when their constituent dislocations are so closely spaced in a complementary way that EXTINCTION width of the cumulative dislocation strain fields is not able to be detected. In the case of small subgrain misorientations where both orientation and extinction effects contribute to determining the appearance of any boundary, the effects can be separated by investigating their different dependences on the diffraction conditions. Adjacent subgrains with a large misorientation between them are able to be observed even in topographic experiments stringently controlled monochromatic radiation and small geometrical divergences if the axis for the misorientation of subgrains is positioned in the plane of diffraction.

The conditions for obtaining stereo-pairs of topographs and the perception of depth which is obtained by simultaneous viewing of the pairs can be matched on a stereographic projection basis with the full three-dimensional range of directions generally taken-up by the defect structures within real crystals. The well-established use of stereo-pairs obtained by the Lang technique for crystal plates on the order of 100 to 1000 microns thick is compared with the more recent employment of stereo-pairs for observing dislocations in Berg-Barrett or Borrman topographs. The relatively shallower crystal surface layers, say, 10 microns or less in depth, typically sampled with the Berg-Barrett (reflection) technique place a severe restriction on the potential of stereo-pairs for resolving depth effects. For the Borrman anomalous transmission of X-rays through crystal plates, say 1000 microns or greater in thickness, a different geometrical consideration occurs for the projection of images to form stereo-pairs because the direction of energy flow for the transmitted beam within the crystal is parallel to the crystal diffracting planes.

The following figures give several examples of the use of stereo pairs and stereographic projections in the analysis of dislocations and boundaries. There are no formal questions, but for each topograph the student should first study the stereographic projection to understand the geometry of the stereo-pair, then examine the topographs (where given) with a stereo viewer to decide what information it contains. Each topograph can in fact be studied in considerable detail, and provides practice material for determining wall planes and dislocation line directions, either by calculation or by use of parallax bars on a photogrammetry stereo viewer. The stereographic projection for Fig. 7 appears as Fig. 1 of chapter 14.

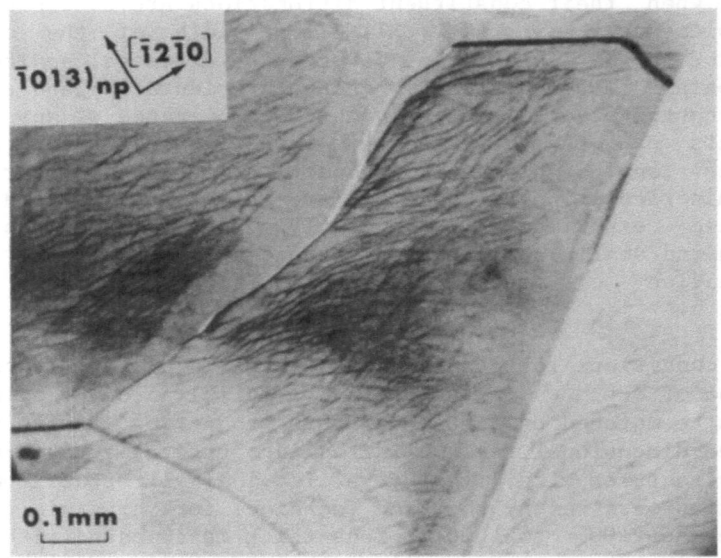

Figure 1. Misorientation contrast for changing boundary directions within (0001) of a zinc single crystal solidified along [0001]. R.W. Armstrong, C.Cm. Wu and E.N. Farabaugh (1977) in Advances in X-ray Analysis, 20, 201

STEREOGRAPHIC PROJECTIONS IN TOPOGRAPHY

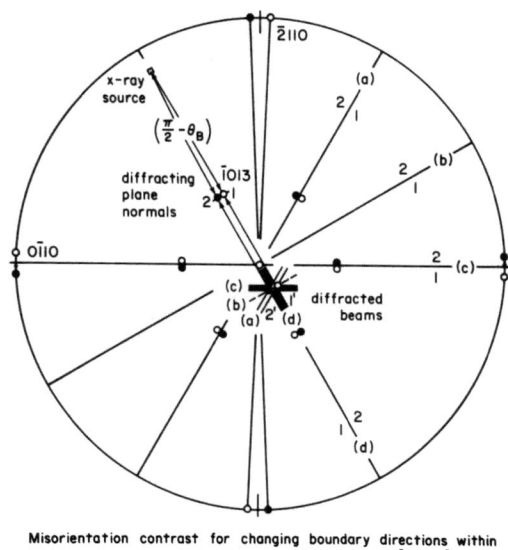

Figure 2. Stereographic projection for determining orientation contrast in Fig. 1

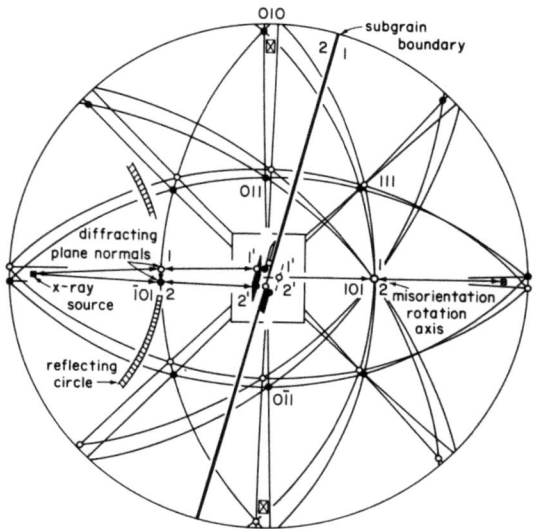

Figure 3. Stereogram of misorientation contrast in {202} reflections obtained with CrKα radiation for a hypothetical subgrain boundary with [101] rotation axis in an MgO crystal.

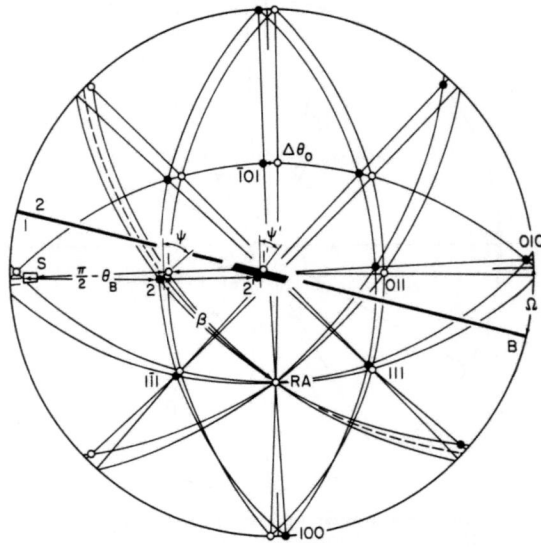

Figure 4. Stereogram of misorientation contrast of subgrain boundary for $0\bar{2}2$ reflection from (001) MgO crystal surface.

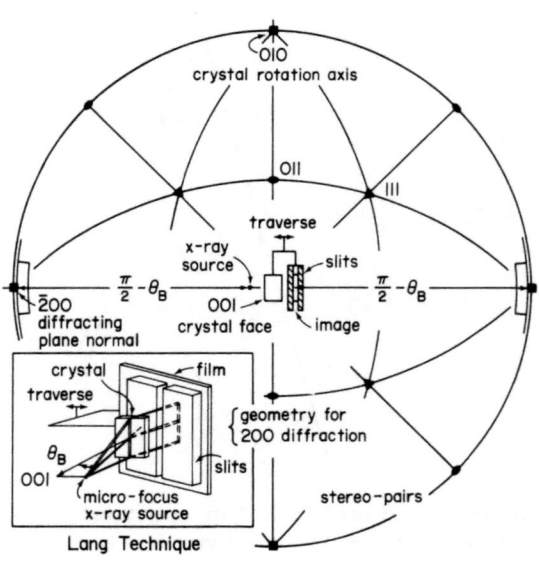

Figure 5. Stereogram for standard Lang topography, showing geometry of Fig. 6.

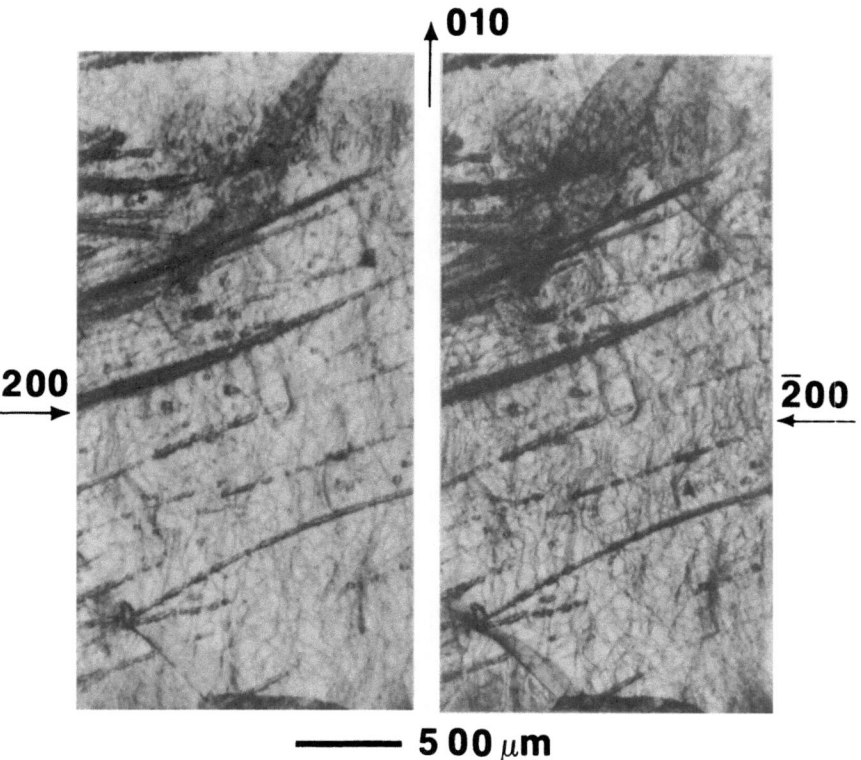

Figure 6. Stereopair Lang topographs of lithium fluoride. Courtesy B. Roessler, P. Kumar and R.J. Clifton (1979) Brown University, unpublished research.

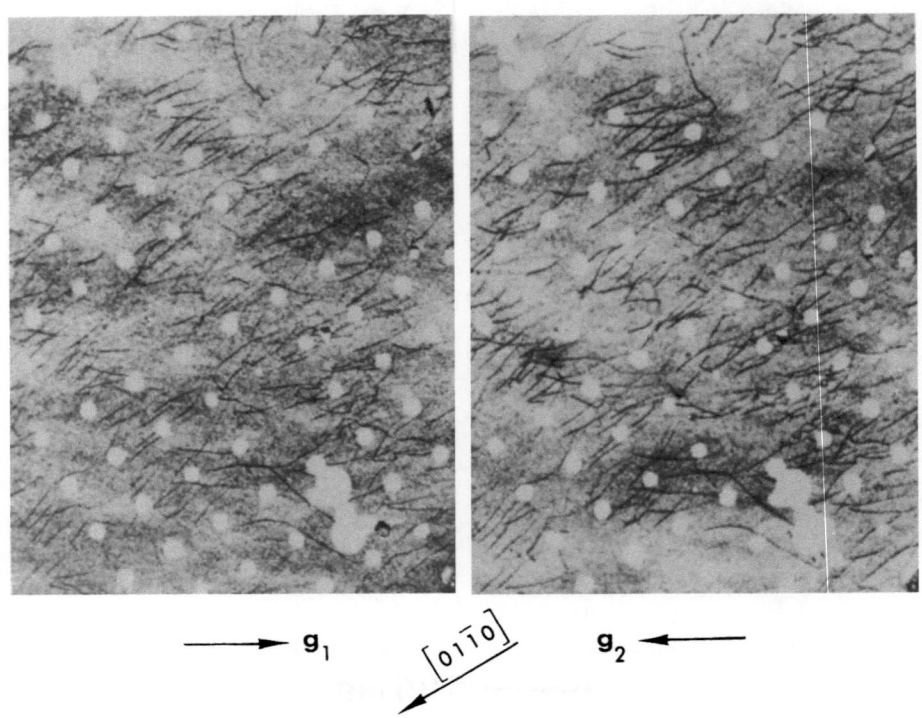

Figure 7. Stereopair Berg-Barrett topographs showing dislocations in zinc; the white dots are a thin gold reference grid. Courtesy T. Vreeland, Jr., (1976) J. Appl. Cryst. 9, 34

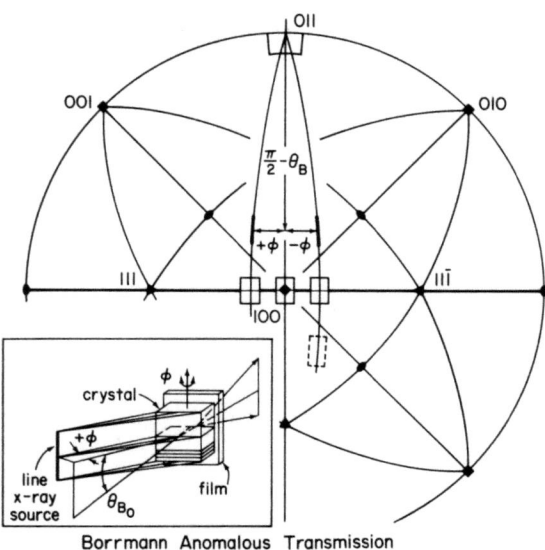

Figure 8. Stereogram of geometry for Borrmann anomalous transmission topographs in Figs. 9 and 10

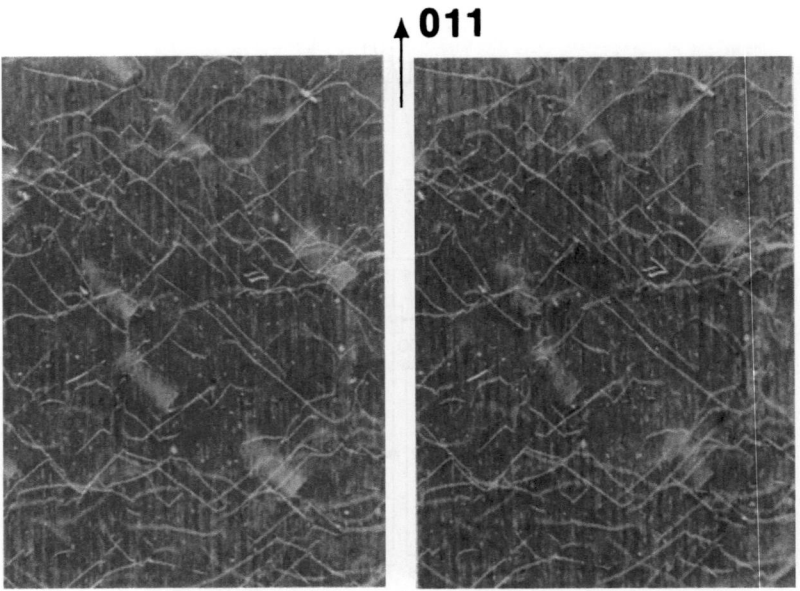

Figure 9. Borrmann anomalous transmission topographs of dislocations (and magnetic domains) in Fe - 3% Si. Courtesy C.Cm. Wu and B. Roessler (1971) Phys. Stat. Sol. (a)8, 571

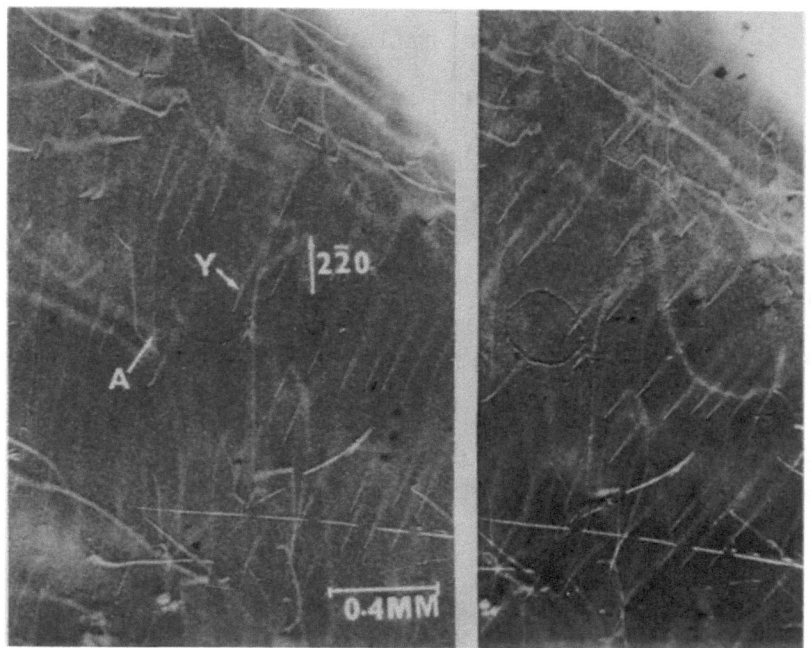

Figure 10. Borrmann anomalous transmission topographs of dislocations in germanium. Courtesy M.C. Narasimhan and B. Roessler (1975) in Microstructural Science (eds. P.M. French, R.J. Gray and J.L. McCall) American Elsevier, New York, p 583

APPENDIX 5

DISPERSION SURFACE EXERCISES

F. BALIBAR

1. Exercise 1 - Bragg diffraction

1.1 The dispersion surface, locus of the extremity of the possible wave-vectors in the crystal, is said to be the analogue of the surface of indices in ordinary optics.
Explain why?
1.2 How is it that Bragg diffraction does not occur in crystal optics of visible light?

2. Exercise 2

2.1 The equation of the dispersion surface (M. Hart Ch. 9 eq. 16) is:

$$\alpha_{oj} \cdot \alpha_{hj} = \frac{k^2}{4} \chi_h \chi_{\bar{h}}$$

(the polarization factor C has been omitted, for simplicity) Give a geometrical interpretation of α_{oj} and α_{hj}.
2.2 Show that the minimum distance between branch 1 and branch 2 of the dispersion surface is $k\sqrt{\chi_h \chi_{\bar{h}}}/\cos\theta_B$ (θ_B = Bragg angle).
2.3 Estimate the angular range of total reflection for Si 220 and MoKα radiation (λ = 0.7A).
e^2/mc^2 = 'classical radius' of the electron = 2.8 Fermi = 2.8 10^{-15}m
    d = crystal parameter for Si = 5.4 A
    f_i = scattering factor Si,220, MoKα radiation
        = 8.6
<u>Hint</u> First calculate the structure factor F_h, then the Fourier component of the electrical susceptibility, for Si 220, MoKα radiation.

3. Exercise 3

From simple qualitative arguments about the dispersion surface and the Poynting vector, give a rough estimation of the dependence of D_{oj} and D_{hj} (j = 1,2) on the departure from exact Bragg angle $\Delta\theta$ in the Laue case. For zero absorption, show that when going from one

end to the other of the domain reflection, there is a reversal of the relative importance of wave-field 1 and wave-field 2.
<u>Hint</u> First determine carefully which end of the domain of reflection corresponds to $\Delta\theta = +\infty$ and which corresponds to $\Delta\theta = -\infty$. Concentrate on these two extremities; first determine which component (o or h) is predominant, then which wave (1 or 2). It can be useful to introduce the parameter $\xi_j = D_{hj}/D_{oj}$.

4. Exercise 4 - Boundary conditions

An incident plane wave in vacuum propagates in a crystal slab made of two identical parts (I and II) separated by an infinitely thin gap. Assume for simplicity a symmetric case where the entrance surface of the crystal is perpendicular to the reflecting planes. Using the dispersion surface construction, determine the tie points of the wave fields which propagate in part II.
<u>Hint</u> : Assume that the gap is filled with air (or vacuum) and describe carefully the structure of the waves which propagate in this gap.

5. Exercise 5 - Spherical wave

Show that the total amplitude of the refracted wave in the plane wave case is proportional to the Fourier transform of the distribution of amplitude on the exit surface due to an incident spherical wave. Applying Parseval's theorem then, show that the integrated intensity received at a point on the exit surface when the crystal is translated (traverse topograph) is equal to the integral of the intensity distribution along the base of the Borrmann fan in a section topograph.

6. Solutions

6.1 Exercise 1

1.1 Let P_j be a point on the dispersion surface and 0 the origin in reciprocal space. Each vector $\underline{OP}_j$ represents a possible wave-vector for an electromagnetic wave in the medium (here a crystalline medium) and:

$$\underline{OP}_j = \underline{K}_{oj} = (\omega/c) n_{\underline{OP}_j}$$

where ω is the angular frequency of the wave, c is the velocity of light in vacuum and n_{OP_j} is the index of refraction of the medium in the $\underline{OP}_j$ direction. The length OP_j in each direction is therefore proportional to the index of refraction in that direction and the dispersion surface is a representation in reciprocal space of the variations of the index of refraction with the direction of the wave vector. This is just what people working in optics of visible light call the surface of indices. Note that for an homogeneous medium (e.g. vacuum or a crystal far from Bragg's incidence), the surface of indices reduces to a sphere of radius n/λ.

1.2 In the case of visible light ($\lambda \sim 10^{-7}$m), $n/\lambda \sim 1/\lambda$ is much smaller that the distance between two reciprocal lattice points O and H (OH = 1/(lattice spacing) $\sim 1/10^{-10}$m); the two surfaces of indices drawn from these two points do not intersect. There is no degeneracy, no split of degeneracy and therefore no Bragg reflection. (Fig. 1). The situation is quite different in the case of X-radiation. Let n be the average index of refraction of the crystalline medium. Since $\lambda = 10^{-10}$m, the two spheres of radius $\bar{n}/\lambda \sim 1/\lambda$ drawn from O and H are such that the separation of their centres OH is of the same order of magnitude as their radius $1/10^{-10}$m); the two spheres intersect and each point of intersection is a point of degeneracy. The split of this degeneracy (due to the crystalline field) then leads to the usual hyperbolic dispersion surfaces (Fig. 1).

6.2 Exercise 2

Let us first recall how the dispersion surface equation is obtained from the basic system of 2 linear equations (see M.Hart Ch. 9 equ. 12) which rules the dependence of the amplitudes of the 2 plane components D_{oj} and D_{hj} for each wave-field $j (j = 1,2)$:

$$\begin{cases} [\underline{K}_{oj} \cdot \underline{K}_{oj} - k^2(1+\chi_o)]D_{oj} - k^2 \chi_{\bar{h}} D_{hj} = 0 \\ - k^2 \chi_h D_{oj} + [\underline{K}_{hj} \cdot \underline{K}_{hj} - k^2(1+\chi_o)]D_{hj} = 0 \end{cases}$$

$$D_j = D_{oj} \exp[-2\pi i \underline{K}_{oj} \cdot \underline{r}] + D_{hj} \exp[-2\pi i \underline{K}_{hj} \cdot \underline{r}]$$

The dispersion surface equation is obtained by equating the characteristic determinant of this system to zero:

$$\left||\underline{K}_{oj}|^2 - k^2(1+\chi_o)\right| \left||\underline{K}_{hj}|^2 - k^2(1+\chi_o)\right| = k^4 \chi_h \chi_{\bar{h}}$$

Introducing α_{oj} and α_{hj} such that:

$$\alpha_{oj} = \frac{|\underline{K}_{oj}|^2 - k^2(1+\chi_o)}{2k} \simeq |\underline{K}_{oj}| - k(1+\frac{\chi_o}{2})$$

$$\alpha_{hj} = \frac{|\underline{K}_{hj}|^2 - k^2(1+\chi_o)}{2k} \simeq |\underline{K}_{hj}| - k(1+\frac{\chi_o}{2})$$

the above equation of the dispersion surface becomes

$$\alpha_{oj} \cdot \alpha_{hj} = \frac{k^2}{4} \chi_h \chi_{\bar{h}}$$

Note, by the way, that the linear system then becomes:

DISPERSION SURFACE EXERCISES

$$2\alpha_{oj} D_{oj} - k\chi_{\bar{h}} D_{hj} = 0$$

$$-k\chi_h D_{oj} + 2\alpha_{hj} D_{hj} = 0$$

The ratio of the amplitudes, for a given wave-field j is thus:

$$\xi_j = \frac{D_{hj}}{D_{oj}} = \frac{k\chi_h}{2\alpha_{hj}} = \frac{2\alpha_{oj}}{k\chi_{\bar{h}}}$$

a result which should be remembered when dealing with Exercise 3.

2.2 Geometrical interpretation of α_{oj} and α_{hj} (Fig. 2).

Let T_o and T_h be the straight lines which approximate the circles centred on O and H with radius $k(1+\chi_o/2)$. Since $\overline{OP_j} = \underline{K}_{oj}$ is a vector of length $|\underline{K}_{oj}|$ perpendicular to T_o, α_{oj}, defined above, measures the distance from the tie point P_j to T_o. Idem for α_{hj} and T_h.

2.2 The <u>minimum distance between branch 1 and branch 2</u> is A_1A_2, A_1 and A_2 being the apices of the dispersion surface. For each of these points $\alpha_{oj} = \alpha_{hj}$ and

$$A_1A_2 = 2 A_1L_o = 2 (k/2 \cdot \sqrt{\chi_h\chi_{\bar{h}}} / \cos\theta_B = k\sqrt{\chi_h\chi_{\bar{h}}}/\cos\theta_B$$

2.3 Structure factor for Si,220

$$F_h = \Sigma f_i \exp 2\pi i \underline{h} \cdot \underline{r}_j$$

For the 220 reflection and the 4 possible values of r_j corresponding to a f.c.c. lattice, $\Sigma \exp 2\pi i h \cdot r_j = 4$. For Si (and 220 reflection), the summation over the 2 atoms of the basis gives $F_h = 8 f_i = 8 \times 8.6$

<u>h-Fourier component of the susceptibility</u>

$$\chi_h = - (e^2/mc^2) \cdot \frac{\lambda^2}{\pi V} F_h,$$

$$\chi_h = - \frac{2.8 \; 10^{-5} \cdot (8.6 \times 8) \cdot (0.7)^2}{3.14 \; (5.4)^2} \simeq - 2 \; 10^{-6}$$

<u>Diameter of the dispersion surface</u>

$A_1A_2 = k\sqrt{\chi_h\chi_{\bar{h}}}/\cos\theta_B$ where $\chi_h = \chi_{\bar{h}}$ for a centrosymmetric crystal. θ_B is given by Bragg's law: $\theta_B = 10°40'$; $A_1A_2 = k \cdot 2 \; 10^{-6}$

Angular width of total reflection (Fig. 3)

$$IJ/k = A_1 A_2/k \sin\theta_B = 2|\chi_h|/\sin 2\theta_B = 10^{-15} \text{radian} = 2''$$

6.3 Exercise 3

3.1. For each wave-field j, the Poynting vector $\underline{P}_j$, along which the energy propagates, is the sum of two terms:

$$\underline{P}_j = |D_{oj}|^2 \underline{s}_o + |D_{hj}|^2 \underline{s}_h$$

where $\underline{s}_o$ and $\underline{s}_h$ are unit vectors along the refracted and reflected directions inside the crystal (Fig. 4).

3.2. Far from Bragg's angle (i.e. $\Delta\theta = \pm\infty$), there is no Bragg reflection; therefore D_{h1} and D_{h2} both tend to zero.

3.3. The boundary conditions at the crystal-vacuum interface at the entrance surface, namely :

$D_{o1} + D_{o2}$ = amplitude of the incident wave;

$D_{h1} + D_{h2} = 0$

imply that for a given value of $\Delta\theta$ (departure from Bragg's law), $|D_{h1}|$ should be equal to $|D_{h2}|$; which in turn implies that the predominant O-wave will be the one for which $\underline{P}_j$ lies closest to $\underline{s}_o$, as is exemplified on Fig. 4 (see reference [1]).

3.4. Applying this rule to both ends of the domain of reflection, one sees that for $\Delta\theta = -\infty$, D_{o1} is the predominant wave, while at $\Delta\theta = +\infty$, D_{o2} predominates. This is shown on Fig. 5 where the arrows perpendicular to the dispersion surface are proportional to the value of P_j.

3.5. For the intermediate region around $\Delta\theta = 0$, the relative value of the h-wave as compared to the O-one is given by

$$\xi_j = \frac{D_{hj}}{D_{oj}} = \frac{k\chi_h}{2\alpha_{hj}} = \frac{2\alpha_{oj}}{k\chi_{\bar{h}}}$$

By looking at the value and sign of α_{oj} and α_{hj} for different values of $\Delta\theta$, it is possible to make a reasonable guess concerning the variations of D_{h1} and D_{h2}. It is thus possible to draw a rough sketch of the variations D_{h1} and D_{h2} as functions of $\Delta\theta$; this is shown on Fig. 6.

6.4 Exercise 4

Let P_o (Fig. 7) represent the incident plane wave in vacuum ($L_a P_o = k\Delta\theta$). The two tie points excited by this wave in part I of the crystal are P_1 and P_2 at the intersection of the parallel to (normal to the entrance surface of the crystal) and the two branches

DISPERSION SURFACE EXERCISES

of the dispersion surface.

In the gap (which we suppose to be filled with 'vacuum'); each wave-field, say 1 for instance, gives rise to 2 plane waves with wave vectors $\underline{OP_{o1}}$ and $\underline{HP_{o1}}$, where P_{o1} and P_{o2} lie at the intersection of the parallel to $\underline{n'}$ drawn from P_1 and the dispersion surfaces in vacuum T_o and T'_h respectively. The same argument, when applied to wave field 2 of part II leads to two other tie-points P_{o2} and P_{h2} which represent the two plane waves induced in the vacuum of the gap by wave-field 2.

Now, the vacuum wave with wave vector $\underline{OP_{o1}}$, when entering part II of the crystal, gives rise to two wave-fields, the tie points of which are at the intersection of the parallel to $\underline{n'}$ drawn from P_{o1} and the two branches of the dispersion surface. Let us call P_{11} and P_{12} these two tie points which are excited by the same wave field P_1 of the first part of the crystal. It is clear that P_{11} is nothing other than P_1. Moreover it is also obvious that the two tie points excited by the vacuum wave which has $\underline{HP_{o1}}$ as wave vector are also $P_{11} \equiv P_1$ and P_{12}. The conclusion is that the wave field P_1 of the first part of the crystal gives rise in part II to only 2 wave-fields, one (P_1) which is just the continuation of the one which propagates in the first part and the other (P_{12}) which is of type 2 and which we call a 'newly created' wave-field. Similarly, wave-field 2 of part I (tie points P_2) gives rise, in part II to 2 wave-fields, one of which has the same tie point P_2 and another one, of type 1, which is newly created and has P_{21} as tie-point.

6.5 Exercise 5

Let $D_h^{sw}(\underline{r})$ the amplitude (at a point ($\underline{r}$) at the exit surface of the crystal) induced by an incident spherical wave.

$$D_h^{sw}(\vec{r}) \not\equiv J_o \left[\frac{\pi t}{\Lambda_o} \sqrt{1 - \frac{x_o^2}{\ell^2}} \right]$$

where J_o = Bessel Function
t = crystal thickness
$x_o = \underline{EP}$, $\ell = BC/2$

$$\Lambda_o = k \sqrt{|\gamma_o \gamma_h|} / kC \sqrt{\chi_h \chi_{\bar{h}}}$$

γ_o and γ_h are the cosines of the angle ($\underline{s_o}, \underline{n}$ and ($\underline{s_h}, \underline{n}$) respectively (see ref. [2])

Applying Huygens' principle, one can say, very simply, that the amplitude distribution on the exit surface due to an incident plane wave (let it be $D_h^{pw}(\underline{r})$) is obtained as the convolution product of

549

- J_o the effect of a single point source located on the entrance surface

- the distribution of point sources on the entrance surface, as 'lit up' by the vacuum wave. This distribution has the same x dependence as the incident plane curve. Let it be $\exp-2\pi i K_x x$ where K_x is proportional to $\Delta\theta$ departure from Bragg's angle.

$$D_h^{pw}(\underline{r}) = \text{Source distribution} \times J_o(MP)$$

$$D_h^{pw}(\vec{r}) = \int_{-\infty}^{+\infty} \exp\left[-2\pi i K_x x_o\right] J_o\left[\frac{\pi t}{\Lambda_o}\sqrt{1-\frac{x_o^2}{\ell^2}}\right] dx_o$$

Now the limits of integration in the above expression do not run from $-\infty$ to $+\infty$, because MP has to lie inside $(\underline{s}_o, \underline{s}_h)$. The limits have to be reduced to $-\ell + \ell$; the 'sources' which lie outside $(-\ell, +\ell)$ do not contribute to the amplitude at P. This means that if we introduce $D_h(x_o)$ such that

$$D_h(x_o) = 0 \text{ if } |x_o| > \ell$$
$$D_h(x_o) = J_o\left(\frac{\pi t}{\Lambda_o}\sqrt{1-\frac{x_o^2}{\ell^2}}\right)$$

then $D_h^{pw}(\underline{r})$ = Fourier Transform of $D_h(x_o)$
the reciprocal variable being x_o, position of the considered point on the exit surface where the spherical wave intensity is calculated and a quantity which is proportional to $\Delta\theta$, the departure from Bragg's law.

Application of Parseval's theorem, then yields:

$$\int_{-\ell}^{+\ell} |D_h(x_o)|^2 dx_o \neq \int_{-\infty}^{+\infty} |D_h(P)|^2 d(\Delta\theta)$$

The left hand side represents the integrated intensity on the base of the Borrmann fan for a spherical wave traverse topograph. The right hand side is the integrated intensity from an impurity of plane waves without any phase relationship. This corresponds to the reciprocity theorem.

REFERENCES

1. E. Dunia, C. Malgrange, and J.F. Petroff (1980) Phil. Mag. A41, 291
2. A. Authier and D. Simon (1968) Acta Cryst. A24, 517

DISPERSION SURFACE EXERCISES

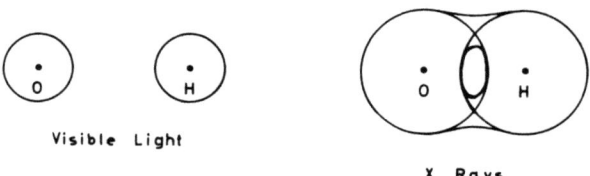

Figure 1. Wave-vector surfaces for visible light and X-rays.

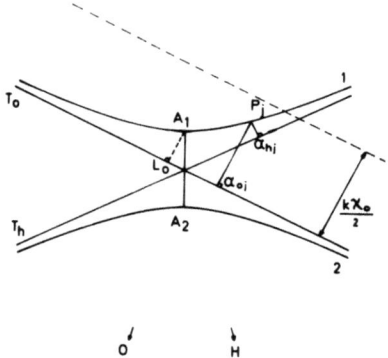

Figure 2. Dispersion surface construction for X-rays.

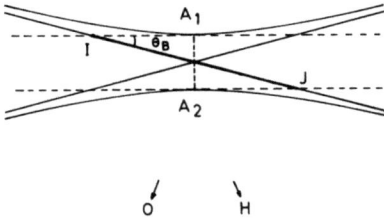

Figure 3. Dispersion surface construction showing diameter region.

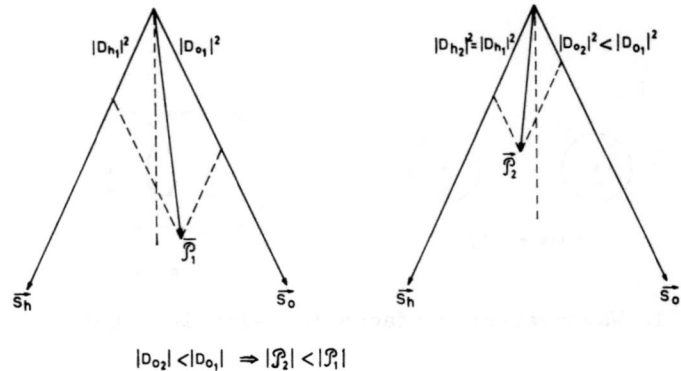

$|D_{o_2}| < |D_{o_1}| \Rightarrow |\vec{\mathcal{P}}_2| < |\vec{\mathcal{P}}_1|$

Figure 4. Poynting vectors giving ray directions.

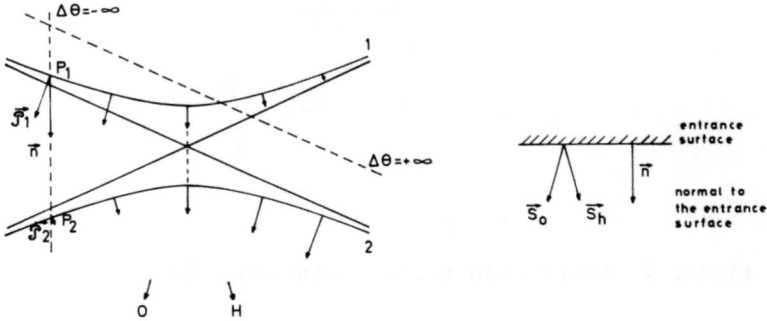

Figure 5. Dispersion surface showing tie points excited in the symmetric Laue geometry.

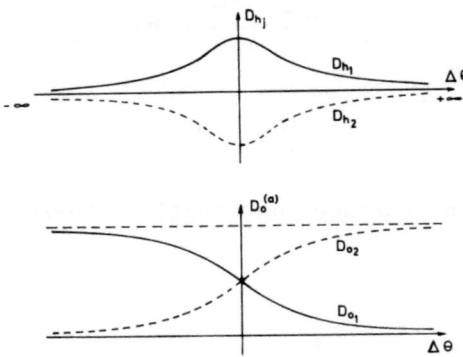

Figure 6. Wave amplitudes as a function of deviation from the Bragg position.

DISPERSION SURFACE EXERCISES

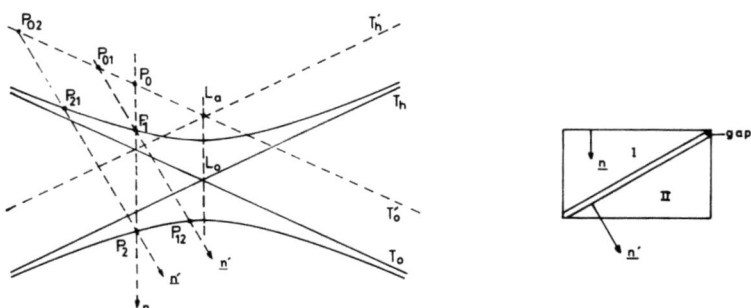

Figure 7. Tie points excited in a crystal containing a gap.

APPENDIX 6

CONTRAST OF STACKING FAULTS

Y. EPELBOIN

1. Introduction

Let us consider a f.c.c. structure such as silicon or a metal. We will draw the various possible stacking faults. For simplicity we will assume a monoatomic motif.

a) Draw the repartition of the atoms in a single (111) plane. Indicate the $\langle 110 \rangle$ rows in the figure – (answer in Fig 5.) Complete the figure, indicating the position of the atoms in the upper layer and calculate the components of the vector which binds an atom in the first plane to one of his nearest neighbours in the upper plane.

b) Draw the pile of successive (111) layers in the case of an extrinsic or intrinsic stacking fault in silicon. What is the fault vector – (answer in Fig 6.) What is the nature of the partial dislocation which limits the fault? Draw the repartition of the atoms near the core of the dislocation. Suppose a circular stacking fault. Is it possible that the contrast of all the partial dislocation may vanish in one single Bragg reflection. Why? – answer below and in Fig. 7.

c) Draw the repartition of the atoms near the core of a Shockley partial dislocation and answer to the same question as before – (answer below and in Fig. 8.)

2. Study of the contrast of the faults

Stacking faults are very common in silicon:

a) Which are the extinction conditions for such a fault? Assume that the fault vector is one of the possible $\langle 111 \rangle$ directions.

b) In which of the following reflections will the contrast of each possible fault vanish? $\underline{g} = \bar{1}\bar{1}1, \bar{1}1\bar{1}, 1\bar{1}\bar{1}, 02\bar{2}$

CONTRAST OF STACKING FAULTS

c) In the following topographs find some examples of such faults (Fig,1),

d) Examine carefully the contrast of the partial dislocation which limits the stacking fault in topograph of Fig. 2. Can you determine qualitatively the inclination of the fault in the crystal?

3. Theory

a) Consider Fig. 3. It may be shown that all wavefields are focussed at point A'. Indicate which paths correspond to normal or newly created wavefields. Explain the formation of the contrast at point p and discuss the visibility of the various fringes when the photoelectric absorption varies.

b) In the following sequence of topographs (Fig. 4.) explain the variation of the contrast. All topographs are taken with MoKα radiation except topograph 4b which has been taken with CuKα radiation. µd increases by about a factor 10. Does it explain the variation of the contrast?

c) Build approximately the shape of the fringes in the plane of incidence, then in the plane of the photograph. - (see Fig. 8.)

d) Draw approximately the shape of the traverse topograph of a stacking fault from the shape of the section topograph. - (see Figs. 10 and 11)

4. Answers

4.1 Structural study

a) The vector which binds two neighbours may be written as:

$$\underline{t} = \underline{n} + \underline{m}$$

where $\underline{n} = \pm 1/3\ [111]$ and $\underline{m} = 1/6\ [\bar{1}2\bar{1}]$ or $1/6\ [2\bar{1}\bar{1}]$ or $1/6\ [\bar{1}\bar{1}2]$ (Verify that $\underline{m}$ vectors are not compatible with the structure).

b) The partial dislocation is an edge dislocation, so for a given reflection $\underline{g}$ its contrast may vanish when $\underline{g} \cdot \underline{f} = 0$
The complete rule needs a second condition $(\underline{g}, \underline{f}, \underline{u}) = 0$ where $\underline{u}$ is a unit vector lying along the line but in some cases the first condition is satisfactory and the contrast may vanish or may have the same contrast all along the line.

555

c) The nature of a Shockley partial dislocation varies along the line, from edge to screw, so is doubtful that it will be invisible in one single reflection except if the condition $\underline{g} \cdot \underline{b} = 0$ is sufficient.

4.2 Study of the contrast of the fault

a) The crystal may be divided in two parts shifted one from the other by a displacement vector $\underline{f}$. The dielectric susceptibility may be written as:

$$\chi = \sum_h \chi_h \exp i\, 2\pi\, \underline{g} \cdot \underline{r}$$

in the first part. In the second part one adds the displacement $\underline{f}$:

$$\chi = \sum_h \chi_h \exp i\, 2\pi\, \underline{g} \cdot (\underline{r}+\underline{f})$$

Thus:

$$\chi_h^{II} = \chi_h^{I} \exp i\, 2\pi\, \underline{g} \cdot \underline{f}$$

which explains the criterion of invisibility:

$$\underline{g} \cdot \underline{f} = m \ (m \text{ is an integer})$$

For a Frank fault $\underline{f} = \pm\, 1/3\, \langle 111 \rangle$ so a condition is (for example):

$$h + k + l = 3m.$$

b) Extinction of contrast

f \ g	$\bar{1}\bar{1}1$	$\bar{1}1\bar{1}$	$1\bar{1}\bar{1}$	$02\bar{2}$
$1/3\ [\bar{1}\bar{1}1]$	x			
$1/3\ [\bar{1}1\bar{1}]$		x		
$1/3\ [1\bar{1}\bar{1}]$			x	x
$1/3\ [111]$				x

4.3 Theory

a) – along Ap normal wavefields 1 and 2
– along qp and rp newly created wavefields

CONTRAST OF STACKING FAULTS

Total intensity at point p may be written as $I = I_1 + I_2 + I_3$
I_1: interference between normal wavefields (extinction fringes)
I_2: interference between newly created wavefields visible only for very low absorption.
I_3: interference between normal and newly created wavefields. They are flat fringes visible also for a strong absorption.

b) Due to the strong absorption fringes I_3 are only visible near the surfaces of the crystal, since wavefields 2 travelling along qp or rp would be absorbed if q lies in the middle of the crystal. Fringes near the entrance surface are blurred in the translation of the traverse topograph. This effect is more sensitive with Cu Kα than MoKα. Notice that since $\underline{f}$ is one of the possible 1/3 <111> fault vectors the contrast of the stacking fault vanishes in the $\overline{3}3\overline{3}$ reflection.
Extinction in $20\overline{2}$ and $1\overline{1}1$ reflections permits the determination of $\underline{f} = 1/3\ [1\overline{1}1]$

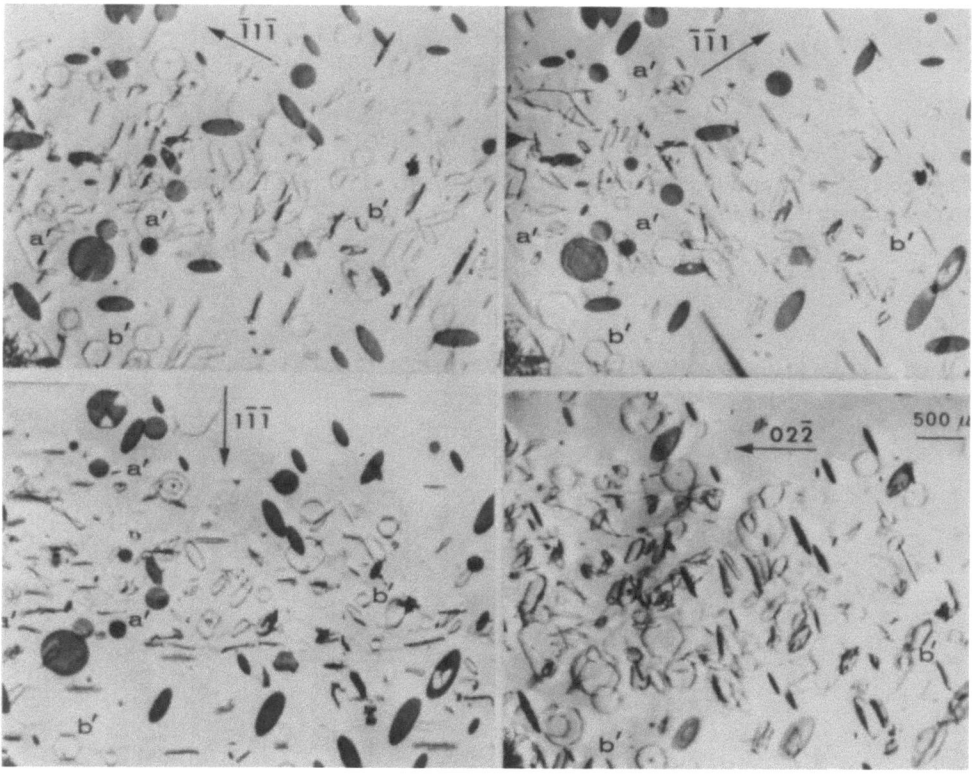

Figure 1. topographs of a silicon wafer. CuKα radiation (after J.R. Patel and A. Authier (1975) J. Appl. Phys. 46,118)

Figure 2. Silicon wafer . CuKα radiation. 111 reflection. (from J.R. Patel unpublished),

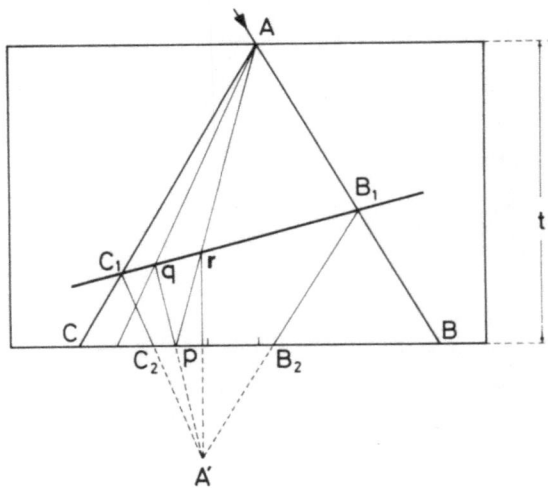

Figure 3. Paths of the wavefields in a plane of incidence.

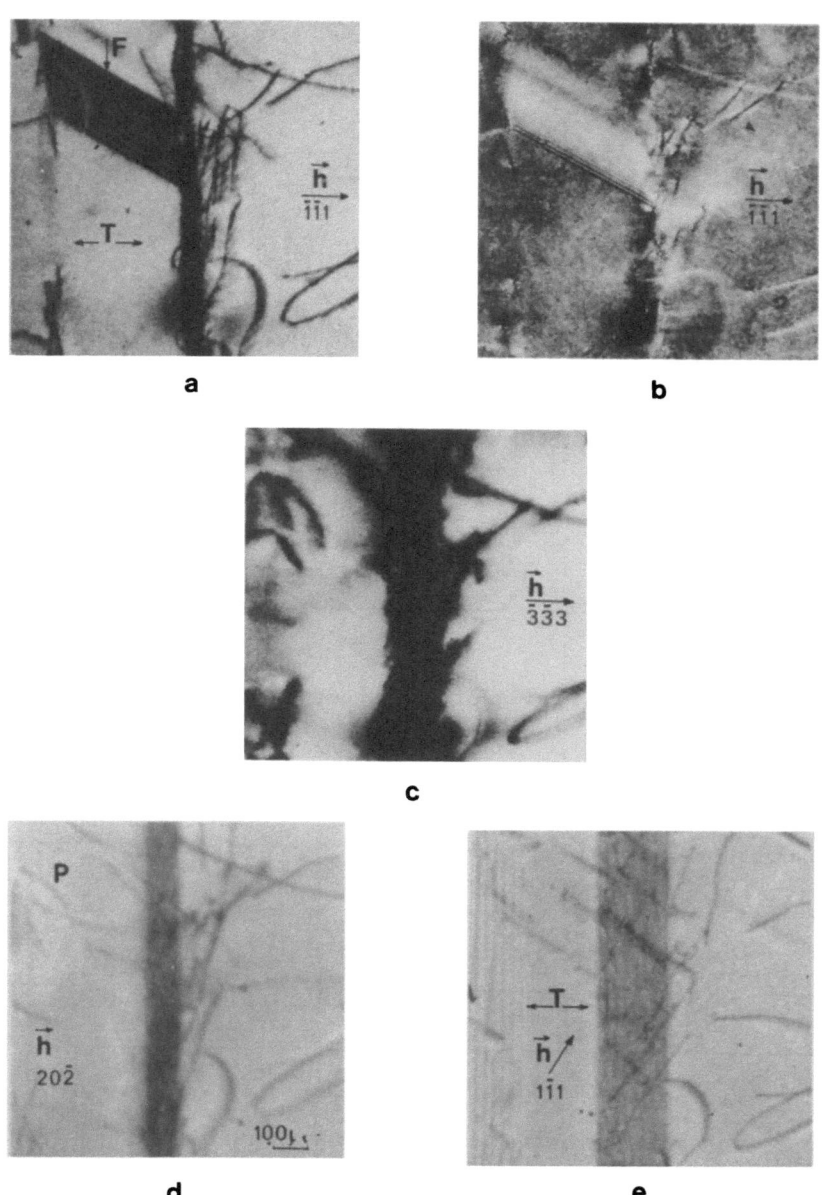

Figure 4. Topographs of a silicon wafer. All recorded with MoKα except b) with CuKα. Vertical fringes are due to a twin boundary. (from J.F. Petroff in M. Sauvage and C. Malgrange (1970)) Phys. Stat. Sol. 37, 759)

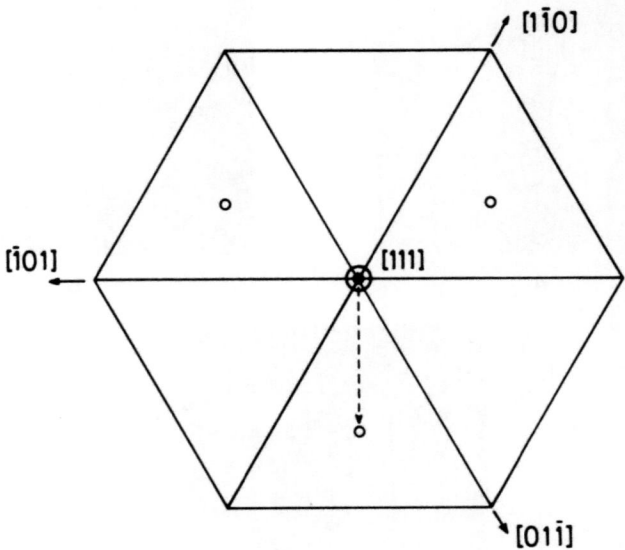

Figure 5. Repartition of the atoms in the (111) plane.

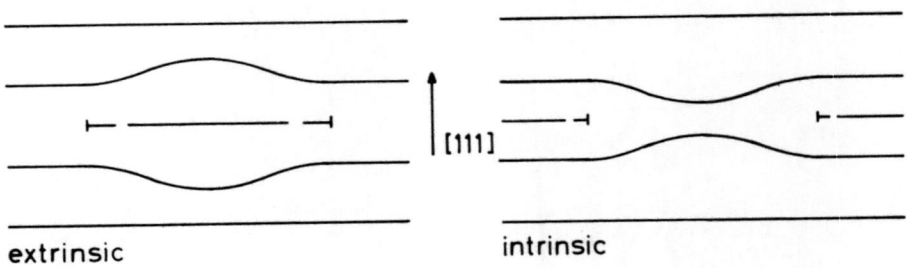

Figure 6. Schematic drawing of a Frank intrinsic and extrinsic stacking faults.

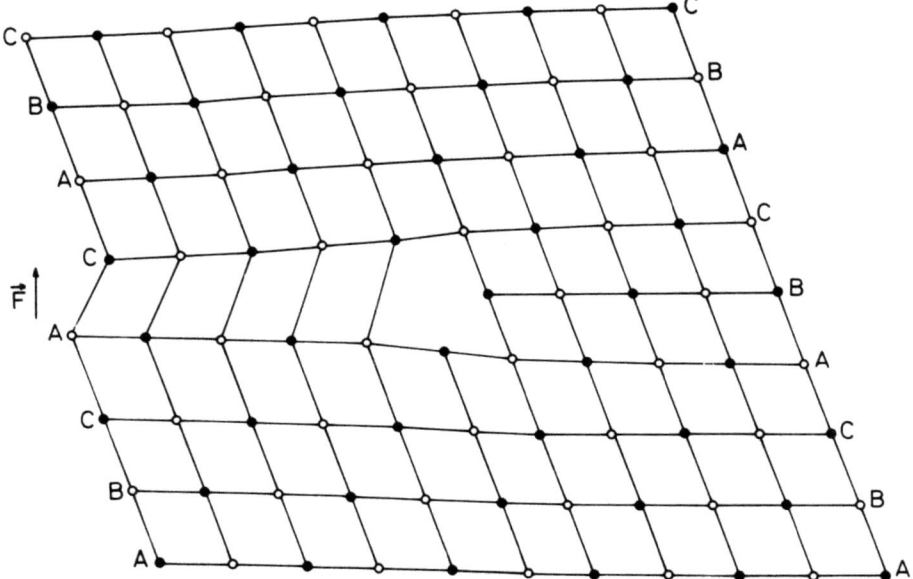

Figure 7. Shape of a partial Frank dislocation (after W.T. Read: Dislocations in crystals, McGraw Hill).

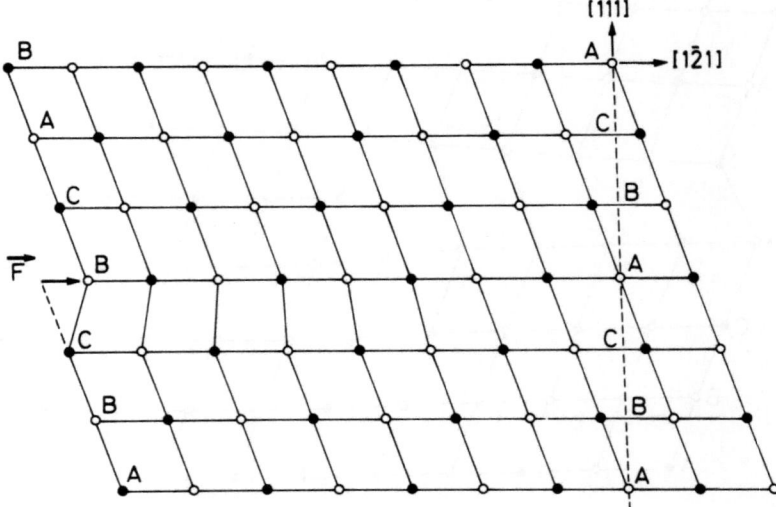

Figure 8. Shape of a Shockley partial dislocation (after W.T. Read).

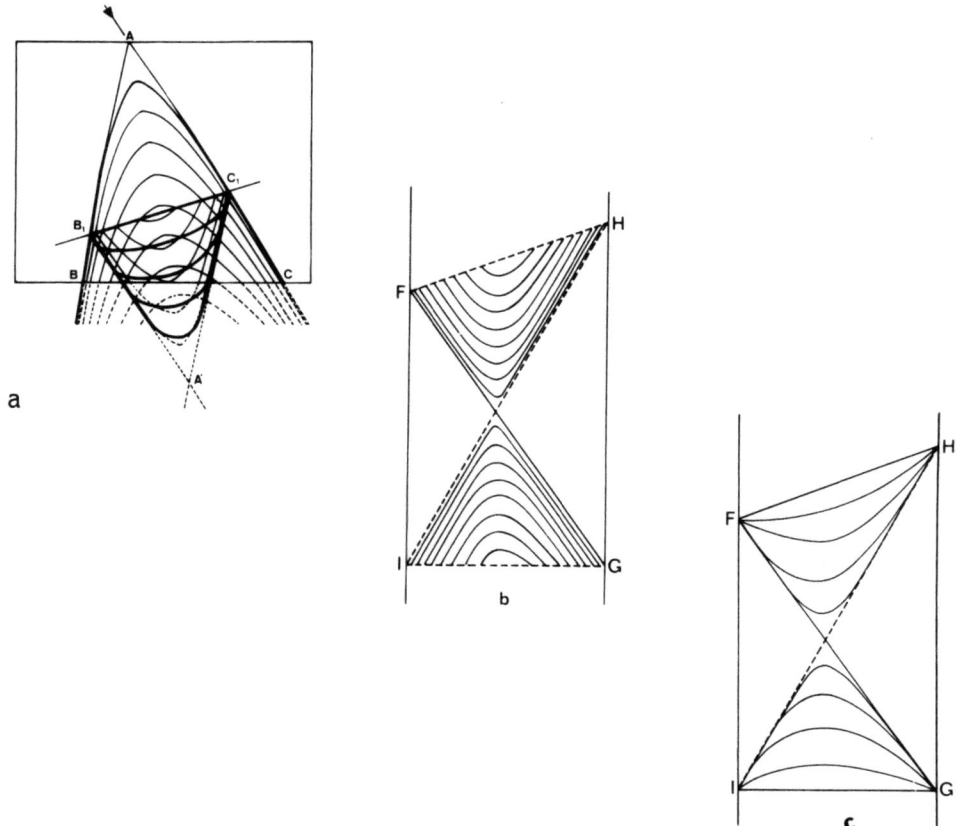

Figure 9. a) shape of the fringes in a plane of incidence. b) I_2 and c) I_3 fringes in the plane of the photograph.

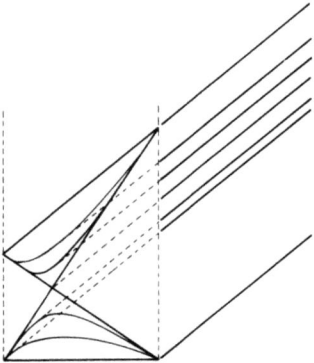

Figure 10. Construction of the traverse from the section topograph.

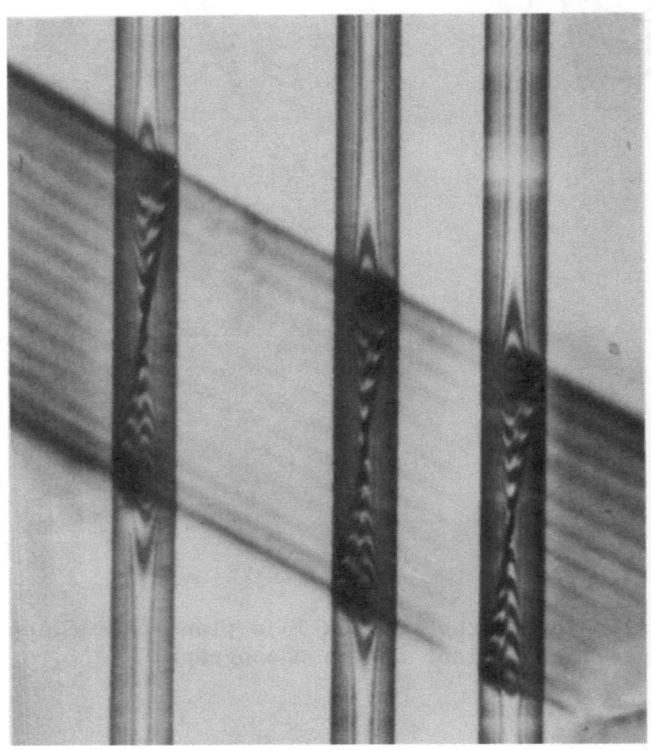

Figure 11. Section and traverse topograph of a quartz crystal MoKα (10$\bar{1}$0) x 45 (courtesty A. Zarka).

APPENDIX 7

MISFIT BOUNDARIES AND JUNCTIONS OF PURELY ROTATIONAL BOUNDARIES

J. MILTAT

1. Introduction

A misfit boundary separating regions I and II of the crystal (Fig.1) is characterized by :
- a rotation $\Delta\omega$ of the lattice planes across the boundary and or
- a change in lattice spacing Δa.

It is assumed that the structure factor is identical in regions I and II. The reciprocal lattice vector $\underline{h}'$ in region II is different from the reciprocal lattice vector $\underline{h}$ in region I, which implies that regions I and II are, in the reciprocal space, characterized by distinct dispersion surfaces.

1.1 Dispersion surface construction (plane wave)

1.1.1 Rotation boundaries

We have $|\underline{h}'| = |\underline{h}|$ and $\underline{h}' \neq \underline{h}$. Fig.2 shows the dispersion surface for region I. Assuming a given deviation parameter, draw on Fig.2a,b, the excited tie-points in region I. The wave-field associated to a given tie point is composed of two waves, O and H. Fig.3a shows the reciprocal lattice in regions I and II for O waves, whereas Fig.3b shows the reciprocal lattice for H waves.
From Fig. 3a draw on Fig.2 the dispersion surface in region II pertaining to O waves.
Similarly, draw on Fig. 2 the dispersion surface in region II pertaining to H waves.
Mark the excited tie points in region II.
How many distinct tie points are excited in region II?
What is the basic difference between the present situation and, for instance, a stacking fault?

Instead of considering that the dispersion surface pertaining to region II, O or H waves, is distinct from the dispersion surface in region I, let us consider the dispersion surface to be unique. Fig. 4 indicates the appropriate construction for O waves. The origins of the wave vectors in region II are now O" and H" : Why? Why are O" and H" situated on the trace of the Ewald sphere (drawn for region I) in the plane of incidence.

Let us call δ_o^R and δ_h^R the rotations, on the Ewald sphere, $O \to O"$ and $H \to H"$, respectively.
What is the relation between δ_o^R and δ_h^R?

A similar construction may be performed for H waves. The origins O" and H" of wave-vectors in region II are obviously identical to those found for O waves.

1.1.2 Dilatation boundaries

We now have $\underline{h}' // \underline{h}$; $|\underline{h}'| \neq |\underline{h}|$. We shall assume that $|\underline{h}'| > |\underline{h}|$

Draw on a figure identical to Fig.2 which represents the dispersion surface in region I, the dispersion surface pertaining to O waves in region II.
How is it possible to let dispersion surfaces in regions I and II coincide ?
The construction pertaining to H waves is shown in Fig. 6. Should O" and H" be situated on the trace of the Ewald sphere in the incidence plane ?

Let us call δ_o^D and δ_h^D the rotations, on the Ewald sphere, $O \to O"$ and $H \to H"$, respectively.
What is the relation between δ_o^D and δ_h^D?
A similar construction applies to H waves.
Consider now a boundary with mixed rotational and dilatational character.
Express the total rotations δ_o^D ($O \to O"$), δ_h^D ($H \to H"$) as a function of δ_o^R, δ_o^D, δ_h^R, δ_h^D.
Finally, in which conditions do O and H waves excited wave-fields reunify in a single wave-field?

1.2 Ray trajectories and fringes

We shall now assume an incident spherical wave and a symmetrical Laue reflection. We shall further assume that $\underline{nf}$ is perpendicular to $\underline{g}$

Let us consider P_{12} wave-fields, that is wave-fields in region II, with tie-points on branch 2 excited by wave-fields with tie points on branch 1 in region I. We have drawn in Fig. 7a, for a purely rotational boundary ($\delta = \delta_o = \delta_h \ll \delta$, where δ is the width of the rocking curve) the positions of tie-points P_{12} excited by wave-fields with tie points P_1. The direction of propagation of wave-fields being normal to the dispersion surface, it is straightforward to draw the trajectories in region II (Fig.7b).

MISFIT BOUNDARIES

Do P_{12} trajectories focalize at any given point in region II?
Draw the caustics determined by the P_{12} trajectories.
What is the behaviour of P_{21} wave-fields?
In analogy to the case of a stacking fault, can you infer the shape of the zone where P_{12} and P_{21} wave-fields contribute to the intensity in a section topograph when the fault is inclined to the entrance surface?
Setting the exit surface along A in Fig.7b, show that interferences between the various wave-fields result in two distinct sets of fringes.

2. Junctions of purely rotational boundaries

Let us consider a straight junction of such boundaries. Each boundary (a) is characterized by a rotation $\Delta\omega_i^a$.
Draw a Burgers circuit around the junction. What happens if $\sum_a \Delta\omega_i^a$ along the circuit does not sum up to 0? In order to simplify the graphical interpretation of the result, you may assume that all rotations $\Delta\omega_i^a$ are rotations around the junction axis.
What type of elastic defect is located along the junction line if $\sum_a \Delta\omega_i^a \neq 0$?

In order to appreciate the contrast properties of such defects, the remainder of this section is devoted to a qualitative comparison of the contrast expected from a twist-quasi-disclination (TqD) (Fig. 8(c)) and a screw dislocation (Sd) both parallel to the X-ray exit surface and relatively close to that surface. The geometry is illustrated in Fig. 9 $\underline{g} \parallel \underline{L}$
In the relaxed state, that is in a state where the defects exert no force on the nearly free surface, the distortions $\beta_{ij} = \frac{\partial u_j}{\partial x_i}$ associated to these defects are:

Screw disl.   Twist quasi-disclination

$$\beta_{13} = \frac{b}{2\pi}\left[\frac{y}{x^2+y^2} - \frac{y}{(x-2c)^2+y^2}\right] \quad \beta_{13} = -\frac{\Omega}{2\pi} \ln\left(\frac{x^2+y^2}{(x-2c)^2+y^2}\right)$$

$$\beta_{23} = \frac{b}{2\pi}\left[\frac{x}{x^2+y^2} - \frac{x-2c}{(x-2c)^2+y^2}\right] \quad \beta_{23} = -\frac{\Omega}{2\pi}\left[\text{arctg}\left(\frac{x}{y}\right) - \text{arctg}\left(\frac{x-2c}{y}\right)\right]$$

All other terms are equal to 0, c is the distance to the free surface.

Let us further define the effective misorientation of the lattice planes around the defect as

$$\delta(\Delta\theta) = -\frac{1}{k \sin 2\theta_B} \frac{\partial}{\partial s_h} (\underline{g} \cdot \underline{u})$$

and the misorientation gradient as

$$G = -\frac{1}{k \sin 2\theta_B} \frac{\partial^2}{\partial s_o \partial s_h} (\underline{g} \cdot \underline{u})$$

Equi $\delta(\Delta\theta)$ and G curves in the $\underline{e}_1 \underline{e}_2$ plane have been plotted in Fig. 10 to 15. The figures apply either to a screw dislocation in silicon, $\underline{g} = \{2\bar{2}0\}$, CuK$\alpha$ radiation, or a twist quasi-disclination in iron, $\underline{g} = \{1\bar{1}0\}$, MoK$\alpha$ radiation. The dotted regions correspond to negative $\delta(\Delta\theta)$ or G. c = 5µm and 20 µm.
Indicate the trace of the incidence plane on the figures.
The direct image : Fig. 10 and 11 show equi $\delta(\Delta\theta) = \alpha\delta$ curves (scale mark: 1µm), in the $e_1 e_2$ plane, drawn for $|\alpha| = 0,5$; 1 ; 1.5, for the Sd and TqD, respectively (δ is the width of the rocking curve). Assuming that the width of the direct image is determined by the width of the zone where $|\delta(\Delta\theta)| > \delta$,
1) estimate the influence of relaxation upon the direct image width of the Sd and the TqD.
2) what happens when c becomes infinite? Is the result obtained for the TqD physically meaningfull?
The intermediary image: Fig. 12 and 13 show equi $G = \beta\delta/\Lambda$ curves, in the $e_1 e_2$ plane, drawn for $|\beta| = 1$; 0.5, for the Sd and TqD, respectively. (Λ is the Pendellosung periodicity). Scale mark : 10µm. Assuming that the width of the intermediary image is approximately determined by the size of the zones where $|G| > \delta/\Lambda$,
i) estimate the influence of relaxation upon the intermediary image width of the Sd and the TqD.
2) what happens when c becomes infinite? Is the result obtained for the TqD physically meaninfull?
The dynamical image : Fig. 14 and 15 show equi $G = \beta\delta/\Lambda$ curves, in the $e_1 e_2$ plane, drawn for $|\beta| = 10^{-1}$; $5 \cdot 10^{-2}$ for the Sd and TqD, respectively. Scale mark : 10 µm.
Knowing that wave-fields belonging to branch I of the dispersion surface are deflected in opposite directions according to the sign of G, and that, in the case of a constant strain gradient (G), under intermediate absorption conditions, a black (white) contrast corresponds to a positive (negative) G, discuss the expected dynamical image properties of the Sd and TqD as a function of distance to the surface.

<u>ANSWERS</u>

AI.1 The tie points in region I are obtained from the condition of continuity of the tangential components of wave-vectors. The tie-points P_1 (branch 1) and P_2 (branch 2) lie on the same normal to the entrance surface (see Fig. 2). Each wave-field is composed of two waves, an 0 wave and an H wave. With respect to the $\underline{s}_o$ direction in region I, the reciprocal lattice for 0 waves is, according to Fig. 3a, rotated by an amount $\Delta\omega$ around the origin 0 of the reciprocal lattice. The dispersion surface for 0 waves undergoes the same rotation around 0.

Similarly, for H waves, the dispersion surface in region II is with respect to the $\underline{s}_h$ direction in region I, deduced from the dispersion surface in region I by a rotation, $-\Delta\omega$ around H. The condition of continuity of the tangential components of wave-vectors implies that excited wave-fields in region II have tie-points situated at the intersection of the branches of the dispersion surfaces pertaining to O and H waves and normals to the fault plane passing through P_1 and P_2. Fig. 2' indicates the appropriate construction for H waves. One gets the following synoptic diagram:

There will be in general 8 different wave-fields propagating in region II of the crystal (for one polarization state). In the case of a stacking fault, since dispersion surfaces in regions I and II coincide, only 4 tie-points are excited in region II, two of them being identical to tie-points P_1 and P_2 in region I.

It is advantageous to work on a single dispersion surface, since it simplifies the mathematical treatment of the problem. The dispersion surfaces pertaining to O or H waves in region II are geometrically identical to the dispersion surface in region I. It is therefore possible to let them coincide via an appropriate rotation (around O or H). The construction is indicated for O waves in Fig. 4. However, the proper propagation directions for the waves constituting a wave-field in region II were those defined by constructions such as shown in Fig. 2'.

Therefore, after letting the dispersion surfaces coincide in regions II and I, it is necessary to define new origins O' and H' to the wave-vectors in region II. O' and H' are necessarily situated on the trace of the Ewald sphere in the plane of incidence since the magnitude of wave-vectors has not changed ($k = n/\lambda$ in regions I and II). After defining a positive direction along the trace of the Ewald sphere in the plane of incidence, one has:

$$\delta_o^R = \delta_h^R$$

A.I.2 : In the case of dilatation boundaries, similar dispersion surface constructions may be performed. Fig. 5 indicates the appropriate construction for O waves. Now, however, the dispersion surfaces in region II are not strictly geometrically equivalent to the dispersion surface in region I. The angle between asymptotes is now $2(\theta_B + \Delta\theta_B)$ where

$$\Delta\theta_B = -1/\cot g\ \theta_B)(\Delta a/a)$$

In practical cases, however, $\Delta\theta_B$ is very small with respect to θ_B and it is a fair approximation to consider dispersion surfaces in region II as geometrically equivalent to the dispersion surface in region I. It is then possible to let dispersion surfaces in region II coincide with the dispersion surface in region I via an appropriate rotation (around O or H). Fig. 6 shows the construction applying to H waves.

As for rotation boundaries, wave-vectors in region II have new origins O" and H" situated on the trace of the Ewald sphere in the plane of incidence. One now however gets:

$$\delta_o^D = -\delta_h^D$$

If the fault has a mixed character (rotation + dilatation), one obtains:

$$\begin{cases} \delta_o = \delta_o^R + \delta_o^D \\ \delta_h = \delta_h^R - \delta_h^D \end{cases}$$

Finally, it may be shown that O and H waves excited wave-fields in region II reunify in a single wave-field if $\Delta a/a = 0$ (no dilatation) and $\underline{n}_e /\!/ \underline{n}_f \perp \underline{h}$.

AII : Assuming an incident spherical wave implies that all points on the dispersion surface in region I are excited. We have further assumed that $\Delta a/a = 0$, $\underline{n}_e /\!/ \underline{n}_f \perp \underline{h}$. Therefore, P_{ij}^Q and P_{ij}^h tie-points coincide. P_1 in Fig. 7a is a tie-point in region I on branch 1 of the dispersion surface. Its corresponding wave-field propagates in a direction normal to the dispersion surface. P_{12} is a tie-point in region II, on branch 2 of the dispersion surface; its corresponding wave-field has been excited by wavefield P_1 (remember there is a rotation). The propagation direction of wave-fields P_{12} is normal to the dispersion surface. It is therefore straightforward to construct the corresponding ray-trajectories in real space (Fig. 7b). It is readily seen that rays in region II do not focus at any one given point. The rays caustics is just the envelope of ray trajectories in region II.

The caustics of P_{21} wave-fields trajectories is symmetrical of that of P_{12} trajectories with respect to P in Fig. 7b (just inspect Fig. 7a). The zone where P_{12} and P_{21} wave-fields contribute to diffracted intensity on a section topograph has the shape of a defocused hour-glass. This may be seen by moving the trace of the fault plane between E and S in Fig. 7b).

Finally, setting the exit surface along A in Fig. 7b, (this is equivalent to placing the trace of the fault plane mid-way between the entrance and exit surfaces) shows that wave-fields P_{21} solely contribute to diffracted intensity along P'P whereas wave-fields P_{12} only contribute to diffracted intensity along PP". It is therefore not surprising to observe two sets of fringes. This is illustrated in Fig. 7c,d.

B.I: Fig. 8a illustrates the nature of the defect to be located at the junction of rotation boundaries when $\sum_a \Delta\omega_i^a \neq 0$. The sector 1 of the crystal has been cut into two parts 1' and 1". One then assembles sector 1' to sector 2 with the proper rotation between sector 1 and 2. The same operation is performed for sectors 2 and 3 and sectors 3 and 1". If $\sum_a \Delta\omega_i^a \neq 0$, the lips A and B do not coincide. For coincidence to be achieved, it is necessary to add an elastic defect of the disclination type (rotation dislocation). Fig. 8b is a wedge disclination. The rotation axis $\underline{\Omega}$ is parallel to the defect line. Fig. 8c and 8d are twist-disclinations. $\underline{\Omega}$ is perpendicular to the defect line. In Fig. 8c, $\underline{\Omega}$ is perpendicular to the cut plane whereas, in Fig. 8d, $\underline{\Omega}$ is in the cut plane.

B.II: The trace of the plane of incidence is drawn a solid vertical straight line.

The direct image: Surface relaxation is unimportant in this case for the Sd direct image width. On the contrary, the TqD direct image width depends strongly on c. If c becomes infinite, the TqD direct image width also becomes infinite, which is equivalent to say that the elastic energy would become infinite, which is physically impossible. Quasi disclinations in massive crystals may only exist in pairs with opposite rotations or very close to free surfaces.

The intermediary image: Surface relaxation affects nearly equally , but moderately, the widths of the intermediary images of the Sd and the TqD. Widths tend to decrease as c increases, tending towards asymptotic values. This behaviour is well known for dislocation.For quasi-disclinations, it expresses the fact that distortions are slowly varying functions of position. The result has however very little physical meaning (see former paragraph).

The dynamical image: Surface relaxation influences rather strongly the Sd and even more the TqD. Width, if determined by, for instance, the curves $G = 5 \times 10^{-2} \delta/\Lambda$ tends to increase as c increases. For the TqD, if one excludes the central region, rays will only pass through regions of negative G. A white contrast is expected. Intermediary absorption conditions are very favourable to the observation of such defects. On the contrary, little may be said about the expected contrast from the Sd since, along most of the traces of incidence planes, rays pass successively through regions of $G > 0$ and < 0. The net contrast may be expected to be small.

BIBLIOGRAPHY

Misfit boundaries : dynamical theory

A. Authier and M. Sauvage (1966), J. de Phys. $\underline{27}$, C3-137
N. Kato (1974), in X-ray diffraction, Ed. L. Azaroff, McGraw-Hill, New York
M. Polcarova (1978), Phys. Stat. Sol. (a) $\underline{46}$, 567; (1978), Ibid, $\underline{47}$, 179

Direct images

A. Authier (1967), Advances in X-ray Analysis, $\underline{10}$, 1
J. Miltat and D. K. Bowen (1975), J Appl. Cryst. $\underline{8}$, 657

limit of validity of geometrical optics

A. Authier and F. Balibar (1970), Acta Cryst. A$\underline{26}$, 647

Constant strain gradient - wave-field curvature

N. Kato (1963). Acta Cryst. $\underline{16}$, 276 and 282; (1963), J. Phys. Soc. Jap. $\underline{18}$, 1785;(1964) Ibid, $\underline{19}$, 67 and 971
Y. Ando and N. Kato (1970), J. Appl. Cryst. $\underline{3}$, 74

Quasi-disclinations : Junctions of rotation boundaries

M. Kleman and M. Schlenker (1972), J. Appl. Phys. $\underline{43}$, 3184
J. Miltat and M. Kleman (1973), Phil. Mag. $\underline{28}$, 1015
M. Kleman (1974), Appl. Phys. $\underline{45}$, 1377
J. Miltat (1976), Phil. Mag. $\underline{33}$, 225
M. Labrune (1976), J. de Phys. $\underline{37}$, 1033
M. Kleman, M. Labrune, J. Miltat, C. Nourtier and D. Taupin (1978), J. Appl. Phys. $\underline{49}$, 1989

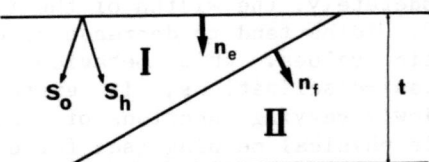

Figure 1. Definition of directions in the plane of incidence.

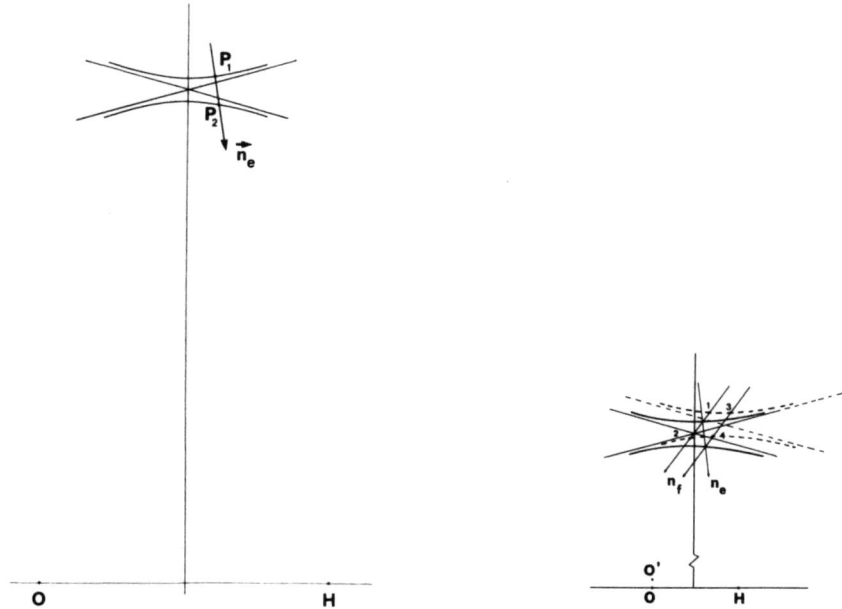

Figure 2. The dispersion surface in region I.
Figure 2'. Dispersion surface construction in region II for a pure rotation boundary and H waves excited wave-fields 1 : P_{11}^H; 1 : P_{12}^H; 3 : P_{21}^H; 4 : P_{22}^H.

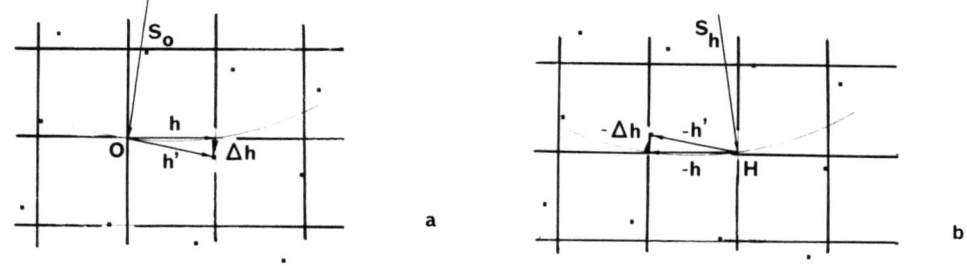

Figure 3. The reciprocal lattice in regions I (square grid) and II square dots). An O wave sees a rotation of the reciprocal lattice as sketched in 3a. An H wave sees a rotation as sketched in 3b.

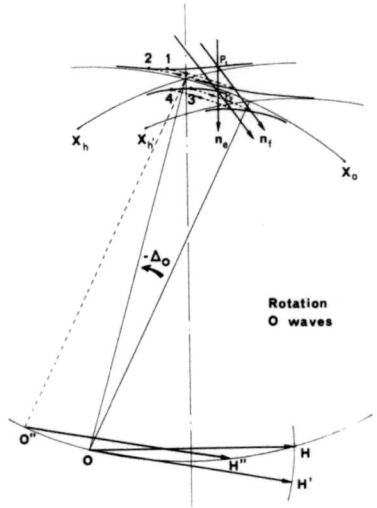

Figure 4. Inverse rotation construction for a pure rotation boundary and O waves excited wave-fields. 1 : P_{11}^O; 1 : P_{21}^O; 3 : P_{12}^O; 4 : P_{22}^O.

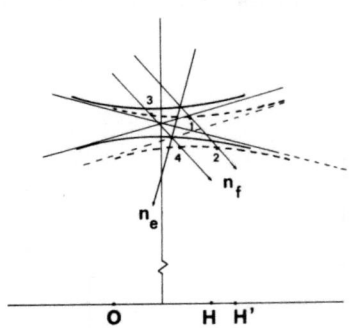

Figure 5. Dispersion surface construction in region 11 for a dilatation boundary and O waves excited wave-fields. 1 : P_{11}^O; 2 : P_{11}^O; 3 : P_{21}^O; 4 : P_{22}^O.

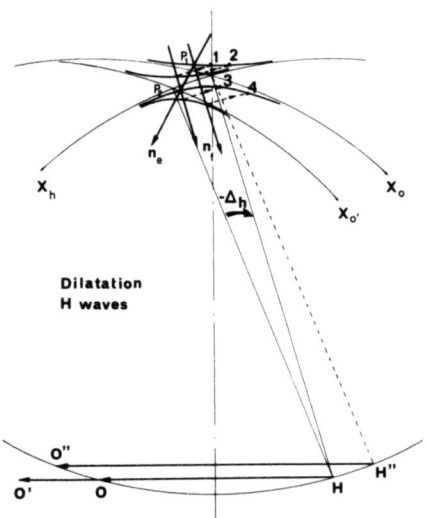

Figure 6. Inverse rotation construction for a dilatation boundary and H waves excited wave-fields. 1 : P^H_{21}; 2 : P^H_{11}; 3 : P^H_{22}; 4 : P^H_{12}.

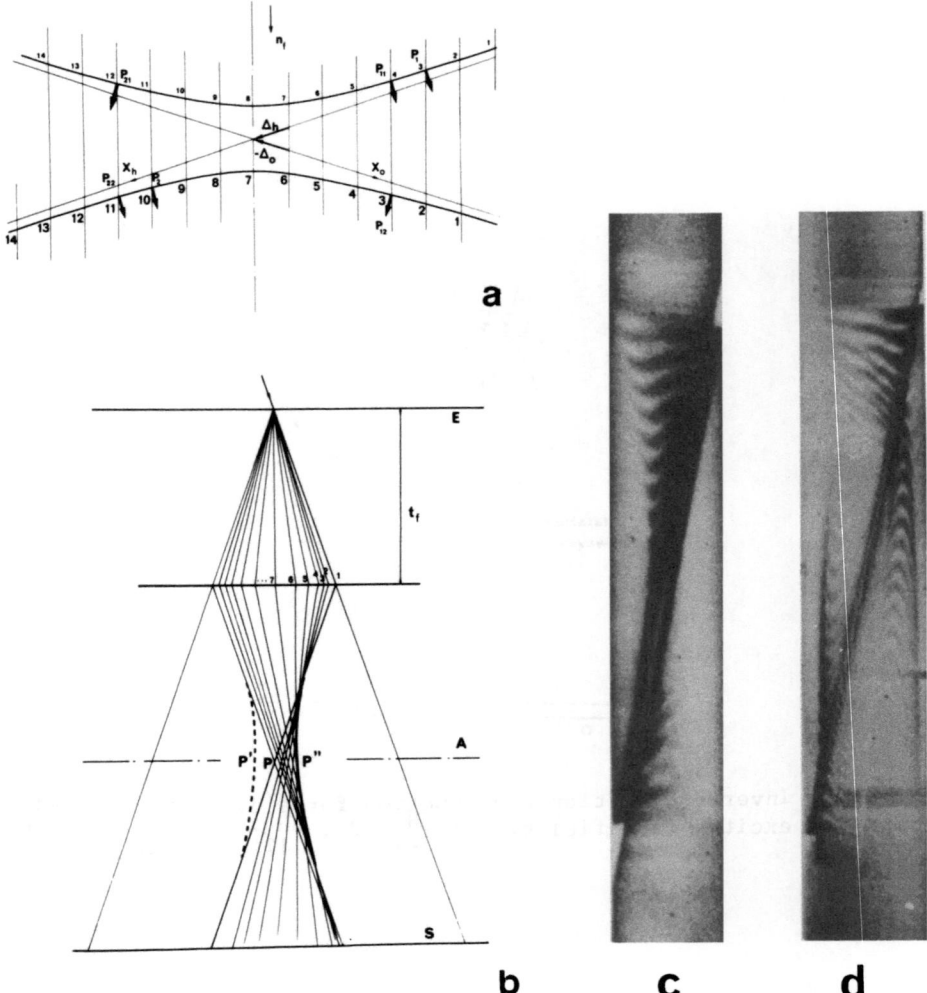

Figure 7. (a) Dispersion surface in regions I and II for $\underline{n}_e \mathbin{/\mkern-6mu/} \underline{n}_f \perp \underline{g}$ and a pure rotation boundary. (b) Ray trajectories deduced from fig. 7a. The solid curve is the trajectories caustics of P_{12} wave-fields. The dotted curve is the trajectories caustics of P_{21} wave-fields (c,d) Section topographs of twin lamellae in calcite. These lamellae induce a rotation of the lattice on each side of the lamella. (c), very small rotation (d), rotation approaching δ. The two parts of the crystal tend to behave independently. (courtesty of M. Sauvage).

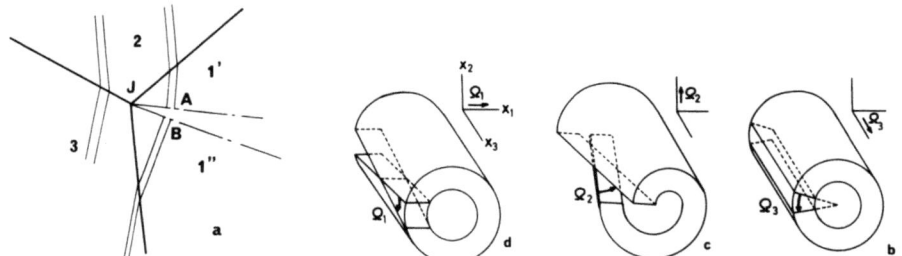

Figure 8. (a) Assembling the various crystals sectors : if $\sum_a \Delta\omega_i^a \neq 0$ the lips A and B do not coincide. (b) wedge disclination (c,d) twist disclinations.

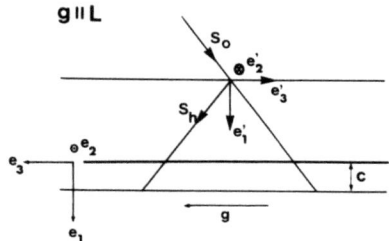

Figure 9. Definition of directions for Sd and TqD contrast analysis

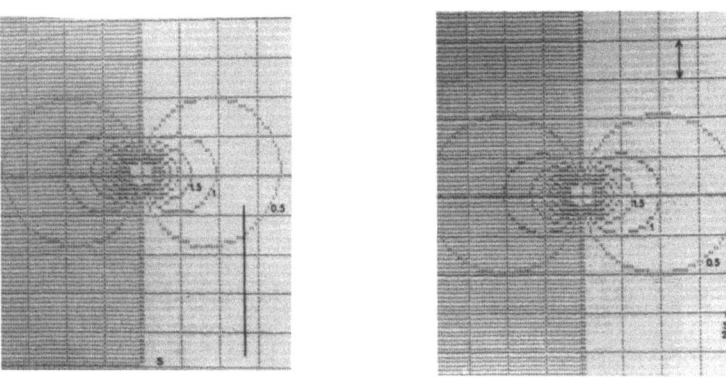

Figure 10. Equi $\delta(\Delta\theta)$ curves. Sd **left**: c = 5 µm **right** c = 20 µm scale mark, 1 µm.

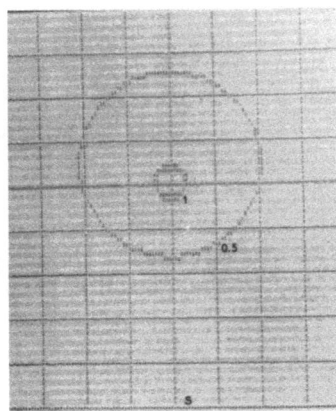 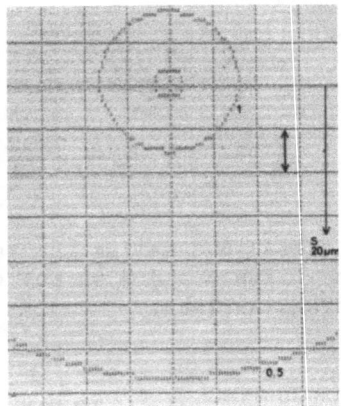

Figure 11. Hqui $\delta(\Delta\theta)$ curves: QD **left**: c = 5µm **right** c = 20 µm scale mark : 1 µm.

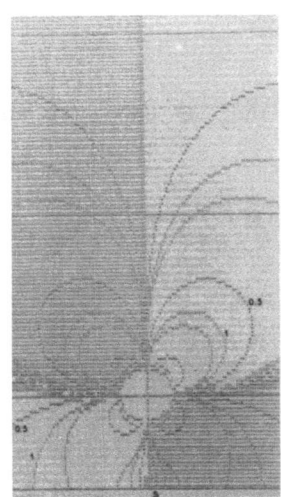 

Figure 12. Equi G curves : Sd β = 1, 0,5 **left** : 5 µm **right** c = 20 µm scale mark : 10 µm.

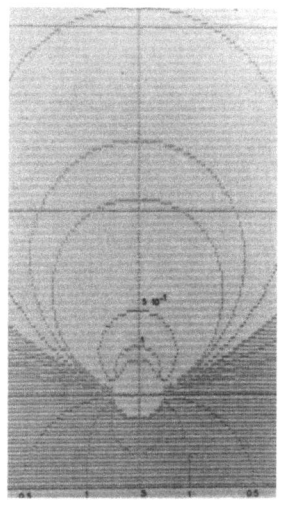 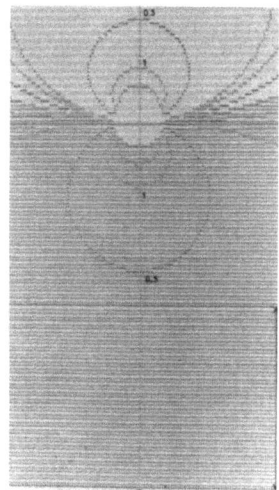

Figure 13. Equi G curves : TqD $\beta = 1$, 0,5 left: $c = 5$ μm right $c = 20$ μm. scale mark : 10 μm.

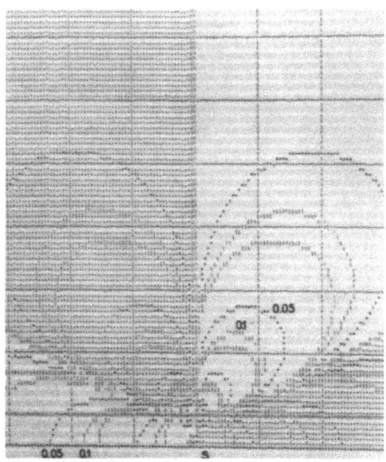 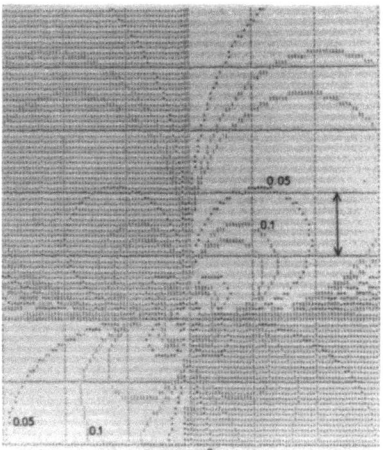

Figure 14. Equi G curves : Sd $\beta = 10^{-1}$, $5\ 10^{-2}$ left: $c = 5$ μm right $c = 20$ μm. scale mark : 10 μm.

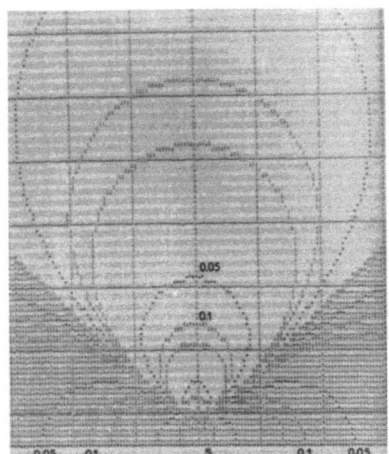 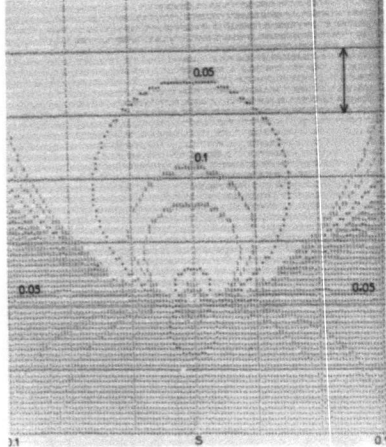

Figure 15. Equi G curves : TqD $\beta = 10^{-1}$, $5\ 10^{-2}$ left c = 5 μm right c = 20 μm. scale mark : 10 μm.

CHARACTERIZATION OF CRYSTAL GROWTH DEFECTS

BY X-RAY METHODS

29th August - 10th September 1979

SPONSORS

North Atlantic Treaty Organization
Texas Instruments Ltd.
Barr and Stroud Ltd.
International Business Machines (UK) Ltd.
Metals Research Ltd.

ORGANISING COMMITTEE

Dr. A.R. Lang   Director
Dr. B.K. Tanner   Executive Secretary
Dr. D.K. Bowen   Programme Secretary
Dr. I.M. Buckley-Golder
Prof. M. Hart
Dr. M. Moore
Prof. J.N. Sherwood
Dr. D.P. Siddons

INTERNATIONAL ADVISORY PANEL

R.W. Armstrong   USA
A. Authier   France
B.W. Batterman   USA
U. Bonse   BRD
G. Champier   France
B. Cockayne   UK
G. Hildebrandt   BRD
N. Kato   Japan
K. Kohra   Japan
H. Klapper   BRD
A.J.R.de Kock   Netherlands
J.R. Patel   USA
W. Stacey   Netherlands

THE CHARACTERIZATION OF CRYSTAL GROWTH DEFECTS BY X-RAY METHODS
29 August – 9 September, 1979
DURHAM, ENGLAND

A NATO Advanced Study Institute

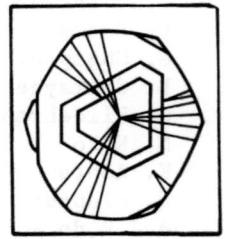

INDEX

Absorption, 227-8,233,263,325, 326,354,406,409,415,424,433, 442,467,507,518,555,571.
Absorption topography, 60,506ff.
Acceptable Quality Level (AQL) 7-8.
Adhesives, 346.
Aluminium iodate, 143.
Ammonium hydrogen oxalate, 135, 136,143,154.
Ammonium hydrogen oxalate (hemi-hydrate), 135,141.
Angular range of reflection, 162,405,422,527,548.
Anode tube, 384.
Anomalous dispersion, 168,272.
Anomalous transmission, 4,6,16, 23,73,186,235,241,272,282-3, 371,381,535,541.
Anthracene, 155.
Argand diagram, 224-5.
Asymmetric diffraction, 354,355, 358,373,435,465.
Atomic displacement, 244,256.
'Aufhellungen', 168.

Background scatter, 167.
Beam conditioning, 424,435.
Beam conditioned micro-radiography, 507.
Beam divergence control, 430.
Bezophenone, 135,147,155,156.
Benzil, 135,144,145,155.
Berg-Barrett technique, 349,351, 356,360,363,364,365,370,371, 465,535,540.
Beryl, 145.
Bicrystals, 255.
Bimodal profiles, 410.
Birefringence, 147,148.
Black and white phenomena, 283,412.
Borrmann effect, 16,177,202, 232ff,248,252,282,406,433, 535.
Boundary conditions, 229,545, 548.

Brazil twin, 183.
Bridgman Technique, 50,98,123, 201,208,464.
Bubble domains, 37.
Buried channel devices, 36.
Burgers vector, 67,77,108,115, 117,136,138,139,140,141,142, 148,151,178,179,182,256,353, 373,383,389,395,410,414,416, 517,527,528ff.
Cadmium mercury telluride, 362.
Calcite, 138,576.
Cathodoluminescence, 170,456.
Caustics, 273,570.
Cellular structures, 55ff.
Characteristic line spectrum, 302ff,314,370.
Charge Coupled Devices (CCD), 29ff.
Charge Transfer Inefficiency, 35.
Chemical etching, 47.
Chemical vapour deposition (CVD) 38.
Chromatic aberration correction method (CAC), 372.
Climb, 98,100,122,126,127,128, 149,153,156,362,364.
Compton scattering, 191,197.
Computer modelling of crystal growth, 503.
Computer simulations, 73-75, 277-8,288,450.
Continuous spectrum (Bremsspectrum) 300,301ff,315,362,401,466.
Contrast reversal, 177.
Cooling rate, 128.
Core energy, 140,142,143.
Cost estimates, 475,495.
Counting electronics, 481,485, 487.
Counting Statistics, 322,325, 382.
Critical wavelength, 307.
Cryostats, 459,460,466,467.
Crystal melting, 393,464,465.
Crystal orientation, 333, 334ff.
Czochralski method, 48,98,104ff, 147,151,363.

INDEX

Dangling bonds, 76.
Dark current, 36.
Dark current spikes, 37.
Dauphine twin, 183,276-8.
Delay lines, 42.
Dendritic Structures, 55ff.
Detectors, 480,485,489.
Diamond, 157,173,180,456,512, 520,521,526.
Die yield, 7ff.
Diffuse reflections, 512,513.
Diffuse scattering, 82,88,187ff.
Digital image processing, 387, 388,499.
Dilation boundaries, 566.
Direct images, 409,476,517,521, 530,568,571.
Dislocation decoration, 156.
Dislocation density, 99ff,111, 115,117,119,120,123,125,126, 173,178,201,527.
Dislocation-free crystals, 104, 106.
Dispersion, 164,217,371,372,373.
Dispersion points, 270,277,281.
Dispersion surface, 87,222ff, 230,235,240,242,252,258,261, 270-1,273,281,378,544ff, 565ff.
Dislocation mobility, 391
Dissolution, 504.
Domain boundary, 183,330.
Double crystal cameras, 483ff.
Double crystal diffraction, 4,6,16,47,49,146,179,236,238, 358,362,372,381,423,447,459, 474,485,491,500,516.
Double crystal spectrometer, 82,370,423,436,470,482.
Du Mond diagram, 405,436ff.
Dynamic experiments, 465ff.
Dynamical images, 521,528,568,571.
Dynamical theory, 216ff,421.

Eikonal approximation, 242
Eikonal theory (see also ray theory), 278,288-9.
Elastic anisotropy, 141.
Electrolytic machining, 341.

Electron impact X-ray sources, 300ff.
Electropolishing, 340,343.
Energy factor, 140,141.
Energy flow, 87.
Energy transfer equations, 290ff.
Epitaxial layers, 84ff,444,482, 501,518.
Etching, 342ff,517.
Ewald sphere, 82,197,566,569,570.
Extinction, 88,89,91,203,205,207, 209,282,290,354,355,356,365, 409,410.
Extinction coefficient, 294,354.
Extinction contrast, 351,353,360, 366.
Extinction distance, 89,161,178, 179,184,198,199,203,293,353, 422,518.

Facets, 47,50ff,137,148,150,176, 180,335.
Faulted growth sector boundaries, 145.
Fault vector, 373,554ff.
Fermat's principle, 245,280.
Ferroelectric domains, 458.
Fick's Law, 47.
Floating zone, 79.
Floating-zone technique, 79,147, 380.
Fluorescence, 87ff.
Flux-grown crystals, 145.
Frank networth, 123.
Frank-Read spirals, 155.
Fraunhofer diffraction, 264,267, 509.
Free energy, 138,139.
Fresnel diffraction, 509.
Friedel's Law, 180.
Furnaces, 461,465,466,467,498.

Gadolinium Gallium Garnet (GGG), 38,147,148,149,151.
'Gamma', 324.
Gamma-ray diffractometry, 187ff, 196ff.
Garnets, 1-2,23,25,44,446.
Gas-filled counters, 320,321.

INDEX

Geiger counter, 321,322,480.
Gettering, 36.
Gettering centres, 3,23.
Glide, 153,154,155.
Grain boundaries, 155.
'Grainless' recorders, 328.
Grown-in dislocations, 134ff, 150ff,153.
Growth accidents, 137,143.
Growth banding, 137,144,145ff, 173,182,503,510,526.
Growth sectors, 134,142.
Growth-sector boundaries, 134, 135,136,143,144,145ff,173,182, 184,510.
Harmonics, 405,406,412,434,446.
Haze, 8,9.
Helices, 113,116,122,149,157,374.
Homogeneously deformed crystals, 247-8,253.
Huang diffuse scattering, 186, 189.
Huygen's principle, 272,290,549.
Ice, 458,464,475.
Image intensification, 321,330, 331,358,497.
Impurity zoning (see growth banding).
Impurity clustering, 79ff.
Impurity location studies, 86.
Inclusions, 136,138,144,150, 155,163,173,181,526.
Infrared absorption, 79,80.
Insitu experiments, 457ff, 497ff.
'Instant' topography, 320,331,382, 390,497ff.
Integrated intensity, 404,405, 545.
Integrated reflecting power, 199,201,207,249,293,354.
Interbranch scattering, 183, 184,412,448.
Intermediary image, 568,571.
Ion beam machining, 341.
Ionization chambers, 321.
Iso defect curves, 7,18,19,25.

KDP, 135,137,138,141,142,143, 145,146,213,335.
Kinematical image technique, 380.
Lang camera, 479,486.
Lang method, 73ff,97ff,363,369, 370,371,409,434,457,474, 486,535.
Lattice phase, 283,285,292.
Laue cameras, 33.
Laue topography, 403.
Limited projection topographs, 168.
Liquid phase epitaxy, (LPE), 38.
Lithium formate hydrate, 143, 154,155.
Lithium fumerate tetrahydrate, 155,156.
Lithium Niobate, 29,41,43.
Lomer-Cottrell locks, 107,108.
Long-range distortions, 278ff,289.
Low-angle boundaries (see also sub grains), 173,178ff, 182,526,527.
Magnetic bubble domain memory, 29,37ff.
Magnetic domains, 414,428,416, 441,458,459,467,499,500, 542.
Maxwell's equations, 217,221ff, 242,261,283.
Metal crystals, 97.
Metal-Oxide semiconductor (MOS) Capacitor, 30.
Microprocessor control systems, 488ff.
Microradiography, 506ff.
Minimum-energy theory, 138,151.
Misfit boundaries, 565ff.
Misfit dislocations, 415,446, 465.
Modulation transfer function (MTF), 386,387.
Monochromator crystals, 208.
Monochromatic synchrotron radiation topography, 433ff.

INDEX

Mosaic crystals, 195,202ff,264, 293,354,356.
Multiple reflection, 421,424,427.

Nacken-Kyropoulos method, 151.
Non-metal crystals, 133ff.
Nuclear emulsion, 320,326,327, 330,465,474,486.

Olivine, 402,403.
Optical Modulators, 29.
Orientation contrast, 352,353.

Pendellosung fringes, 13,85,163, 168,176,177,178,184,238,239ff, 248,250,252,257,274,277,295, 376,389,393,406,526.
Pendellosung fringes (Bragg case) 275.
Pendellosung length, 80,91,238, 422.
Penetration depth, 518.
Phase increments, 269.
Phase integral, 280.
Phase transformation method, 117, 396,511.
Phase transitions, 213,414,433.
Photoelectric absorption, 86,87.
Phosphors, 320,321,331,397,497.
Photographic densitometry, 323, 447,448.
Photographic recording, 323,368ff.
Photographic sensitivity, 323.
Photomicrography, 329.
Planar defects, 73,254,513,570.
Plane-wave imaging, 372,435, 444ff.
Plastic deformation, 392,414,416, 461,465.
Platelike crystals, 101.
Plate-like defects, 275,277,289.
Point defects, 73,34ff.187ff,202, 380.
Point-defect aggregates, 46.
Polarisation 'beat', 177.
Polarisation factor, 162,224,267, 353,405.
Polishing, 342ff.

Poly - [1,2-bis-(p tolylsulphon-oxymethylene)] - 1 - butane - 3 inylene, 365-6.
Post-growth defects, 153ff.
Potash alum, 138.
Poynting vector, 226,236,262, 544,548.
Precipitates, 13,46,60,73,76, 79ff,157,173,181,526.
Preferred directions of dislocations, 138.
Projection topograph, 161,165, 167ff,368,369ff,479,521ff, 526.
Proportional counters, 321,322, 480,485.
Pulse compression filters, 42.

Quantum detection efficiency, 383.
Quartz, 29,41,43,77,135,145,146, 183,276,335,362,452,454,458.

Ratemeters, 323.
Ray approximation (see also Eikonal approximation), 254.
Ray optics, 226,266ff,273.
Ray tracing, 235,250,566,576.
Recrystallization, 113,115,414, 466,467,468.
Reflection profile analysis, 239.
Reflection section topographs, 170.
Reflection topography, 6,349ff, 515ff.
Refraction of dislocation lines, 151.
Relps, 512.
Residual contrast, 531.
Resolution, 351,370,402,403,404, 450,457,466,476,498-9,506,508, 513,520.
Resolution limits, 163,324,331, 382,509.
Rocking curve, 236,250,440,457, 482,485,486,489,516.
Rotating anode tubes, 298,305, 312-3,316,397,465,477,478, 482.

INDEX

Rotational boundaries 565ff.
Rutherford backscattering, 91.

Salol, 136,138,144,146.
Sample preparation, 333,361.
Sapphire, 1-2,23,25,69,151,157, 350,363.
Scanning electron microscopy (SEM) 76.
Scanning reflection topographs, 161,169.
Schulz technique, 349,356,362.
Scintillation counters, 320,321, 322,480.
Sealed X-ray tubes, 305,312,476, 477,483.
Section topography, 4,6,13,73-75, 77,146,161,163ff,239ff,252, 277-8,289,368,369,372,373,374, 381,461,479,520ff,545,555,569, 576.
Segregation, 46,47,51,511.
Self interstitials, 156,192, 193ff.
Semi conductor detectors, 320, 321,322.
Shaping and forming, 338ff.
Signal to noise ratio (SNR), 383,388,389,402.
Silicon, 2,49,68,73ff.
Silicon web dendrite, 11.
Skew reflection, 350,351.
Slip, 126,128,416,526.
Slip planes, 99.
Solid state growth, 110ff.
Solid State Imaging, 33,34.
Solute striations, 46,48ff.
Solute trails, 60,63,64,70.
Sonar signal retrieval, 34.
Spark erosion, 341.
Specimen mounting, 345.
Spherical waves, 231ff,265,268, 368ff,372,441,545,549,570.
Spike topography, 187,512ff.
Spinel, 49.
Spiral Dislocations, 68.
Stacking faults, 73ff,146,157 173,176,178,182,254,255,275-6, 369,378,515,526,554ff,565.

Standing wave field, 73,90-93, 217,234.
Stationary phase method, 274.
Statistical theory of dynamical diffraction, 289ff.
Stereo methods, 170ff,371,521ff, 530,535ff.
Stereographic projection, 165, 166,167,520,535ff.
Storage ring sources, 307,310, 312,317,430.
Strain-anneal method, 110ff,363.
Striation, 147,149.
Stroboscopic techniques, 428, 434,441.
Structure factor, 162,219,240, 268,353,405,422,424,544.
Sub-boundaries (also sub-grain boundaries), 99,106,116, 351,356ff,360,362,364,371, 408,527,535ff.
Sublimation-condensation, 117.
Supercooling, 55-9.
Superjogs, 122.
Surface Acoustic Wave (SAW) Devices, 29ff.
Surface damage, 12,13,135,155, 173,180ff,526.
Swirl, 8ff,22,151,380,381,395, 465.
Synchrotron radiation, 97,113, 128,164,166,187,195,232,258, 298,300,306ff,401ff,466,474, 491,506ff,513.
Synchrotron radiation sources, 306ff,331,401,469,491ff.

Takagi's equations (Takagi-Taupin equations), 78,283ff, 290.
Temperature gradient, 99,104, 109.
Tensile Stages, 462,463.
Tetraoxan, 155.
Textured grains, 114.
TGS triglycene sulphate), 135,154. 154.
Thermal gradients, 151.
Thiourea, 135,155.

Tie points (see also dispersion points), 87,270,545,549,565, 568ff.
Topaz, 145.
Transmission electron microscopy (TEM), 76,82,364,379,391, 393,450,513.
Twin lamellae, 255,576.
Twist quasi-disclination, 568, 569.

Undulators, 300,310,312.

Vacancies, 98,153,156,194,365.
Vacancy sink, 125.
Vacancy supersaturation, 107,113, 114,116,127,128.
Vapour-grown crystals, 146.
Vegard's law, 84.
Veils, 137.
Video display, 382,466,497ff.
Verneuille method, 147,152.
Voids, 69ff.

Wafer diameter, 2.
Wafer distortion, 23,24,371.
Wafer flatness, 2.
Weak-beam topography, 379,450.
Whiskers, 117,118-9,468.
White beam topography, 401ff,423, 433,434,469,491.
White radiation microradiography, 506.
Wigglers, 300,310,491.

X-ray detectors, 320.
X-ray films, 326.
X-ray polarizer, 427.
X-ray sources, 298ff,397,457, 476,482.

Yield, 1ff,18ff.

"Zwischenreflex scattering", 186, 189,191.

CHEMICAL FORMULA INDEX

Aℓ, 104,105,111,123,124,126, 191,204,208,341,414,461.
$Aℓ_2O_3$, 364.
Ag, 105,108.
Ag-Sn, 116.

$BaTiO_3$, 364,458,467.
Be, 207.
BeO, 250,369.

CaF_2, 66.
$CaWO_4$, 62,67.
Cd, 100,102,117,122,127,341.
CdS, 375,376,379.
Cu, 98,104,105,116,118,200,201, 206,340,341,362,381.
Cu-Ag, 119.
Cu-Zn, 119.
Fe, 115,118.
Fe-Si, 100,115,341,442,443,467, 468,499,542.

Ga, 103,122,465.
GaAs, 66,67,68,381,501.
GaAs - $Ga_xAℓ_{1-x}As$, 84.
$Gd_3Ga_5O_{12}$, 38,53-5,68,86.
$Gd_2(MoO_4)_3$, 433,434.
Ge, 56,63,64,68,87,88,150,354-6, 370,381,475,543

III-V Compounds, 444.
In, 101,103,120.
InP, 67,70.
InSb, 51,58.

KCℓ, 151,153.
$KCoF_3$, 142.
$KNiF_3$, 135,142,145,467.
Kr, 468.

LiF, 171,173,354-6,3600,408,526, 539.
$LiRF_4$, 70.

Mg, 100,102,117,128.
MgO, 343.
Mo, 116.

NaCl, 146,283.
$NaClO_3$, 154,155,156,283,528.
Ni, 105,108,354,357,359.
NiO, 459.
Nb, 105,108,362,363.

RbI, 214.

Si, 2,49,68,73ff,157,199,250,277, 283,340,381,389,394,407,413, 475,501,520,530,554.
SiC, 135,155.
Sn, 115.
$Sr(NO_3)_2$, 452,453.
S, 155.

Ti, 117.
$TbVO_4$, 214.
$TmAsO_4$, 214.

W, 359.

Xe, 468.

$Y_3Aℓ_5O_{12}$, 52,53,62,135,145.

Zn, 101,103,117,349,352,354-6, 360,361,536,540.
ZnS, 467.

CHEMICAL FORMULA INDEX

Ag, 104,109,111,113,124,126,
191,204,208,341,414,461.
Ag_2O, 354.
As, 105,106.
As-Sn, 110.

BaTiO, 366,458,467.
Be, 267.
BeO, 250,359.

CdS, 68.
$CdWO_4$, 65,67.
Cd, 100,101,115,122,127,341.
CdS, 373,379,379.
Co, 98,104,105,115,116,200,201,
206,340,341,361,367.
Cr-Ag, 119.
Co-Zn, 119.
Fe, 115,118.
Fe-Ni, 100,115,141,142,143,167,
468,469,54.

Ga, 107,122,465.
GaAs, 65,67,66,381,501.
GaAs — Ga,Al, A, B,.
Gd-Ge, 2, ,58,53,5,68,69.
Gd(OH)₃, 433,434.
Ge, 56,67,82,68,87,83,150,354-6,
370,381,475,545.

III-V Compounds, 468.
In, 101,103,120.
InSb, 67,70.
InP, 51,52.

KCl, 151,151.
$KCoF_3$, 161.
KMnF₃, 135,162,164,467.
K_2S, 465.

LiF, 171,175,154-6,3800,408,520.
, 530.
$LiHF_4$, 70.

Hg, 000,102,113,71.
M_2O, 343.
Mn, 116.

NaCl, 146,288.
$NaClO_3$, 154,155,179 ?, 181 ?,
Ni, 105,108,354, 7, 519.
P(Ni, 458?
3b, 105,108,362,58.

PbI, 214.

Si, 7,48,86, 791,1, 280,271,
181,240,181,155, , ,41, ,415,
475,501,420,56,
510, 356,165.
Sn, 115.
Sr(No)₂, 652,65.
Se, 153 ?.

Ti, 117,4.
UVO_3, 214.
$UgAg_2$, 21.

W, 559.

Ze, 464 ?

$Y_3As_5O_2$, 52,55.

Zn, 101,103,117, , , 354-1
360, 381, 358, , ,.
ZnS, 464.

MIX
Papier aus verantwortungsvollen Quellen
Paper from responsible sources
FSC® C105338

If you have any concerns about our products,
you can contact us on
ProductSafety@springernature.com

In case Publisher is established outside the EU,
the EU authorized representative is:
**Springer Nature Customer Service Center GmbH
Europaplatz 3, 69115 Heidelberg, Germany**

Printed by Libri Plureos GmbH
in Hamburg, Germany